WILEY'S ENGLISH-SPANISH
SPANISH-ENGLISH
CHEMISTRY DICTIONARY

DICCIONARIO DE QUÍMICA
INGLÉS-ESPAÑOL
ESPAÑOL-INGLÉS WILEY

By the same author

English-Spanish, Spanish-English Legal Dictionary,
Second Edition published 1997.

Wiley's English-Spanish, Spanish-English Business Dictionary.
Published 1996.

*English-Spanish, Spanish-English Electrical and Computer
Engineering Dictionary.* Published 1996.

*Wiley's English-Spanish, Spanish-English Dictionary of
Psychology and Psychiatry.* Published 1995.

Por el mismo autor

Diccionario jurídico inglés-español, español-inglés,
Segunda Edición publicada en 1997.

Diccionario de negocios inglés-español, español-inglés Wiley.
Publicado en 1996.

*Diccionario de ingeniería eléctrica y de computadoras
inglés-español, español-inglés.* Publicada en 1996.

*Diccionario de psicología y psiquiatría inglés-español,
español-inglés Wiley.* Publicada en 1995.

WILEY'S ENGLISH-SPANISH SPANISH-ENGLISH CHEMISTRY DICTIONARY

DICCIONARIO DE QUÍMICA INGLÉS-ESPAÑOL ESPAÑOL-INGLÉS WILEY

Steven M. Kaplan
Lexicographer

Editorial Advisors
Dr. Rolf Altschul
Dr. Sylvia Márquez-Pirazzi

A WILEY-INTERSCIENCE PUBLICATION
JOHN WILEY & SONS, INC.
New York • Chichester • Weinheim • Brisbane • Singapore • Toronto

This text is printed on acid-free paper. ⊗

Copyright © 1998 by John Wiley & Sons, Inc. All rights reserved.

Published simultaneously in Canada.

Library of Congress Cataloging in Publication Data:

Kaplan, Steven M.
    Wiley's English-Spanish, Spanish-English chemistry dictionary = Diccionario de química inglés-español, español-inglés Wiley / Steven M. Kaplan, lexicographer ; editorial advisory board, Rolf Altschul, Sylvia Márquez-Pirazzi.
        p. cm
    "A Wiley-Interscience publication."
    Includes index.
    ISBN 0-471-19288-0 (cloth : alk. paper). — ISBN 0-471-24923-8 (paper : alk. paper)
    1. Chemistry—Dictionaries. 2. English language—Dictionaries—Spanish. 3. Chemistry—Dictionaries—Spanish. 4. Spanish language—Dictionaries—English. I. Title.
QD5.K292 1998
540'.3—dc21   97-34936
                                                              CIP

Printed in the United States of America

10 9 8 7 6 5 4 3 2 1

PREFACE

Throughout the world, over 350 million people speak Spanish and over 450 million speak English. Most of them live in America. Over the Internet, dozens of countries are providing a massive amount of information concerning chemistry. This has further increased the need for precise communication between users of chemical terminology in the two languages.

The primary objective of this dictionary is to provide accurate equivalents for the words and phrases most commonly encountered in the chemistry literature. There are over 40,000 total entries, approximately 20,000 in each section of the dictionary.

Prefixes such as cis, trans, ortho, meta, para, 1, 2, 3, D, L, alpha, beta, N, O, and sec, among many others, are not placed in front of compound names in the dictionary. There are three main reasons for this. First, this eliminates the possible confusion concerning alphabetization. It also avoids wasting space with certain groups of entries which would have almost identical equivalents. For example, to have three entries for

1,2,3-trimethylbenzene
1,2,4-trimethylbenzene, and
1,3,5-trimethylbenzene,

instead of just **trimethylbenzene,** would add to the bulk of the dictionary without increasing usefulness. Finally, when users of the dictionary encounter compounds of this nature, simply placing the prefix in the target language in the same spot it was found in the source language is all that is needed for transliteration. So **1,3,5-trimethylbenzene** would be **1,3,5-trimetilbenceno.** Another example will illustrate how this works with another prefix in a different location: **m-sulfobenzoic acid** has **ácido m-sulfobenzoico** as its equivalent.

Many thousands of chemical compounds are included in this dictionary, allowing for quick and simple reference. Unfortunately, many chemical compounds have names that are either too long or are not encountered frequently enough in the literature to warrant inclusion. To help users in these cases, most every prefix and suffix that can be an integral part of a chemical name is included. For example, cyclo, di, ethyl, and ate, among over one hundred others, will help piece together even the most involved name.

I am very grateful to John Wiley Editor Carla A. Fjerstad who, after recognizing the merits of this project, enlisted the prominent advisors and supervised the work. The editorial advisors, Dr. Rolf Altschul and Dr. Sylvia Márquez-Pirazzi, made significant contributions to the quality of the dictionary, which is greatly appreciated. Special thanks go to

John Wiley Senior Editor George J. Telecki, who kindly insured that the idea for this book was communicated to the appropriate person. I also wish to thank Wiley Editorial Assistant Jennifer Campaniolo who helped things march smoothly throughout the project.

Steven M. Kaplan

Miami, Florida

PREFACIO

A través del mundo, más de 350 millones de personas hablan español y otras 450 millones hablan inglés. La mayoría de estas viven en América. A través de la Interred, docenas de países están proveyendo una cantidad masiva de información concerniente a la química. Esto ha aumentado aun más la necesidad para la comunicación precisa entre los usuarios de la terminología química en los dos idiomas.

El objetivo principal de este diccionario es proveer equivalentes precisos para las palabras y frases más comúnmente encontradas en la literatura de la química. Hay más de 40,000 voces de entradas en total, aproximadamente 20,000 en cada sección del diccionario.

Prefijos tales como cis, trans, orto, meta, para, 1, 2, 3, D, L, alfa, beta, N, O, y sec, entre muchos otros, no están colocados delante de los nombres de compuestos en el diccionario. Hay tres razones principales para esto. Primero, esto elimina la posible confusión referente a la alfabetización. Además evita desperdiciar espacio con ciertos grupos de voces de entradas las cuales tendrían equivalentes casi idénticos. Por ejemplo, el tener tres voces de entrada para

1,2,3-trimethylbenzene
1,2,4-trimethylbenzene, y
1,3,5-trimethylbenzene,

en vez de sólo **trimethylbenzene,** añadiría tamaño al diccionario sin aumentar la utilidad. Finalmente, cuando los usuarios del diccionario encuentren compuestos de esta naturaleza, el sencillamente colocar el prefijo en el idioma objeto en el mismo lugar en que se encontró en el idioma de origen, es todo lo que se necesita para la transliteración. Así que, **1,3,5-trimethylbenzene** sería **1,3,5-trimetilbenceno.** Otro ejemplo ilustrará cómo esto funciona con otro prefijo en otro lugar: **m-sulfobenzoic acid** tiene **ácido m-sulfobenzoico** como su equivalente.

Muchos miles de compuestos químicos están incluidos en este diccionario, permitiendo referencia sencilla y rápida. Desafortunadamente, muchos compuestos químicos tienen nombres los cuales son o muy largos, o que no se encuentran suficientemente frecuentemente como para justificar su inclusión. Para ayudar a los usuarios en estos casos, casi cualquier prefijo y sufijo el cual podría ser una parte integral de un nombre químico está incluido. Por ejemplo, ciclo, di, etil, y ato, entre más de cien otros, ayudarán a formar hasta el nombre más envuelto.

Le estoy muy agradecido a la editora de John Wiley Carla A. Fjerstad, quien tras reconocer los méritos de este proyecto, reclutó los asesores prominentes, y supervisó el trabajo. Los asesores editoriales, el Dr. Rolf Altschul y la Dra. Sylvia Márquez-Pirazzi, hicieron contribuciones significativas a la calidad del diccionario, lo cual se aprecia grandemente. Gracias especiales van al editor de John Wiley George J. Telecki, quien amablemente se aseguró de que la idea para este diccionario fuera comunicada a la persona apropiada. También quiero dar mis gracias a la asistente de editor de Wiley Jennifer Campaniolo quien ayudó a que todo marchara sobre ruedas a través del proyecto.

Steven M. Kaplan

Miami, Florida

ENGLISH-SPANISH
INGLÉS-ESPAÑOL

A

abaca abacá
Abbegg rule regla de Abbegg
Abel reagent reactivo de Abel
aberration aberración
abherent antiadherente
abieninic acid ácido abienínico
abietadienoic acid ácido abietadienoico
abietic acid ácido abiético
abietic anhydride anhídrido abiético
ablation ablación
ablative material material ablativo
abrasion abrasión
abrasive abrasivo
abscisic acid ácido abscísico
absinthe absintio
absinthium absintio, ajenjo
absolute absoluto
absolute alcohol alcohol absoluto
absolute boiling point punto de
    ebullición absoluto
absolute configuration configuración
    absoluta
absolute density densidad absoluta
absolute error error absoluto
absolute humidity humedad absoluta
absolute stereochemistry estereoquímica
    absoluta
absolute system sistema absoluto
absolute temperature temperatura
    absoluta
absolute temperature scale escala de
    temperatura absoluta
absolute unit unidad absoluta
absolute zero cero absoluto
absolute zero temperature temperatura
    de cero absoluto
absorbance absorbancia
absorbancy absorbancia
absorbent absorbente
absorptiometer absorciómetro
absorptiometric analysis análisis
    absorciométrico
absorption absorción

absorption apparatus aparato de
    absorción
absorption band banda de absorción
absorption cell célula de absorción
absorption coefficient coeficiente de
    absorción
absorption constant constante de
    absorción
absorption edge borde de absorción
absorption indicator indicador de
    absorción
absorption limit límite de absorción
absorption line línea de absorción
absorption of light absorción de luz
absorption paper papel de absorción
absorption peak pico de absorción
absorption spectroscopy espectroscopia
    de absorción
absorption spectrum espectro de
    absorción
absorption tower torre de absorción
absorption tube tubo de absorción
absorptive power poder de absorción
absorptivity absortividad
abundance abundancia
abundance ratio razón de abundancia
acacetin acacetina
acacia acacia
acacia gum goma de acacia
acadesine acadesina
acanthite acantita
accelerating anode ánodo acelerador
accelerating electrode electrodo
    acelerador
acceleration aceleración
accelerator acelerador
accelofilter filtro acelerador
acceptable explosive explosivo aceptable
acceptable risk riesgo aceptable
acceptor aceptor
accessory accesorio
accumulator acumulador
accuracy precisión, exactitud

acenaphthene acenafteno
acenaphthenequinone acenaftenoquinona
acenaphthylene acenaftileno
acetal acetal
acetal resin resina acetálica
acetaldehyde acetaldehído
acetaldehyde cyanohydrin cianhidrina
  de acetaldehído
acetaldol acetaldol
acetamide acetamida
acetamide chloride cloruro de acetamida
acetamide hydrochloride clorhidrato de
  acetamida
acetamide nitrate nitrato de acetamida
acetamidine acetamidina
acetamidine hydrochloride clorhidrato
  de acetamidina
acetaminophen acetaminofeno,
  acetaminofén
acetaminosalol acetaminosalol
acetanilide acetanilida
acetate acetato
acetate film película de acetato
acetazolamide acetazolamida
acetiamine acetiamina
acetic acid ácido acético
acetic acid amide amida de ácido acético
acetic acid amine amina de ácido acético
acetic acid glacial ácido acético glacial
acetic aldehyde aldehído acético
acetic anhydride anhídrido acético
acetic ester éster acético
acetic ether éter acético
acetic oxide óxido acético
acetin acetina
acetoacetanilide acetoacetanilida
acetoacetate acetoacetato
acetoacetic acid ácido acetoacético
acetoacetic ester éster acetoacético
acetoin acetoína
acetol acetol
acetolysis acetólisis
acetone acetona
acetone bromoform bromoformo de
  acetona
acetone chloroform cloroformo de
  acetona
acetone cyanohydrin cianhidrina de
  acetona

acetone dicarboxylic acid ácido
  acetonadicarboxílico
acetone monocarboxylic acid ácido
  acetonamonocarboxílico
acetone number número de acetona
acetone semicarbazone semicarbazona
  de acetona
acetone sodium bisulfate bisulfato
  sódico de acetona
acetonitrile acetonitrilo
acetonylacetone acetonilacetona
acetonylidene acetonilideno
acetophenetidine acetofenetinina
acetophenone acetofenona
acetophenone acetone acetona de
  acetofenona
acetophenone oxime oxima de
  acetofenona
acetostearin acetoestearina
acetotoluide acetotoluida
acetotoluidide acetotoluidida
acetoxime acetoxima
acetoxylation acetoxilación
acetozone acetozona
acetyl acetilo
acetyl bromide bromuro de acetilo
acetyl chloride cloruro de acetilo
acetyl coenzyme acetilcoenzima
acetyl fluoride fluoruro de acetilo
acetyl iodide yoduro de acetilo
acetyl ketene acetilcetena
acetyl nitrate nitrato de acetilo
acetyl oxide óxido de acetilo
acetyl peroxide peróxido de acetilo
acetyl value valor de acetilo
acetylacetic acid ácido acetilacético
acetylacetonate acetilacetonato
acetylacetone acetilacetona
acetylamino acetilamino
acetylating agent agente acetilante
acetylation acetilación
acetylbenzene acetilbenceno
acetylcarbromal acetilcarbromal
acetylcholine acetilcolina
acetylcholinesterase acetilcolinesterasa
acetylene acetileno
acetylene black negro de acetileno
acetylene complex complejo de acetileno

acetylene dibromide dibromuro de acetileno

acetylene dichloride dicloruro de acetileno

acetylene hydrocarbon hidrocarburo de acetileno

acetylene tetrabromide tetrabromuro de acetileno

acetylenogen acetilenógeno

acetylenyl acetilenilo

acetylferrocene acetilferroceno

acetylformic acid ácido acetilfórmico

acetylide acetiluro

acetylmethylcarbinol acetilmetilcarbinol

acetylphenol acetilfenol

acetylpropionic acid ácido acetilpropiónico

acetylsalicylic acid ácido acetilsalicílico

acetylurea acetilurea

Acheson process proceso de Acheson

achiral aquiral

achromatic acromático

achromatic indicator indicador acromático

acicular acicular

acid ácido

acid acceptor aceptor de ácido

acid alcohol alcohol ácido

acid amide amida de ácido

acid ammonium sulfate sulfato amónico ácido

acid ammonium tartrate tartrato amónico ácido

acid anhydride anhídrido de ácido

acid-base catalysis catálisis ácido-base

acid-base equilibrium equilibrio ácido-base

acid-base indicator indicador ácido-base

acid-base pair par ácido-base

acid-base titration titulación ácido-base, valoración ácido-base

acid chloride cloruro de ácido

acid content contenido ácido

acid dissociation disociación de ácido

acid dye colorante ácido

acid fuchsin fucsina ácida

acid group grupo ácido

acid halide haluro ácido

acid magnesium citrate citrato ácido magnésico

acid methyl sulfate sulfato ácido de metilo

acid number número de ácido

acid oxide óxido ácido

acid phosphatase fosfatasa ácida

acid phosphate fosfato ácido

acid potassium sulfate sulfato ácido potásico

acid precipitation precipitación ácida

acid radical radical de ácido

acid rain lluvia ácida

acid reaction reacción ácida

acid-resistant resistente al ácido

acid salt sal ácida

acid solution solución ácida

acid solvent solvente ácido

acid value valor de ácido

acidic acídico

acidic group grupo acídico

acidic hydrogen hidrógeno acídico

acidification acidificación

acidify acidificar

acidimeter acidímetro

acidimetry acidimetría

acidity acidez

acidity coefficient coeficiente de acidez

acidity constant constante de acidez

acidolysis acidólisis

acidproof a prueba de ácido

acidulant acidulante

Acker process proceso de Acker

aconic acid ácido acónico

aconite acónito

aconitic acid ácido aconítico

aconitine aconitina

Acree-Rosenheim reaction reacción de Acree-Rosenheim

acrid acre

acridine acridina

acridine orange anaranjado de acridina

acridinic acid ácido acridínico

acriflavine acriflavina

acrimonious acrimonioso

acroleic acid ácido acroleico

acrolein acroleína

acrolein dimer dímero de acroleína

acrolein polymer polímero de acroleína

acrolein test prueba de acroleína

acrometer acrómetro
acrylaldehyde acrilaldehído
acrylamide acrilamida
acrylamide copolymer copolímero de
    acrilamida
acrylamide polymer polímero de
    acrilamida
acrylate acrilato
acrylate plastic plástico de acrilato
acrylate resin resina de acrilato
acrylic acrílico
acrylic acid ácido acrílico
acrylic acid polymer polímero de ácido
    acrílico
acrylic aldehyde aldehído acrílico
acrylic ester éster acrílico
acrylic polymer polímero acrílico
acrylic resin resina acrílica
acrylonitrile acrilonitrilo
acrylonitrile copolymer copolímero de
    acrilonitrilo
acrylonitrile dimer dímero de
    acrilonitrilo
acrylonitrile polymer polímero de
    acrilonitrilo
acryloyl chloride cloruro de acriloilo
actinic actínico
actinic radiation radiación actínica
actinide actínido
actinide series serie actínida
actinism actinismo
actinium actinio
actinium series serie de actinio
actinochemistry actinoquímica
actinoid actinoide
actinoid contraction contracción de
    actinoides
actinometer actinómetro
actinometry actinometría
actinomycin actinomicina
actinon actinón
actinoquinol actinquinol
action acción
action spectrum espectro de acción
activated activado
activated adsorption adsorción activada
activated alumina alúmina activada
activated carbon carbón activado
activated charcoal carbón activado

activated complex complejo activado
activated molecule molécula activada
activated state estado activado
activation activación
activation analysis análisis por
    activación
activation energy energía de activación
activator activador
active activo
active carbon carbón activo
active center centro activo
active component componente activo
active element elemento activo
active group grupo activo
active mass masa activa
active site lugar activo
active transport transporte activo
activity actividad
activity coefficient coeficiente de
    actividad
activity curve curva de actividad
activity series serie de actividad
actuator actuador
acyclic acíclico
acyclic compound compuesto acíclico
acyl acilo
acyl group grupo acilo
acyl halide haluro de acilo
acylation acilación
acyloin aciloína
acyloin condensation condensación
    aciloínica
adamantane adamantano
Adamkiewicz reaction reacción de
    Adamkiewicz
Adams catalyst catalizador de Adams
adamsite adamsita
adapter adaptador
adatom adátomo
addition adición
addition agent agente de adición
addition complex complejo de adición
addition polymer polímero de adición
addition reaction reacción de adición
addition resin resina de adición
additive aditivo
adduct aducto
adenine adenina
adenosine adenosina

adenosine diphosphate difosfato de adenosina

adenosine monophosphate monofosfato de adenosina

adenosine phosphate fosfato de adenosina

adenosine triphosphate trifosfato de adenosina

adenylic acid ácido adenílico

adherence adherencia

adhesion adhesión

adhesion agent agente de adhesión

adhesive adhesivo

adiabatic adiabático

adiabatic calorimeter calorímetro adiabático

adiabatic change cambio adiabático

adiabatic compression compresión adiabática

adiabatic cooling enfriamiento adiabático

adiabatic expansion expansión adiabática

adiabatic heating calentamiento adiabático

adiabatic process proceso adiabático

adipate adipato

adipic acid ácido adípico

adipinketone adipincetona

adiponitrile adiponitrilo

adjacent position posición adyacente

adjuvant adyuvante

Adkins-Peterson reaction reacción de Adkins-Peterson

adrenaline adrenalina

adsorbate adsorbato

adsorbent adsorbente

adsorption adsorción

adsorption analysis análisis de adsorción

adsorption chromatography cromatografía por adsorción

adsorption coefficient coeficiente de adsorción

adsorption indicator indicador de adsorción

adulterant adulterante

adulterated adulterado

adulteration adulteración

advection advección

aerate airear

aerated water agua aireada

aerobic aerobio, aeróbico

aerogel aerogel

aerometer aerómetro

aerometry aerometría

aerosol aerosol

affinity afinidad

affinity chromatography cromatografía por afinidad

affinity curve curva de afinidad

aflatoxin aflatoxina

afterglow luminiscencia persistente, luminiscencia residual

agar agar

agaric acid ácido agárico

agaricin agaricina

agate ágata

agent agente

agglomeration aglomeración

agglutinate aglutinar

agglutination aglutinación

aggregate agregado

aggregation agregación

aging envejecimiento

agitator agitador

aglucone aglucona

aglycone aglicona

agmatine agmatina

agricultural chemical producto químico agrícola

agricultural waste desperdicios agrícolas

Agulhon reagent reactivo de Agulhon

Aich metal metal de Aich

air aire

air bath baño de aire

air compressor compresor de aire

air filter filtro de aire

air oxidation oxidación por aire

air pollution contaminación de aire

air pump bomba de aire

air regulator regulador de aire

alabandite alabandina

alabaster alabastro

alacreatine alacreatina

alanine alanina

alantic acid ácido alántico

alanyl alanilo

alazopeptin alazopeptina

albaspidin albaspidina

albata albata

albedo albedo
albertite albertita
albite albita
albomycin albomicina
albumen albumen
albumin albúmina
albuminoid albuminoide
albumitate albuminato
albutoin albutoína
alclophenac alclofenac
alcogel alcogel
alcohol alcohol
alcoholate alcoholato
alcoholic alcohólico
alcoholic fermentation fermentación
  alcohólica
alcoholysis alcohólisis
alcosol alcosol
alcyl alcilo
aldehyde aldehído
aldehyde condensation condensación de
  aldehído
aldehyde ketone cetona de aldehído
aldehyde polymer polímero de aldehído
aldehydine aldehidina
Alder-Rickert rule regla de Alder-
  Rickert
Alder-Stein rules reglas de Alder-Stein
aldicarb aldicarb
aldohexose aldohexosa
aldoketene aldocetena
aldol aldol
aldol condensation condensación aldólica
aldolase aldolasa
aldonic acid ácido aldónico
aldopentose aldopentosa
aldose aldosa
aldosterone aldosterona
aldoxime aldoxima
aldrin aldrina
aleuritic acid ácido aleurítico
alexandrite alexandrita
alfin alfin
alfin catalyst catalizador alfin
algae algas
algarose algarosa
algicide algicida
algin algina
alginate alginato

alginic acid ácido algínico
algulose algulosa
alicyclic alicíclico
alicyclic compound compuesto alicíclico
alignment alineación
aliphatic alifático
aliphatic acid ácido alifático
aliphatic compound compuesto alifático
aliphatic group grupo alifático
aliphatic hydrocarbon hidrocarburo
  alifático
aliphatic polycyclic hydrocarbon
  hidrocarburo policíclico alifático
aliphatic series serie alifática
alipite alipita
aliquot alícuota
alite alita
alizaramide alizaramida
alizarate alizarato
alizaric acid ácido alizárico
alizarin alizarina
alizarin black negro de alizarina
alizarin blue azul de alizarina
alizarin cyanine cianina de alizarina
alizarin dye colorante de alizarina
alizarin red rojo de alizarina
alizarin sulfonate sulfonato de alizarina
alizarin yellow amarillo de alizarina
alkadiene alcadieno
alkalamide alcalamida
alkalescence alcalescencia
alkalescent alcalescente
alkali álcali
alkali alcoholate alcoholato alcalino
alkali blue azul alcalino
alkali metal metal alcalino
alkali reaction reacción alcalina
alkalide alcalida
alkalimeter alcalímetro
alkalimetry alcalimetría
alkaline alcalino
alkaline earth tierra alcalina
alkaline-earth metal metal alcalinotérreo
alkaline-earth oxide óxido alcalinotérreo
alkaline reaction reacción alcalina
alkaline solution solución alcalina
alkalinity alcalinidad
alkalize alcalizar
alkaloid alcaloide

alkaloidal alcaloidal
alkalometry alcalometría
alkamine alcamina
alkanal alcanal
alkane alcano
alkanesulfonic acid ácido alcanosulfónico
alkanization alcanización
alkannin alcanina
alkanol alcanol
alkanolamine alcanolamina
alkasal alcasal
alkene alqueno
alkene series serie de alquenos
alkoxide alcóxido
alkoxy alcoxi
alkoxylate alcoxilato
alkyd resin resina alquídica
alkyl alquilo
alkyl halide haluro de alquilo
alkyl nitrite nitrito de alquilo
alkyl oxide óxido de alquilo
alkylamine alquilamina
alkylate alquilato
alkylation alquilación
alkylbenzene alquilbenceno
alkylbenzene sulfonate sulfonato de alquilbenceno
alkyle alquilo
alkylene alquileno
alkylide alquiluro
alkylidene alquilideno
alkylolamide alquilolamida
alkylolamine alquilolamina
alkylphenol alquilfenol
alkyne alquino
Allan-Robinson reaction reacción de Allan-Robinson
allanic acid ácido alánico
allanite alanita
allantoic acid ácido alantoico
allantoin alantoína
allantoxanic acid ácido alantoxánico
allanturic acid ácido alantúrico
allelochemical aleloquímico
allelomorph alelomorfo
allelopathic chemical producto químico alelopático
allelotrope alelótropo

allelotropism alelotropismo
allemontite alemontita
allene aleno
allethrin aletrina
allicin alicina
alligation aligación
allo alo
allochromatic alocromático
allocinnamic acid ácido alocinámico
alloclasite aloclasita
alloisomerism aloisomerismo
allomaleic acid ácido alomaleico
allomeric alomérico
allomerism alomerismo
allomerization alomerización
allomorphism alomorfismo
allophanamide alofanamida
allophanate alofanato
allophane alófana
allophanic alofánico
allophanic acid ácido alofánico
allose alosa
allosteric effect efecto alostérico
allosteric enzyme enzima alostérica
allotoxin alotoxina
allotriomorphism alotriomorfismo
allotrope alótropo
allotropic alotrópico
allotropism alotropismo
allotropy alotropía
allowable permisible
allowed permitido
allowed band banda permitida
allowed transition transición permitida
alloxan aloxana
alloxanic acid ácido aloxánico
alloxanthin aloxantina
alloy aleación
alloy steel acero aleado
allulose alulosa
allyl alilo
allyl acetate acetato de alilo
allyl acrylate acrilato de alilo
allyl alcohol alcohol alílico
allyl aldehyde aldehído alílico
allyl benzoate benzoato de alilo
allyl bromide bromuro de alilo
allyl butyrate butirato de alilo
allyl carbamate carbamato de alilo

allyl chloride cloruro de alilo
allyl cinnamate cinamato de alilo
allyl cyanide cianuro de alilo
allyl ether éter alílico
allyl fluoride floruro de alilo
allyl group grupo alilo
allyl iodide yoduro de alilo
allyl isocyanate isocianato de alilo
allyl isosulfocyanate isosulfocianato de alilo
allyl isothiocyanate isotiocinato de alilo
allyl oxalate oxalato de alilo
allyl plastic plástico alílico
allyl polymer polímero de alilo
allyl resin resina alílica
allyl sulfide sulfuro de alilo
allyl thioether tioéter alílico
allyl tribromide tribromuro de alilo
allylacetone alilacetona
allylacetonitrile alilacetonitrilo
allylamine alilamina
allylaniline alilanilina
allylene alileno
allylic alílico
allylic rearrangement reordenamiento alílico
allylin alilina
allylthiourea aliltiourea
allyltrichlorosilane aliltriclorosilano
almandite almandita
Almen test prueba de Almen
almond oil aceite de almendra
alneon alneón
alnico álnico
aloe áloe
aloeresic acid ácido aloerésico
aloin aloína
aloxiprin aloxiprina
alpax alpax
alpha alfa
alpha cellulose celulosa alfa
alpha decay desintegración alfa
alpha disintegration desintegración alfa
alpha emission emisión alfa
alpha emitter emisor alfa
alpha helix hélice alfa
alpha particle partícula alfa
alpha position posición alfa
alpha radiator radiador alfa

alpha rays rayos alfa
alphahydroxy acid alfahidroxiácido
alphatopic alfatópico
alphyl alfilo
alstonidine alstonidina
alstonine alstonina
alternating polymer polímero alternante
alternation alternación
alternation law ley de alternación
altheine alteína
altrose altrosa
aludel aludel
alum alumbre
alumina alúmina
alumina gel gel de alúmina
alumina trihydrate trihidrato de alúmina
aluminate aluminato
aluminic acid ácido alumínico
aluminiferous aluminífero
aluminite aluminita
aluminium aluminio
aluminized aluminizado
aluminoferric aluminoférrico
aluminosilicate aluminosilicato
aluminum aluminio
aluminum acetate acetato alumínico, acetato de aluminio
aluminum acetotartrate acetotartrato alumínico
aluminum alkoxide alcóxido alumínico
aluminum alloy aleación alumínica
aluminum ammonium sulfate sulfato amónico alumínico
aluminum antimonide antimoniuro alumínico
aluminum benzoate benzoato alumínico
aluminum borate borato alumínico
aluminum boride boruro alumínico
aluminum boroformate boroformiato alumínico
aluminum borohydride borohidruro alumínico
aluminum brass latón de aluminio
aluminum bromate bromato alumínico
aluminum bromide bromuro alumínico
aluminum bronze bronce de aluminio
aluminum butoxide butóxido alumínico
aluminum calcium hydride hidruro cálcico alumínico

aluminum carbide carburo alumínico
aluminum carbonate carbonato alumínico
aluminum chlorate clorato alumínico
aluminum chloride cloruro alumínico
aluminum chlorohydrate clorohidrato alumínico
aluminum diacetate diacetato alumínico
aluminum dichromate dicromato alumínico
aluminum diformate diformiato alumínico
aluminum distearate diestearato alumínico
aluminum ethanoate etanoato alumínico
aluminum ethoxide etóxido alumínico
aluminum ethylate etilato alumínico
aluminum fluoride fluoruro alumínico
aluminum foil papel de aluminio
aluminum formate formiato alumínico
aluminum gluconate gluconato alumínico
aluminum hydrate hidrato alumínico
aluminum hydride hidruro alumínico
aluminum hydroxide hidróxido alumínico
aluminum iodide yoduro alumínico
aluminum isopropoxide isopropóxido alumínico
aluminum isopropylate isopropilato alumínico
aluminum lactate lactato alumínico
aluminum lithium hydride hidruro de litio alumínico
aluminum metaphosphate metafosfato alumínico
aluminum monostearate monoestearato alumínico
aluminum nitrate nitrato alumínico
aluminum nitride nitruro alumínico
aluminum oleate oleato alumínico
aluminum orthophosphate ortofosfato alumínico
aluminum oxalate oxalato alumínico
aluminum oxide óxido alumínico
aluminum palmitate palmitato alumínico
aluminum phenolate fenolato alumínico
aluminum phosphate fosfato alumínico
aluminum phosphide fosfuro alumínico

aluminum phosphinate fosfinato alumínico
aluminum potassium chloride cloruro potásico alumínico
aluminum potassium sulfate sulfato potásico alumínico
aluminum potassium tartrate tartrato potásico alumínico
aluminum propoxide propóxido alumínico
aluminum resinate resinato alumínico
aluminum rubidium sulfate sulfato de rubidio alumínico
aluminum salicylate salicilato alumínico
aluminum silicate silicato alumínico
aluminum sodium chloride cloruro sódico alumínico
aluminum sodium fluoride fluoruro sódico alumínico
aluminum sodium sulfate sulfato sódico alumínico
aluminum stearate estearato alumínico
aluminum sulfate sulfato alumínico
aluminum sulfide sulfuro alumínico
aluminum tartrate tartrato alumínico
aluminum thallium sulfate sulfato de talio alumínico
aluminum thiocyanate tiocianato alumínico
aluminum triacetate triacetato alumínico
aluminum zinc sulfate sulfato de cinc alumínico
alumite alumita
alunite alunita
alunogen alonógeno
alunogenite alunogenita
alvite alvita
Amadori rearrangement reordenamiento de Amadori
amalgam amalgama
amalgamation amalgamación
amalgamation process proceso de amalgamación
amalgamator amalgamador
amalinic acid ácido amalínico
amanitine amanitina
amaranth amaranto
amarine amarina
amaroid amaroide

amatol amatol
amazonite amazonita
amber ámbar
ambident ambidente
ambident anion anión ambidente
ambient air aire ambiente
ambient pressure presión ambiente
ambient temperature temperatura
    ambiente
amblygonite ambligonita
ambrite ambrita
ameboid ameboide
americium americio
amesite amesita
amethocaine ametocaína
amethyst amatista
amicron amicrón
amidase amidasa
amidation amidación
amide amida
amide chloride cloruro de amida
amide group grupo amida
amidine amidina
amidino amidino
amido amido
amidol amidol
amidone amidona
amidrazone amidrazona
amination aminación
amine amina
aminic acid ácido amínico
aminimide aminimida
amino amino
amino acid aminoácido
amino acid oxidase aminoácido oxidasa
amino alcohol aminoalcohol
amino group grupo amino
amino oxidase aminooxidasa
amino resin aminorresina
aminoacetanilide aminoacetanilida
aminoacetic acid ácido aminoacético
aminoacetone aminoacetona
aminoacetophenone aminoacetofenona
aminoanthraquinone aminoantraquinona
aminoantipyrine aminoantipirina
aminoazo aminoazo
aminoazobenzene aminoazobenceno
aminoazobenzene hydrochloride
    clorhidrato de aminoazobenceno

aminoazonaphthalene
    aminoazonaftaleno
aminoazotoluene aminoazotolueno
aminobenzene aminobenceno
aminobenzenesulfonic acid ácido
    aminobencenosulfónico
aminobenzoic acid ácido aminobenzoico
aminobenzothiazole aminobenzotiazol
aminobiphenyl aminobifenilo
aminobutane aminobutano
aminobutanoic acid ácido
    aminobutanoico
aminobutyric acid ácido aminobutírico
aminocaproic acid ácido aminocaproico
aminocyclohexane aminociclohexano
aminodiethylaniline aminodietilanilina
aminodimethylaniline
    aminodimetilanilina
aminodimethylbenzene
    aminodimetilbenceno
aminodiphenyl aminodifenilo
aminodiphenylamine aminodifenilamina
aminodithioformic acid ácido
    aminoditiofórmico
aminoethane aminoetano
aminoethanol aminoetanol
aminoethyl alcohol alcohol aminoetílico
aminoethylsulfuric acid ácido
    aminoetilsulfúrico
aminoform aminoformo
aminoglutaric acid ácido aminoglutárico
aminohexane aminohexano
aminohexanoic acid ácido
    aminohexanoico
aminohippuric acid ácido aminohipúrico
aminoid aminoide
aminoisobutyric acid ácido
    aminoisobutírico
aminoketone aminocetona
aminomethane aminometano
aminomethylation aminometilación
aminonaphthalene aminonaftaleno
aminonaphthol aminonaftol
aminonaphtholsulfonic acid ácido
    aminonaftolsulfónico
aminonicotinic acid ácido
    aminonicotínico
aminophenol aminofenol
aminophenylacetic acid ácido

aminofenilacético

aminophosphoric acid ácido aminofosfórico

aminophylline aminofilina

aminopicoline aminopicolina

aminopropane aminopropano

aminopropanoic acid ácido aminopropanoico

aminopropanol aminopropanol

aminopropene aminopropeno

aminopropyl alcohol alcohol aminopropílico

aminopropylmorpholine aminopropilmorfolina

aminopterin aminopterina

aminopurine aminopurina

aminopyridine aminopiridina

aminopyrine aminopirina

aminoquinoline aminoquinolina

aminosalicylic acid ácido aminosalicílico

aminothiazole aminotiazol

aminothiopene aminotiopeno

aminothiourea aminotiourea

aminotoluene aminotolueno

aminotriazole aminotriazol

aminourea aminourea

aminoxylene aminoxileno

aminyl aminilo

amithiozone amitiozona

amitrole amitrol

ammeline amelina

ammine amina

ammonation amonación

ammonia amoníaco

ammonia nitrogen nitrógeno amoniacal

ammonia water agua amoniacal

ammoniacal amoniacal

ammoniated iron hierro amoniacal

ammoniated mercury mercurio amoniacal

ammoniated mercury chloride cloruro mercúrico amoniacal

ammoniated superphosphate superfosfato amoniacal

ammonification amonificación

ammonium amonio

ammonium acetate acetato amónico, acetato de amonio

ammonium acid carbonate carbonato

ácido amónico

ammonium acid fluoride fluoruro ácido amónico

ammonium acid phosphate fosfato ácido amónico

ammonium acid tartrate tartrato ácido amónico

ammonium aldehyde aldehído amónico

ammonium alginate alginato amónico

ammonium alum alumbre amónico

ammonium aluminum sulfate sulfato de aluminio amónico

ammonium aminosulfonate aminosulfonato amónico

ammonium antimoniate antimoniato amónico

ammonium arsenate arseniato amónico

ammonium arsenite arsenito amónico

ammonium benzenesulfonate bencenosulfonato amónico

ammonium benzoate benzoato amónico

ammonium bicarbonate bicarbonato amónico

ammonium bichromate bicromato amónico

ammonium bifluoride bifluoruro amónico

ammonium bioxalate bioxalato amónico

ammonium biphosphate bifosfato amónico

ammonium bisulfate bisulfato amónico

ammonium bisulfite bisulfito amónico

ammonium bitartrate bitartrato amónico

ammonium borate borato amónico

ammonium bromide bromuro amónico

ammonium caprylate caprilato amónico

ammonium carbamate carbamato amónico

ammonium carbazotate carbazotato amónico

ammonium carbonate carbonato amónico

ammonium chlorate clorato amónico

ammonium chloride cloruro amónico

ammonium chromate cromato amónico

ammonium chromium sulfate sulfato de cromo amónico

ammonium citrate citrato amónico

ammonium cuprate cuprato amónico

ammonium cyanide cianuro amónico

ammonium dichromate dicromato amónico

ammonium dithiocarbamate ditiocarbamato amónico

ammonium dithionate ditionato amónico

ammonium ferrocyanide ferrocianuro amónico

ammonium fluoride fluoruro amónico

ammonium fluoroborate fluoroborato amónico

ammonium fluosilicate fluosilicato amónico

ammonium formate formiato amónico

ammonium gallate galato amónico

ammonium gluconate gluconato amónico

ammonium glutamate glutamato amónico

ammonium hexabromoplatinate hexabromoplatinato amónico

ammonium hexachloroplatinate hexacloroplatinato amónico

ammonium hexachlorostannate hexacloroestannato amónico

ammonium hexacyanoferrate hexacianoferrato amónico

ammonium hexafluoroaluminate hexafluoroaluminato amónico

ammonium hexafluorophosphate hexafluorofosfato amónico

ammonium hexafluorosilicate hexafluorosilicato amónico

ammonium hydrate hidrato amónico

ammonium hydrogen carbonate carbonato de hidrógeno amónico

ammonium hydrogen difluoride difluoruro de hidrógeno amónico

ammonium hydrogen fluoride fluoruro de hidrógeno amónico

ammonium hydrogen sulfate sulfato de hidrógeno amónico

ammonium hydrogen sulfide sulfuro de hidrógeno amónico

ammonium hydrogen tartrate tartrato de hidrógeno amónico

ammonium hydrosulfide hidrosulfuro amónico

ammonium hydroxide hidróxido

amónico

ammonium hypophosphate hipofosfato amónico

ammonium iodate yodato amónico

ammonium iodide yoduro amónico

ammonium ion ion amónico

ammonium iron bromide bromuro de hierro amónico

ammonium iron chloride cloruro de hierro amónico

ammonium iron chromate cromato de hierro amónico

ammonium iron citrate citrato de hierro amónico

ammonium iron oxalate oxalato de hierro amónico

ammonium iron sulfate sulfato de hierro amónico

ammonium iron tartrate tartrato de hierro amónico

ammonium lactate lactato amónico

ammonium laurate laurato amónico

ammonium linoleate linoleato amónico

ammonium magnesium arsenate arseniato magnésico amónico

ammonium magnesium chloride cloruro magnésico amónico

ammonium magnesium phosphate fosfato magnésico amónico

ammonium magnesium sulfate sulfato magnésico amónico

ammonium manganese phosphate fosfato de manganeso amónico

ammonium manganese sulfate sulfato de manganeso amónico

ammonium mercuric chloride cloruro mercúrico amónico

ammonium metaphosphate metafosfato amónico

ammonium metavanadate metavanadato amónico

ammonium molybdate molibdato amónico

ammonium nickel chloride cloruro de níquel amónico

ammonium nickel sulfate sulfato de níquel amónico

ammonium nitrate nitrato amónico

ammonium nitrite nitrito amónico

ammonium oleate oleato amónico
ammonium oxalate oxalato amónico
ammonium oxamate oxamato amónico
ammonium palmitate palmitato amónico
ammonium pentaborate pentaborato amónico
ammonium perchlorate perclorato amónico
ammonium perchromate percromato amónico
ammonium permanganate permanganato amónico
ammonium peroxoborate peroxoborato amónico
ammonium peroxodisulfate peroxodisulfato amónico
ammonium persulfate persulfato amónico
ammonium phenolsulfonate fenolsulfonato amónico
ammonium phosphate fosfato amónico
ammonium phosphite fosfito amónico
ammonium phosphomolybdate fosfomolibdato amónico
ammonium phosphotungstate fosfotungstato amónico
ammonium picrate picrato amónico
ammonium polyphosphate polifosfato amónico
ammonium polysulfide polisulfuro amónico
ammonium ricinoleate ricinoleato amónico
ammonium salicylate salicilato amónico
ammonium salt sal amónica
ammonium selenate seleniato amónico
ammonium selenite selenito amónico
ammonium sodium phosphate fosfato sódico amónico
ammonium sodium sulfate sulfato sódico amónico
ammonium stearate estearato amónico
ammonium succinate succinato amónico
ammonium sulfamate sulfamato amónico
ammonium sulfate sulfato amónico
ammonium sulfide sulfuro amónico
ammonium sulfite sulfito amónico
ammonium tartrate tartrato amónico

ammonium tellurate telurato amónico
ammonium thioacetate tioacetato amónico
ammonium thiocyanate tiocianato amónico
ammonium thiostanate tioestannato amónico
ammonium thiosulfate tiosulfato amónico
ammonium trinitrophenolate trinitrofenolato amónico
ammonium tungstate tungstato amónico
ammonium valerate valerato amónico
ammonium vanadate vanadato amónico
ammonium zirconyl carbonate carbonato de zirconilo amónico
ammonolysis amonólisis
amobarbital amobarbital
amorphism amorfismo
amorphous amorfo
amorphous polymer polímero amorfo
amorphous state estado amorfo
amount of substance cantidad de substancia
ampelite ampelita
amperometric titration titulación amperométrica, valoración amperométrica
amperometry amperometría
amphetamine anfetamina
amphibole anfibol
amphibolite anfibolita
amphichroic anficroico
amphichromatic anficromático
amphiphatic anfifático
amphiphile anfífilo
amphiphilic anfifílico
amphiprotic anfiprótico
ampholyte anfolito
ampholytoid anfolitoide
amphoteric anfótero
amphoteric electrolyte electrólito anfótero
amphoteric hydroxide hidróxido anfótero
amphoteric oxide óxido anfótero
amphoterism anfoterismo
ampicillin ampicilina
ampul ampolla

ampule ampolla
amsonic acid ácido amsónico
amyctic amíctico
amygdalic acid ácido amigdálico
amygdalin amigdalina
amyl amilo
amyl acetate acetato de amilo
amyl acid phosphate fosfato ácido de amilo
amyl alcohol alcohol amílico
amyl aldehyde aldehído amílico
amyl benzoate benzoato de amilo
amyl butyrate butirato de amilo
amyl chloride cloruro de amilo
amyl ethanoate etanoato de amilo
amyl ether éter amílico
amyl formate formiato de amilo
amyl hydrate hidrato de amilo
amyl hydride hidruro de amilo
amyl nitrate nitrato de amilo
amyl nitrite nitrito de amilo
amyl propionate propionato de amilo
amyl salicylate salicilato de amilo
amyl sulfide sulfuro de amilo
amyl xanthate xantato de amilo
amylacetic ester éster amilacético
amylamine amilamina
amylase amilasa
amylate amilato
amylbenzene amilbenceno
amylcarbinol amilcarbinol
amylcinnamic alcohol alcohol amilcinámico
amylcinnamic aldehyde aldehído amilcinámico
amylene amileno
amylene dichloride dicloruro de amileno
amylene hydrate hidrato de amileno
amyloid amiloide
amylolysis amilólisis
amylolytic amilolítico
amylopectin amilopectina
amylose amilosa
amylphenol amilfenol
amyltrichlorosilane amiltriclorosilano
amyrin amirina
anabasine anabasina
anabolic anabólico
anabolism anabolismo

anacardic acid ácido anacárdico
anaerobe anaerobio
anaerobic anaerobio
analog display visualizador analógico
analysis análisis
analytical analítico
analytical balance balanza analítica
analytical chemistry química analítica
analytical distillation destilación analítica
analytical extraction extracción analítica
analytical reaction reacción analítica
analyzer analizador
anamorphism anamorfismo
anaphoresis anaforesis
anaphylaxis anafilaxis
anarcotine anarcotina
anatase anatasa
anchoic acid ácido ancoico
andalusite andalucita
andesine andesina
androgen andrógeno
androstane androstano
androsterone androsterona
anemometer anemómetro
anemonic acid ácido anemónico
anemosite anemosita
anesthesia anestesia
anesthetic anestésico
anethole anetol
angelic acid ácido angélico
angelica lactone lactona de angélica
angle ángulo
anglesite anglesita
angosturine angosturina
angstrom angstrom
angstrom unit unidad angstrom
angular angular
anharmonic anarmónico
anharmonicity anarmonicidad
anhydride anhídrido
anhydrite anhidrita
anhydro anhidro
anhydrone anhidrona
anhydrosynthesis anhidrosíntesis
anhydrous anhidro
anhydrous alcohol alcohol anhidro
anhydrous ammonia amoníaco anhidro
anhydrous hydrogen chloride cloruro de

hidrógeno anhidro
anhydrous phosphoric acid ácido
  fosfórico anhidro
anhydrous sodium sulfate sulfato sódico
  anhidro
anibine anibina
anilide anilida
anilinate anilinato
aniline anilina
aniline acetate acetato de anilina
aniline black negro de anilina
aniline blue azul de anilina
aniline chloride cloruro de anilina
aniline dye colorante de anilina
aniline fluoride fluoruro de anilina
aniline hydrochloride clorhidrato de
  anilina
aniline oxalate oxalato de anilina
aniline point punto de anilina
aniline salt sal de anilina
aniline sulfate sulfato de anilina
aniline yellow amarillo de anilina
anilinesulfonic acid ácido
  anilinosulfónico
anilinium anilinio
anilino anilino
anion anión
anionic aniónico
anionic polymerization polimerización
  aniónica
anionic resin resina aniónica
anionotropy anionotropía
anisal anisal
anisaldehyde anisaldehído
anise alcohol alcohol anísico
anise oil aceite de anís
anisic anísico
anisic acid ácido anísico
anisic alcohol alcohol anísico
anisic aldehyde aldehído anísico
anisidine anisidina
anisole anisol
anisomeric anisomérico
anisonitrile anisonitrilo
anisotonic anisotónico
anisotonic solution solución anisotónica
anisotropic anisótropo, anisotrópico
anisoyl anisoilo
anisoyl chloride cloruro de anisoilo

anisyl anisilo
anisyl acetate acetato de anisilo
anisyl alcohol alcohol anisílico
anisylacetone anisilacetona
annealing recocido
annular anular
annulene anuleno
anode ánodo
anode coating revestimiento anódico
anode effect efecto anódico
anode film película anódica
anode layer capa anódica
anode mud residuos anódicos, lodo
  anódico
anode oxidation oxidación anódica
anode polarization polarización anódica
anode slime residuos anódicos, lodo
  anódico
anode solution solución anódica
anodic anódico
anodic coating revestimiento anódico
anodic effect efecto anódico
anodic film película anódica
anodic layer capa anódica
anodic mud residuos anódicos, lodo
  anódico
anodic oxidation oxidación anódica
anodic polarization polarización anódica
anodic slime residuos anódicos, lodo
  anódico
anodic solution solución anódica
anodize anodizar
anodizing anodización
anolyte anolito
anomalous anómalo
anomalous dispersion dispersión
  anómala
anomaly anomalía
anomer anómero
anomeric anomérico
anorganic anorgánico
anorthic anórtico
anorthite anortita
antacid anitácido
antagonism antagonismo
antagonist antagonista
antalkaline antalcalino
anthemol antemol
anthion antión

anthocyanidin antocianidina
anthocyanin antocianina
anthophyllite antofilita
anthracene antraceno
anthracene blue azul de antraceno
anthracene oil aceite de antraceno
anthracene violet violeta de antraceno
anthracite antracita
anthradiol antradiol
anthragallol antragalol
anthrahydroquinone antrahidroquinona
anthraldehyde antraldehído
anthralin antralina
anthranil antranil
anthranilic acid ácido antranílico
anthranol antranol
anthranone antranona
anthraparazene antraparaceno
anthrapurpurin antrapurpurina
anthraquinone antraquinona
anthraquinone dye colorante de
    antraquinona
anthrarufin antrarrufina
anthratetrol antratetrol
anthratriol antratriol
anthroic acid ácido antroico
anthrol antrol
anthrone antrona
anthryl antrilo
anthrylene antrileno
antiaromatic antiaromático
antibacterial antibacterial, antibacteriano
antibiotic antibiótico
antiblock agent agente anibloqueo
antibody anticuerpo
antibonding antienlazado
antibonding orbital orbital de
    antienlazado, orbital antienlazante
anticatalyzer anticatalizador
anticathode anticátodo
antichlor anticloro
anticholingeric anticolinérgico
anticoagulant anticoagulante
anticomplement anticomplemento
anticorrosive anticorrosivo
antidote antídoto
antienzyme antienzima
antiferromagnetic antiferromagnético
antiferromagnetism

antiferromagnetismo
antifluorite structure estructura
    antifluorita
antifoam agent agente antiespumante
antifoaming agent agente antiespumante
antiformin antiformina
antifreeze anticongelante
antigen antígeno
antiglobulin antiglobulina
antigorite antigorita
antihistamine antihistamina
antiisomer antiisómero
antiisomorphism antiisomorfismo
antiknock antigolpeteo, antidetonante
antiknock additive aditivo antigolpeteo
antiknock agent agente antigolpeteo
antimatter antimateria
antimer antímero
antimetabolite antimetabolito
antimonate antimoniato
antimonic antimónico
antimonic acid ácido antimónico
antimonic anhydride anhídrido
    antimónico
antimonic chloride cloruro antimónico
antimonic compound compuesto
    antimónico
antimonic fluoride fluoruro antimónico
antimonic sulfide sulfuro antimónico
antimonite antimonita
antimonous antimonioso
antimonous bromide bromuro
    antimonioso
antimonous chloride cloruro
    antimonioso
antimonous compound compuesto
    antimonioso
antimonous fluoride fluoruro
    antimonioso
antimonous hydride hidruro antimonioso
antimonous iodide yoduro antimonioso
antimonous oxalate oxalato antimonioso
antimonous oxide óxido antimonioso
antimonous oxychloride oxicloruro
    antimonioso
antimonous oxysulfide oxisulfuro
    antimonioso
antimonous sulfate sulfato antimonioso
antimonous sulfide sulfuro antimonioso

antimony antimonio
antimony anhydride anhídrido de antimonio
antimony arsenide arseniuro de antimonio
antimony black negro de antimonio
antimony bromide bromuro de antimonio
antimony chloride cloruro de antimonio
antimony dichlorotrifluoride diclorotrifluoruro de antimonio
antimony fluoride fluoruro de antimonio
antimony hydride hidruro de antimonio
antimony iodide yoduro de antimonio
antimony lactate lactato de antimonio
antimony orange anaranjado de antimonio
antimony oxide óxido de antimonio
antimony oxychloride oxicloruro de antimonio
antimony pentachloride pentacloruro de antimonio
antimony pentafluoride pentafluoruro de antimonio
antimony pentasulfide pentasulfuro de antimonio
antimony pentoxide pentóxido de antimonio
antimony persulfide persulfuro de antimonio
antimony potassium oxalate oxalato potásico de antimonio
antimony potassium tartrate tartrato potásico de antimonio
antimony red rojo de antimonio
antimony salt sal de antimonio
antimony sodium tartrate tartrato sódico de antimonio
antimony sulfate sulfato de antimonio
antimony sulfide sulfuro de antimonio
antimony tribromide tribromuro de antimonio
antimony trichloride tricloruro de antimonio
antimony trifluoride trifluoruro de antimonio
antimony triiodide triyoduro de antimonio
antimony trioxide trióxido de antimonio

antimony trisulfate trisulfato de antimonio
antimony trisulfide trisulfuro de antimonio
antimony yellow amarillo de antimonio
antimonyl antimonilo
antimycin antimicina
antineutron antineutrón
antioxidant antioxidante
antioxygen antioxígeno
antiozonant antiozonizante
antiparallel spin espín antiparalelo
antiparasitic antiparasítico
antiparticle antipartícula
antiperspirant antiperspirante
antiproton antiprotón
antipsychotic antipsicótico
antipyrene antipireno
antipyrene benzoate benzoato de antipireno
antipyrene monobromide monobromuro de antipireno
antipyretic antipirético
antipyrine antipirina
antipyryl antipirilo
antiquinol antiquinol
antiradiation antirradiación
antiseptic antiséptico
antistatic antiestático
antistatic agent agente antiestático
antitoxic antitóxico
antitoxin antitoxina
antitussive antitusivo
antiviral antiviral
antizymotic antizimótico
antozone antozono
anytol anitol
apatite apatita
aphrodine afrodina
apiolic acid ácido apiólico
apionol apionol
apoatropine apoatropina
apocarotenal apocarotenal
apocholic acid ácido apocólico
apocodeine apocodeína
apolar adsorption adsorción apolar
apomorphine apomorfina
apophyllite apofilita
apparatus aparato

apparent density densidad aparente
apple acid ácido de manzana
apple oil aceite de manzana
applied chemistry química aplicada
applied research investigación aplicada
approved name nombre aprobado
approximate aproximado
aprotic aprótico
aprotic solvent solvente aprótico
aprotic substance substancia aprótica
aqua regia agua regia
aquation acuación
aqueous acuoso
aqueous solution solución acuosa
aqueous solvent solvente acuoso
aqueous tension tensión acuosa
arabic gum goma arábiga
arabin arabina
arabinose arabinosa
arabinoside arabinosida
arabite arabita
arabitol arabitol
arabonic acid ácido arabónico
arachidic acid ácido araquídico
arachidonic acid ácido araquidónico
aragonite aragonita
aralkyl aralquilo
aramid aramida
arborescent arborescente
arbutin arbutina
arc spectrum espectro de arco
areametric analysis análisis areamétrico
arecaidine arecaidina
arecaine arecaína
arecoline arecolina
arene areno
areometer areómetro
argenate argenato
argentamine argentamina
argentan argentán
argentic argéntico
argentic compound compuesto argéntico
argentic oxide óxido argéntico
argentite argentita
argentocyanide argentocianuro
argentometry argentometría
argentopyrite argentopirita
arginase arginasa
arginine arginina

argol argol
argon argón
argyrodite argirodita
aribine aribina
armored thermometer termómetro armado
Armstrong acid ácido de Armstrong
Arndt alloy aleación de Arndt
Arndt-Eistert synthesis síntesis de Arndt-Eistert
Arndt tube tubo de Arndt
arnicin arnicina
aromatic aromático
aromatic acid ácido aromático
aromatic alcohol alcohol aromático
aromatic aldehyde aldehído aromático
aromatic amine amina aromática
aromatic compound compuesto aromático
aromatic hydrocarbon hidrocarburo aromático
aromatic ketone cetona aromática
aromatic nucleus núcleo aromático
aromatic series serie aromática
aromaticity aromaticidad
aromatization aromatización
aroyl aroilo
arrhenic acid ácido arrénico
Arrhenius equation ecuación de Arrhenius
arrowroot arrurruz
arsacetin arsacetina
arsanilate arsanilato
arsanilic acid ácido arsanílico
arsenate arseniato
arsenenic acid ácido arsenénico
arsenic arsénico
arsenic acid ácido arsénico
arsenic anhydride anhídrido arsénico
arsenic bromide bromuro de arsénico
arsenic chloride cloruro de arsénico
arsenic diiodide diyoduro de arsénico
arsenic disulfide disulfuro de arsénico
arsenic fluoride fluoruro de arsénico
arsenic hydride hidruro de arsénico
arsenic iodide yoduro de arsénico
arsenic oxide óxido de arsénico
arsenic pentafluoride pentafluoruro de arsénico

arsenic pentasulfide pentasulfuro de arsénico

arsenic pentoxide pentóxido de arsénico

arsenic selenide seleniuro de arsénico

arsenic sulfide sulfuro de arsénico

arsenic tetrasulfide tetrasulfuro de arsénico

arsenic thioarsenate tioarseniato de arsénico

arsenic thiocyanate tiocianato de arsénico

arsenic tribromide tribromuro de arsénico

arsenic trichloride tricloruro de arsénico

arsenic trifluoride trifluoruro de arsénico

arsenic trioxide trióxido de arsénico

arsenic trisulfide trisulfuro de arsénico

arsenical arsenical

arsenide arseniuro

arsenidine arsenidina

arsenious arsenioso

arsenite arsenito

arseno arseno

arsenobenzene arsenobenceno

arsenolite arsenolita

arsenopyrite arsenopirita

arsenous arsenioso

arsenous acid ácido arsenioso

arsenous bromide bromuro arsenioso

arsenous chloride cloruro arsenioso

arsenous fluoride fluoruro arsenioso

arsenous hydride hidruro arsenioso

arsenous iodide yoduro arsenioso

arsenous oxide óxido arsenioso

arsenous oxychloride oxicloruro arsenioso

arsenous phosphide fosfuro arsenioso

arsenous selenide seleniuro arsenioso

arsenous sulfide sulfuro arsenioso

arsenyl arsenilo

arsine arsina

arsinic acid ácido arsínico

arsinoso arsinoso

arsonate arsonato

arsonic acid ácido arsónico

arsonium arsonio

arsono arsono

arsonoso arsonoso

arsonoyl arsonoilo

arthranitin artranitín

artificial artificial

artificial element elemento artificial

artificial radioactive element elemento radiactivo artificial

artificial radioactivity radiactividad artificial

artificial rubber caucho artificial

artinite artinita

aryl arilo

arylalkyl arilalquilo

arylamine arilamina

arylarsonate arilarsonato

arylene arileno

arylide ariluro

aryne arino

asarone asarona

asaronic acid ácido asarónico

asaryl asarilo

asbestine asbestina

asbestos asbesto

asbolane asbolana

asbolite asbolita

ascaridic acid ácido ascarídico

ascaridole ascaridol

ascaridolic acid ácido ascaridólico

ascorbic acid ácido ascórbico

ascorbic acid oxidase oxidasa de ácido ascórbico

aseptol aseptol

aseptoline aseptolina

ash ceniza

ashing incineración

asiatic acid ácido asiático

asiminine asiminina

askarel askarel

asparagic acid ácido asparágico

asparaginase asparaginasa

asparagine asparagina

asparaginic acid ácido asparagínico

asparamide asparamida

aspariginyl aspariginilo

aspartame aspartama

aspartamic acid ácido aspartámico

aspartamide aspartamida

aspartase aspartasa

aspartate aspartato

aspartic acid ácido aspártico

aspartoyl aspartoilo
aspartyl aspartilo
aspergillic acid ácido aspergílico
asphalt asfalto
asphaltene asfalteno
asphaltic bitumen bitumen asfáltico
asphaltite asfaltita
asphyxiant gas gas asfixiante
aspidinol aspidinol
aspidosamine aspidosamina
aspidospermine aspidospermina
aspirator aspirador
aspirin aspirina
assay ensayo
assimilation asimilación
associated liquids líquidos asociados
association asociación
astacene astaceno
astatic astátco
astatine astatino, ástato
asterisk asterisco
Aston rule regla de Aston
Aston spectrum espectro de Aston
astorism astorismo
astrochemistry astroquímica
astrophyllite astrofilita
asymmetric asimétrico
asymmetric atom átomo asimétrico
asymmetric carbon carbono asimétrico
asymmetric carbon atom átomo de
    carbono asimétrico
asymmetric compound compuesto
    asimétrico
asymmetric induction inducción
    asimétrica
asymmetric structure estructura
    asimétrica
asymmetric synthesis síntesis asimétrica
asymmetrical asimétrico
asymmetry asimetría
atacamite atacamita
atactic atáctico
atactic polymer polímero atáctico
ate ato
atenolol atenolol
atisine atisina
atmolysis atmólisis
atmometer atmómetro
atmosphere atmósfera

atmospheric atmosférico
atmospheric gas gas atmosférico
atmospheric humidity humedad
    atmosférica
atmospheric pollution contaminación
    atmosférica
atmospheric pressure presión
    atmosférica
atom átomo
atomic atómico
atomic absorption absorción atómica
atomic absorption coefficient coeficiente
    de absorción atómica
atomic absorption spectroscopy
    espectroscopia de absorción atómica
atomic attraction atracción atómica
atomic charge carga atómica
atomic constant constante atómica
atomic diamagnetism diamagnetismo
    atómico
atomic diameter diámetro atómico
atomic disintegration desintegración
    atómica
atomic distance distancia atómica
atomic electricity electricidad atómica
atomic emission spectroscopy
    espectroscopia de emisión atómica
atomic energy energía atómica
atomic energy level nivel de energía
    atómica
atomic excitation excitación atómica
atomic field campo atómico
atomic fission fisión atómica
atomic force microscope microscopio de
    fuerza atómica
atomic fragment fragmento atómico
atomic frequency frecuencia atómica
atomic fusion fusión atómica
atomic group grupo atómico
atomic heat calor atómico
atomic hydrogen hidrógeno atómico
atomic hydrogen welding soldadura con
    hidrógeno atómico
atomic level nivel atómico
atomic magnetic moment momento
    magnético atómico
atomic mass masa atómica
atomic mass number número de masa
    atómica

atomic mass unit unidad de masa atómica
atomic migration migración atómica
atomic model modelo atómico
atomic nucleus núcleo atómico
atomic number número atómico
atomic orbit órbita atómica
atomic orbital orbital atómico
atomic oscillation oscilación atómica
atomic particle partícula atómica
atomic plane plano atómico
atomic polarization polarización atómica
atomic potential potencial atómico
atomic properties propiedades atómicas
atomic radius radio atómico
atomic refraction refracción atómica
atomic resonance resonancia atómica
atomic scale escala atómica
atomic scattering factor factor de dispersión atómica
atomic species especie atómica
atomic spectroscopy espectroscopia atómica
atomic spectrum espectro atómico
atomic structure estructura atómica
atomic susceptibility susceptibilidad atómica
atomic symbol símbolo atómico
atomic theory teoría atómica
atomic transformation transformación atómica
atomic transmutation transmutación atómica
atomic unit unidad atómica
atomic vibration vibración atómica
atomic volume volumen atómico
atomic weight peso atómico
atomic weight unit unidad de peso atómico
atomically atómicamente
atomicity atomicidad
atomization atomización
atomizer atomizador
atomology atomología
atomsite atomsita
atopite atopita
atrazine atrazina
atrolactic acid ácido atroláctico
atronene atroneno

atronic acid ácido atrónico
atropic acid ácido atrópico
atropine atropina
atropine nitrate nitrato de atropina
atropine salicylate salicilato de atropina
atropine stearate estearato de atropina
atropine sulfate sulfato de atropina
atropine valerate valerato de atropina
atropoyl atropoilo
attapulgite attapulgita
attar aceite esencial
attenuation atenuación
attenuation coefficient coeficiente de atenuación
attenuation factor factor de atenuación
attraction atracción
Atwater calorimeter calorímetro de Atwater
aubepine aubepina
Auger effect afecto de Auger
Auger electron electrón de Auger
augite augita
auramine auramina
aurate aurato
aurelite aurelita
aureolin aureolina
auribromhydric acid ácido auribromhídrico
auribromide auribromuro
auric áurico
auric acid ácido áurico
auric bromide bromuro áurico
auric chloride cloruro áurico
auric compound compuesto áurico
auric cyanide cianuro áurico
auric hydroxide hidróxido áurico
auric iodide yoduro áurico
auric oxide óxido áurico
auric sulfate sulfato áurico
auric sulfide sulfuro áurico
aurichalcite auricalcita
aurichloride auricloruro
auriferous aurífero
aurin aurina
auroauric auroáurico
aurobromide aurobromuro
aurochloride aurocloruro
auromine auromina
aurone aurona

aurous auro
aurous bromide aurobromuro
aurous chloride aurocloruro
aurous cyanide aurocianuro
aurous iodide auroyoduro
aurous sulfide aurosulfuro
austenite austenita
autocatalysis autocatálisis
autocatalytic autocatalítico
autoclave autoclave
autodecomposition autodescomposición
autoexcitation autoexcitación
autoignition point punto de autoignición
autoionization autoionización
autoluminescence autoluminiscencia
autolysis autólisis
automatic automático
automatic buret bureta automática
automatic control control automático
automatic pipet pipeta automática
automatic titration titulación
    automática, valoración automática
automation automatización
autooxidation autooxidación
autoprotolysis autoprotólisis
autunite autunita
auxiliary auxiliar
auxin auxina
auxochrome auxocromo
availability disponibilidad
available disponible
available electron electrón disponible
avalite avalita
avenine avenina
average molecular weight peso
    molecular medio
avidin avidina
avocado oil aceite de aguacate
Avogadro constant constante de
    Avogadro
Avogadro hypothesis hipótesis de
    Avogadro
Avogadro law ley de Avogadro
Avogadro number número de Avogadro
Avogadro theory teoría de Avogadro
axial axial
axis eje
axis of symmetry eje de simetría
azaguanine azaguanina

azauridine azauridina
azelaic acid ácido azelaico
azelate azelato
azeotrope azeótropo
azeotropic azeotrópico
azeotropic distillation destilación
    azeotrópica
azeotropic mixture mezcla azeotrópica
azetidine azetidina
azide azida
azido azido
azimide azimida
azimido azimido
azimidobenzene azimidobenceno
azimino azimino
aziminobenzene aziminobenceno
azimuthal quantum number número
    cuántico acimutal
azine azina
azine dye colorante de azina
azino azino
aziridine aziridina
azlon azlón
azo azo
azo dye colorante azoico, azocolorante
azobenzene azobenceno
azobenzoic acid ácido azobenzoico
azobisisobutyronitrile
    azobisisobutironitrilo
azocyanide azocianuro
azodicarbonamide azodicabornamida
azodine azodina
azoic azoico
azoimide azoimida
azole azol
azomethine azometina
azonaphthalene azonaftaleno
azonium azonio
azophenol azofenol
azophenyl azofenilo
azophenylene azofenileno
azorite azorita
azosulfamide azosulfamida
azotoluene azotolueno
azotometer azotómetro
azoxy azoxi
azoxybenzene azoxibenceno
azoxynaphthalene azoxinaftaleno
azoxytoluidine azoxitoluidina

azulene azuleno
azulin azulina

azurite azurita

B

babassu oil aceite de babasú
Babbitt metal metal de Babbitt
Babcock test prueba de Babcock
Babo absorption tube tubo de absorción de Babo
Babo law ley de Babo
baccarin bacarina
bacillus bacilo
bacitracin bacitracina
back donation retrodonación
back-extraction retroextracción
Back-Goudsmit effect efecto de Back-Goudsmit
back titration retrotitulación, retrovaloración
background radiation radiación de fondo
backscatter retrodispersión
bacteria bacterias
bacterial bacterial, bacteriano
bactericidal batericida
bactericide bactericida
bacterin bacterina
bacteriocidal batericida
bacteriological bacteriológico
bacteriolysis bacteriólisis
bacteriophage bacteriófago
bacteriostatic bacteriostático
bacterium bacteria
baddeckite badequita
baddeleyite badeleyita
Badger rule regla de Badger
baeumlerite beumlerita
Baeyer strain theory teoría de deformación de Baeyer
Baeyer-Villiger reactions reacciones de Baeyer-Villiger
baffle bafle
bagasse bagazo
bagrationite bagrationita
baicalein baicaleína
baicalin baicalina

baikalite baikalita
bainite bainita
Baker-Nathan effect efecto de Baker-Nathan
Baker-Venkataraman effect efecto de Baker-Venkataraman
bakerite baquerita
baking finish acabado con horneado
baking powder polvo de hornear
baking soda bicarbonato sódico, bicarbonato de sosa
balance balance, equilibrio, balanza
balanced balanceado
balanced reaction reacción balanceada
balata balata
Baldwin phosphorus fósforo de Baldwin
Baldwin rules reglas de Baldwin
ball mill molino de bolas
balloelectric baloeléctrico
Balmer formula fórmula de Balmer
Balmer lines líneas de Balmer
Balmer series serie de Balmer
balometer balómetro
balsam bálsamo
balsam of Peru bálsamo de Perú
balsam of Tolu bálsamo de Tolú
Bamberger formula fórmula de Bamberger
banalsite banalsita
banana oil aceite de plátano
Banbury mixer mezclador de Banbury
band banda
band spectrum espectro de bandas
Bandrowski base base de Bandrowski
banisterine banisterina
baptisin baptisina
baptitoxine baptitoxina
bar barra
barbaloin barbaloína
barban barbana
barberite barberita
Barbier-Wieland degradation degradación de Barbier-Wieland

barbital barbital
barbitone barbitona
barbiturate barbiturato
barbituric acid ácido barbitúrico
barcenite barcenita
Barff process proceso de Barff
Barfoed reagent reactivo de Barfoed
Barfoed test prueba de Barfoed
barite barita
barium bario
barium acetate acetato bárico, acetato de
    bario
barium aluminate aluminato bárico
barium arsenate arseniato bárico
barium azide azida bárica
barium benzoate benzoato bárico
barium bicarbonate bicarbonato bárico
barium bichromate bicromato bárico
barium binoxide binóxido bárico
barium bioxalate bioxalato bárico
barium bisulfate bisulfato bárico
barium borate borato bárico
barium boride boruro bárico
barium borotungstate borotungstato
    bárico
barium bromate bromato bárico
barium bromide bromuro bárico
barium butyrate butirato bárico
barium carbide carburo bárico
barium carbonate carbonato bárico
barium chlorate clorato bárico
barium chloride cloruro bárico
barium chloroplatinate cloroplatinato
    bárico
barium chromate cromato bárico
barium citrate citrato bárico
barium cyanate cianato bárico
barium cyanide cianuro bárico
barium cyanoplatinite cianoplatinita
    bárica
barium dichromate dicromato bárico
barium dioxide dióxido bárico
barium diphenylamine sulfonate
    difenilaminosulfonato bárico
barium dithionate ditionato bárico
barium diurinate diuriniato bárico
barium ethylsulfate etilsulfato bárico
barium ferrate ferrato bárico
barium ferrite ferrita bárica

barium ferrocyanide ferrocianuro bárico
barium fluoride fluoruro bárico
barium fluosilicate fluosilicato bárico
barium formate formiato bárico
barium hexachloroplatinate
    hexacloroplatinato bárico
barium hexacyanoferrate
    hexacianoferrato bárico
barium hexafluorogermantane
    hexafluorogermantano bárico
barium hexanitride hexanitruro bárico
barium hydrate hidrato bárico
barium hydride hidruro bárico
barium hydrosulfide hidrosulfuro bárico
barium hydroxide hidróxido bárico
barium hypophosphate hipofosfato
    bárico
barium hypophosphite hipofosfito
    bárico
barium hyposulfate hiposulfato bárico
barium hyposulfite hiposulfito bárico
barium iodate yodato bárico
barium iodide yoduro bárico
barium lactate lactato bárico
barium malate malato bárico
barium malonate malonato bárico
barium manganate manganato bárico
barium mercuric iodide yoduro
    mercúrico bárico
barium mercury bromide bromuro
    mercúrico bárico
barium mercury iodide yoduro
    mercúrico bárico
barium metaphosphate metafosfato
    bárico
barium metasilicate metasilicato bárico
barium methylsulfate metilsulfato bárico
barium molybdate molibdato bárico
barium monohydrate monohidrato
    bárico
barium monophosphate monofosfato
    bárico
barium monosulfide monosulfuro bárico
barium monoxide monóxido bárico
barium nitrate nitrato bárico
barium nitrite nitrito bárico
barium octahydrate octahidrato bárico
barium oleate oleato bárico
barium orthoperiodate ortoperyodato

bárico

barium orthophosphate ortofosfato bárico

barium oxalate oxalato bárico

barium oxide óxido bárico

barium pentahydrate pentahidrato bárico

barium pentylsulfate pentilsulfato bárico

barium perchlorate perclorato bárico

barium periodate peryodato bárico

barium permanganate permanganato bárico

barium peroxide peróxido bárico

barium peroxodisulfate peroxodisulfato bárico

barium phosphate fosfato bárico

barium phosphide fosfuro bárico

barium phosphinate fosfinato bárico

barium phosphosilicate fosfosilicato bárico

barium potassium chlorate clorato potásico bárico

barium potassium chromate cromato potásico bárico

barium propionate propionato bárico

barium protoxide protóxido bárico

barium pyrophosphate pirofosfato bárico

barium salicylate salicilato bárico

barium selenate seleniato bárico

barium selenide seleniuro bárico

barium silicate silicato bárico

barium silicide siliciuro bárico

barium silicofluoride silicofluoruro bárico

barium stannate estannato bárico

barium stearate estearato bárico

barium succinate succinato bárico

barium sulfate sulfato bárico

barium sulfide sulfuro bárico

barium sulfite sulfito bárico

barium sulfocyanide sulfocianuro bárico

barium superoxide superóxido bárico

barium tartrate tartrato bárico

barium tetrasulfide tetrasulfuro bárico

barium thiocyanate tiocianato bárico

barium thiosulfate tiosulfato bárico

barium titanate titanato bárico

barium triphosphate trifosfato bárico

barium trisulfide trisulfuro bárico

barium trithionate tritionato bárico

barium tungstate tunsgstato bárico

barium value valor de bario

barium zirconate zirconato bárico

barkevite barkevita

Barlow rule regla de Barlow

barn barn

Barnett acetylation method método de acetilación de Barnett

baroluminescence baroluminiscencia

barometer barómetro

barometric pressure presión barométrica

barophoresis baroforesis

baroscope baroscopio

barosimin barosimina

barostat barostato

barrandite barrandita

barrel barril

barrier barrera

barrier layer capa de barrera

Bart reaction reacción de Bart

barthrin bartrina

Barton reaction reacción de Barton

barye baria

baryon barión

barysilite barisilita

baryte barita

basanite basanita

base base

base dissociation constant constante de dissociación básica

base metal metal de base

base number número de base

base unit unidad básica

basic básico

basic anhydride anhídrido básico

basic capacity capacidad básica

basic group grupo básico

basic lining revestimiento básico

basic oxide óxido básico

basic salt sal básica

basic solution solución básica

basic solvent solvente básico

basicity basicidad

basicity constant constante de basicidad

basis metal metal base

bassanite basanita

bassisterol basisterol

bassorin basorina
bastnasite bastnasita
bath baño
bathochromatic batocromático
bathochromatic shift deslizamiento
  batocromático
bathochromic batocrómico
bathocuproine batocuproína
bathophenanthroline batofenantrolina
batrachotoxinin batracotoxinina
battery batería
battery acid ácido de batería
battery electrolyte electrólito de batería
batu batu
batyl alcohol alcohol batílico
Baudisch reaction reacción de Baudisch
bauxite bauxita
Bayer acid ácido de Bayer
Bayer process proceso de Bayer
bayerite bayerita
bead test prueba de perla
beaker vaso de laboratorio
beam attenuator atenuador de haz
beam balance balanza de astil
Beattie-Bridgman equation ecuación de
  Beattie-Bridgman
beccarite becarita
Béchamp reaction reacción de Béchamp
Béchamp reduction reducción de
  Béchamp
Bechhold filter filtro de Bechhold
Becke line línea de Becke
beckelite bequelita
Beckmann reaction reacción de
  Beckmann
Beckmann rearrangement
  reordenamiento de Beckmann
Beckmann thermometer termómetro de
  Beckmann
becquerel becquerel
Becquerel effect efecto de Becquerel
Becquerel rays rayos de Becquerel
becquerelite becquerelita
beehive colmena
beer cerveza
Beer-Lambert law ley de Beer-Lambert
Beer law ley de Beer
beeswax cera de abejas
beet sugar azúcar de remolacha

behenic acid ácido behénico
behenolic acid ácido behenólico
behenolyl behenolilo
behenone behenona
behenyl alcohol alcohol behenílico
Beilby layer capa de Beilby
Beilstein test prueba de Beilstein
bel bel
beldongrite beldongrita
belite belita
bell glass campana
bell jar campana
bell metal bronce de campana
belladonna belladona
belladonnine belladonina
Bellier index índice de Bellier
Bellier number número de Bellier
belonesite belonesita
Belousov-Zhabotinskii reaction reacción
  de Belousov-Zhabotinskii
bemberg bemberg
Bénard cell célula de Bénard
Benary reaction reacción de Benary
benazolin benazolina
bench test prueba de banco
bendiocarb bendiocarb
Benedict solution solución de Benedict
Benedict test prueba de Benedict
beneficiation beneficio
benefin benefina
Benfield process proceso de Benfield
benomyl benomilo
bensulide bensulida
bentazone bentazona
bentonite bentonita
benzacetin benzacetina
benzaconine benzaconina
benzacridine benzacridina
benzal benzal
benzal chloride benzalcloruro
benzalacetone benzalacetona
benzalazine benzalazina
benzalcohol benzalcohol
benzaldehyde benzaldehído
benzaldehyde azide azida de
  benzaldehído
benzaldehyde cyanohydrin cianhidrina
  de benzaldehído
benzaldehyde green verde de

benzaldehído

benzaldehyde phenylhydrazone
fenilhidrazona de benzaldehído

benzaldoxime benzaldoxima

benzaldoxime acetate acetato de
benzaldoxima

benzalkonium benzalconio

benzalkonium chloride cloruro de
benzalconio

benzalphthalide benzalftalida

benzamide benzamida

benzamide oxime oxima de benzamida

benzamidine benzamidina

benzamido benzamido

benzamidoacetic acid ácido
benzamidoacético

benzaminoacetic acid ácido
benzaminoacético

benzanilide benzanilida

benzanthracene benzantracina

benzanthrone benzantrona

benzathine benzatina

benzazide benzazida

benzazimide benzazimida

benzazine benzazina

benzene benceno

benzene azimide bencenoazimida

benzene dibromide dibromuro de
benceno

benzene dichloride dicloruro benceno

benzene hexabromide hexabromuro de
benceno

benzene hexachloride hexacloruro de
benceno

benzene ring anillo bencénico

benzene series serie de benceno

benzene structure estructura de benceno

benzeneazoanilide bencenoazoanilida

benzeneazoaniline bencenoazoanilina

benzeneazobenzene bencenoazobenceno

benzenecarbaldehyde
bencenocarbaldehído

benzenecarbinol bencenocarbinol

benzenecarbonitrile bencenocarbonitrilo

benzenecarbonyl chloride cloruro de
bencenocarbonilo

benzenecarboxylate bencenocarboxilato

benzenecarboxylic acid ácido
bencenocarboxílico

benzenediazonium chloride cloruro de
bencenodiazonio

benzenediazonium hydroxide hidróxido
de bencenodiazonio

benzenedicarbinol bencenodicarbinol

benzenedicarbonal bencenodicarbonal

benzenedicarbonitrile
bencenodicarbonitrilo

benzenedicarboxylic acid ácido
bencenodicarboxílico

benzenedisulfonic acid ácido
bencenodisulfónico

benzenedisulfoxide bencenodisulfóxido

benzenedithiol bencenoditiol

benzenemonosulfonic acid ácido
bencenomonosulfónico

benzenephosphinic acid ácido
bencenofosfínico

benzenephosphonic acid ácido
bencenofosfónico

benzenesulfamide bencenosulfamida

benzenesulfide bencenosulfuro

benzenesulfinic acid ácido
bencenosulfínico

benzenesulfonamide
bencenosulfonamida

benzenesulfonic acid ácido
bencenosulfónico

benzenesulfonyl bencenosulfonilo

benzenesulfoxide bencenosulfóxido

benzenetetracarboxylic acid ácido
bencenotetracarboxílico

benzenetriol bencenotriol

benzenetrisulfonic acid ácido
bencenotrisulfónico

benzenide bencenida

benzenium bencenio

benzenoid bencenoide

benzenone bencenona

benzenyl bencenilo

benzenyl trichloride tricloruro de
bencenilo

benzethonium chloride cloruro de
bencetonio

benzfuran benzfurano

benzhydrazoin benzhidrazoína

benzhydrol benzhidrol

benzhydryl benzhidrilo

benzhydryl bromide bromuro de

benzhidrilo
benzhydryl chloride cloruro de benzhidrilo
benzidine bencidina
benzidine conversion conversión de bencidina
benzidine dye colorante de bencidina
benzidine sulfate sulfato de bencidina
benzidine test prueba de bencidina
benzidinedicarboxylic acid ácido bencidinodicarboxílico
benzidino bencidino
benzil bencilo
benzilic acid ácido bencílico
benzilidene bencilideno
benzimidazole bencimidazol
benzimidazolyl bencimidazolilo
benzimidoyl bencimidoilo
benzin bencina
benzindole bencindol
benzine bencina
benzisoxasole bencisoxasol
benzoate benzoato
benzocaine benzocaína
benzocarbolic acid ácido benzocarbólico
benzodianthrone benzodiantrona
benzodiazine benzodiazina
benzodiazole benzodiazol
benzodihydropyrone benzodihidropirona
benzodioxazine benzodioxazina
benzofluorene benzofluoreno
benzofuran benzofurano
benzofuranone benzofuranona
benzofuranyl benzofuranilo
benzoglycolic acid ácido benzoglicólico
benzoguanamine benzoguanamina
benzohydrol benzohidrol
benzohydroximic acid ácido benzohidroxímico
benzohydryl benzohidrilo
benzoic acid ácido benzoico
benzoic acid anhydride anhídrido de ácido benzoico
benzoic alcohol alcohol benzoico
benzoic aldehyde aldehído benzoico
benzoic anhydride anhídrido benzoico
benzoic sulfimide sulfimido benzoico
benzoic trichloride tricloruro benzoico
benzoilacetone benzoilacetona

benzoin benzoína
benzoin condensation condensación de benzoína
benzoin oxime oxima de benzoína
benzol benzol
benzol bichloride bicloruro de benzol
benzol bromide bromuro de benzol
benzol chloride cloruro de benzol
benzomorpholine benzomorfolina
benzonaphthene benzonafteno
benzonaphthol benzonaftol
benzonitrile benzonitrilo
benzopentazole benzopentazol
benzoperoxide benzoperóxido
benzophenol benzofenol
benzophenone benzofenona
benzophenone oxide óxido de benzofenona
benzophenone sulfide sulfuro de benzofenona
benzopinacol benzopinacol
benzopurpurine benzopurpurina
benzopyran benzopirano
benzopyranyl benzopiranilo
benzopyrene benzopireno
benzopyridine benzopiridina
benzopyrone benzopirona
benzopyrrole benzopirrol
benzoquinol benzoquinol
benzoquinone benzoquinona
benzoquinone monoxime monoxima de benzoquinona
benzoselenofuran benzoselenofurano
benzosulfimide benzosulfimida
benzotetrazine benzotetrazina
benzotetrazole benzotetrazol
benzothiazine benzotiazina
benzothiazole benzotiazol
benzothiazyl benzotiacilo
benzothiazyl disulfide disulfuro de benzotiacilo
benzothiofuran benzotiofurano
benzothiopyran benzotiopirano
benzothiopyrone benzotiopirona
benzotoluide benzotoluida
benzotriazepine benzotriazepina
benzotriazine benzotriazina
benzotriazole benzotriazol
benzotrichloride benzotricloruro

benzotrifluoride benzotrifluoruro
benzotrifuran benzotrifurano
benzoxanthene benzoxanteno
benzoxazine benzoxazina
benzoxazole benzoxazol
benzoxazolyl benzoxazolilo
benzoxozolone benzoxazolona
benzoxtetrazine benzoxtetrazina
benzoxtriazine benzoxtriazina
benzoyl benzoilo
benzoyl acetate acetato de benzoilo
benzoyl benzoate benzoato de benzoilo
benzoyl bromide bromuro de benzoilo
benzoyl chloride cloruro de benzoilo
benzoyl cyanide cianuro de benzoilo
benzoyl disulfide disulfuro de benzoilo
benzoyl fluoride fluoruro de benzoilo
benzoyl iodide yoduro de benzoilo
benzoyl oxide óxido de benzoilo
benzoyl peroxide peróxido de benzoilo
benzoyl sulfide sulfuro de benzoilo
benzoylacetaldehyde benzoilacetaldehído
benzoylacetic acid ácido benzoilacético
benzoylacrylic acid ácido benzoilacrílico
benzoylamide benzoilamida
benzoylamino benzoilamino
benzoylaminoacetic acid ácido
 benzoilaminoacético
benzoylaniline benzoilanilina
benzoylation benzoilación
benzoylazide benzoilazida
benzoylbenzoic acid ácido
 benzoilbenzoico
benzoylcyclopropane
 benzoilciclopropano
benzoylferrocene benzoilferroceno
benzoylglycin benzoilglicina
benzoylhydrazine benzoilhidrazina
benzoyloxy benzoiloxi
benzoylpropionic acid ácido
 benzoilpropiónico
benzoylpyridine benzoilpiridina
benzozone benzozono
benzpyrazole benzpirazol
benzthiophene benztiofeno
benztropine mesylate mesilato de
 benztropina
benzyl bencilo
benzyl abietate abietato de bencilo

benzyl acetate acetato de bencilo
benzyl alcohol alcohol bencílico
benzyl azide azida de bencilo
benzyl benzoate benzoato de bencilo
benzyl bichloride bicloruro de bencilo
benzyl bromide bromuro de bencilo
benzyl butyrate butirato de bencilo
benzyl carbinol bencilcarbinol
benzyl chloride cloruro de bencilo
benzyl chlorocarbonate clorocarbonato
 de bencilo
benzyl chloroformate cloroformiato de
 bencilo
benzyl cinnamate cinamato de bencilo
benzyl cyanide cianuro de bencilo
benzyl dichloride dicloruro de bencilo
benzyl disulfide disulfuro de bencilo
benzyl ether éter bencílico
benzyl ethyl ether éter etílico bencílico
benzyl ethyl ketone cetona etílica
 bencílica
benzyl ethylsalicylate etilsalicilato de
 bencilo
benzyl fluoride fluoruro de bencilo
benzyl formate formiato de bencilo
benzyl fumarate fumarato de bencilo
benzyl iodide yoduro de bencilo
benzyl isoeugenol bencilisoeugenol
benzyl isothiocyanate bencilisotiocianato
benzyl mercaptan bencilmercaptano
benzyl nitrile nitrilo de bencilo
benzyl pelargonate pelargonato de
 bencilo
benzyl phenylacetate fenilacetato de
 bencilo
benzyl propionate propionato de bencilo
benzyl salicylate salicilato de bencilo
benzyl succinate succinato de bencilo
benzyl sulfide sulfuro de bencilo
benzyl thiocyanate tiocianato de bencilo
benzylacetamide bencilacetamida
benzylamine bencilamina
benzylaniline bencilanilina
benzylation bencilación
benzylbenzene bencilbenceno
benzylcyanamide bencilcianamida
benzyldianil bencildianilo
benzyldimethylamine bencildimetilamina
benzylene bencileno

benzylethanolamine benciletanolamina
benzylformamide bencilformamida
benzylhydrazine bencilhidrazina
benzylhydroquinone bencilhidroquinona
benzylidene bencilideno
benzylidene dibromide dibromuro de
     bencilideno
benzylidene dichloride dicloruro de
     bencilideno
benzylideneacetone bencilidenacetona
benzylideneacetophenone
     bencilidenacetofenona
benzylidiethanolamine
     bencilidietanolamina
benzylidine bencilidina
benzylisopropylamine
     bencilisopropilamina
benzylmethylamine bencilmetilamina
benzylnaphthalene bencilnaftaleno
benzyloxyphenol benciloxifenol
benzylphenanthrene bencilfenantreno
benzylphenol bencilfenol
benzylphenylamine bencilfenilamina
benzylpyridine bencilpiridina
benzylsulfonic acid ácido bencilsulfónico
benzylthiol benciltiol
benzyltrimethylammonium chloride
     cloruro de benciltrimetilamonio
benzyltrimethylammonium methoxide
     metóxido de benciltrimetilamonio
benzyne bencino
berberine berberina
berberonic acid ácido berberónico
beresovite beresovita
bergamot oil aceite de bergamota
Bergius process proceso de Bergius
Bergmann degradation degradación de
     Bergmann
Berkefeld filter filtro de Berkefeld
berkelium berkelio
Berthelot equation ecuación de Berthelot
Berthelot reaction reacción de Berthelot
berthierine bertierina
bertholite bertolita
berthollide bertolida
bertrandite bertrandita
beryl berilo
beryllate berilato
beryllia berilia

beryllide beriliuro
beryllium berilio
beryllium acetate acetato de berilio
beryllium acetylacetonate
     acetilacetonato de berilio
beryllium bromide bromuro de berilio
beryllium carbide carburo de berilio
beryllium carbonate carbonato de berilio
beryllium chloride cloruro de berilio
beryllium-copper berilio-cobre
beryllium fluoride fluoruro de berilio
beryllium hydrate hidrato de berilio
beryllium hydroxide hidróxido de berilio
beryllium iodide yoduro de berilio
beryllium metaphosphate metafosfato de
     berilio
beryllium nitrate nitrato de berilio
beryllium nitride nitruro de berilio
beryllium oxalate oxalato de berilio
beryllium oxide óxido de berilio
beryllium potassium fluoride fluoruro
     potásico berílico
beryllium potassium sulfate sulfato
     potásico berílico
beryllium silicate silicato de berilio
beryllium sodium fluoride fluoruro
     sódico berílico
beryllium sulfate sulfato de berilio
beryllonite berilonita
Bessemer process proceso de Bessemer
Bessonoff reagent reactivo de Bessonoff
beta beta
beta acid ácido beta
beta activity actividad beta
beta decay desintegración beta
beta disintegration desintegración beta
beta emitter emisor beta
beta iron hierro beta
beta lactone lactona beta
beta particle partícula beta
beta position posición beta
beta radiation radiación beta
beta ray rayo beta
beta-ray spectrometer espectrómetro de
     rayos beta
beta sheet lámina beta
betaine betaína
betaine hydrochloride clorhidrato de
     betaína

betaine phosphate fosfato de betaína
betamethasone betametasona
betatopic betatópico
betatron betatrón
Bettendorf reagent reactivo de Bettendorf
Bettendorf test prueba de Bettendorf
Betterton-Kroll process proceso de Betterton-Kroll
Betti reaction reacción de Betti
Betts process proceso de Betts
betula oil aceite de betula
betulin betulina
betulinic acid ácido betulínico
betulinol betulinol
beyrichite beiriquita
bi bi
biacene biaceno
biacetylene biacetileno
biallyl bialilo
bianiline bianilina
bianisidine bianisidina
bianthryl biantrilo
biaxial biaxial
bibenzyl bibencilo
biborate biborato
bicalcium phosphate fosfato bicálcico
bicarbonate bicarbonato
bicarbonate of potash bicarbonato potásico
bicarbonate of soda bicarbonato sódico, bicarbonato de sosa
bichloride bicloruro
bicyclic bicíclico
bicyclic compound compuesto bicíclico
bicyclo numbering numeración bicíclica
bicyclodecane biciclodecano
bicyclohexyl biciclohexilo
bieberite bieberita
biethylene bietileno
biflavonyl biflavonilo
biformin biformina
bifurcation bifurcación
Biginelli reaction reacción de Biginelli
biguanide biguanida
bihexyl bihexilo
biimino biimino
bilateral bilateral
bile acid ácido biliar

bile salt sal biliar
bilifuscin bilifuscina
bilirubin bilirrubina
bilirubinic acid ácido bilirrubínico
bimetal bimetal
bimolecular bimolecular
bimolecular reaction reacción bimolecular
binapacryl binapacrilo
binaphthalene binaftaleno
binaphthalenyl binaftalenilo
binaphthyl binaftilo
binary binario
binary acid ácido binario
binary alloy aleación binaria
binary compound compuesto binario
binary mixture mezcla binaria
binary salt sal binaria
binary system sistema binario
bind fijar, aglutinar
binder aglutinante, ligador
binding aglutinación, ligadura
binding energy energía de separación
binding site lugar de unión
binitro binitro
binnite binita
binoxide binóxido
binuclear binuclear
bioaccumulation bioacumulación
bioactivation bioactivación
bioanalysis bioanálisis
bioassay bioensayo
biocatalyst biocatalizador
biochemical bioquímico
biochemistry bioquímica
biochrome biocromo
biocide biocida
biocolloid biocoloide
bioconversion bioconversión
biocytin biocitina
biodegradable biodegradable
bioelectrochemistry bioelectroquímica
bioelement bioelemento
bioengineering bioingeniería
bioethics bioética
bioflavonoid bioflavonoide
biogas biogas
biogenic biogénico
biogeochemistry biogeoquímica

bioinorganic chemistry química
  bioinorgánica
biological biológico
biological availability disponibilidad
  biológica
biology biología
bioluminescence bioluminiscencia
biomass biomasa
biomaterial biomaterial
biomimetic chemistry química
  biomimética
biomineral biomineral
biomolecule biomolécula
biophyl biofila
biopolymer biopolímero
bioreactor biorreactor
bioresmethrin biorresmetrina
biorization biorización
biose biosa
biosphere biosfera
biostat biostático
biosterol biosterol
biosynthesis biosíntesis
biotechnology biotecnología
biotic biótico
biotin biotina
biotite biotita
bioxyl bioxilo
biperiden biperiden
biphenyl bifenilo
biphenylamine bifenilamina
biphenyldiamine bifenildiamina
biphenylene bifenileno
biphenylene oxide óxido de bifenileno
biphenylyl bifenililo
biphenylylamine bifenililamina
bipropenyl bipropenilo
bipyridyl bipiridilo
biquinoline biquinolina
biquinolyl biquinolilo
biradical birradical
birectification birrectificación
birefringence birrefrigencia
birefringent birrefringente
Birge-Sponer extrapolation
  extrapolación de Birge-Sponer
Birkeland-Eyde process proceso de
  Birkeland-Eyde
birotation birrotación

bis bis
bisabolene bisaboleno
bisamide bisamida
bisazo bizaso
bisdiazo bisdiazo
bishydroxymethylurea
  bishidroximetilurea
bismanol bismanol
bismite bismita
bismuth bismuto
bismuth acetate acetato de bismuto
bismuth antimonide antimoniuro de
  bismuto
bismuth borate borato de bismuto
bismuth bromide bromuro de bismuto
bismuth carbonate carbonato de bismuto
bismuth chloride cloruro de bismuto
bismuth chromate cromato de bismuto
bismuth citrate citrato de bismuto
bismuth dichloride dicloruro de bismuto
bismuth ditannate ditanato de bismuto
bismuth gallate galato de bismuto
bismuth hydrate hidrato de bismuto
bismuth hydroxide hidróxido de bismuto
bismuth iodide yoduro de bismuto
bismuth nitrate nitrato de bismuto
bismuth oleate oleato de bismuto
bismuth oxalate oxalato de bismuto
bismuth oxide óxido de bismuto
bismuth oxobromide oxobromuro de
  bismuto
bismuth oxochloride oxocloruro de
  bismuto
bismuth oxofluoride oxofluoruro de
  bismuto
bismuth oxoiodide oxoyoduro de
  bismuto
bismuth oxybromide oxibromuro de
  bismuto
bismuth oxycarbonate oxicarbonato de
  bismuto
bismuth oxychloride oxicloruro de
  bismuto
bismuth oxyhydrate oxihidrato de
  bismuto
bismuth oxynitrate oxinitrato de bismuto
bismuth pentafluoride pentafluoruro de
  bismuto
bismuth pentoxide pentóxido de bismuto

bismuth permanganate permanganato de bismuto
bismuth peroxide peróxido de bismuto
bismuth phosphate fosfato de bismuto
bismuth potassium iodide yoduro potásico de bismuto
bismuth potassium tartrate tartrato potásico de bismuto
bismuth propionate propionato de bismuto
bismuth selenide seleniuro de bismuto
bismuth stannate estannato de bismuto
bismuth subcarbonate subcarbonato de bismuto
bismuth subchloride subcloruro de bismuto
bismuth subgallate subgalato de bismuto
bismuth subnitrate subnitrato de bismuto
bismuth subsalicylate subsalicilato de bismuto
bismuth sulfate sulfato de bismuto
bismuth sulfide sulfuro de bismuto
bismuth sulfite sulfito de bismuto
bismuth tannate tanato de bismuto
bismuth telluride telururo de bismuto
bismuth tetraoxide tetróxido de bismuto
bismuth tribromide tribromuro de bismuto
bismuth trichloride tricloruro de bismuto
bismuth trihydrate trihidrato de bismuto
bismuth trihydroxide trihidróxido de bismuto
bismuth triiodide triyoduro de bismuto
bismuth trinitrate trinitrato de bismuto
bismuth trioxide trióxido de bismuto
bismuth trisulfide trisulfuro de bismuto
bismuth tritelluride tritelururo de bismuto
bismuth tungstate tungstato de bismuto
bismuth yellow amarillo de bismuto
bismuthide bismutida
bismuthine bismutina
bismuthinite bismutinita
bismuthite bismutita
bismuthous bismutoso
bismuthyl bismutilo
bismuthyl bromide bromuro de bismutilo
bismuthyl carbonate carbonato de bismutilo
bismuthyl chloride cloruro de bismutilo
bismuthyl fluoride fluoruro de bismutilo
bismuthyl hydroxide hidróxido de bismutilo
bismuthyl nitrate nitrato de bismutilo
bisphenol bisfenol
bistability biestabilidad
bistetrazole bistetrazol
bistriazole bistriazol
bistridecyl phthalate ftalato de bistridecilo
bisulfate bisulfato
bisulfite bisulfito
bitartrate bitartrato
bithionol bitionol
bitrex bitrex
bitter amargo
bitter principle principio amargo
bittern bittern
bitumen bitumen
bituminous bituminoso
bituminous coal carbón bituminoso
bituminous plastic plástico bituminoso
biurea biurea
biuret biuret
biuret reaction reacción de biuret
biuret test prueba de biuret
bivalence bivalencia
bivalent bivalente
bivinyl bivinilo
black alum alumbre negro
black cyanide cianuro negro
black lead plomo negro
black powder polvo negro
Blagden law ley de Blagden
Blaise reaction reacción de Blaise
blanc fixe blanco fijo
Blanc reaction reacción de Blanc
Blanc rule regla de Blanc
blast furnace alto horno
blasting gelatin gelatina explosiva
bleach blanquear
bleaching blanqueo
bleaching material material de blanqueo
bleaching powder polvo de blanqueo
bleed sangrar, drenar

blend mezcla, combinación
blende blenda
bleomycin bleomicina
blister packaging empaquetado
    ampollado
Bloch theorem teorema de Bloch
block bloque, bloqueo
block copolymer copolímero de bloques
bloom capa fina superficial
blooming capa fina superficial
blown oil aceite soplado
blowpipe soplete
blue copper cobre azul
blue copperas caparrosa azul
blue lead plomo azul
blue powder polvo azul
blue salt sal azul
blue verdigris cardenillo azul
blue vitriol vitriolo azul
blueing azulamiento
blush capa fina superficial
boat artesa
boat conformation conformación en bote
Bodroux reaction reacción de Bodroux
body cuerpo
body-centered cubic cúbico centrado en
    el cuerpo
body-centered cubic structure
    estructura cúbica centrada en el
    cuerpo
body-centered lattice red centrada en el
    cuerpo
body-centered structure estructura
    centrada en el cuerpo
boehmite boehmita
Boeseken method método de Boeseken
Boettger test prueba de Boettger
Bogert-Cook synthesis síntesis de
    Bogert-Cook
boghead coal boghead
Bohn-Schmidt reaction reacción de
    Bohn-Schmidt
Bohr atom átomo de Bohr
Bohr magneton magnetón de Bohr
Bohr orbit órbita de Bohr
Bohr radius radio de Bohr
Bohr-Sommerfeld theory teoría de
    Bohr-Sommerfeld
Bohr theory teoría de Bohr

boil hervir
boiled oil aceite cocido
boiler incrustation incrustación de
    caldera
boiler scale incrustaciones de caldera
boiler stone incrustación de caldera
boiling ebullición
boiling chips gravilla de ebullición
boiling point punto de ebullición
boiling-point apparatus aparato de punto
    de ebullición
boiling-point diagram diagrama de
    puntos de ebullición
boiling-point elevation elevación de
    punto de ebullición
boiling-point range intervalo de puntos
    de ebullición
boiling-point reduction reducción de
    punto de ebullición
boiling stones gravilla de ebullición
boleite boleíta
bolometer bolómetro
Bolton reagent reactivo de Bolton
Boltzmann constant constante de
    Boltzmann
Boltzmann equation ecuación de
    Boltzmann
Boltzmann law ley de Boltzmann
bomb bomba
bomb calorimeter bomba calorimétrica
bombardment bombardeo
bond enlace
bond angle ángulo de enlace
bond distance distancia de enlace
bond energy energía de enlace
bond hybridization hibridización de
    enlace
bonding enlazado
bonding distance distancia de enlazado
bonding electron electrón de enlazado
bonding orbital orbital de enlazado,
    orbital enlazante
bondur bondur
bone ash ceniza de huesos
bone black negro de huesos
bone oil aceite de huesos
bone phosphate fosfato de huesos
boracic acid ácido borácico
boracite boracita

borane borano
boranediyl boranodiilo
boranetriyl broranotriilo
borate borato
borax bórax
borax-bead test prueba de perla con
    bórax
borax glass vidrio de bórax
borazine borazina
borazole borazol
Borcher metal metal de Borcher
Borcher process proceso de Borcher
boric acid ácido bórico
boric acid ester éster de ácido bórico
boric anhydride anhídrido bórico
boric oxide óxido bórico
boride boruro
borine borina
Born-Haber cycle ciclo de Born-Haber
Born-Oppenheimer approximation
    aproximación de Born-Oppenheimer
Born-Oppenheimer method método de
    Born-Oppenheimer
bornane bornano
bornene borneno
borneol borneol
bornite bornita
bornyl bornilo
bornyl acetate acetato de bornilo
bornyl alcohol alcohol bornílico
bornyl formate formiato de bornilo
bornyl isovalerate isovalerato de bornilo
bornyl salicylate salicilato de bornilo
boroethane boroetano
borohydride borohidruro
borol borol
boron boro
boron alloy aleación de boro
boron bromide bromuro de boro
boron carbide carburo de boro
boron chloride cloruro de boro
boron fluoride fluoruro de boro
boron fuel combustible de boro
boron hydride hidruro de boro
boron hydroxide hidróxido de boro
boron nitride nitruro de boro
boron oxide óxido de boro
boron phosphate fosfato de boro
boron phosphide fosfuro de boro

boron steel acero al boro
boron sulfide sulfuro de boro
boron tribromide tribromuro de boro
boron trichloride tricloruro de boro
boron trifluoride trifluoruro de boro
boron trisulfide trisulfuro de boro
borophosphoric acid ácido borofosfórico
borosalicylate borosalicilato
borosilicate borosilicato
borosilicate glass vidrio de borosilicato
borotungstic acid ácido borotúngstico
boroxine boroxina
bort bort
Bosch process proceso de Bosch
Bose-Einstein statistics estadística de
    Bose-Einstein
bottle botella
Bouchardat solution solución de
    Bouchardat
Bouguer-Lambert law ley de Bouguer-
    Lambert
Bouin fluid fluido de Bouin
bound electron electrón ligado
bound state estado ligado
bound water agua retenida
Bouveault aldehyde synthesis síntesis de
    aldehídos de Bouveault
Bouveault-Blanc method método de
    Bouveault-Blanc
Bouveault-Blanc reduction reduccion de
    Bouveault-Blanc
Bowen tube tubo de Bowen
Boyce burner mechero de Boyce
Boyle law ley de Boyle
Brackett series serie de Brackett
brackish water agua salobre
Bradsher reaction reacción de Bradsher
Brady reagent reactivo de Brady
Bragg angle ángulo de Bragg
Bragg crystallogram cristalograma de
    Bragg
Bragg curve curva de Bragg
Bragg equation ecuación de Bragg
Bragg ionization curve curva de
    ionización de Bragg
Bragg law ley de Bragg
Bragg method método de Bragg
Bragg peak pico de Bragg
Bragg rule regla de Bragg

Bragg spectrometer espectrómetro de Bragg
bran oil aceite de salvado
branched ramificado
branched chain cadena ramificada
branched-chain reaction reacción de cadena ramificada
branching ramificación
brasilin brasilina
brass latón
brassidic acid ácido brasídico
brassylic acid ácido brasílico
braunite braunita
Brazil wax cera de Brasil
brazing soldadura fuerte
Bredt rule regla de Bredt
breeder reactor engendrador
breeze cisco
bretonite bretonita
breunnerite breunerita
brewer's yeast levadura de cerveza
brewing fabricación de cerveza
Brewster angle ángulo de Brewster
Brewster process proceso de Brewster
bridge puente
bridgehead molecule molécula con puente
Briet-Wigner formula fómula de Briet-Wigner
bright-line spectrum espectro de líneas brillantes
brightener abrillantador
brightening abrillantamiento
brightening agent agente de abrillantamiento
brilliant brillante
brilliant crocein croceína brillante
brilliant green verde brillante
brilliant yellow amarillo brillante
brimstone piedra de azufre
Brin process proceso de Brin
brine salmuera
Brinell hardness dureza de Brinell
Brinell hardness test prueba de dureza de Brinell
britholite britolita
brittle quebradizo
brittle point punto de destrozo
Brix degree grado de Brix

brochanite brocanita
Brodie solution solución de Brodie
Broenner acid ácido de Broenner
bromacetate bromacetato
bromacil bromacil
bromal bromal
bromalin bromalina
bromate bromato
bromated brominado
bromelain bromelaína
bromelin bromelina
bromeosin bromeosina
bromethyl brometilo
bromic acid ácido brómico
brominated brominado
brominated camphor alcanfor brominado
bromination brominación
bromine bromo
bromine azide azida de bromo
bromine chloride cloruro de bromo
bromine cyanide cianuro de bromo
bromine fluoride fluoruro de bromo
bromine hydride hidruro de bromo
bromine iodide yoduro de bromo
bromine pentafluoride pentafluoruro de bromo
bromine trifluoride trifluoruro de bromo
bromine water agua de bromo
bromoacetamide bromoacetamida
bromoacetate bromoacetato
bromoacetic acid ácido bromoacético
bromoacetone bromoacetona
bromoacetone cyanohydrin cianhidrina de bromoacetona
bromoallylene bromoalileno
bromoaniline bromoanilina
bromoanthrallic acid ácido bromoantrálico
bromoauric acid ácido bromoaúrico
bromobenzaldehyde bromobenzaldehído
bromobenzene bromobenceno
bromobenzenesulfonic acid ácido bromobencenosulfónico
bromobenzoic acid ácido bromobenzoico
bromobenzyl cyanide cianuro de bromobencilo
bromobutane bromobutano
bromobutyric acid ácido bromobutírico

bromochloroethane bromocloroetano
bromochloromethane bromoclorometano
bromochloropropane bromocloropropano
bromocresol green verde de bromocresol
bromocresol purple púrpura de bromocresol
bromocyclopentane bromociclopentano
bromododecane bromododecano
bromoethane bromoetano
bromoethyl alcohol alcohol bromoetílico
bromofluorobenzene bromofluorobenceno
bromoform bromoformo
bromohexane bromohexano
bromol bromol
bromomethane bromometano
bromometry bromometría
bromonaphthalene bromonaftaleno
bromopentane bromopentano
bromophenol bromofenol
bromophenol blue azul de bromofenol
bromophenol red rojo de bromofenol
bromophenylacetonitrile bromofenilacetonitrilo
bromophosgene bromofosgeno
bromopicrin bromopicrina
bromopropene bromopropeno
bromopropionic acid ácido bromopropiónico
bromopyridine bromopiridina
bromopyrine bromopirina
bromostyrene bromoestireno
bromosuccinic acid ácido bromosuccínico
bromosuccinimide bromosuccinimida
bromothiophenol bromotiofenol
bromothymol blue azul de bromotimol
bromotoluene bromotolueno
bromotrichloromethane bromotriclorometano
bromotrifluoroethylene bromotrifluoroetileno
bromotrifluoromethane bromotrifluorometano
Brönner acid ácido de Brönner
Brönsted acid ácido de Brönsted
Brönsted-Lowry theory teoría de Brönsted-Lowry

Brönsted theory teoría de Brönsted
bronze bronce
bronze blue azul de bronce
bronze orange anaranjado de bronce
brosylate ester éster de brosilato
Brown-Boverti test prueba de Brown-Boverti
brown-ring test prueba de anillo pardo
Brownian motion movimiento browniano
Brownian movement movimiento browniano
browning reaction reacción de dorado
brucine brucina
brucine nitrate nitrato de brucina
brucine sulfate sulfato de brucina
brucite brucita
Bruehl receiver receptor de Bruehl
bubble burbuja
bubble counter contador de burbujas
Bucherer-Bergs reaction reacción de Bucherer-Bergs
Bucherer reaction reacción de Bucherer
Buchner funnel embudo de Buchner
buckyballs fulereno
buclizine hydrochloride clorhidrato de buclizina
buffer tampón, amortiguador
buffer action acción amortiguadora
buffer capacity capacidad amortiguadora
buffer index índice amortiguador
buffer range intervalo amortiguador
buffer salt sal amortiguadora
buffer solution solución amortiguadora
buffer value valor de amortiguador
buffering amortiguamiento
bufotanine bufotanina
bufotenine bufotenina
bulb matraz, cubeta
bumping ebullición dispareja
bunamiodyl bunamiodilo
Bunsen burner mechero de Bunsen
Bunsen cell pila de Bunsen
Bunsen clamp abrazadera de Bunsen
Bunsen eudiometer eudiómetro de Bunsen
Bunsen flame llama de Bunsen
Bunsen funnel embudo de Bunsen
Bunsen-Kirchhoff law ley de Bunsen-

Kirchhoff
bunsenite bunsenita
buoyancy flotabilidad
buphanine bufanina
buret bureta
buret clamp abrazadera de bureta
burette bureta
burner mechero
burner attachment aditamento de mechero
burnt quemado, cocido
burnt lime cal viva
bursting disk disco de estallido
butadiene butadieno
butaldehyde butaldehído
butanal butanal
butane butano
butanedial butanodial
butanediamide butanodiamida
butanediamine butanodiamina
butanediol butanodiol
butanediolamine butanodiolamina
butanedione butanodiona
butanenitrile butanonitrilo
butanethiol butanotiol
butanetriol butanotriol
butanoic acid ácido butanoico
butanol butanol
butanol acetate acetato de butanol
butanone butanona
butanoyl chloride cloruro de butanoilo
butenal butenal
butene buteno
butenoic acid ácido butenoico
butenyl butenilo
butenylidene butenilideno
butesin butesina
butethal butetal
butethamine hydrochloride clorhidrato de butetamina
Butler-Volmer equation ecuación de Butler-Volmer
butonate butonato
butopyronoxyl butopironoxilo
butoxy butoxi
butoxyethanol butoxietanol
butoxyethyl laurate laurato de butoxietilo
butoxyethyl oleate oleato de butoxietilo

butoxyethyl stearate estearato de butoxietilo
butoxyphenol butoxifenol
butoxypropanol butoxipropanol
butoxytriglycol butoxitriglicol
butter mantequilla
butter yellow amarillo de mantequilla
butyl butilo
butyl acetate acetato de butilo
butyl acetoacetate acetoacetato de butilo
butyl acetoxystearate acetoxiestearato de butilo
butyl acetylene acetileno de butilo
butyl acrylate acrilato de butilo
butyl alcohol alcohol butílico
butyl aldehyde aldehído butílico
butyl anthranilate antralinato de butilo
butyl benzoate benzoato de butilo
butyl borate borato de butilo
butyl bromide bromuro de butilo
butyl butanoate butanoato de butilo
butyl butyrate butirato de butilo
butyl carbamate carbamato de butilo
butyl carbinol butilcarbinol
butyl chloral hydrate hidrdato de butilcloral
butyl chloride cloruro de butilo
butyl chromate cromato de butilo
butyl citrate citrato de butilo
butyl crotonate crotonato de butilo
butyl cyanide cianuro de butilo
butyl diethanolamine dietanolamina de butilo
butyl diglyme diglime de butilo
butyl dodecanoate dodecanoato de butilo
butyl epoxystearate epoxiestearato de butilo
butyl ether éter butílico
butyl formate formiato de butilo
butyl furoate furoato de butilo
butyl iodide yoduro de butilo
butyl lactate lactato de butilo
butyl myristate miristato de butilo
butyl nitrate nitrato de butilo
butyl nitrite nitrito de butilo
butyl nonanoate nonanoato de butilo
butyl octadecanoate octadecanoato de butilo
butyl oleate oleato de butilo

butyl pelargonate pelargonato de butilo
butyl peracetate peracetato de butilo
butyl perbenzoate perbenzoato de butilo
butyl peroxide peróxido de butilo
butyl peroxyacetate peroxiacetato de butilo
butyl peroxybenzoate peroxibenzoato de butilo
butyl peroxyisobutyrate peroxiisobutirato de butilo
butyl phthalate ftalato de butilo
butyl propionate propionato de butilo
butyl ricinoleate ricinoleato de butilo
butyl rubber caucho butilo
butyl sebacate sebacato de butilo
butyl sorbate sorbato de butilo
butyl stearamide estearamida de butilo
butyl stearate estearato de butilo
butyl sulfide sulfuro de butilo
butyl thiocyanate tiocianiato de butilo
butyl titanate titanato de butilo
butyl valerate valerato de butilo
butylacrylamide butilacrilamida
butylamine butilamina
butylaminoethanol butilaminoetanol
butylaniline butilanilina
butylanthraquinone butilantraquinona
butylated hydroxyanisole hidroxianisol butilado
butylated hydroxytoluene hidroxitolueno butilado
butylbenzene butilbenceno
butylbenzenesulfonamide butilbencenosulfonamida
butylbenzyl phthalate ftalato de butilbencilo
butyldiamylamine butildiamilamina
butylene butileno
butylene dimethacrylate dimetracrilato de butileno
butylene glycol butilenglycol

butylene oxide óxido de butileno
butylethanolamine butiletanolamina
butylethylacetaldehyde butiletilacetaldehído
butylformamide butilformamida
butylhydroquinone butilhidroquinona
butylidene butilideno
butylin trichloride tricloruro de butilina
butylisocyanate butilisocianato
butyllithium butil-litio
butylmagnesium chloride cloruro de butilmagnesio
butylmercaptan butilmercaptano
butylmethacrylate butilmetacrilato
butylparaben butilparabeno
butylperoxypivalate butilperoxipivalato
butylphenol butilfenol
butylphenyl ether éter butilfenílico
butylstannic acid ácido butilestánnico
butylthiophenol butiltiofenol
butyltoluene butiltolueno
butyltrichlorosilane butiltriclorosilano
butylurea butilurea
butyne butino
butynediol butinodiol
butyraldehyde butiraldehído
butyric butírico
butyric acid ácido butírico
butyric alcohol alcohol butírico
butyric aldehyde aldehído butírico
butyric anhydride anhídrido butírico
butyrin butirina
butyrolactam butirolactama
butyrolactone butirolactona
butyrone butirona
butyronitrile butironitrilo
butyroyl chloride cloruro de butiroilo
butyryl butirilo
butyryl chloride cloruro de butirilo
by-product producto secundario

C

Cabannes factor factor de Cabannes
cabrerite cabrerita
cacao butter manteca de cacao
cacao syrup jarabe de cacao
cacodyl cacodilo
cacodyl chloride cloruro de cacodilo
cacodyl oxide óxido de cacodilo
cacodylate cacodilato
cacodylic acid ácido cacodílico
cacotheline cacotelina
cactine cactina
cadalene cadaleno
cadaverine cadaverina
cadinene cadineno
cadion cadión
cadmium cadmio
cadmium acetate acetato de cadmio
cadmium amalgam amalgama de cadmio
cadmium ammonium bromide bromuro amónico de cadmio
cadmium antimonide antimoniuro de cadmio
cadmium borotungstate borotungstato de cadmio
cadmium bromate bromato de cadmio
cadmium bromide bromuro de cadmio
cadmium carbonate carbonato de cadmio
cadmium cell pila de cadmio
cadmium chlorate clorato de cadmio
cadmium chloride cloruro de cadmio
cadmium chloroacetate cloroacetato de cadmio
cadmium cinnamate cinamato de cadmio
cadmium cyanide cianuro de cadmio
cadmium diethyldithiocarbamate dietilditiocarbamato de cadmio
cadmium electrode electrodo de cadmio
cadmium fluoride fluoruro de cadmio
cadmium formate formiato de cadmio
cadmium fumarate fumarato de cadmio
cadmium halide haluro de cadmio
cadmium hexafluorosilicate hexafluorurosilicato de cadmio
cadmium hydroxide hidróxido de cadmio
cadmium iodate yodato de cadmio
cadmium iodide yoduro de cadmio
cadmium lactate lactato de cadmio
cadmium molybdate molybdato de cadmio
cadmium nitrate nitrato de cadmio
cadmium oxalate oxalato de cadmio
cadmium oxide óxido de cadmio
cadmium permanganate permanganato de cadmio
cadmium phosphate fosfato de cadmio
cadmium pigment pigmento de cadmio
cadmium potassium cyanide cianuro potásico de cadmio
cadmium potassium iodide yoduro potásico de cadmio
cadmium propionate propionato de cadmio
cadmium red rojo de cadmio
cadmium ricinoleate ricinoleato de cadmio
cadmium selenate seleniato de cadmio
cadmium selenide seleniuro de cadmio
cadmium stearate estearato de cadmio
cadmium succinate succinato de cadmio
cadmium sulfate sulfato de cadmio
cadmium sulfide sulfuro de cadmio
cadmium tartrate tartrato de cadmio
cadmium telluride telururo de cadmio
cadmium tungstate tungstato de cadmio
cadmium valerate valerato de cadmio
cadmium yellow amarillo de cadmio
cadmous cadmoso
caesium cesio
caffeic acid ácido cafeico
caffeine cafeína
caffeine benzoate benzoato de cafeína
caffeine citrate citrato de cafeína
caffeine hydrochloride clorhidrato de cafeína

caffeine salicylate salicilato de cafeína
caffetannic acid ácido cafetánico
cage compound compuesto en jaula
cage zeolite zeolita en jaula
calabarine calabarina
calamine calamina
calamus oil aceite de cálamo
calaverite calaverita
calcein calceína
calciferol calciferol
calcimeter calcímetro
calcimine calcimina
calcination calcinación
calcine calcina
calcined calcinado
calcined bauxite bauxita calcinada
calcinite calcinita
calcinol calcinol
calcite calcita
calcium calcio
calcium abietate abietato cálcico
calcium acetate acetato cálcico, acetato de calcio
calcium acetylide acetiluro cálcico
calcium acetylsalicylate acetilsalicilato cálcico
calcium acrylate acrilato cálcico
calcium alginate alginato cálcico
calcium aluminate aluminato cálcico
calcium aluminum hydride hidruro alumínico cálcico
calcium aluminum nitrate nitrato alumínico cálcico
calcium arsenate arseniato cálcico
calcium arsenite arsenito cálcico
calcium ascorbate ascorbato cálcico
calcium benzoate benzoato cálcico
calcium biophosphate biofosfato cálcico
calcium bisulfide bisulfuro cálcico
calcium bisulfite bisulfito cálcico
calcium borate borato cálcico
calcium bromate bromato cálcico
calcium bromide bromuro cálcico
calcium butanoate butanoato cálcico
calcium carbide carburo cálcico
calcium carbonate carbonato cálcico
calcium caseinate caseinato cálcico
calcium chlorate clorato cálcico
calcium chloride cloruro cálcico

calcium chlorite clorito cálcico
calcium chromate cromato cálcico
calcium cinnamate cinamato cálcico
calcium citrate citrato cálcico
calcium cyanamide cianamida cálcica
calcium cyanide cianuro cálcico
calcium cyclamate ciclamato cálcico
calcium dehydroacetate deshidroacetato cálcico
calcium diacetate diacetato cálcico
calcium dicarbide dicarburo cálcico
calcium dichromate dicromato cálcico
calcium dioxide dióxido cálcico
calcium disodium edetate edetato disódico cálcico
calcium ethanoate etanoato cálcico
calcium ethylhexoate etilhexoato cálcico
calcium ferrocyanide ferrocianuro cálcico
calcium fluoride fluoruro cálcico
calcium fluorophosphate fluorofosfato cálcico
calcium fluorosilicate fluorosilicato cálcico
calcium formate formiato cálcico
calcium fumarate fumarato cálcico
calcium gluconate gluconato cálcico
calcium glutamate glutamato cálcico
calcium glycerophosphate glicerofosfato cálcico
calcium glycolate glicolato cálcico
calcium hexafluorosilicate hexafluorosilicato cálcico
calcium hydrate hidrato cálcico
calcium hydride hidruro cálcico
calcium hydrogen sulfite sulfito de hidrógeno cálcico
calcium hydrosulfide hidrosulfuro cálcico
calcium hydroxide hidróxido cálcico
calcium hypochlorite hipoclorito cálcico
calcium hypophosphite hipofosfito cálcico
calcium iodate yodato cálcico
calcium iodide yoduro cálcico
calcium iodobehenate yodobehenato cálcico
calcium isobutyrate isobutirato cálcico
calcium lactate lactato cálcico

calcium malate malato cálcico
calcium metaphosphate metafosfato
  cálcico
calcium metasilicate metasilicato cálcico
calcium molybdate molibdato cálcico
calcium naphthenate naftenato cálcico
calcium nitrate nitrato cálcico
calcium nitride nitruro cálcico
calcium nitrite nitrito cálcico
calcium octoate octoato cálcico
calcium oleate oleato cálcico
calcium orthophosphate ortofosfato
  cálcico
calcium orthotungstate ortotungstato
  cálcico
calcium oxalate oxalato cálcico
calcium oxide óxido cálcico
calcium oxychloride oxicloruro cálcico
calcium palmitate palmitato cálcico
calcium pantothenate pantotenato
  cálcico
calcium perborate perborato cálcico
calcium perchlorate perclorato cálcico
calcium permanganate permanganato
  cálcico
calcium peroxide peróxido cálcico
calcium phenolate fenolato cálcico
calcium phenylate fenilato cálcico
calcium phosphate fosfato cálcico
calcium phosphide fosfuro cálcico
calcium phosphinate fosfinato cálcico
calcium phosphite fosfito cálcico
calcium phytate fitato cálcico
calcium plumbate plumbato cálcico
calcium polysilicate polisilicato cálcico
calcium propionate propionato cálcico
calcium pyrophosphate pirofosfato
  cálcico
calcium resinate resinato cálcico
calcium ricinoleate ricinoleato cálcico
calcium selenate seleniato cálcico
calcium silicate silicato cálcico
calcium silicide siliciuro cálcico
calcium silicofluoride silicofluoruro
  cálcico
calcium sorbate sorbato cálcico
calcium stannate estannato cálcico
calcium stearate estearato cálcico
calcium succinate succinato cálcico

calcium sulfamate sulfamato cálcico
calcium sulfate sulfato cálcico
calcium sulfhydrate sulfhidrato cálcico
calcium sulfide sulfuro cálcico
calcium sulfite sulfito cálcico
calcium sulfocyanate sulfocianato
  cálcico
calcium superoxide superóxido cálcico
calcium superphosphate superfosfato
  cálcico
calcium tannate tanato cálcico
calcium tartrate tartrato cálcico
calcium thiocyanate tiocianato cálcico
calcium titanate titanato cálcico
calcium tungstate tungstato cálcico
calcium undecylenate undecilenato
  cálcico
calcium valerate valerato cálcico
calcium zirconate zirconato cálcico
caledonite caledonita
calender calandria
calibrate calibrar
calibrated calibrado
calibration calibración
calibration error error de calibración
caliche caliche
californite californita
californium californio
calixarene calixareno
calomel calomel
calomel electrode electrodo de calomel
calorie caloría
calorific value valor calorífico
calorimeter calorímetro
calorimetric calorimétrico
calorimetry calorimetría
calorizing calorización
calotropin calotropina
calutron calutrón
calycanthine calicantina
cambopinic acid ácido cambopínico
Campbell synthesis síntesis de Campbell
camphane canfano
camphanol canfanol
camphanone canfanona
camphene canfeno
camphocarboxylic acid ácido
  canfocarboxílico
campholic acid ácido canfólico

camphor alcanfor
camphor bromate bromato de alcanfor
camphor imide imida de alcanfor
camphor oxime oxima de alcanfor
camphoric acid ácido canfórico
camphoroquinone canforoquinona
camphoroyl canforoilo
Camps quinoline synthesis síntesis de
    quinolinas de Camps
canadine canadina
canal rays rayos canales
cananga oil aceite de cananga
canavanine canavanina
cancrinite cancrinita
candelilla wax cera de candelilla
cane sugar azúcar de caña
cannabidiol cannabidiol
cannabinol cannabinol
cannabis cannabis
cannel coal carbón de bujía
Cannizzaro reaction reacción de
    Cannizzaro
cannonite canonita
canola aceite de colza
canonical form forma canónica
canphane canfano
cantharidin cantaridina
canthaxanthin cantaxantina
capacity capacidad
capillarity capilaridad
capillary condensation condensación
    capilar
capillary pipet pipeta capilar
capraldehyde capraldehído
caprate caprato
capreomycin capreomicina
capric acid ácido cáprico
capric aldehyde aldehído cáprico
caproic acid ácido caproico
caproic aldehyde aldehído caproico
caprolactam caprolactama
caprolactone caprolactona
capronyl capronilo
caproyl caproilo
capryl caprilo
caprylamide caprilamida
caprylate caprilato
caprylic acid ácido caprílico
caprylic anhydride anhídrido caprílico

caprylyl caprililo
capsule cápsula
captafol captafol
captan captano
capture captura
caracolite caracolita
caramel caramelo
caramelan caramelano
caramelization caramelización
carane carano
carat quilate
caraway oil aceite de alcaravea
carbaborane carbaborano
carbachol carbacol
carbaldehyde carbaldehído
carbamate carbamato
carbamic carbámico
carbamic acid ácido carbámico
carbamide carbamida
carbamide chloride cloruro de
    carbamida
carbamide peroxide peróxido de
    carbamida
carbamidine carbamidina
carbamido carbamido
carbamite carbamita
carbamoyl carbamoilo
carbamyl carbamilo
carbamylurea carbamilurea
carbanil carbanilo
carbanilic acid ácido carbanílico
carbanilide carbanilida
carbanion carbanión
carbaryl carbarilo
carbate carbato
carbazic acid ácido carbázico
carbazide carbazida
carbazole carbazol
carbazolyl carbazolilo
carbazone carbazona
carbazotic acid ácido carbazótico
carbazoyl carbazoilo
carbene carbeno
carbenium carbenio
carbenium ion ion carbenio
carbenoid carbenoide
carbethoxycyclohexanone
    carbetoxiciclohexanona
carbethoxycyclopentanone

carbetoxiciclopentanona
carbethoxypiperazine
carbetoxipiperazina
carbide carburo
carbinol carbinol
carbitol carbitol
carbobenzoyl carbobenzoilo
carbocation carbocatión
carbocyclic carbocíclico
carbodiazone carbodiazona
carbodihydrazide carbodihidrazida
carbodiimide carbodiimida
carbodithioic carboditioico
carbofuran carbofurano
carbohydrase carbohidrasa
carbohydrate carbohidrato, hidrato de carbono
carbohydrazide carbohidrazida
carbohydride carbohidruro
carboid carboide
carbolfuchsin carbolfucsina
carbolic acid ácido carbólico
carbolic oil aceite carbólico
carbomer carbómero
carbometer carbonómetro
carbomethene carbometeno
carbomethoxy carbometoxi
carbomycin carbomicina
carbon carbono, carbón
carbon assimilation asimilación de carbono
carbon atom átomo de carbono
carbon bisulfide bisulfuro de carbono
carbon black negro de carbón
carbon bond enlace de carbono
carbon chain cadena de carbonos
carbon compound compuesto de carbono
carbon cycle ciclo de carbono
carbon dating datación por carbono
carbon dichloride dicloruro de carbono
carbon dioxide dióxido de carbono
carbon dioxide snow nieve de dióxido de carbono
carbon disulfide disulfuro de carbono
carbon electrode electrodo de carbono
carbon fiber fibra de carbono
carbon fluoride fluoruro de carbono
carbon hexachloride hexacloruro de carbono

carbon monosulfide monosulfuro de carbono
carbon monoxide monóxido de carbono
carbon nitride nitruro de carbono
carbon-nitrogen cycle ciclo de carbono-nitrógeno
carbon oxide óxido de carbono
carbon oxybromide oxibromuro de carbono
carbon oxychloride oxicloruro de carbono
carbon oxycyanide oxicianuro de carbono
carbon oxyfluoride oxifluoruro de carbono
carbon oxysulfide oxisulfuro de carbono
carbon ratio razón de carbono
carbon steel acero al carbono
carbon suboxide subóxido de carbono
carbon tetrabromide tetrabromuro de carbono
carbon tetrachloride tetracloruro de carbono
carbon tetrafluoride tetrafluoruro de carbono
carbon tetraiodide tetrayoduro de carbono
carbon trichloride tricloruro de carbono
carbonado carbonado
carbonatation carbonatación
carbonate carbonato
carbonate mineral mineral de carbonato, carbonatomineral
carbonate rock roca de carbonato, roca carbonatada
carbonation carbonación
carbonic carbónico
carbonic acid ácido carbónico
carbonic anhydrase anhidrasa carbónica
carbonic ester éster carbónico
carbonic ether éter carbónico
carbonimidoyl carbonimidoilo
carbonite carbonita
carbonitrile carbonitrilo
carbonium carbonio
carbonium ion ion carbonio
carbonization carbonización
carbonize carbonizar
carbonized carbonizado

carbonohydrazide carbonohidrazida
carbonohydrazido carbonohidrazido
carbonyl carbonilo
carbonyl bromide bromuro de carbonilo
carbonyl chloride cloruro de carbonilo
carbonyl cyanide cianuro de carbonilo
carbonyl fluoride fluoruro de carbonilo
carbonyl group grupo carbonilo
carbonyl halide haluro de carbonilo
carbonyl hydride hidruro de carbonilo
carbonyl sulfide sulfuro de carbonilo
carbonylate carbonilato
carbonylation carbonilación
carbonyldiimidazole carbonildiimidazol
carbonyldioxy carbonldioxi
carbophenothion carbofenotión
carborane carborano
carbostyril carboestirilo
carbothioic carbotioico
carbothionic carbotiónico
carboxamide carboxamida
carboxamidine carboxamidina
carboxy carboxi
carboxybenzene carboxibenceno
carboxyl carboxilo
carboxyl group grupo carboxilo
carboxylase carboxilasa
carboxylation carboxilación
carboxylic carboxílico
carboxylic acid ácido carboxílico
carboxymethylcellulose
    carboximetilcelulosa
carboxypeptidase carboxipeptidasa
carburite carburita
carburization carburización
carbylamine carbilamina
carbylic acid ácido carbílico
carbyloxime carbiloxima
carbyne carbino
carcinogen carcinógeno
carcinomic acid ácido carcinómico
cardamom oil aceite de cardamomo
cardenolide cardenolida
carene careno
Carius method método de Carius
Carius tube tubo de Carius
Carl solution solución de Carl
carlic acid ácido cárlico
carmine carmín

carminic acid ácido carmínico
carminite carminita
carnallite carnalita
carnauba wax cera de carnauba
carnaubic acid ácido carnáubico
carnegine carnegina
carnine carnina
carnitine carnitina
carnosine carnosina
Carnot cycle ciclo de Carnot
Carnot reagent reactivo de Carnot
carnotite carnotita
Carnoy fluid fluido de Carnoy
Caro acid ácido de Caro
Caro reagent reactivo de Caro
carobic acid ácido caróbico
carobine carobina
carolic acid ácido carólico
carone carona
carotene caroteno
carotenoid carotenoide
carpain carpaína
carrageenan carraginina
carreen carreno
carrier portador
carrier gas gas portador
Carroll reaction reacción de Carroll
carthamic acid ácido cartámico
carthamin cartamina
carvacrol carvacrol
carveol carveol
carvestrene carvestreno
carvone carvona
caryophyllene cariofileno
caryophyllic acid ácido cariofílico
cascade cascada
cascade process proceso en cascada
cascarilla oil aceite de cascarilla
casein caseína
caseinase caseinasa
cashew oil aceite de anacardo
Cassella acid ácido de Cassella
cassia oil aceite de casia
cassiterite casiterita
cast iron hierro fundido
Castner-Kellner cell pila de Castner-
    Kellner
Castner process proceso de Castner
castor oil aceite de castor

castor oil acid ácido de aceite de castor
catabiotic catabiótico
catabolic catabólico
catabolism catabolismo
catalase catalasa
catalysis catálisis
catalyst catalizador
catalyst carrier portador de catalizador
catalyst promotor promotor de catalizador
catalyst selectivity selectividad de catalizador
catalytic catalítico
catalytic action acción catalítica
catalytic converter convertidor catalítico
catalytic cracking craqueo catalítico
catalytic force fuerza catalítica
catalytic poison veneno catalítico
catalytic polymerization polimerización catalítica
catalytic reforming reformación catalítica
catalyzer catalizador
cataphoresis cataforesis
catechin catequina
catechol catecol
catecholborane catecolborano
catena catena
catenane catenano
catenary catenario
catenation catenación
catenyl catenilo
catharometer catarómetro
cathartin catartina
cathetometer catetómetro
cathine catina
cathode cátodo
cathode coating revestimiento catódico
cathode evaporation evaporación catódica
cathode layer capa catódica
cathode polarization polarización catódica
cathode protection protección catódica
cathode rays rayos catódicos
cathode sputtering pulverización catódica
cathodic catódico
cathodic coating revestimiento catódico

cathodic evaporation evaporación catódica
cathodic layer capa catódica
cathodic polarization polarización catódica
cathodic protection protección catódica
cathodic rays rayos catódicos
cathodic sputtering pulverización catódica
catholyte católito
cation catión
cation exchange intercambio de cationes
cationic catiónico
cationic polymerization polimerización catiónica
cationic reagent reactivo catiónico
cationotropy cationotropía
cationtrophy cationtrofía
catlinite catlinita
caustic cáustico
caustic alcohol alcohol cáustico
caustic antimony antimonio cáustico
caustic baryta barita cáustica
caustic cracking fragilización cáustica
caustic embrittlement fragilización cáustica
caustic lime cal cáustica
caustic potash potasa cáustica
caustic soda sosa cáustica
caustic wash lavado cáustico
causticity causticidad
causticized ash ceniza caustificada
cavitation cavitación
cedrin cedrina
cedrol cedrol
cedryl acetate acetato de cedrilo
celestine celestina
celestite celestita
cell célula, pila
cell constant constante de célula
cellase celasa
cellobiose celobiosa
cellodextrin celodextrina
celloidin celoidina
cellophane celofán
cellosteric effect efecto celostérico
cellular plastic plástico celular
cellulase celulasa
cellulose celulosa

cellulose acetate acetato de celulosa
cellulose acetate butyrate acetatobutirato de celulosa
cellulose acetatobutyrate acetatobutirato de celulosa
cellulose diacetate diacetato de celulosa
cellulose ether éter celulósico
cellulose fiber fibra de celulosa
cellulose gum goma de celulosa
cellulose monoacetate monoacetato de celulosa
cellulose nitrate nitrato de celulosa
cellulose propionate propionato de celulosa
cellulose triacetate triacetato de celulosa
cellulose trinitrate trinitrato de celulosa
cellulose xanthate xantato de celulosa
cellulosic celulósico
cellulosic plastic plástico celulósico
cellulosic resin resina celulósica
cellulosic thiocarbonate tiocarbonato celulósico
Celsius scale escala Celsius
cement cemento
cementation cementación
cementite cementita
center of symmetry centro de simetría
centi centi
centigrade centígrado
centigrade scale escala centígrada
centinormal centinormal
centinormal solution solución centinormal
centrifugal centrífugo
centrifugal force fuerza centrífuga
centrifugal pump bomba centrífuga
centrifugal separation separación centrífuga
centrifugation centrifugación
centrifuge centrífuga
centripetal centrípeto
centroid diagram diagrama centroide
centron centrón
cephalin cefalina
cephalosporin cefalosporina
cephamycin cefamicina
ceramic cerámica
ceramic alloy aleación cerámica
ceramide ceramida

cerate cerato
cerebronic acid ácido cerebrónico
cerebrose cerebrosa
cerebrosides cerebrósidos
Cerenkov effect efecto de Cerenkov
ceresin ceresina
ceresin wax cera de ceresina
ceria ceria
ceric cérico
ceric ammonium nitrate nitrato amónico cérico
ceric fluoride fluoruro cérico
ceric hydroxide hidróxido cérico
ceric nitrate nitrato cérico
ceric oxide óxido cérico
ceric sulfate sulfato cérico
ceric sulfide sulfuro cérico
cerin cerina
ceritamic acid ácido ceritámico
cerite cerita
cerium cerio
cerium carbonate carbonato de cerio
cerium chloride cloruro de cerio
cerium dioxide dióxido de cerio
cerium hydrate hidrato de cerio
cerium nitrate nitrato de cerio
cerium oxalate oxalato de cerio
cerium oxide óxido de cerio
cerium stearate estearato de cerio
cerium sulfate sulfato de cerio
cermet cermet
ceroplastic acid ácido ceroplástico
cerosin cerosina
cerosinyl cerosinilo
cerotene ceroteno
cerotic acid ácido cerótico
cerotinic acid ácido cerotínico
cerous ceroso
cerous acetate acetato ceroso
cerous benzoate benzoato ceroso
cerous bromate bromato ceroso
cerous bromide bromuro ceroso
cerous carbonate carbonato ceroso
cerous chloride cloruro ceroso
cerous citrate citrato ceroso
cerous fluoride fluoruro ceroso
cerous hydroxide hidróxido ceroso
cerous iodide yoduro ceroso
cerous lactate lactato ceroso

cerous nitrate nitrato ceroso
cerous oxalate oxalato ceroso
cerous phosphate fosfato ceroso
cerous sulfate sulfato ceroso
cerous valerate valerato ceroso
certified color color certificado
cerulean blue azul cerúleo
cerussite cerusita
cervantite cervantita
ceryl cerilo
ceryl alcohol alcohol cerílico
cerylic cerílico
cesium cesio
cesium acetate acetato de cesio
cesium alum alumbre de cesio
cesium aluminum sulfate sulfato
     alumínico de cesio
cesium antimonide antimoniuro de cesio
cesium arsenide arseniuro de cesio
cesium benzoate benzoato de cesio
cesium bicarbonate bicarbonato de cesio
cesium bromate bromato de cesio
cesium bromide bromuro de cesio
cesium carbonate carbonato de cesio
cesium chloride cloruro de cesio
cesium chromate cromato de cesio
cesium cyanide cianuro de cesio
cesium dichromate dicromato de cesio
cesium dioxide dióxido de cesio
cesium fluoride fluoruro de cesio
cesium hexachloroplatinate
     hexacloroplatinato de cesio
cesium hexafluorosilicate
     hexafluorosilicato de cesio
cesium hydrate hidrato de cesio
cesium hydrogensulfate
     hidrogenosulfato de cesio
cesium hydroxide hidróxido de cesio
cesium iodate yodato de cesio
cesium iodide yoduro de cesio
cesium nitrate nitrato de cesio
cesium oxalate oxalato de cesio
cesium perchlorate perclorato de cesio
cesium periodate peryodato de cesio
cesium permanganate permanganato de
     cesio
cesium peroxide peróxido de cesio
cesium phosphide fosfuro de cesio
cesium salicylate salicilato de cesio

cesium sulfate sulfato de cesio
cesium sulfite sulfito de cesio
cesium tetraoxide tetróxido de cesio
cesium trioxide trióxido de cesio
cesium trisulfide trisulfuro de cesio
cespitine cespitina
cetane cetano
cetane number número de cetano
cetene ceteno
cetin cetina
cetol cetol
cetolic acid ácido cetólico
cetrimide cetrimida
cetyl cetilo
cetyl alcohol alcohol cetílico
cetyl bromide bromuro de cetilo
cetyl ether éter cetílico
cetyl mercaptan cetilmercaptano
cetyl palmitate palmitato de cetilo
cetylate cetilato
cetylbetamine cetilbetamina
cetylene cetileno
cetylic acid ácido cetílico
cetylic alcohol alcohol cetílico
cetylpyridinium bromide bromuro de
     cetilpiridinio
cetylpyridinium chloride cloruro de
     cetilpiridinio
cetyltrimethylammonium bromide
     bromuro de cetiltrimetilamonio
cetyltrimethylammonium chloride
     cloruro de cetiltrimetilamonio
chabazite chabasita
Chadwick-Goldhaber effect efecto de
     Chadwick-Goldhaber
chain cadena
chain conformation conformación en
     cadena
chain decay desintegración en cadena
chain disintegration desintegración en
     cadena
chain initiation inicio de cadena
chain isomerization isomerización de
     cadena
chain mechanism mecanismo en cadena
chain polymer polímero de cadena
chain reaction reacción en cadena
chain termination terminación de cadena
chain transfer transferencia de cadena

chair conformation conformación en silla
chalcanthite calcantita
chalcedony calcedonia
chalcocite calcosina
chalcogen calcógeno
chalcone calcona
chalcophanite calcofanita
chalcopyrite calcopirita
chalcosiderite calcosiderita
chalk creta
chamber cámara
chamber acid ácido de cámara
chamomile oil aceite de manzanilla
Chance-Claus process proceso de Chance-Claus
change of phase cambio de fase
change of state cambio de estado
channeling canalización
Channing solution solución de Channing
chaos caos
chaotic reaction reacción caótica
Chapman rearrangement reordenamiento de Chapman
charcoal carbón
charge carga
charge density densidad de carga
charge independence independencia de carga
charge-mass ratio razón de carga a masa
charge of electron carga de electrón
charge transfer transferencia de carga
charged cargado
charged particle partícula cargada
Charles law ley de Charles
chaulmoogric acid ácido chaulmúgrico
chavicol chavicol
chebulinic acid ácido quebulínico
cheiramidine queiramidina
chelate quelato
chelate effect efecto de quelato
chelating agent agente quelante
chelation quelación
chelerythrine queleritrina
chelidonic acid ácido quelidónico
chemical substancia química, producto químico
chemical action acción química
chemical activity actividad química

chemical affinity afinidad química
chemical agent agente químico
chemical analysis análisis químico
chemical bond enlace químico
chemical burn quemadura química
chemical change cambio químico
chemical combination combinación química
chemical compound compuesto químico
chemical constant constante química
chemical constitution constitución química
chemical corrosion corrosión química
chemical data datos químicos
chemical dating datación química
chemical deposition deposición química
chemical detector detector químico
chemical effect efecto químico
chemical efficiency eficiencia química
chemical element elemento químico
chemical energy energía química
chemical engineering ingeniería química
chemical entity entidad química
chemical equation ecuación química
chemical equilibrium equilibrio químico
chemical equivalent equivalente químico
chemical erosion erosión química
chemical exchange intercambio químico
chemical flooding inundación química
chemical formula fórmula química
chemical garden jardín químico
chemical incompatibility incompatibilidad química
chemical indicator indicador químico
chemical inhibitor inhibidor químico
chemical kinetics cinética química
chemical laws leyes químicas
chemical light luz química
chemical load carga química
chemical microscopy microscopia química
chemical milling molienda química
chemical modification modificación química
chemical nomenclature nomenclatura química
chemical oxygen demand demanda química de oxígeno
chemical passivation pasivación química

chemical passivity pasividad química
chemical polarity polaridad química
chemical pollution contaminación química
chemical potential potencial químico
chemical process proceso químico
chemical processing procesamiento químico
chemical product producto químico
chemical property propiedad química
chemical protector protector químico
chemical purity pureza química
chemical reaction reacción química
chemical reactivity reactividad química
chemical reduction reducción química
chemical research investigación química
chemical sediment sedimento químico
chemical series serie química
chemical shift deslizamiento químico
chemical smoke humo químico
chemical solution solución química
chemical solvent solvente químico
chemical specialty especialidad química
chemical spectrum espectro químico
chemical standard patrón químico
chemical substance substancia química
chemical symbol símbolo químico
chemical synthesis síntesis química
chemical technology tecnología química
chemical testing pruebas químicas
chemical thermodynamics
    termodinámica química
chemical warfare guerra química
chemical waste desperdicios químicos
chemically induced químicamente inducido
chemically pure químicamente puro
chemiluminescence quimioluminiscencia
chemisorption quimisorción
chemist químico
chemistry química
chemodynamics quimiodinámica
chemolysis quimiólisis
chemometrics quimiometría
chemonuclear quimionuclear
chemonuclear production producción quimionuclear
chemosmosis quimiósmosis
chemosmotic quimiosmótico

chemosorption quimiosorción
chemostat quimióstato
chemosynthesis quimiosíntesis
chemotaxis quimiotaxis
chemotaxonomy quimiotaxonomía
chemotherapy quimioterapia
chemotropism quimiotropismo
chenopodium oil aceite de quenopodio
chevkinite quevkinita
chi chi
chi acid ácido chi
Chicago acid ácido de Chicago
Chichibabin pyridine synthesis síntesis de piridinas de Chichibabin
Chichibabin reaction reacción de Chichibabin
chicle chicle
Chinese blue azul de China
Chinese wax cera de China
Chinese white blanco de China
chinic acid ácido quínico
chinoidine quinoidina
chinone quinona
chiral quiral
chiral center centrol quiral
chirality quiralidad
chirooptical quiroóptico
chirooptical spectroscopy espectroscopia quiroóptica
chitin quitina
chitinase quitinasa
chitosan quitosana
chloral cloral
chloral amide amida de cloral
chloral hydrate hidrato de cloral
chloralkali clorálcali
chloralose cloralosa
chloramine cloramina
chloramphenicol cloranfenicol
chloranil cloranil
chloranilate cloranilato
chlorapatite clorapatita
chlorate clorato
chloraurite cloraurita
chlorazine clorazina
chlorbenside clorbensida
chlorbutanol clorbutanol
chlordane clordano
chlordiazepoxide hydrochloride

clorhidrato de clordiazepóxido
chlordimeform clordimeformo
chlorendic anhydride anhídrido
  cloréndico
chlorfenvinphos clorfenvinfos
chlorhexidine clorhexidina
chloric acid ácido clórico
chloride cloruro
chloridion cloridión
chlorimide clorimida
chlorinated clorado
chlorinated acetone acetona clorada
chlorinated camphene canfeno clorado
chlorinated diphenyl difenilo clorado
chlorinated hydrocarbon hidrocarburo
  clorado
chlorinated isocyanuric acid ácido
  isocianúrico clorado
chlorinated lime cal clorada
chlorinated naphthalene naftaleno
  clorado
chlorinated paraffin parafina clorada
chlorinated polyether poliéter clorado
chlorinated polyolefin poliolefina
  clorada
chlorinated rubber caucho clorado
chlorinated solvent solvente clorado
chlorinated trisodium phosphate fosfato
  trisódico clorado
chlorination clorinación
chlorine cloro
chlorine bromide bromuro de cloro
chlorine cyanide cianuro de cloro
chlorine dioxide dióxido de cloro
chlorine halide haluro de cloro
chlorine heptoxide heptóxido de cloro
chlorine hydrate hidrato de cloro
chlorine monofluoride monofluoruro de
  cloro
chlorine monoxide monóxido de cloro
chlorine oxide óxido de cloro
chlorine trifluoride trifluoruro de cloro
chlorine water agua de cloro
chlorinity clorinidad
chlorinolysis clorinólisis
chlorite clorito
chlorition cloritión
chlormethazanone clormetazanona
chloroacetaldehyde cloroacetaldehído

chloroacetamide cloroacetamida
chloroacetate cloroacetato
chloroacetic acid ácido cloroacético
chloroacetic anhydride anhídrido
  cloroacético
chloroacetoacetanilide
  cloroacetoacetanilida
chloroacetone cloroacetona
chloroacetonitrile cloroacetonitrilo
chloroacetophenone cloroacetofenona
chloroacetyl cloroacetilo
chloroacetyl chloride cloruro de
  cloroacetilo
chloroacetylurethane cloroacetiluretano
chloroacrolein cloroacroleína
chloroacrylate cloroacrilato
chloroacrylonitrile cloroacrilonitrilo
chloroaluminum diisopropoxide
  diisopropóxido de cloroaluminio
chloroaniline cloroanilina
chloroanthraquinone cloroantraquinona
chloroargentate cloroargentato
chloroauric acid ácido cloroáurico
chloroazotic acid ácido cloroazótico
chlorobenzal clorobenzal
chlorobenzaldehyde clorobenzaldehído
chlorobenzanthrone clorobenzantrona
chlorobenzene clorobenceno
chlorobenzenesulfonamide
  clorobencenosulfonamida
chlorobenzenethiol clorobencenotiol
chlorobenzhydrol clorobenzhidrol
chlorobenzoate clorobenzoato
chlorobenzohydrol clorobenzohidrol
chlorobenzoic acid ácido clorobenzoico
chlorobenzophenone clorobenzofenona
chlorobenzotrichloride
  clorobenzotricloruro
chlorobenzotrifluoride
  clorobenzotrifluoruro
chlorobenzoyl clorobenzoilo
chlorobenzoyl chloride cloruro de
  clorobenzoilo
chlorobenzoyl peroxide peróxido de
  clorobenzoilo
chlorobenzyl chloride cloruro de
  clorobecilo
chlorobenzyl cyanate cianato de
  clorobecilo

chlorobenzylpyridine clorobecilpiridina
chlorobromo clorobromo
chlorobutadiene clorobutadieno
chlorobutane clorobutano
chlorobutanol clorobutanol
chlorobutyryl clorobutirilo
chlorocarbon clorocarburo
chlorocarbonate clorocarbonato
chlorocarbonic clorocarbónico
chlorocarbonyl ferrocene ferroceno de clorocarbonilo
chlorochromate clorocromato
chlorochromic anhydride anhídrido clorocrómico
chlorocosane clorocosano
chlorocoumarin clorocumarina
chlorocyanide clorocianuro
chlorocyanogen clorocianógeno
chlorodecone clorodecona
chlorodifluoroacetic acid ácido clorodifluoroacético
chlorodifluoroethane clorodifluoroetano
chlorodifluoromethane clorodifluorometano
chlorodinitrobenzene clorodinitrobenceno
chlorodiphenyl clorodifenilo
chloroethanamide cloroetanamida
chloroethane cloroetano
chloroethanoic acid ácido cloroetanoico
chloroethanoic anhydride anhídrido cloroetanoico
chloroethanol cloroetanol
chloroethene cloroeteno
chloroethyl alcohol alcohol cloroetílico
chloroethylchloroformate cloroetilcloroformiato
chloroethylene cloroetileno
chloroethylphosphonic acid ácido cloroetilfosfónico
chlorofenethol clorofenetol
chlorofluoride clorofluoruro
chlorofluorocarbon clorofluorocarburo
chloroform cloroformo
chloroformate cloroformiato
chloroformic acid ácido clorofórmico
chloroformoxime cloroformoxima
chloroformyl chloride cloruro de cloroformilo

chlorogenic acid ácido clorogénico
chlorohydrin clorohidrina
chlorohydrin rubber caucho de clorohidrina
chlorohydroquinone clorohidroquinona
chlorohydroxybenzene clorohidroxibenceno
chlorohydroxytoluene clorohidroxitolueno
chloroiodide cloroyoduro
chloroisopropyl alcohol alcohol cloroisopropílico
chloromadinone acetate acetato de cloromadinona
chloromaleic anhydride anhídrido cloromaleico
chloromalonic acid ácido cloromalónico
chloromercuriferrocene clomercuriferroceno
chloromethane clorometano
chloromethyl clorometilo
chloromethyl cyanide cianuro de clorometilo
chloromethylaniline clorometilanilina
chloromethylated diphenyl oxide óxido de difenilo clorometilado
chloromethylation clorometilación
chloromethylbenzene clorometilbenceno
chloromethylchloroformate clorometilcloroformiato
chloromethylchlorosulfonate clorometilclorosulfonato
chloromethylnaphthalene clorometilnaftaleno
chloromethylphosphonic acid ácido clorometilfosfónico
chloromethylphosphonic dichloride dicloruro clorometilfosfónico
chloronaphthalene cloronaftaleno
chloronaphthalene oil aceite de cloronaftaleno
chloronitric acid ácido cloronítrico
chloronitroaniline cloronitroanilina
chloronitrobenzene cloronitrobenceno
chloronitrobenzoic acid ácido cloronitrobenzoico
chloronitrotoluene cloronitrotolueno
chloronitrous acid ácido cloronitroso
chloronium ion ion cloronio

chloropentafluoroacetone cloropentafluoroacetona

chloropentane cloropentano

chloroperbenzoic acid ácido cloroperbenzoico

chlorophenol clorofenol

chlorophenol red rojo de clorofenol

chlorophenyl isocyanate isocianato de clorofenilo

chlorophenyltrichlorosilane clorofeniltriclorosilano

chlorophoenicite clorofenicita

chlorophyll clorofila

chlorophyllin clorofilina

chloropicrin cloropicrina

chloroplatinate cloroplatinato

chloroplatinic acid ácido cloroplatínico

chloroprene cloropreno

chloropropane cloropropano

chloropropene cloropropeno

chloropropionic acid ácido cloropropiónico

chloropropionitrile cloropropionitrilo

chloropropylene cloropropileno

chloropyridine cloropiridina

chloroquinaldine cloroquinaldina

chloroquine cloroquina

chlorosalicylanilide clorosalicilanilida

chlorosalicylic acid ácido clorosalicílico

chlorostyrene cloroestireno

chlorosuccinimide clorosuccinimida

chlorosulfonic acid ácido clorosulfónico

chlorosulfonylbenzoic acid ácido clorosulfonilbenzoico

chlorosulfuric acid ácido clorosulfúrico

chlorosyl clorosilo

chlorothiazide clorotiazida

chlorothiophenol clorotiofenol

chlorothymol clorotimol

chlorotoluene clorotolueno

chlorotrifluoromethane clorotrifluorometano

chlorotrifluoroethylene clorotrifluoroetileno

chlorotrifluoroethylene polymer polímero de clorotrifluoroetileno

chlorous cloroso

chlorous acid ácido cloroso

chlorovaleric acid ácido clorovalérico

chlorovinyldichloroarsine clorovinildicloroarsina

chloroxine cloroxina

chloroxylene cloroxileno

chlorpromazine clorpromazina

chlorquinaldol clorquinaldol

chlortetracycline clortetraciclina

chloryl clorilo

cholaic acid ácido colaico

cholane colano

cholanthrene colantreno

cholecalciferol colecalciferol

choleic acid ácido coleico

cholesteric colestérico

cholesterol colesterol

cholestyramine colestiramina

cholic acid ácido cólico

choline colina

choline base base de colina

choline bicarbonate bicarbonato de colina

choline bitartrate bitartrato de colina

choline chloride cloruro de colina

cholinesterase colinesterasa

cholinesterase inhibitor inhibidor de colinesterasa

chorionic gonadotropin gonadotropina coriónica

chorismic acid ácido corísmico

choritoid coritoide

chroma croma

chroman cromano

chromate cromato

chromatic cromático

chromatic aberration aberración cromática

chromatic scale escala cromática

chromaticity cromaticidad

chromatin cromatina

chromatogram cromatograma

chromatographic adsorption adsorción cromatográfica

chromatography cromatografía

chrome cromo

chrome alum alumbre de cromo

chrome dye colorante de cromo

chrome green verde de cromo

chrome orange anaranjado de cromo

chrome pigment pigmento de cromo

chrome red rojo de cromo
chrome steel acero al cromo
chrome yellow amarillo de cromo
chromic crómico
chromic acetate acetato crómico
chromic acid ácido crómico
chromic anhydride anhídrido crómico
chromic bromide bromuro crómico
chromic carbide carburo crómico
chromic chloride cloruro crómico
chromic fluoride fluoruro crómico
chromic hydroxide hidróxido crómico
chromic iodide yoduro crómico
chromic nitrate nitrato crómico
chromic oxide óxido crómico
chromic oxychloride oxicloruro crómico
chromic phosphate fosfato crómico
chromic silicide siliciuro crómico
chromic sulfate sulfato crómico
chromic tartrate tartrato crómico
chrominance crominancia
chromite cromita
chromitite cromitita
chromium cromo
chromium acetate acetato de cromo
chromium acetylacetonate
    acetilacetonato de cromo
chromium ammonium sulfate sulfato
    amónico de cromo
chromium arsenide arseniuro de cromo
chromium boride boruro de cromo
chromium bromide bromuro de cromo
chromium carbide carburo de cromo
chromium carbonyl carbonilo de cromo
chromium chloride cloruro de cromo
chromium-copper cromo-cobre
chromium dioxide dióxido de cromo
chromium fluoride fluoruro de cromo
chromium hexacarbonyl hexacarbonilo
    de cromo
chromium hydrate hidrato de cromo
chromium hydroxide hidróxido de
    cromo
chromium manganese antimonide
    antimoniuro de manganeso crómico
chromium naphthenate naftenato de
    cromo
chromium nitrate nitrato de cromo
chromium oxide óxido de cromo

chromium oxychloride oxicloruro de
    cromo
chromium oxyfluoride oxifluoruro de
    cromo
chromium phosphate fosfato de cromo
chromium phosphide fosfuro de cromo
chromium potassium oxalate oxalato
    potásico crómico
chromium potassium sulfate sulfato
    potásico crómico
chromium sesquichloride sesquicloruro
    de cromo
chromium steel acero al cromo
chromium sulfate sulfato de cromo
chromium tetrasulfide tetrasulfuro de
    cromo
chromium trichloride tricloruro de
    cromo
chromium trifluoride trifluoruro de
    cromo
chromium trioxide trióxido de cromo
chromo cromo
chromogen cromógeno
chromogenic cromogénico
chromoisomer cromoisómero
chromone cromona
chromophilic cromofílico
chromophore cromóforo
chromophoric cromofórico
chromosan cromosana
chromosome cromosoma
chromotropic acid ácido cromotrópico
chromotropy cromotropía
chromous cromoso
chromous acetate acetato cromoso
chromous acid ácido cromoso
chromous bromide bromuro cromoso
chromous carbonate carbonato cromoso
chromous chloride cloruro cromoso
chromous fluoride fluoruro cromoso
chromous formate formiato cromoso
chromous hydroxide hidróxido cromoso
chromous oxalate oxalato cromoso
chromous oxide óxido cromoso
chromous sulfate sulfato cromoso
chromyl cromilo
chromyl chloride cloruro de cromilo
chromyl fluoride fluoruro de cromilo
chronopotentiometry

cronopotenciometría
chrysalicic acid ácido crisalícico
chrysaniline crisanilina
chrysanisic acid ácido crisanísico
chrysanthemummonocarboxylic acid
    ácido crisantemomonocarboxílico
chrysene criseno
chrysolite crisólito
chrysophanic acid ácido crisofánico
chrysoquinone crisoquinona
chrysotile crisotilo
Chugaev reaction reacción de Chugaev
chymosin quimosina
chymotrypsin quimotripsina
chymotrypsinogen quimotripsinógeno
Ciamician-Dennstedt rearrangement
    reordenamiento de Ciamician-
    Dennstedt
cicutoxine cicutoxina
ciguatoxin ciguatoxina
cimetidine cimetidina
cinchocain cincocaína
cinchomeronic acid ácido cincomerónico
cinchona bark quina
cinchonamine cinconamina
cinchonidin cinconidina
cinchonine cinconina
cinchonine nitrate nitrato de cinconina
cinchoninic acid ácido cinconínico
cinene cineno
cineol cineol
cinerin cinerina
cinerolone cinerolona
cinnabar cinabrio
cinnamaldehyde cinamaldehído
cinnamamide cinamamida
cinnamate cinamato
cinnamein cinameína
cinnamene cinameno
cinnamenyl cinamenilo
cinnamic acid ácido cinámico
cinnamic alcohol alcohol cinámico
cinnamic aldehyde aldehído cinámico
cinnamic anhydride anhídrido cinámico
cinnamide cinamida
cinnamilidene cinamilideno
cinnamon oil aceite de canela
cinnamoyl cinamoilo
cinnamoyl chloride cloruro de cinamoilo

cinnamyl acetate acetato de cinamilo
cinnamyl alcohol alcohol cinamílico
cinnamyl aldehyde aldehído cinamílico
cinnamyl cinnamate cinamato de
    cinamilo
cinnamyl hydride hidruro de cinamilo
cinnamylidene cinamilideno
cinnoline cinolina
circular birefringence birrefrigencia
    circular
circular chromatography cromatografía
    circular
circular dichroism dicroismo circular
circular polarization polarización
    circular
cis cis
cis-isomer isómero cis
cis-trans isomerism isomerismo cis-trans
cisoid cisoide
citiolone citiolona
citraconic acid ácido citracrónico
citraconic anhydride anhídrido
    citacrónico
citraconoyl citraconoilo
citral citral
citramalic acid ácido citramálico
citramide citramida
citrate citrato
citrene citreno
citresia citresia
citric acid ácido cítrico
citric acid cycle ciclo de ácido cítrico
citridic acid ácido citrídico
citrinin citrinina
citronella oil aceite de citronela
citronellal citronelal
citronellic acid ácido citronélico
citronellol citronelol
citronyl citronilo
citrulline citrulina
civet civeto
civettal civetal
cladding chapado, revestimiento
Claisen condensation condensación de
    Claisen
Claisen flask matraz de Claisen
Claisen reaction reacción de Claisen
Claisen rearrangement reordenamiento
    de Claisen

Claisen-Schmidt condensation condensación de Claisen-Schmidt

clamp abrazadera

Clapeyron-Clausius equation ecuación de Clapeyron-Clausius

Clapeyron equation ecuación de Clapeyron

clarification clarificación

clarifier clarificador

Clark cell pila de Clark

Clark electrode electrodo de Clark

Clark-Lubs indicators indicadores de Clark-Lubs

Clark process proceso de Clark

classical electron radius radio de electrón clásico

classification clasificación

clathrate clatrato

Claude process proceso de Claude

Claude system sistema de Claude

Clausius law ley de Clausius

Clausius-Mosotti equation ecuación de Clausius-Mosotti

Clausius-Mosotti law ley de Clausius-Mosotti

clay arcilla

clean room sala limpia

cleavage división, hendidura

clebrium clebrio

Cleland reagent reactivo de Cleland

Clemmensen reduction reducción de Clemmensen

Cleve acid ácido de Cleve

clinical chemistry química clínica

clinohedrite clinohedrita

clinoptilolite clinoptilolita

clioquinol clioquinol

clomiphene clomifeno

clonazepam clonazepam

clopidol clopidol

close-packed structure estructura compacta

closed chain cadena cerrada

closo closo

cloudy turbio

clove oil aceite de clavero

Clusius column columna de Clusius

cluster compound compuesto de grupos

coacervation coacervación

coagulant coagulante

coagulation coagulación

Coahran process proceso de Coahran

coal hulla, carbón

coal gas gas de hulla

coal oil aceite de hulla

coal tar alquitrán de hulla

coal-tar distillate destilado de alquitrán de hulla

coal-tar dye colorante de alquitrán de hulla

coal-tar oil aceite de alquitrán de hulla

coal-tar pitch brea de alquitrán de hulla

coalescence coalescencia

coalescent coalescente

coated cubierto, revestido

coating revestimiento

cobalamin cobalamina

cobalt cobalto

cobalt acetate acetato de cobalto

cobalt alloy aleación de cobalto

cobalt aluminate aluminato de cobalto

cobalt ammonium sulfate sulfato amónico de cobalto

cobalt arsenate aresniato de cobalto

cobalt black negro de cobalto

cobalt blue azul de cobalto

cobalt bromide bromuro de cobalto

cobalt carbonate carbonato de cobalto

cobalt carbonyl carbonilo de cobalto

cobalt chloride cloruro de cobalto

cobalt chromate cromato de cobalto

cobalt cyanide cianuro de cobalto

cobalt difluoride difluoruro de cobalto

cobalt ethylhexoate etilhexoato de cobalto

cobalt fluorosilicate fluorosilicato de cobalto

cobalt green verde de cobalto

cobalt halide haluro de cobalto

cobalt hydrate hidrato de cobalto

cobalt hydrocarbonyl hidrocarbonilo de cobalto

cobalt hydroxide hidróxido de cobalto

cobalt iodide yoduro de cobalto

cobalt linoleate linoleato de cobalto

cobalt molybdate molibdato de cobalto

cobalt monoxide monóxido de cobalto

cobalt naphthenate naftenato de cobalto

cobalt nitrate nitrato de cobalto
cobalt nitride nitruro de cobalto
cobalt oleate oleato de cobalto
cobalt oxide óxido de cobalto
cobalt phosphate fosfato de cobalto
cobalt phosphide fosfuro de cobalto
cobalt potassium nitrate nitrato potásico de cobalto
cobalt potassium sulfate sulfato potásico de cobalto
cobalt resinate resinato de cobalto
cobalt selenite selenito de cobalto
cobalt silicide siliciuro de cobalto
cobalt sulfate sulfato de cobalto
cobalt tetracarbonyl tetracarbonilo de cobalto
cobalt trifluoride trifluoruro de cobalto
cobalt tungstate tungstato de cobalto
cobalt violet violeta de cobalto
cobalt yellow amarillo de cobalto
cobalt zincate cincato cobalto
cobaltammine cobaltamina
cobaltic cobáltico
cobaltic boride boruro cobáltico
cobaltic chloride cloruro cobáltico
cobaltic fluoride fluoruro cobáltico
cobaltic hydroxide hidróxido cobáltico
cobaltic oxide óxido cobáltico
cobaltic potassium nitrite nitrito potásico cobáltico
cobaltic sulfate sulfato cobáltico
cobaltic sulfide sulfuro cobáltico
cobaltite cobaltita
cobaltocene cobaltoceno
cobaltous acetate acetato cobaltoso
cobaltous aluminate aluminato cobaltoso
cobaltous ammonium phosphate fosfato amónico cobaltoso
cobaltous ammonium sulfate sulfato amónico cobaltoso
cobaltous arsenate arseniato cobaltoso
cobaltous benzoate benzoato cobaltoso
cobaltous bromate bromato cobaltoso
cobaltous bromide bromuro cobaltoso
cobaltous butyrate butirato cobaltoso
cobaltous carbonate carbonato cobaltoso
cobaltous chlorate clorato cobaltoso
cobaltous chloride cloruro cobaltoso
cobaltous chromate cromato cobaltoso

cobaltous citrate citrato cobaltoso
cobaltous cyanide cianuro cobaltoso
cobaltous ferrite ferrita cobaltosa
cobaltous fluoride fluoruro cobaltoso
cobaltous formate formiato cobaltoso
cobaltous hexafluorosilicate hexafluorosilicato cobaltoso
cobaltous hydroxide hidróxido cobaltoso
cobaltous iodide yoduro cobaltoso
cobaltous linoleate linoleato cobaltoso
cobaltous naphthenate naftenato cobaltoso
cobaltous nitrate nitrato cobaltoso
cobaltous oleate oleato cobaltoso
cobaltous oxalate oxalato cobaltoso
cobaltous oxide óxido cobaltoso
cobaltous perchlorate perclorato cobaltoso
cobaltous phosphate fosfato cobaltoso
cobaltous propionate propionato cobaltoso
cobaltous resinate resinato cobaltoso
cobaltous silicate silicato cobaltoso
cobaltous stannate estannato cobaltoso
cobaltous succinate succinato cobaltoso
cobaltous sulfate sulfato cobaltoso
cobaltous sulfide sulfuro cobaltoso
cobaltous tartrate tartrato cobaltoso
cobaltous tungstate tungstato cobaltoso
cobaltouscobaltic oxide óxido colbatosocobáltico
coca coca
cocaine cocaína
cocaine benzoate benzoato de cocaína
cocaine borate borato de cocaína
cocaine chloride cloruro de cocaína
cocaine hydrochloride clorhidrato de cocaína
cocaine nitrate nitrato de cocaína
cocaine sulfate sulfato de cocaína
cocarboxylase cocarboxilasa
cocculin coculina
coccus coco
cochineal cochinilla
cochinilin cochinilina
cochrome cocromo
cocoa cacao
cocoa butter manteca de cacao
cocoa oil aceite de cacao

coconut acid ácido de nuez de coco
coconut oil aceite de nuez de coco
cod-liver oil aceite de hígado de bacalao
codeine codeína
codeine citrate citrato de codeína
codeine hydrochloride clorhidrato de
    codeína
codeine phosphate fosfato de codeína
codeine salicylate salicilato de codeína
codeine sulfate sulfato de codeína
coefficient coeficiente
Coehn law ley de Coehn
coenzyme coenzima
coercivity coercitividad
cofactor cofactor
coherence coherencia
coherent coherente
coherent light luz coherente
coherent precipitate precipitado
    coherente
cohesion cohesión
cohobation cohobación
coil condenser condensador en espiral
coiled condenser condensador en espiral
coke coque
coke gas gas de coque
coke oven horno de coque
coking coquización
cola cola
colamine colamina
colchicine colquicina
cold flow flujo en frío
cold light luz fría
cold rubber caucho frío
colemanite colemanita
colestipol colestipol
colistin colistina
collagen colágeno
collagenase colagenasa
collargol colargol
collateral series serie colateral
collector colector
collidine colidina
colligate coligar
colligative coligativo
colligative property propiedad coligativa
collision colisión
collision density densidad de colisiones
collision excitation excitación por
    colisión
collision ionization ionización por
    colisión
Collman reagent reactivo de Collman
collodion colodión
colloid coloide
colloid equivalent equivalente coloidal
colloid mill molino coloidal
colloidal coloidal
colloidal dispersion dispersión coloidal
colloidal electrolyte electrólito coloidal
colloidal metal metal coloidal
colloidal solution solución coloidal
colloidal state estado coloidal
colloidal suspension suspensión coloidal
colloidal system sistema coloidal
colpholic acid ácido colofólico
colophonic acid ácido colofónico
colophonite colofonita
colophony colofonia
color comparator comparador de colores
color indicator indicador de color
color standard patrón de color
colorant colorante
coloration coloración
colorimeter colorímetro
colorimetric colorimétrico
colorimetric analysis análisis
    colorimétrico
colorimetric purity pureza colorimétrica
colorimetry colorimetría
colorization colorización
colorless dye colorante incoloro
columbate columbato
columbite columbita
columbium columbio
column columna
colza colza
Combes quinoline synthesis síntesis de
    quinolinas de Combes
combination combinación
combined combinado
combined carbon carbono combinado
combining number número de
    combinación
combining volume volumen de
    combinación
combining weight peso de combinación
combustibility combustibilidad

combustible combustible
combustible material material
    combustible
combustion combustión
combustion boat artesa de combustión
combustion capsule cápsula de
    combustión
combustion furnace horno de
    combustión
combustion rate velocidad de
    combustión
combustion tube tubo de combustión
comenic acid ácido coménico
comicellization comicelización
comirin comirina
comminution pulverización
common-ion effect efecto de ion común
common salt sal común
comparator comparador
compatibility compatibilidad
compatible compatible
complex complejo
complex acid ácido complejo
complex compound compuesto complejo
complex ion ion complejo
complex reaction reacción compleja
complex salt sal compleja
complexation complejación
complexing complejación
complexing agent agente complejante
complexometric analysis análisis
    complejométrico
complexometric titration titulación
    complejométrica, valoración
    complejométrica
complexone complexona
component componente
composite compuesto
composition composición
compost abono orgánico
compound compuesto
compound molecule molécula compuesta
compound nucleus núcleo compuesto
compressed comprimido
compressed gas gas comprimido
Compton effect efecto de Compton
Compton rule regla de Compton
computational chemistry química
    computacional

concave cóncavo
concentrate concentrar
concentrated concentrado
concentrated solution solución
    concentrada
concentration concentración
concentration gradient gradiente de
    concentración
concentration limit límite de
    concentración
concentrator concentrador
conchiolin conquiolina
conchoidal concoidal
concrete concreto
condensation condensación
condensation polymer polímero de
    condensación
condensation reaction reacción de
    condensación
condensation temperature temperatura
    de condensación
condensed condensado
condenser condensador
condistillation condestilación
conductance conductancia
conduction conducción
conduction band banda de conducción
conduction electron electrón de
    conducción
conductivity conductividad
conductivity apparatus aparato de
    conductividad
conductometric conductométrico
conductometric analysis análisis
    conductométrico
conductometric method método
    conductométrico
conductometric titration titulación
    conductiométrica, valoración
    conductométrica
conductor conductor
Condy fluid fluido de Condy
cone cono
conessine conesina
configuration configuración
configuration space espacio de
    configuración
conformation conformación
conformational analysis análisis

coordinated complex   63

conformacional
congealing congelación
congelation congelación
conglomerate conglomerado
congo blue azul congo
congo red rojo congo
congo yellow amarillo congo
conicein coniceína
coniferin coniferina
coniferol coniferol
coniferyl coniferilo
coniferyl alcohol alcohol coniferilo
conine conina
conjuctive nomenclature nomenclatura
    conjuntiva
conjugate acid ácido conjugado
conjugate base base conjugada
conjugated conjugado
conjugated double bonds enlaces dobles
    conjugados
conjugated fatty acid ácido graso
    conjugado
conjugated system sistema conjugado
conjugation conjugación
conproportionation conproporción
Conrad-Limpach reaction reacción de
    Conrad-Limpach
conservation law ley de conservación
conservation of energy conservación de
    energía
conservation of matter conservación de
    materia
consistence consistencia
consistency consistencia
consistometer consistómetro
consolute consoluto
constant constante
constant-boiling mixture mezcla de
    ebullición constante
constant composition composición
    constante
constant current corriente constante
constant pressure presión constante
constant proportion proporción
    constante
constant temperature temperatura
    constante
constantan constantán
constituent constituyente

constitutional constitucional
constitutional formula fórmula
    constitucional
contact contacto
contact acid ácido de contacto
contact potential potencial de contacto
contact process proceso de contacto
contact resin resina de contacto
contact substance substancia de contacto
container contenedor, receptáculo
contaminant contaminante
contamination contaminación
continuity of state continuidad de estado
continuous continuo
continuous distillation destilación
    continua
continuous phase fase continua
continuous spectrum espectro continuo
contraction contracción
contrast contraste
control control
controlled atmosphere atmósfera
    controlada
controlled substance substancia
    controlada
convection convección
convergence limit límite de convergencia
conversion conversión
conversion coefficient coeficiente de
    conversión
conversion electron electrón de
    conversión
conversion factor factor de conversión
conversion process proceso de
    conversión
conversion ratio razón de conversión
converter convertidor
convex convexo
Cook formula fórmula de Cook
cooking cocido
coolant refrigerante
Coolidge tube tubo de Coolidge
cooling enfriamiento
coordinate coordinado
coordinate bond enlace coordinado
coordinate covalence covalencia
    coordinada
coordinated complex complejo
    coordinado

coordination coordinación

coordination compound compuesto de coordinación

coordination isomerism isomerismo por coordinación

coordination number número de coordinación

coordination polymer polímero de coordinación

copaiba oil aceite de copaiba

copaiba resin resina de copaiba

copal copal

Cope elimination reaction reacción de eliminación de Cope

Cope rearrangement reordenamiento de Cope

copigment copigmento

copolymer copolímero

copolymerization copolimerización

copper cobre

copper abietate abietato de cobre

copper acetate acetato de cobre

copper acetoarsenite acetoarsenito de cobre

copper alloy aleación de cobre

copper amalgam amalgama de cobre

copper arsenate arseniato de cobre

copper arsenite arsenito de cobre

copper benzoate benzoato de cobre

copper blue azul de cobre

copper borate borato de cobre

copper bromide bromuro de cobre

copper carbonate carbonato de cobre

copper chloride cloruro de cobre

copper chromate cromato de cobre

copper cyanide cianuro de cobre

copper ethylacetoacetate etilacetoacetato de cobre

copper ferrocyanide ferrocianuro de cobre

copper fluoride fluoruro de cobre

copper gluconate gluconato de cobre

copper glycinate glicinato de cobre

copper green verde de cobre

copper hexacyanoferrate hexacianoferrato de cobre

copper hydroxide hidróxido de cobre

copper iodide yoduro de cobre

copper lactate lactato de cobre

copper metaborate metaborato de cobre

copper molybdate molibdato de cobre

copper monoxide monóxido de cobre

copper naphthenate naftenato de cobre

copper nitrate nitrato de cobre

copper nitrite nitrito de cobre

copper number número de cobre

copper oleate oleato de cobre

copper oxalate oxalato de cobre

copper oxide black óxido negro de cobre

copper oxychloride oxicloruro de cobre

copper phenolsulfonate fenolsulfonato de cobre

copper phosphate fosfato de cobre

copper phosphide fosfuro de cobre

copper resinate resinato de cobre

copper silicate silicato de cobre

copper stearate estearato de cobre

copper sulfate sulfato de cobre

copper sulfide sulfuro de cobre

copper trifluoroacetylacetonate trifluoroacetilacetonato de cobre

copper tungstate tungstato de cobre

copperas caparrosa

Coppet law ley de Coppet

coprastanol coprastanol

coprecipitation coprecipitación

coprosterol coprosterol

copyrine copirina

coral coral

cordite cordita

coriandrol coriandrol

cork corcho

corn oil aceite de maíz

corn sugar azúcar de almidón de maíz

corn syrup jarabe de maíz

cornstarch almidón de maíz

corona corona

coronene coroneno

correlation correlación

correlation diagram diagrama de correlación

correlation function función de correlación

correlation spectroscopy espectroscopia de correlación

correspondence principle principio de correspondencia

corresponding state estado

correspondiente
corresponding temperature temperatura
   correspondiente
corrosion corrosión
corrosive corrosivo
corrosive material material corrosivo
corrosive substance substancia corrosiva
corsite corsita
corticoid corticoide
corticoid hormone hormona corticoide
corticosterone corticosterona
corticotropin corticotropina
cortisol cortisol
cortisone cortisona
corundum corindón
cosanoic acid ácido cosanoico
cosanol cosanol
cosmetic cosmético
cosyl cosilo
cotton algodón
Cotton effect efecto de Cotton
cottonseed semilla de algodón
cottonseed meal harina de semilla de
   algodón
cottonseed oil aceite de semilla de
   algodón
coulomb coulombio
Coulomb barrier barrera de Coulomb
Coulomb excitation excitación de
   Coulomb
Coulomb explosion explosión de
   Coulomb
coumalic acid ácido cumálico
coumaric acid ácido cumárico
coumarin cumarina
coumarone cumarona
count cuenta
countercurrent contracorriente
counterion contraion
counterstain contracolorante
coupling acoplamiento
coupling constant constante de
   acoplamiento
covalence covalencia
covalent covalente
covalent bond enlace covalente
covalent crystal cristal covalente
covalent radius radio covalente
coveline covelina

covellite covelita
Cox chart diagrama de Cox
crack craquear, fraccionar
cracking craqueo, fraccionamiento
Craig method método de Craig
Cram rule regla de Cram
cream of tartar crémor tártaro
creatinase creatinasa
creatine creatina
creatininase creatininasa
creatinine creatinina
creeping precipitación ascendente
Creighton process proceso de Creighton
creosol creosol
creosote creosota
cresol cresol
cresol purple púrpura de cresol
cresol red rojo de cresol
cresolphthalein cresolftaleína
cresorcin cresorcina
cresotic acid ácido cresótico
cresyl cresilo
cresyl acetate acetato de cresilo
cresyl blue azul de cresilo
cresyl phosphate fosfato de cresilo
cresylate cresilato
cresylic acid ácido cresílico
Criegee reaction reacción de Criegee
cristobalite cristobalita
critical crítico
critical assembly conjunto crítico
critical coefficient coeficiente crítico
critical conditions condiciones críticas
critical constant constante crítica
critical density densidad crítica
critical humidity humedad crítica
critical mass masa crítica
critical point punto crítico
critical potential potencial crítico
critical pressure presión crítica
critical solution temperature
   temperatura de solución crítica
critical state estado crítico
critical temperature temperatura crítica
critical volume volumen crítico
crocetin crocetina
crocidolite crocidolita
croconic acid ácido crocónico
Crookes tube tubo de Crookes

cross-linkage encadenamiento cruzado
cross-linking encadenamiento cruzado
cross-section sección transversal
crossed-beam reaction reacción de haces cruzadas
croton oil aceite de crotón
crotonal crotonal
crotonaldehyde crotonaldehído
crotonic acid ácido crotónico
crotonic aldehyde aldehído crotónico
crotonoid crotonoide
crotonoyl crotonoilo
crotonyl crotonilo
crotonylene crotonileno
crotoxin crotoxina
crotyl alcohol alcohol crotílico
crown ether éter en corona
crucible crisol
crucible holder sujetador de crisol
crucible tongs tenazas de crisol
crucible triangle triángulo de crisol
crude crudo, bruto
crude oil petróleo bruto
crude petroleum petróleo bruto
cryochemistry crioquímica
cryogenic criogénico
cryogenic pump bomba criogénica
cryogenics criogénica, criogenia
cryogenin criogenina
cryohydrate criohidrato
cryolite criolita
cryometer criómetro
cryoscopic constant constante crioscópica
cryoscopy crioscopia
cryostat crióstato
cryptand criptando
cryptate criptato
cryptidine criptidina
cryptocyanine criptocianina
cryptohalite criptohalita
cryptoxanthin criptoxantina
crystal cristal
crystal-field theory teoría de campo cristalino
crystal form forma cristalina
crystal growth crecimiento de cristal
crystal lattice red cristalina
crystal structure estructura cristalina

crystal symmetry simetría de cristal
crystal system sistema cristalográfico
crystalline cristalino
crystalline state estado cristalino
crystallite cristalito
crystallization cristalización
crystallize cristalizar
crystallogram cristalograma
crystallographic system sistema cristalográfico
crystallography cristalografía
crystalloid cristaloide
crystalloluminescence cristaloluminiscencia
cube cubo
cubic-centered centrado en el cubo
cubic close packing estructura compacta cúbica
cubic crystal cristal cúbico
cubic system sistema cúbico
cubical cúbico
cumal cumal
cumaldehyde cumaldehído
cumene cumeno
cumene hydroperoxide hidroperóxido de cumeno
cumene process proceso de cumeno
cumenyl cumenilo
cumic acid ácido cúmico
cumic alcohol alcohol cúmico
cumic aldehyde aldehído cúmico
cumidine cumidina
cuminic acid ácido cumínico
cuminic alcohol alcohol cumínico
cuminic aldehyde aldehído cumínico
cuminol cuminol
cumol cumol
cumulalive double bonds enlaces dobles acumulativos
cumulene cumuleno
cumyl cumilo
cupellation copelación
cupellation process proceso de copelación
cupferron cupferrón
cuprate cuprato
cupreine cupreína
cuprene cupreno
cupric cúprico

cupric acetate acetato cúprico
cupric arsenate arseniato cúprico
cupric benzoate benzoato cúprico
cupric bromide bromuro cúprico
cupric carbonate carbonato cúprico
cupric chloride cloruro cúprico
cupric chromate cromato cúprico
cupric citrate citrato cúprico
cupric cyanide cianuro cúprico
cupric fluoride fluoruro cúprico
cupric hexafluorosilicate
    hexafluorosilicato cúprico
cupric hydroxide hidróxido cúprico
cupric iodide yoduro cúprico
cupric nitrate nitrato cúprico
cupric nitrite nitrito cúprico
cupric oxalate oxalato cúprico
cupric oxide óxido cúprico
cupric phosphate fosfato cúprico
cupric selenate seleniato cúprico
cupric subcarbonate subcarbonato
    cúprico
cupric sulfate sulfato cúprico
cupric sulfide sulfuro cúprico
cupric sulfite sulfito cúprico
cupric tartrate tartrato cúprico
cuprite cuprita
cuprocupric cuprocúprico
cuprocyanide cuprocianuro
cuproine cuproína
cupronickel cuproníquel
cuprotungsten cuprotungsteno
cuprous cuproso
cuprous acetylide acetiluro cuproso
cuprous bromide bromuro cuproso
cuprous carbonate carbonato cuproso
cuprous chloride cloruro cuproso
cuprous cyanide cianuro cuproso
cuprous hexafluorosilicate
    hexafluorosilicato cuproso
cuprous hydroxide hidróxido cuproso
cuprous iodide yoduro cuproso
cuprous oxide óxido cuproso
cuprous phosphide fosfuro cuproso
cuprous potassium cyanide cianuro
    potásico cuproso
cuprous selenide seleniuro cuproso
cuprous sulfate sulfato cuproso
cuprous sulfide sulfuro cuproso

cuprous sulfite sulfito cuproso
cuprous thiocyanate tiocianato cuproso
curare curare
curative curativo
curcumin curcumina
curie curie
Curie law ley de Curie
Curie point punto de Curie
Curie temperature temperatura de Curie
Curie-Weiss law ley de Curie-Weiss
curing curado
curing agent agente de curado
curium curio
current density densidad de corriente
Curtius reaction reacción de Curtius
Curtius rearrangement reordenamiento
    de Curtius
Curtius transformation transformación
    de Curtius
cusparine cusparina
cyanamide cianamida
cyanamide process proceso de cianamida
cyanate cianato
cyanato cianato
cyanaurite cianaurita
cyanazine cianazina
cyanoacetamide cianoacetamida
cyanic acid ácido ciánico
cyanide cianuro
cyanide process proceso de cianuro
cyanidine cianidina
cyanine cianina
cyanine dye colorante de cianina
cyanite cianita
cyano ciano
cyanoacetic acid ácido cianoacético
cyanoacetic ester éster cianoacético
cyanoacrylate cianoacrilato
cyanoaldehyde cianoaldehído
cyanoaniline cianoanilina
cyanobenzamide cianobenzamida
cyanobenzene cianobenceno
cyanobenzyl cianobencilo
cyanocarbon cianocarburo
cyanocarbonic acid ácido cianocarbónico
cyanocobalamin cianocobalamina
cyanoethanoic acid ácido cianoetanoico
cyanoethyl acryate acrilato de cianoetilo
cyanoethylation cianoetilación

cyanoferrate cianoferrato
cyanoformate cianoformiato
cyanoformic chloride cloruro cianofórmico
cyanogen cianógeno
cyanogen azide azida de cianógeno
cyanogen bromide bromuro de cianógeno
cyanogen chloride cloruro de cianógeno
cyanogen fluoride fluoruro de cianógeno
cyanogen iodide yoduro de cianógeno
cyanogen sulfide sulfuro de cianógeno
cyanohydrin cianhidrina
cyanomethyl acetate acetato de cianometilo
cyanoplatinate cianoplatinato
cyanopyridine cianopiridina
cyanuric cianúrico
cyanuric acid ácido cianúrico
cyanuric chloride cloruro cianúrico
cyanuric ester éster cianúrico
cybotactic cibotáctico
cyclamate ciclamato
cyclamic acid ácido ciclámico
cycle ciclo
cyclene cicleno
cyclic cíclico
cyclic action acción cíclica
cyclic adenosine monophosphate monofosfato de adenosina cíclico
cyclic amide amida cíclica
cyclic anhydride anhídrido cíclico
cyclic compound compuesto cíclico
cyclic hydrocarbon hidrocarburo cíclico
cyclic process proceso cíclico
cyclitol ciclitol
cyclization ciclización
cyclizine hydrochloride clorhidrato de ciclizina
cyclo ciclo
cycloaddition cicloadición
cycloaliphatic cicloalifático
cycloalkane cicloalcano
cycloalkene cicloalqueno
cycloalkyl cicloalquilo
cyclobarbital ciclobarbital
cyclobarbitone ciclobarbitona
cyclobutadiene ciclobutadieno
cyclobutane ciclobutano

cyclobutanol ciclobutanol
cyclobutene ciclobuteno
cyclobutyl ciclobutilo
cycloheptadecene cicloheptadeceno
cycloheptane cicloheptano
cycloheptanol cicloheptanol
cycloheptanone cicloheptanona
cycloheptene ciclohepteno
cycloheptyl cicloheptilo
cyclohexane ciclohexano
cyclohexanecarboxylic acid ácido ciclohexanocarboxílico
cyclohexanediol ciclohexanodiol
cyclohexanol ciclohexanol
cyclohexanol acetate acetato de ciclohexanol
cyclohexanone ciclohexanona
cyclohexanone peroxide peróxido de ciclohexanona
cyclohexene ciclohexeno
cyclohexene oxide óxido de ciclohexeno
cyclohexenol ciclohexenol
cyclohexenyltrichlorosilane ciclohexeniltriclorosilano
cycloheximide cicloheximida
cyclohexyl bromide bromuro de ciclohexilo
cyclohexyl chloride cloruro de ciclohexilo
cyclohexyl isocyanate isocianato de ciclohexilo
cyclohexyl stearate estearato de ciclohexilo
cyclohexyl trichlorosilane triclorosilano de ciclohexilo
cyclohexylamine ciclohexilamina
cyclohexylbenzene ciclohexilbenceno
cyclohexylidene ciclohexilideno
cyclohexylphenol ciclohexilfenol
cyclohexylpiperdine ciclohexilpiperdina
cyclone ciclón
cyclonite ciclonita
cyclononane ciclononano
cyclooctadiene ciclooctadieno
cyclooctane ciclooctano
cyclooctatetraene ciclooctatetraeno
cyclooctene cicloocteno
cycloolefin cicloolefina
cycloparaffin cicloparafina

cyclopentadecanone ciclopentadecanona
cyclopentadiene ciclopentadieno
cyclopentane ciclopentano
cyclopentanol ciclopentanol
cyclopentanone ciclopentanona
cyclopentene ciclopenteno
cyclopentenyl ciclopentenilo
cyclopentenylacetone
    ciclopentenilacetona
cyclopentyl ciclopentilo
cyclopentylacetone ciclopentilacetona
cyclopentylene ciclopentileno
cyclopentylpropionic acid ácido
    ciclopentilpropiónico
cyclophane ciclofano
cyclophosphamide ciclofosfamida
cyclopropane ciclopropano
cyclopropyl ciclopropilo
cyclosilane ciclosilano

cyclotetraene ciclotetraeno
cyclotron ciclotrón
cylinder cilindro
cylindrite cilindrita
cymene cimeno
cymyl cimilo
cyprite ciprita
cystamine cistamina
cysteine cisteína
cystine cistina
cytidine citidina
cytidylic acid ácido citidílico
cytochemistry citoquímica
cytochrome citocromo
cytoplasm citoplasma
cytosine citosina
cytotoxic citotóxico
cytotoxic agent agente citotóxico
cytotoxin citotoxina

D

dacite dacita
dactinomycin dactinomicina
dactylin dactilina
Dahl acid ácido de Dahl
dahlin dalina
dahllite dahllita
dahmenite damenita
Dakin reaction reacción de Dakin
Dakin solution solución de Dakin
dalapon dalapón
dalton dalton
Dalton atomic theory teoría atómica de Dalton
Dalton law ley de Dalton
damascinine damascinina
dambonite dambonita
damourite damourita
damping period periodo de amortiguamiento
danaite danaíta
danalite danalita
Daniell cell pila de Daniell
dansyl dansilo
daphnetin dafnetina
daphnite dafnita
dapsone dapsona
darapskite darapskita
dark reaction reacción obscura
Darzens condensation condensación de Darzens
Darzens procedure procedimiento de Darzens
Darzens reaction reacción de Darzens
Darzens synthesis síntesis de Darzens
dating datación
dating technique técnica de datación
dative bond enlace dativo
dative covalent bond enlace covalente dativo
datolite datolita
daturic acid ácido datúrico
daubreelite daubreelita
daunomycin daunomicina

daviesite daviesita
dawsonite dawsonita
de Broglie equation ecuación de de Broglie
de Broglie wavelength longitud de onda de de Broglie
De Mayo reaction reacción de De Mayo
deacetylation desacetilación
Deacon process proceso de Deacon
deactivating collision colisión desactivante
deactivation desactivación
deactivator desactivador
deamidation desamidación
deamination desaminación
deanol deanol
debye debye
Debye characteristic temperature temperatura característica de Debye
Debye-Hückel theory teoría de Debye-Hückel
Debye-Scherrer method método de Debye-Scherrer
decaborane decaborano
decahydrate decahidrato
decahydronaphthalene decahidronaftaleno
decalcification descalcificación
decalin decalina
decamethrin decametrina
decamethyltetrasiloxane decametiltetrasiloxano
decanal decanal
decanamide decanamida
decane decano
decanecarboxylic acid ácido decanocarboxílico
decanedioic acid ácido decanodioico
decanediol decanodiol
decanenitrile decanonitrilo
decanoate decanoato
decanoic acid ácido decanoico
decanoic anhydride anhídrido decanoico

decanol decanol
decanoyl decanoilo
decanoyl chloride cloruro de decanoilo
decanoyl peroxide peróxido de decanoilo
decant decantar
decantation decantación
decarbonize descarbonizar
decarboxylase descarboxilasa
decarboxylate descarboxilar
decarboxylation descarboxilación
decarboxyrissic acid ácido
     descarboxirrísico
decarburization descarburización
decarburize descarburizar
decatoic acid ácido decatoico
decatyl decatilo
decatyl alcohol alcohol decatílico
decay desintegración, decaimiento
decay chain cadena de desintegración
decay constant constante de
     desintegración
decay family familia de desintegración
decay mode modo de desintegración
decay particle partícula de
     desintegración
decay product producto de
     desintegración
decay rate velocidad de desintegración
decay series serie de desintegración
decene deceno
dechlorination desclorinación
decicaine decicaína
decimolar decimolar
decimolar solution solución decimolar
decinormal decinormal
decinormal solution solución decinormal
decoction decocción
decoic acid ácido decoico
decolorant decolorante
decoloration decoloración
decolorization decolorización
decolorizing agent agente decolorante
decompose descomponer
decomposition descomposición
decomposition point punto de
     descomposición
decontamination descontaminación
decouple desacoplar
decrepitate decrepitar

decrepitation decrepitación
decyclic acid ácido decíclico
decyl decilo
decyl acetate acetato de decilo
decyl alcohol alcohol decílico
decyl aldehyde aldehído decílico
decyl hydride hidruro de decilo
decyl mercaptan decilmercaptano
decylamide decilamida
decylamine decilamina
decylene decileno
decylethylene deciletileno
decylic acid ácido decílico
decyltrichlorosilane deciltriclorosilano
decyne decino
defect defecto
defect structure estructura con defectos
defervescence defervescencia
definite composition law ley de
     composición definida
definite proportions proporciones
     definidas
definite proportions law ley de
     proporciones definidas
deflagration deflagración
deflocculant desfloculante
deflocculation desfloculación
defoamer desespumante
defoaming agent agente desespumante
deformability deformabilidad
deformation energy energía de
     deformación
degasification desgasificación
degassing desgaseamiento
Degener indicator indicador de Degener
degenerate degenerado
degenerate orbital orbital degenerado
degenerate system sistema degenerado
degeneration degeneración
degradation degradación
degrade degradar
degree grado
degree of dilution grado de dilución
degree of dissociation grado de
     disociación
degree of hardness grado de dureza
degree of hydrolysis grado de hidrólisis
degree of polymerization grado de
     polimerización

degree of substitution grado de substitución

degree of temperature grado de temperatura

degree of vacuum grado de vacío

degrees of freedom grados de libertad

dehumidification deshumidificación

dehydrant deshidratante

dehydrated deshidratado

dehydration deshidratación

dehydrator deshidratador

dehydro deshidro

dehydroabietic acid ácido deshidroabiético

dehydroacetic acid ácido deshidroacético

dehydroascorbic acid ácido deshidroascórbico

dehydrocholesterol deshidrocolesterol

dehydrocholic acid ácido deshidrocólico

dehydrocyclization deshidrociclización

dehydrogenase deshidrogenasa

dehydrogenation deshidrogenación

dehydroisoandrosterone deshidroisoandrosterona

dehydrolysis deshidrólisis

deionization desionización

deionized desionizado

deionizing desionización

delanium delanio

delayed neutron neutrón retardado

delayed proton protón retardado

Delbrück scattering dispersión de Delbrück

delcosine delcosina

Delepine reaction reacción de Delepine

delessite delessita

deliquescence delicuescencia

deliquescent delicuescente

deliquescent substance substancia delicuescente

delocalization deslocalización

delphinic acid ácido delfínico

delphinin delfinina

delta delta

delta acid ácido delta

delta bond enlace delta

delta bonding enlazado delta

delta iron hierro delta

delta orbital orbital delta

delta ray rayo delta

delta value valor delta

delvauxite delvauxita

demagnetization desmagnetización

demal demal

demeclocycline demeclociclina

demethylation desmetilación

demethylene demetileno

demeton demetona

demineralization desmineralización

Demjanov rearrangement reordenamiento de Demjanov

demolization demolización

demulsibility demulsibilidad

demulsification desemulsificación

demuriation desmuriación

denatonium benzoate benzoato de denatonio

denaturant desnaturalizante

denaturation desnaturalización

denature desnaturalizar

denatured desnaturalizado

denatured alcohol alcohol desnaturalizado

dendrite dendrita

dendrobine dendrobina

Denigè reagent reactivo de Denigè

denitration desnitración

denitrification desnitrificación

denitrogenation desnitrogenación

densitometer densitómetro

density densidad

density comparator comparador de densidad

density-gradient column columna de gradiente de densidad

dental amalgam amalgama dental

deodorant desodorante

deoxidant desoxidante

deoxidation desoxidación

deoxidize desoxidar

deoxidizer desoxidante

deoxy desoxi, deoxi

deoxyanisoin desoxianisoína

deoxybenzoin desoxibenzoína

deoxycholic acid ácido desoxicólico

deoxygenation desoxigenación

deoxyribonuclease desoxirribonucleasa

deoxyribonucleic acid ácido

desoxirribonucleico
deoxyribose desoxirribosa
dephlegmation deflegmación
depleted uranium uranio agotado
depolarization despolarización
depolarizer despolarizador
depolymerization despolmerización
deposit depósito, precipitado
deposition deposición
deposition potential potencial de
     deposición
depression depresión
depression of freezing point depresión
     de punto de congelación
depressor depresor
depside depsida
depurator depurador
derivation derivación
derivative derivado
derived unit unidad derivada
derric acid ácido dérrico
derritol derritol
des des
desalination desalinación
descending chromatography
     cromatografía descendente
desclizite desclizita
desiccant desecante
desiccated desecado
desiccating desecante
desiccator desecador
deslanoside deslanosida
desmin desmina
desmotrope desmótropo
desmotropism desmotropismo
desorption desorción
desoxy desoxi
desoxycholic acid ácido desoxicólico
Despretz law ley de Despretz
destructive distillation destilación
     destructiva
desulfuration desulfuración
desyl desilo
detergent detergente
determinate error error determinado
determination determinación
detonating gas gas detonante
detonation detonación
detonator detonador

detoxification destoxificación
deuterate deuterar
deuterated deuterado
deuteration deuteración
deuteric acid ácido deutérico
deuteride deuteruro
deuterio deuterio
deuterium deuterio
deuterium oxide óxido de deuterio
deutero deutero
deuteron deuterón
deuteroxyl deuteroxilo
Devarda alloy aleación de Devarda
Devarda metal metal de Devarda
developed dye colorante desarrollado
developer revelador, desarrollador
development desarrollo
devitrification desvitrificación
dew point punto de rocío
Dewar flask recipiente de Dewar
Dewar structure estructura de Dewar
deweylite deweylita
dexamphetamine dexanfetamina
dextran dextrano
dextranase dextranasa
dextrin dextrina
dextrinose dextrinosa
dextro dextro
dextrogyric dextrógiro
dextrorotary dextrógiro
dextrorotatory dextrógiro
dextrose dextrosa
di di
diabase diabasa
diacetamide diacetamida
diacetanilide diacetanilida
diacetate diacetato
diacetic acid ácido diacético
diacetic ether éter diacético
diacetin diacetina
diacetonamine diacetonamina
diacetone diacetona
diacetone acrylamide
     diacetonaacrilamida
diacetone alcohol diacetonaalcohol
diacetyl diacetilo
diacetylamide diacetilamida
diacetylanilide diacetilanilida
diacetylene diacetileno

diacetylethane diacetiletano
diacetylmorphine diacetilmorfina
diacetylperoxide diacetilperóxido
diacetylurea diacetilurea
diacid diácido
diacolation diacolación
diadochite diadoquita
diagenesis diagénesis
diagonal relationship relación diagonal
dialdehyde dialdehído
dialkene dialqueno
dialkyl dialquilo
dialkylene dialquileno
diallyl dialilo
diallyl adipate adipato de dialilo
diallyl chlorendate clorendato de dialilo
diallyl cyanamide cianamida de dialilo
diallyl isophthalate isoftalato de dialilo
diallyl maleate maleato de dialilo
diallyl phosphite fosfito de dialilo
diallyl phthalate ftalato de dialilo
diallyl sulfide sulfuro de dialilo
diallylamine dialilamina
diallylmelamine dialilmelanina
diallylurea dialilurea
dialozite dialozita
dialuric acid ácido dialúrico
dialysis diálisis
dialyzed dializado
dialyzer dializador
diamagnetism diamagnetismo
diamide diamida
diamide hydrate hidrato de diamida
diamidine diamidina
diamido diamido
diamine diamina
diamino diamino
diaminoacridine diaminoacridina
diaminoazobenzene diaminoazobenceno
diaminoazoxytoluene
     diaminoazoxitolueno
diaminobenzene diaminobenceno
diaminobenzidine diaminobencidina
diaminobenzophenone
     diaminobenzofenona
diaminobutane diaminobutano
diaminocaproic acid ácido
     diaminocaproico
diaminodiphenic acid ácido

diaminodifénico
diaminodiphenyl diaminodifenilo
diaminodiphenylamine
     diaminodifenilamina
diaminodiphenylmethane
     diaminodifenilmetano
diaminoethane diaminoetano
diaminohexane diaminohexano
diaminopentane diaminopentano
diaminophenazine diaminofenazina
diaminophenol diaminofenol
diaminophenol hydrochloride
     clorhidrato de diaminofenol
diaminopropane diaminopropano
diaminotoluene diaminotolueno
diammonium diamonio
diammonium phosphate fostato
     diamónico
diamond diamante
diamorphine diamorfina
diamyl diamilo
diamyl phenol diamilfenol
diamyl sulfide sulfuro de diamilo
diamylamine diamilamina
diamylaniline diamilanilina
diamylene diamileno
dianiline dianilina
dianilinoethane dianilinoetano
dianisidine dianisidina
diaphorite diaforita
diaphragm diafragma
diarabinose diarabinosa
diarsane diarsano
diarsenite diarsenito
diarsyl diarsilo
diasolysis diasólisis
diaspore diásporo
diastase diastasa
diastereoisomer diaestereoisómero
diastereomer diaestereoisómero
diastereotopic diaestereotópico
diathermy diatermia
diatomaceous earth tierra de diatomeas
diatomic diatómico
diatomic molecule molécula diatómica
diatomite diatomita
diazacyclo diazaciclo
diazepam diazepam
diazine diazina

diazinon diazinón
diazo diazo
diazo dye colorante diazoico
diazoacetate diazoacetato
diazoacetic ester éster diazoacético
diazoamine diazoamina
diazoamino diazoamino
diazoaminobenzene diazoaminobenceno
diazoaminonaphthalene
   diazoaminonaftaleno
diazoate diazoato
diazobenzene diazobenceno
diazobenzene chloride cloruro de
   diazobenceno
diazobenzene hydroxide hidróxido de
   diazobenceno
diazobenzene nitrate nitrato de
   diazobenceno
diazobenzenesulfonic acid ácido
   diazobencenosulfónico
diazoethane diazoetano
diazohydrate diazohidrato
diazohydroxide diazohidróxido
diazoic acid ácido diazoico
diazoimide diazoimida
diazoimido diazoimido
diazole diazol
diazomethane diazometano
diazonitrophenol diazonitrofenol
diazonium diazonio
diazonium salt sal de diazonio
diazophenol diazofenol
diazosulfonate diazosulfonato
diazosulfonic acid ácido diazosulfónico
diazotization diazotización
diazoxide diazóxido
diazoxy diazoxi
dibasic dibásico
dibasic acid ácido dibásico
dibenalacetone dibenalacetona
dibenzamide dibenzamida
dibenzanthracene dibenzantraceno
dibenzanthrone dibenzantrona
dibenzanthronyl dibenzantronilo
dibenzene dibenceno
dibenzenechromium dibencenocromo
dibenzenyl dibencenilo
dibenzo dibenzo
dibenzoanthracene dibenzoantraceno

dibenzocyclohepatadienone
   dibenzocicloheptadienona
dibenzopyrone dibenzopirona
dibenzopyrrole dibenzopirrol
dibenzothiazine dibenzotiacina
dibenzoyl dibenzoilo
dibenzoyl peroxide peróxido de
   dibenzoilo
dibenzoylethane dibenzoiletano
dibenzoylethylene dibenzoiletileno
dibenzyl dibencilo
dibenzyl disulfide disulfuro de dibencilo
dibenzylamine dibencilamina
dibenzylether dinenciléter
dibenzylidene dibencilideno
dibenzylidyne dibencilidino
dibenzylmethylamine dibencilmetilamina
diborane diborano
dibromide dibromuro
dibromo dibromo
dibromoacetylene dibromoacetileno
dibromoanthracene dibromoantraceno
dibromobenzene dibromobenceno
dibromochloromethane
   dibromoclorometano
dibromochloropropane
   dibromocloropropano
dibromodiethyl sulfide sulfuro de
   dibromodietilo
dibromodifluoromethane
   dibromodifluorometano
dibromoethane dibromoetano
dibromofluorobenzene
   dibromofluorobenceno
dibromomalonic acid ácido
   dibromomalónico
dibromomethane dibromometano
dibromopentane dibromopentano
dibromopropane dibromopropano
dibromopropanol dibromopropanol
dibucaine dibucaína
dibutoxyaniline dibutoxianilina
dibutyl dibutilo
dibutyl chlorophosphate clorofosfato de
   dibutilo
dibutyl fumarate fumarato de dibutilo
dibutyl maleate maleato de dibutilo
dibutyl oxalate oxalato de dibutilo
dibutyl phosphate fosfato de dibutilo

dibutyl phosphite fosfito de dibutilo
dibutyl phthalate ftalato de dibutilo
dibutyl tartrate tartrato de dibutilo
dibutylamine dibultilamina
dibutylbutyl phosphonate fosfonato de dibutilbutilo
dibutylthiourea dibutiltiourea
dibutyltin dibutilestaño
dibutyltin diacetate diacetato de dibutilestaño
dibutyltin dichloride dicloruro de dibutilestaño
dibutyltin dilaurate dilaurato de dibutilestaño
dibutyltin oxide óxido de dibutilestaño
dibutyltin sulfide sulfuro de dibutilestaño
dicalcium orthophosphate ortofosfato dicálcico
dicalcium phosphate fosfato dicálcico
dicapryl dicaprilo
dicarbide dicarburo
dicarbocyanine dicarbocianina
dicarbonate dicarbonato
dicarboxyl dicarboxilo
dicarboxylic acid ácido dicarboxílico
dicarbylamine dicarbilamina
dicetyl dicetilo
dichlone diclona
dichloramine dicloramina
dichloride dicloruro
dichlorine dicloro
dichlorine oxide óxido de dicloro
dichloro dicloro
dichloroacetal dicloroacetal
dichloroacetic acid ácido dicloroacético
dichloroacetyl chloride cloruro de dicloroacetilo
dichloroacetylene dicloroacetileno
dichloroaniline dicloroanilina
dichloroanthracene dicloroantraceno
dichlorobenzaldehyde diclorobenzaldehído
dichlorobenzene diclorobenceno
dichlorobenzidine diclorobencidina
dichlorobenzoic acid ácido diclorobenzoico
dichlorobenzoyl chloride cloruro de diclorobenzoilo

dichlorobenzyl chloride cloruro de diclorobencilo
dichlorobutane diclorobutano
dichlorobutene diclorobuteno
dichlorocarbene diclorocarbeno
dichlorodiethyl sulfide sulfuro de diclorodietilo
dichlorodifluoromethane diclorodifluorurometano
dichloroethane dicloroetano
dichloroethanoic acid ácido dicloroetanoico
dichloroethene dicloroeteno
dichloroethyl acetate acetato de dicloroetilo
dichloroethyl ether éter dicloroetílico
dichloroethyl oxide óxido de dicloroetilo
dichloroethylene dicloroetileno
dichlorohydrin diclorohidrina
dichloroisopropyl alcohol alcohol dicloroisopropílico
dichloromethane diclorometano
dichloromethylsilane diclorometilsilano
dichloronaphthalene dicloronaftaleno
dichloronitrobenzene dicloronitrobenceno
dichloropentane dicloropentano
dichlorophene diclorofeno
dichlorophenol diclorofenol
dichlorophenoxyacetic acid ácido diclorofenoxiacético
dichloropropane dicloropropano
dichloropropene dicloropropeno
dichloropropionic acid ácido dicloropropiónico
dichloropropyl alcohol alcohol dicloropropílico
dichlorothiophene diclororiofeno
dichlorotoluene diclorotolueno
dichroic dicroico
dichroism dicroismo
dichromate dicromato
dichromatic dicromático
dichromic dicrómico
dichromic acid ácido dicrómico
dicryl dicrilo
dicyandiamide diciandiamida
dicyanide dicianuro
dicyano diciano

dicyanogen dicianógeno
dicyclic dicíclico
dicyclohexyl diciclohexilo
dicyclohexylamine diciclohexilamina
dicyclohexylcarbodiimide
    diciclohexilcarbodiimida
dicyclopentadiene diciclopentadieno
dicyclopentadienyl diciclopentadienilo
didecyl adipate adipato de didecilo
didecyl ether éter didecílico
didecyl sulfide sulfuro de didecilo
didymium didimio
didymium chloride cloruro de didimio
didymium nitrate nitrato de didimio
didymium oxide óxido de didimio
didymium sulfate sulfato de didimio
didymolite didimolita
Dieckmann reaction reacción de
    Dieckmann
diectyl diectilo
diectyl ether éter diectílico
diectyl sulfide sulfuro de diectilo
dieldrin dieldrina
dielectric dieléctrico
dielectric cohesion cohesión dieléctrica
dielectric constant constante dieléctrica
dielectric strength fuerza dieléctrica
Diels-Alder reaction reacción de Diels-
    Alder
diene dieno
dienestrol dienestrol
dienone-phenol rearrangement
    reordenamiento de dienona-fenol
dienophile dienófilo
diester diéster
Dieterici equation ecuación de Dieterici
Dieterici rule regla de Dieterici
diethanolamine dietanolamina
diether diéter
diethoxyaniline dietoxianilina
diethoxyethane dietoxietano
diethyl dietilo
diethyl carbonate carbonato de dietilo
diethyl chlorophosphate clorosfosfato de
    dietilo
diethyl diethylmalonate dietilmalonato
    de dietilo
diethyl disulfide dislfuro de dietilo
diethyl ether éter dietílico

diethyl ethylmalonate etilmalonato de
    dietilo
diethyl ketone dietilcetona
diethyl malate malato de dietilo
diethyl maleate maleato de dietilo
diethyl oxalate oxalato de dietilo
diethyl oxide oxido de dietilo
diethyl peroxide peroxido de dietilo
diethyl phosphine fosfina de dietilo
diethyl phosphite fosfito de dietilo
diethyl phthalate ftalato de dietilo
diethyl selenide seleniuro de dietilo
diethyl succinate succinato de dietilo
diethyl sulfate sulfato de dietilo
diethyl sulfide sulfuro de dietilo
diethyl sulfite sulfito de dietilo
diethyl tartrate tartrato de dietilo
diethylacetal dietilacetal
diethylacetic acid ácido dietilacético
diethylaluminum chloride cloruro
    dietilalumínico
diethylamine dietilamina
diethylamino dietilamino
diethylaminobenzaldehyde
    dietilaminobenzaldehído
diethylaminoethanol dietilaminoetanol
diethylaminopropylamine
    dietilaminopropilamina
diethylaniline dietilanilina
diethylbarbituric acid ácido
    dietilbarbitúrico
diethylbenzene dietilbenceno
diethylcadmium dietilcadmio
diethylcarbamazine dietilcarbamazina
diethylcarbinol dietilcarbinol
diethylcyclohexylamine
    dietilciclohexilamina
diethylene dietileno
diethylene dioxide dióxido de dietileno
diethylene glycol dietilenglicol
diethylene glycol dinitrate dinitrato de
    dietilenglicol
diethylene glycol ethyl ether éter etílico
    de dietilenglicol
diethylene glycol stearate estearato de
    dietilenglicol
diethylenediamine dietilendiamina
diethylenetriamine dietilenotriamina
diethylethylene dietiletileno

diethylidene dietilideno
diethylmethylmethane dietilmetilmetano
diethyloxamide dietiloxamida
diethylstilbestrol dietilestilbestrol
diethylurea dietilurea
diethylzinc dietilcinc
dietzeite dietzeita
differential adsorption adsorción
   diferencial
differential ebulliometer ebullómetro
   diferencial
differential reduction reducción
   diferencial
differential scanning calorimetry
   calorimetría de exploración diferencial
differential spectrophotometry
   espectrofotometría diferencial
differential thermal analysis análisis
   térmico diferencial
differential titration titulación
   diferencial, valoración diferencial
diffraction difracción
diffraction grating rejilla de difracción
diffraction pattern patrón de difracción
diffraction spectrum espectro de
   difracción
diffuse difuso
diffusion difusión
diffusion analysis análisis por difusión
diffusion constant constante de difusión
diffusion current corriente de difusión
diffusion law ley de difusión
diffusion layer capa de difusión
diffusion limited aggregation agregación
   limitada por difusión
diffusion pump bomba de difusión
difluoride difluoruro
difluoro difluoro
difluoroamino difluoroamino
difluorobenzene difluorobenceno
difluorodiazine difluorodiazina
difluoroethane difluoroetano
difluoromethane difluorometano
digallic acid ácido digálico
digestion digestión
digitalin digitalina
digitalis digitalis
digitalose digitalosa
digitonin digitonina

digitoxigenin digitoxigenina
digitoxin digitoxina
diglyceride diglicérido
diglycerol diglicerol
diglycol diglicol
diglycol carbamate carbamato de
   diglicol
diglycol laurate laurato de diglicol
diglycol oleate oleato de diglicol
diglycol stearate estearato de diglicol
diglycolic acid ácido diglicólico
diglycolide diglicólido
diglyme diglime
digol digol
digoxin digoxina
dihedral dihedral
dihedral angle ángulo dihedral
dihedron dihedrón
dihexyl dihexilo
dihexylamine dihexilamina
dihydrate dihidrato
dihydrazine sulfate sulfato de
   dihidrazina
dihydrazone dihidrazona
dihydric alcohol alcohol dihídrico
dihydride dihidruro
dihydro dihidro
dihydroazirine dihidroazirina
dihydrobenzene dihidrobenceno
dihydrobromide dibromhidrato
dihydrochalcone dihidrocalcona
dihydrochloride diclorhidrato
dihydrocholesterol dihidrocolesterol
dihydronaphthalene dihidronaftaleno
dihydropropene dihidropropeno
dihydropyran dihidropirano
dihydroxy dihidroxi
dihydroxyacetone dihidroxiacetona
dihydroxyamine dihidroxiamina
dihydroxyanthraquinone
   dihidroxiantraquinona
dihydroxybenzene dihidroxibenceno
dihydroxybenzoic acid ácido
   dihidroxibenzoico
dihydroxybutanedioic acid ácido
   dihidroxibutanodioico
dihydroxycinnamic acid ácido
   dihidroxicinámico
dihydroxydiphenylsulfone

dihidroxidifenilsulfona

dihydroxyethane dihidroxietano

dihydroxymalonic acid ácido dihidroximalónico

dihydroxynaphthalene dihidroxinaftaleno

dihydroxypropane dihidroxipropano

dihydroxystearic acid ácido dihidroxiesteárico

dihydroxysuccinic acid ácido dihidroxisuccínico

diimide diimida

diimine diimina

diimino diimino

diiodide diyoduro

diiodo diyodo

diiodoacetic acid ácido diyodoacético

diiodoacetylene diyodoacetileno

diiodomethane diyodometano

diisoamyl diisoamilo

diisobutyl diisobutilo

diisobutyl ketone diisobutilcetona

diisobutylamine diisobutilamina

diisobutylene diisobutileno

diisobutylphenol diisobutilofenol

diisocyanate diisocianato

diisooctyl adipate adipato de diisooctilo

diisopentyl diisopentilo

diisopropyl diisopropilo

diisopropyl ether éter diisopropílico

diisopropylbenzene diisopropilbenceno

diketene dicetena

diketone dicetona

diketopiperazine dicetopiperazina

diketopyrrolidine dicetopirrolidina

dilactone dilactona

dilatation dilatación

dilation dilatación

dilatometer dilatómetro

dilauryl dilaurilo

dilauryl peroxide dilaurilperóxido

dilauryl phosphite dilaurilfosfito

dilauryl sulfide dilaurilsulfuro

dilaurylamine dilaurilamina

dilead diplomo

diluent diluyente

dilute diluido

dilute acid ácido diluido

dilute base base diluida

dilute solution solución diluida

dilution dilución

dilution law ley de dilución

dilution ratio razón de dilución

dilution rule regla de dilución

dimedone dimedona

dimer dímero

dimeric dimérico

dimerization dimerización

dimethano dimetano

dimethoate dimetoato

dimethoxy dimetoxi

dimethoxyaniline dimetoxianilina

dimethoxybenzene dimetoxibenceno

dimethoxyethane dimetoxietano

dimethoxymethane dimetoximetano

dimethoxytetraglycol dimetoxitetraglicol

dimethyl dimetilo

dimethyl cadmium cadmio de dimetilo

dimethyl cyanamide cianamida de dimetilo

dimethyl disulfide disulfuro de dimetilo

dimethyl ether éter dimetílico

dimethyl malonate malonato de dimetilo

dimethyl phosphate fosfato de dimetilo

dimethyl phthalate ftalato de dimetilo

dimethyl sebacate sebacato de dimetilo

dimethyl selenide seleniuro de dimetilo

dimethyl sulfate sulfato de dimetilo

dimethyl sulfide sulfuro de dimetilo

dimethyl sulfoxide sulfóxido de dimetilo

dimethyl telluride telururo de dimetilo

dimethyl terephthalate tereftalato de dimetilo

dimethylacetal dimetilacetal

dimethylamine dimetilamina

dimethylamino dimetilamino

dimethylaminobenzoic acid ácido dimetilaminobenzoico

dimethylaminoethanol dimetilaminoetanol

dimethylaminopropanol dimetilaminopropanol

dimethylaniline dimetilanilina

dimethylanthracene dimetilantraceno

dimethylarsine dimetilarsina

dimethylarsinic acid ácido dimetilarsínico

dimethylbenzene dimetilbenceno

dimethylbenzyl chloride cloruro de dimetilbencilo
dimethylberyllium dimetilberilio
dimethylbutadiene dimetilbutadieno
dimethylbutane dimetilbutano
dimethylbutene dimetilbuteno
dimethylcarbinol dimetilcarbinol
dimethylchloroacetal dimetilcloroacetal
dimethylcyclohexane dimetilciclohexano
dimethylcyclopentane dimetilciclopentano
dimethyldichlorosilane dimetildiclorosilano
dimethylene dimetileno
dimethylethylene dimetiletileno
dimethylformamide dimetilformamida
dimethylglyoxime dimetilglioxima
dimethylheptene dimetilhepteno
dimethylhexadiene dimetilhexadieno
dimethylhexanediol dimetilhexanodiol
dimethylhexynediol dimetilhexinodiol
dimethylhydantoin dimetilhidantoína
dimethylhydrazine dimetilhidrazina
dimethylhydroxybenzene dimetilhidroxibenceno
dimethylisopropanolamine dimetilisopropanolamina
dimethylmercury dimetilmercurio
dimethylmethane dimetilmetano
dimethylnitrosamine dimetilnitrosamina
dimethyloctadiene dimetiloctadieno
dimethylpentaldehyde dimetilpentaldehído
dimethylpentane dimetilpentano
dimethylphenol dimetilfenol
dimethylphosphine dimetilfosfina
dimethylpropane dimetilpropano
dimethyltin dichloride dicloruro de dimetilestaño
dimethylurea dimetilurea
dimethylxanthine dimetilxantina
dimetilan dimetilán
dimetric dimétrico
dimorphic dimórfico
dimorphism dimorfismo
dimyristyl sulfide sulfuro de dimiristilo
dinaphtho dinafto
dinaphthol dinaftol
dinaphthyl dinaftilo

dinaphthylene dinaftileno
dineric dinérico
dineutron dineutrón
dinitrate dinitrato
dinitrite dinitrito
dinitro dinitro
dinitroaniline dinitroanilina
dinitroanthraquinone dinitroantraquinona
dinitrobenzene dinitrobenceno
dinitrobenzoyl chloride cloruro de dinitrobenzoilo
dinitrocresol dinitrocresol
dinitrofluorobenzene dinitrofluorobenceno
dinitrogen dinitrógeno
dinitrogen oxide óxido de dinitrógeno
dinitrogen tetraoxide tetróxido de dinitrógeno
dinitronaphthalene dinitronaftaleno
dinitronaphthol dinitronaftol
dinitrophenol dinitrofenol
dinitrophenylhydrazine dinitrofenilhidrazina
dinitrosalicylic acid ácido dinitrosalicílico
dinitroso dinitroso
dinitrotoluene dinitrotolueno
dinonyl phthalate ftalato de dinonilo
dinucleotide dinucleótido
dioctyl dioctilo
dioctyl chlorophosphate clorofosfato de dioctilo
dioctyl ether éter dioctílico
dioctyl phosphite fosfito de dioctilo
dioctyl phthalate ftalato de dioctilo
dioctyl sebacate sebacato de dioctilo
dioctylamine dioctilamina
dioctylmethane dioctilmetano
diol diol
diolefin diolefina
dione diona
diopside diópsido
dioptase dioptasa
diorite diorita
dioxadiene dioxadieno
dioxalane dioxalano
dioxan dioxana
dioxane dioxano

dioxazole dioxazol
dioxdiazole dioxdiazol
dioxide dióxido
dioxime dioxima
dioxin dioxina
dioxo dioxo
dioxolane dioxolano
dioxonitric acid ácido dioxonítrico
dioxy dioxi
dioxygenyl dioxigenilo
dipalmitate dipalmitato
dipalmitylamine dipalmitilamina
dipentaerythritol dipentaeritritol
dipentene dipenteno
dipentene dioxide dióxido de dipenteno
dipentene glycol dipentenoglicol
dipentene hydrochloride clorhidrato de
    dipenteno
dipentene monoxide monóxido de
    dipenteno
dipentyl dipentilo
dipentyl sulfide sulfuro de dipentilo
dipeptide dipéptido
diphenic acid ácido difénico
diphenol difenol
diphenolate difenolato
diphenyl difenilo
diphenyl carbonate carbonato de difenilo
diphenyl ether éter difenílico
diphenyl ketone difenilcetona
diphenyl oxide óxido de difenilo
diphenyl phosphite fosfito de difenilo
diphenyl phthalate ftalato de difenilo
diphenylamine difenilamina
diphenylamino difenilamino
diphenylbenzene difenilbenceno
diphenylcarbazide difenilcarbazida
diphenylchloroarsine difenilcloroarsina
diphenyldichlorosilane
    difenildiclorosilano
diphenyldiimide difenildiimida
diphenylene difenileno
diphenylethane difeniletano
diphenylethylene difeniletileno
diphenylglyoxal difenilglioxal
diphenylguanidine difenilguanidina
diphenylmethane difenilmetano
diphenylmethyl difenilmetilo
diphenylmethyl bromide bromuro de

difenilmetilo
diphenylsulfone difenilsulfona
diphenyltin difenilestaño
diphenylurea difenilurea
diphosgene difosgeno
diphosphane difosfano
diphosphate difosfato
diphosphine difosfina
diphosphoglyceric acid ácido
    difosfoglicérico
diphosphoric acid ácido difosfórico
dipolar dipolar
dipolar ion ion dipolar
dipole dipolo
dipole moment momento dipolar
dipole radiation radiación dipolar
dipropyl dipropilo
dipropyl ether éter dipropílico
dipropyl sulfide sulfuro de dipropilo
dipropylberyllium dipropilberilio
dipropylene glycol dipropilenglicol
diproton diprotón
dipyridine dipiridina
dipyridyl dipiridilo
diradical dirradical
direct dye colorante directo
direct nuclear reaction reacción nuclear
    directa
diresorcinol dirresorcina
disaccharide disacárido
disagglomeration desaglomeración
disalicylic acid ácido disalicílico
disassociation disociación
discaryl disacrilo
discharge descarga
Dische reaction reacción de Dische
discontinuous phase fase discontinua
discrete spectrum espectro discreto
diselane diselano
diselenide deseleniuro
disilane disilano
disilanyl disinalilo
disilicate disilicato
disilicide disiliciuro
disiloxane disiloxano
disilyl disililo
disinfectant desinfectante
disinfection desinfección
disintegration desintegración

disintegration chain cadena de desintegración
disintegration constant constante de desintegración
disintegration family familia de desintegración
disintegration mode modo de desintegración
disintegration particle partícula de desintegración
disintegration product producto de desintegración
disintegration rate velocidad de desintegración
disintegration series serie de desintegración
dislocation dislocación
dismutation dismutación
disodium citrate citrato disódico
disodium diphosphate difosfato disódico
disodium hydrogenphosphate hidrógenofosfato disódico
disodium phosphate fosfato disódico
dispersator dispersador
disperse dispersar
dispersed disperso
dispersed phase fase dispersa
dispersed system sistema disperso
dispersing dispersante
dispersing agent agente dispersante
dispersion dispersión
dispersion forces fuerzas de dispersión
dispersion medium medio de dispersión
dispersoid dispersoide
displacement desplazamiento
displacement chromatography cromatografía por desplazamiento
displacement law ley de desplazamiento
displacement reaction reacción de desplazamiento
displacement series serie de desplazamiento
display indicador visual, indicación visual
disproportionation desproporción
dissipation disipación
dissipative system sistema disipativo
dissociate disociar
dissociated disociado

dissociated state estado disociado
dissociation disociación
dissociation constant constante de disociación
dissociation energy energía de disociación
dissociation pressure presión de disociación
dissociation reaction reacción de disociación
dissolution disolución
dissolve disolver
dissolved disuelto
dissolved oxygen oxígeno disuelto
dissolvent disolvente
dissolving disolvente
dissolving agent agente disolvente
dissymmetrical disimétrico
dissymmetry disimetría
distance distancia
distearyl sulfide sulfuro de diestearilo
distillate destilado
distillation destilación
distillation apparatus aparato de destilación
distillation column columna de destilación
distillation flask matraz de destilación
distillation residue residuo de destilación
distillation tower torre de destilación
distilled destilado
distilled water agua destilada
distiller destilador
distribution coefficient coeficiente de distribución
distribution law ley de distribución
distribution principle principio de distribución
distyrene diestireno
disulfate disulfato
disulfide disulfuro
disulfiram disulfiram
disulfite disulfito
disulfo disulfo
disulfonate disulfonato
disulfonic acid ácido disulfónico
disulfuric acid ácido disulfúrico
diterpene diterpeno
diterpenoid diterpenoide

dithiane ditiano
dithiazine ditiazina
dithiene ditieno
dithienyl ditienilo
dithio ditio
dithiobenzoic acid ácido ditiobenzoico
dithiocarbamate ditiocarbamato
dithiocarbamic acid ácido ditiocarbámico
dithiocarbonate ditiocarbonato
dithiocarboxy ditiocarboxi
dithiol ditiol
dithiolene ditioleno
dithionate ditionato
dithionic acid ácido ditiónico
dithionite ditionita
dithiooxamide ditiooxamida
dithizone ditizona
ditolyl ditolilo
ditungsten carbide carburo de ditungsteno
divalence divalencia
divalent divalente
divalent carbon carbono divalente
divinyl divinilo
divinyl ether éter divinílico
divinyl oxide óxido de divinilo
divinyl sulfide sulfuro de divinilo
divinylacetylene divinilacetileno
divinylbenzene divinilbenceno
dixylyl dixililo
dixylylethane dixililetano
Dobbin reagent reactivo de Dobbin
docosane docosano
docosanoic acid ácido docosanoico
docosanol docosanol
docosenoic acid ácido docosenoico
dodecahedral dodecaedral
dodecahedro dodecaedro
dodecahedron dodecaedro
dodecahydrate dodecahidrato
dodecanal dodecanal
dodecane dodecano
dodecanoic acid ácido dodecanoico
dodecanol dodecanol
dodecanoyl dodecanoilo
dodecenal dodecenal
dodecene dodeceno
dodecenylsuccinic acid ácido dodecenilsuccinico
dodecyl dodecilo
dodecyl alcohol alcohol dodecílico
dodecylamine dodecilamina
dodecylbenzene dodecilbenceno
dodecylene dodecileno
dodecylphenol dodecilfenol
dodecyltrichlorosilane dodeciltriclorosilano
Doebner-Miller reaction reacción de Doebner-Miller
Doebner reaction reacción de Doebner
dolerite dolerita
dolichol dolicol
dolomite dolomita
dolomol dolomol
dome cúpula
domesticine domesticina
domingite dominguita
donarite donarita
Donnan equilibrium equilibrio de Donnan
Donnan hydrolysis hidrólisis de Donnan
donor donador
donor-acceptor pair par donador-aceptor
donor atom átomo donador
dope dopante
doped dopado
doping dopado
dotriacontane dotriacontano
double-beam spectrophotometer espectrofotómetro de doble haz
double bond enlace doble
double-bond isomerism isomerismo de enlace doble
double decomposition descomposición doble
double displacement desplazamiento doble
double layer capa doble
double refraction refracción doble
double replacement reemplazo doble
double salt sal doble
double weighing pesaje doble
doublet doblete
Downs process proceso de Downs
draconyl draconilo
Dragendorff reaction reacción de Dragendorff

Dragendorff reagent reactivo de Dragendorff
Draper effect efecto de Draper
Draper law ley de Draper
Drew number número de Drew
drimin drimina
dropping bottle botella para gotear
drug droga
dry seco
dry acid ácido seco
dry cell pila seca
dry chemical producto químico seco
dry deposition deposición seca
dry distillation destilación seca
dry ice hielo seco
dry point punto seco
dryer secador
drying secado
drying agent agente secante
drying oil aceite secante
drying tube tubo secante
ductile dúctil
ductility ductilidad
Duff reaction reacción de Duff
dufrenite dufrenita
dufrenoysite dufrenoisita
Duhem equation ecuación de Duhem
Dühring rule regla de Dühring
dulcitol dulcitol
Dumas method método de Dumas
dumortierite dumortierita

dunnite dunita
duplet duplete
duralumin duraluminio
durene dureno
durol durol
duryl durilo
durylene durileno
durylic acid ácido durílico
dust polvo
Dutt-Wormall reaction reacción de Dutt-Wormall
dye colorante, tinte
dye-retarding agent agente retardante de teñido
dyeing assistant asistente de teñido
dyestuff materia colorante
dynad dinada
dynamic equilibrium equilibrio dinámico
dynamic isomerism isomerismo dinámico
dynamic system sistema dinámico
dynamite dinamita
dypnone dipnona
dyscrasite discrasita
Dyson notation notación de Dyson
dysprosium disprosio
dysprosium nitrate nitrato de disprosio
dysprosium oxide óxido de disprosio
dysprosium sulfate sulfato de disprosio
dystectic mixture mezcla distéctica

E

earth tierra
earth alkali metal metal alcalinotérreo
earth wax cera de tierra, ozocerita
easin easina
ebonite ebonita
ebullator ebullidor
ebulliometer ebulliómetro
ebulliometry ebullometría
ebullioscope ebulloscopio
ebullioscopic constant constante
    ebulloscópica
ebullioscopy ebulloscopia
ebullition ebullición
eccentric excéntrico
ecdysone ecdisona
ecgonine ecgonina
echelon grating rejilla escalonada
echinomycin equinomicina
echinopsine equinopsina
eclipsed eclipsado
eclipsed conformation conformación
    eclipsada
eclogite eclogita
ecology ecología
economizer economizador
ecosystem ecosistema
edeate edeato
Edeleanu process proceso de Edeleanu
edenite edenita
Eder solution solución de Eder
edestin edestina
edetate edetato
edetate disodium edetato disódico
edge filtration filtración por bordes
edible oil aceite comestible
edingtonite edingtonita
Edison cell pila de Edison
Edman degradation degradación de
    Edman
effect efecto
effective atomic number número
    atómico efectivo
effective charge carga efectiva

effective molecular diameter diámetro
    molecular efectivo
effective permeability permeabilidad
    efectiva
effective temperature temperatura
    efectiva
effervescence efervescencia
effervescent efervescente
effervescent mixture mezcla
    efervescente
efflorescence eflorescencia
effluent efluente
effusiometer efusiómetro
effusion efusión
egg oil aceite de huevo
Ehrenfest classification clasificación de
    Ehrenfest
Ehrlich-Sachs reaction reacción de
    Ehrlich-Sachs
eicosane eicosano
eicosanoic acid ácido eicosanoico
eicosanoid eicosanoide
eicosanol eicosanol
eigenfunction autofunción
eigenvalue autovalor
Einchluss thermometer termómetro de
    Einchluss
Einhorn-Brunner reaction reacción de
    Einhorn-Brunner
einstein einstein
Einstein coefficient coeficiente de
    Einstein
Einstein law ley de Einstein
einsteinium einstenio
eka eka
ekahafnium ekahafnio
elaboration elaboración
elaidic acid ácido elaídico
elaidinization elaidinización
elastase elastasa
elastic elástico
elastic collision colisión elástica
elastic constant constante elástica

elasticity　elasticidad
elastin　elastina
elastomer　elastómero
elateric acid　ácido elatérico
elaterite　elaterita
Elbs reaction　reacción de Elbs
electret　electreto
electric　eléctrico
electric conductivity　conductividad eléctrica
electric current　corriente eléctrica
electric double layer　capa doble eléctrica
electric field effect　efecto de campo eléctrico
electric furnace　horno eléctrico
electric steel　acero de horno eléctrico
electride　electruro
electrification　electrificación
electroanalysis　electroanálisis
electrochemical　electroquímico
electrochemical cell　célula electroquímica
electrochemical corrosion　corrosión electroquímica
electrochemical deposition　deposición electroquímica
electrochemical deterioration　deterioro electroquímico
electrochemical diffusion　difusión electroquímica
electrochemical effect　efecto electroquímico
electrochemical equivalent　equivalente electroquímico
electrochemical hardening　endurecimiento electroquímico
electrochemical measurement　medida electroquímica
electrochemical migration　migración electroquímica
electrochemical oxidation　oxidación electroquímica
electrochemical passivation　pasivación electroquímica
electrochemical polarization　polarización electroquímica
electrochemical potential　potencial electroquímico
electrochemical process　proceso electroquímico
electrochemical reduction　reducción electroquímica
electrochemical series　serie electroquímica
electrochemical spectrum　espectro electroquímico
electrochemistry　electroquímica
electrochromatography　electrocromatografía
electrocoating　electrorrevestimiento
electrocratic　electrocrático
electrocyclic reaction　reacción electrocíclica
electrode　electrodo
electrode characteristic　característica de electrodo
electrode conductance　conductancia de electrodo
electrode current　corriente de electrodo
electrode dissipation　disipación de electrodo
electrode potential　potencial de electrodo
electrode voltage　tensión de electrodo
electrodeposit　electrodepósito
electrodeposition　electrodeposición
electrodeposition analysis　análisis por electrodeposición
electrodialysis　electrodiálisis
electrodialyzer　electodializador
electrodispersion　electrodispersión
electrodissolution　electrodisolución
electroflotation　electroflotación
electrofocusing　electroenfoque
electroforming　electroformación
electrographic analysis　análisis electrográfico
electrographite　electrografito
electrogravimetry　electrogravimetría
electrokinetic potential　potencial electrocinético
electrokinetics　electrocinética
electroluminescence　electroluminiscencia
electrolysis　electrólisis
electrolyte　electrólito
electrolyte acid　ácido electrolítico
electrolytic　electrolítico
electrolytic action　acción electrolítica

electrolytic analysis análisis electrolítico
electrolytic cathode cátodo electrolítico
electrolytic cell celda electrolítica
electrolytic cleaning limpieza electrolítica
electrolytic conductance conductancia electrolítica
electrolytic conduction conducción electrolítica
electrolytic conductivity conductividad electrolítica
electrolytic corrosion corrosión electrolítica
electrolytic decomposition descomposición electrolítica
electrolytic deposition deposición electrolítica
electrolytic dissociation disociación electrolítica
electrolytic gas gas electrolítico
electrolytic ionization ionización electrolítica
electrolytic iron hierro electrolítico
electrolytic metallization metalización electrolítica
electrolytic oxidation oxidación electrolítica
electrolytic polarization polarización electrolítica
electrolytic potential potencial electrolítico
electrolytic process proceso electrolítico
electrolytic reduction reducción electrolítica
electrolytic separation separación electrolítica
electrolytic solution solución electrolítica
electrolyze electrolizar
electrolyzer electrolizador
electromagnetic electromagnético
electromagnetic radiation radiación electromagnética
electromagnetic separation separación electromagnética
electromagnetic spectrum espectro electromagnético
electromerism electromerismo
electrometallurgy electrometalurgia
electrometer electrómetro

electrometric titration titulación electrométrica, valoración electrométrica
electromigration electromigración
electromotive force fuerza electromotriz
electromotive series serie electromotriz
electron electrón
electron accelerator acelerador electrónico, acelerador de electrones
electron acceptor aceptor electrónico
electron affinity afinidad electrónica
electron avalanche avalancha electrónica
electron band banda electrónica
electron beam haz electrónico
electron bombardment bombardeo electrónico
electron bunching agrupamiento electrónico
electron capture captura electrónica
electron charge carga electrónica
electron-charge density densidad de carga electrónica
electron cloud nube electrónica
electron collection recolección electrónica
electron collision colisión electrónica
electron concentration concentración electrónica
electron conduction conducción electrónica
electron conductor conductor electrónico
electron configuration configuración electrónica
electron crystallography cristalografía electrónica
electron current corriente electrónica
electron deficiency deficiencia electrónica
electron density densidad electrónica
electron diffraction difracción electrónica
electron dipole moment momento dipolar electrónico
electron distribution distribución electrónica
electron donor donador electrónico
electron drift flujo electrónico
electron emission emisión electrónica
electron energy energía electrónica

electron energy level nivel de energía electrónica

electron exchange intercambio electrónico

electron flow flujo electrónico

electron flux flujo electrónico

electron formula fórmula electrónica

electron gas gas electrónico

electron image imagen electrónica

electron impact impacto electrónico

electron linear accelerator acelerador lineal electrónico

electron magnetic moment momento magnético electrónico

electron magnetic resonance resonancia magnética electrónica

electron mass masa de electrón

electron microanalyzer microanalizador electrónico

electron microprobe microsonda electrónica

electron microscope microscopio electrónico

electron microscopy microscopia electrónica

electron motion moción electrónica

electron multiplication multiplicación electrónica

electron multiplicity multiplicidad electrónica

electron multiplier multiplicador electrónico

electron octet octeto electrónico

electron orbit órbita electrónica

electron orbital orbital electrónico

electron oscillator oscilador electrónico

electron pair par de electrones

electron-pair bond enlace de par de electrones

electron paramagnetic resonance resonancia paramagnética electrónica

electron physics física electrónica

electron probe sonda electrónica

electron probe analysis análisis con sonda electrónica

electron resonance resonancia electrónica

electron rest mass masa en reposo de electrón

electron shell capa electrónica

electron source fuente de electrones

electron specific charge carga específica de electrón

electron spectroscopy espectroscopia electrónica

electron spectrum espectro electrónico

electron spin espín de electrón

electron-spin resonance resonancia de espín electrónico

electron stream corriente electrónica

electron trajectory trayectoria electrónica

electron transfer transferencia electrónica

electron transport chain cadena de transporte electrónico

electron velocity velocidad de electrón

electron-volt electrón-volt

electronegative electronegativo

electronegative element elemento electronegativo

electronegative potential potencial electronegativo

electronegative radical radical electronegativo

electronegativity electronegatividad

electronic configuration configuración electrónica

electronic energy curve curva de energía electrónica

electronic transition transición electrónica

electroorganic reaction reacción electroorgánica

electroosmosis electroósmosis

electrophile electrófilo

electrophilic electrofílico

electrophilic addition adición electrofílica

electrophilic reagent reactivo electrofílico

electrophilic substitution substitución electrofílica

electrophoresis electroforesis

electrophoretic electroforético

electrophoretic effect efecto electroforético

electrophoretogram electroforetograma

electrophorogram electroforograma
electroplating galvanoplastia,
  electrorrecubrimiento
electropolishing electropulido
electropositive electropositivo
electropositive element elemento
  electropositivo
electropositive potential potencial
  electropositivo
electropotential electropotencial
electroscope electroscopio
electroscopy electroscopia
electrosol electrosol
electrostatic bond enlace electrostático
electrostatic precipitator precipitador
  electrostático
electrostriction electroestricción
electrosynthesis electrosíntesis
electrovalence electrovalencia
electrovalent electrovalente
electrovalent bond enlace electrovalente
electrovalent compound compuesto
  electrovalente
electroviscosity electroviscocidad
electrowinning extracción electrolítica
electrum electrum
element elemento
element of symmetry elemento de
  simetría
elementary analysis análisis elemental
elementary charge carga elemental
elementary molecule molécula elemental
elementary particle partícula elemental
elementary process proceso elemental
elementary space espacio elemental
elemonic acid ácido elemónico
eleonorite eleonorita
eleostearic acid ácido eleoesteárico
elevation of boiling point elevación de
  punto de ebullición
elimination reaction reacción de
  eliminación
elixir elíxir
ellagic acid ácido elágico
Ellingham diagram diagrama de
  Ellingham
elpidite elpidita
Eltekoff reaction reacción de Eltekoff
eluant eluyente

eluate eluido
elution elución
elutriation elutriación
emanation emanación
embelic acid ácido embélico
embolite embolita
embonic acid ácido embónico
embrittlement fragilización
Emde degradation degradación de Emde
emerald esmeralda
emerald green verde esmeralda
emery esmeril
emetamine emetamina
emetine emetina
emission emisión
emission lines líneas de emisión
emission spectroscopy espectroscopia de
  emisión
emission spectrum espectro de emisión
Emmert reaction reacción de Emmert
emodin emodina
empirical empírico
empirical formula fórmula empírica
emplectite emplectita
emulgator emulagador
emulsification emulsionamiento
emulsifier emulsionador
emulsify emulsionar
emulsifying agent agente emulsionante
emulsion emulsión
emulsion breaker rompedor de emulsión
emulsion breaking rompimiento de
  emulsión
emulsion polymerization polimerización
  en emulsión
emulsion stabilizer estabilizador de
  emulsión
enamel esmalte
enamine enamina
enanthic acid ácido enántico
enanthol enantol
enanthyl enantilo
enantiomer enantiómero
enantiomeric enantiomérico
enantiomorph enantiomorfo
enantiomorphic enantiomórfico
enantiomorphism enantiomorfismo
enantiotopic enantiotópico
enantiotropic enantiótropo,

enantiotrópico
enantiotropy enantiotropía
enargite enargita
encapsulation encapsulación
enclosure compound compuesto de
 encerramiento
end point punto final
endo endo
endogenous endógeno
endomycin endomicina
endopeptidase endopeptidasa
endorphin endorfina
endosmosis endósmosis
endothermic endotérmico
endothermic reaction reacción
 endotérmica
endrin endrina
ene eno
enediol enodiol
energy energía
energy balance equilibrio energético
energy band banda de energía
energy change cambio de energía
energy conservation conservación de
 energía
energy conversion conversión de energía
energy converter convertidor de energía
energy density densidad de energía
energy level nivel de energía
energy-level diagram diagrama de
 niveles de energía
energy of activation energía de
 activación
energy quantum cuanto de energía
energy source fuente de energía
enflurane enflurano
Engel salt sal de Engel
enhanced line línea realzada
enkephalin encefalina
enol enol
enolase enolasa
enriched uranium uranio enriquecido
enrichment enriquecimiento
enstatite enstatita
enthalpy entalpía
enthalpy of evaporation entalpía de
 evaporación
enthalpy of fusion entalpía de fusión
enthalpy of reaction entalpía de reacción

enthalpy titration titulación entálpica,
 valoración entálpica
entropy entropía
entropy of activation entropía de
 activación
entropy of mixing entropía de mezcla
entropy of transition entropía de
 transición
environmental chemistry química
 ambiental
environmental conditions condiciones
 ambientales
environmental contamination
 contaminación ambiental
environmental factors factores
 ambientales
environmental protection protección
 ambiental
environmental requirements requisitos
 ambientales
environmental stability estabilidad
 ambiental
enyl enilo
enzyme enzima
enzyme kinetics cinética de enzima
enzyme-substrate complex complejo
 enzima-substrato
eosin eosina
ephedrine efedrina
ephedrine hydrochloride clorhidrato de
 efedrina
ephedrine sulfate sulfato de efedrina
epi epi
epicamphor epialcanfor
epichlorite epiclorita
epichlorohydrin epiclorohidrina
epicyanohydrin epicianhidrina
epididymite epididimita
epidioxide epidióxido
epidioxy epidioxi
epidithio epiditio
epigenite epigenita
epihydrin epihidrina
epimer epímero
epimerase epimerasa
epimeric epimérico
epimeride epimerida
epimerism epimerismo
epimerization epimerización

epinephrine epinefrina
epistilbite epiestilbita
epitaxy epitaxia
epithio epitio
epithioximino epitioximino
epoxidation epoxidación
epoxide epóxido
epoxy epoxi
epoxy polymer polímero epóxido
epoxy resin resina epóxica
epoxybutane epoxibutano
epoxyethane epoxietano
epoxyimino epoxiimino
epoxynitrilo epoxinitrilo
epoxypropane epoxipropano
epoxypropanol epoxipropanol
epoxythio epoxitio
epsilon epsilón
epsilon acid ácido epsilón
Epsom salts sales de Epsom
epsomite epsomita
equation ecuación
equation of state ecuación de estado
equatorial ecuatorial
equilenin equilenina
equilibrium equilibrio
equilibrium constant constante de equilibrio
equilibrium diagram diagrama de equilibrio
equilibrium dialysis diálisis de equilibrio
equilibrium law ley de equilibrio
equilibrium point punto de equilibrio
equilibrium potential potencial de equilibrio
equilibrium ratio razón de equilibrio
equilibrium solubility solubilidad de equilibrio
equipartition equipartición
equipartition of energy equipartición de energía
equipotential energy energía equipotencial
equivalence equivalencia
equivalence point punto de equivalencia
equivalent equivalente
equivalent charge carga equivalente
equivalent conductance conductancia equivalente

equivalent conductivity conductividad equivalente
equivalent electrons electrones equivalentes
equivalent nuclei núcleos equivalentes
equivalent proportions proporciones equivalentes
equivalent weight peso equivalente
erbia erbia
erbium erbio
erbium acetate acetato de erbio
erbium halide haluro de erbio
erbium nitrate nitrato de erbio
erbium oxalate oxalato de erbio
erbium oxide óxido de erbio
erbium sulfate sulfato de erbio
erbon erbón
Erdmann reagent reactivo de Erdmann
Erdmann salt sal de Erdmann
ergine ergina
ergocalciferol ergocalciferol
ergodic ergódico
ergometrine ergometrina
ergonovine ergonovina
ergosinine ergosinina
ergosterol ergosterol
ergot cornezuelo
ergotamine ergotamina
ergotinine ergotinina
ergotoxine ergotoxina
erilite erilita
erinite erinita
eriodictyol eriodictiol
Erlenmeyer flask matraz de Erlenmeyer
Erlenmeyer rule regla de Erlenmeyer
Erlenmeyer synthesis síntesis de Erlenmeyer
erosion erosión
erubescite erubescita
erucic acid ácido erúcico
erucyl alcohol alcohol erucílico
erythorbic acid ácido eritórbico
erythrin eritrina
erythrite eritrita
erythritol eritritol
erythritol anhydride anhídrido de eritritol
erythrityl eritritilo
erythro eritro

erythroidine eritroidina
erythrol eritrol
erythromycin eritromicina
erythrose eritrosa
erythrosiderite eritrosiderita
erythrosin eritrosina
Eschka mixture mezcla de Eschka
Eschweiler-Clarke modification
     modificación de Eschweiler-Clarke
eserine eserina
essential esencial
essential oil aceite esencial
essential salt sal esencial
ester éster
ester gum goma de éster
ester interchange cambio de éster
ester number número de éster
esterase esterasa
esterification esterificación
esterification law ley de esterificación
estersil estersil
estradiol estradiol
estragole estragol
estrane estrano
estriol estriol
estrogen estrógeno
estrone estrona
eta eta
Étard reaction reacción de Étard
Étard salt sal de Étard
ethacetic acid ácido etacético
ethal etal
ethaldehyde etaldehído
ethamine etamina
ethanal etanal
ethanaldehyde etanaldehído
ethanamide etanamida
ethanamidine etanamidina
ethane etano
ethane hydrate hidrato de etano
ethanedial etanodial
ethanediamide etanodiamida
ethanediamine etanodiamina
ethanedinitrile etanodinitrilo
ethanedioic acid ácido etanodioico
ethanedithiol etanoditiol
ethanenitrile etanonitrilo
ethanesulfonic acid ácido etanosulfónico
ethanethial etanotial

ethanethiol etanotiol
ethanethiolic acid ácido etanotiólico
ethanite etanita
ethanoate etanoato
ethanoic acid ácido etanoico
ethanoic anhydride anhídrido etanoico
ethanol etanol
ethanol formamide formamida de etanol
ethanol hydrazine hidrazina de etanol
ethanolamine etanolamina
ethanolate etanolato
ethanolurea etanolurea
ethanoyl etanoilo
ethanoyl chloride cloruro de etanoilo
ethanoylation etanoilación
ethanyl etanilo
ethanyl alcohol alcohol etanílico
ethene eteno
ethene polymer polímero de eteno
ethenium etenio
ethenol etenol
ethenone etenona
ethenyl etenilo
ether éter
ethereal etéreo
etheric acid ácido etérico
etherification eterificación
ethide etida
ethidine etidina
ethine etina
ethinoic acid ácido etinoico
ethinyl etinilo
ethinylation etinilación
ethion etión
ethionamide etionamida
ethisterone etisterona
ethohexadiol etohexadiol
etholide etolida
ethoxalyl etoxalilo
ethoxide etóxido
ethoxy etoxi
ethoxyacetanilide etoxiacetanilida
ethoxyacetone etoxiacetona
ethoxyaniline etoxianilina
ethoxycarbonyl etoxicarbonilo
ethoxycarbonyl isothiocyanate
     isotiocianato de etoxicarbonilo
ethoxyethanol etoxietanol
ethoxyl etoxilo

ethoxymethylenemalononitrile
  etoximetilenomalononitrilo
ethoxyphenol etoxifenol
ethoxytriglycol etoxitriglicol
ethyl etilo
ethyl abietate abietato de etilo
ethyl acetate acetato de etilo
ethyl acetoacetate acetoacetato de etilo
ethyl acrylate acrilato de etilo
ethyl alcohol alcohol etílico
ethyl amyl ketone etilamilcetona
ethyl benzoate benzoato de etilo
ethyl benzoylacetate benzoilacetato de
  etilo
ethyl bromide bromuro de etilo
ethyl bromoacetate bromoacetato de
  etilo
ethyl butanoate butanoato de etilo
ethyl butyrate butirato de etilo
ethyl caprate caprato de etilo
ethyl caproate caproato de etilo
ethyl caprylate caprilato de etilo
ethyl carbamate carbamato de etilo
ethyl carbazate carbazato de etilo
ethyl carbonate carbonato de etilo
ethyl chloride cloruro de etilo
ethyl chloroacetal cloroacetal de etilo
ethyl chloroacetate cloroacetato de etilo
ethyl chloroacetoacetate
  cloroacetoacetato de etilo
ethyl chlorocarbonate clorocarbonato de
  etilo
ethyl chloroformate cloroformiato de
  etilo
ethyl chloropropionate cloropropionato
  de etilo
ethyl chlorosulfonate clorosulfonato de
  etilo
ethyl cinnamate cinamato de etilo
ethyl citrate citrato de etilo
ethyl crotonate crotonato de etilo
ethyl cyanate cianato de etilo
ethyl cyanide cianuro de etilo
ethyl cyanoacetate cianoacetato de etilo
ethyl cyanoethanoate cianoetanoato de
  etilo
ethyl decanoate decanoato de etilo
ethyl diazoacetate diazoacetato de etilo
ethyl diazoethanoate diazoetanoato de

  etilo
ethyl dibromoacetate dibromoacetato de
  etilo
ethyl dichloroacetate dicloroacetato de
  etilo
ethyl ethanoate etanoato de etilo
ethyl ether éter etílico
ethyl ferrocyanate ferrocianato de etilo
ethyl formamide formamida de etilo
ethyl formate formiato de etilo
ethyl furoate furoato de etilo
ethyl glycerate glicerato de etilo
ethyl glycolate glicolato de etilo
ethyl heptanoate heptanoato de etilo
ethyl hexanoate hexanoato de etilo
ethyl hydrosulfide hidrosulfuro de etilo
ethyl hydroxyethylcellulose
  hidroxietilcelulosa de etilo
ethyl iodide yoduro de etilo
ethyl iodoacetate yodoacetato de etilo
ethyl isobutyrate isobutirato de etilo
ethyl isocyanate isocianato de etilo
ethyl isocyanide isocianuro de etilo
ethyl isophthalate isoftalato de etilo
ethyl isovalerate isovalerato de etilo
ethyl lactate lactato de etilo
ethyl malate malato de etilo
ethyl malonate malonato de etilo
ethyl mercaptan etilmercaptano
ethyl methacrylate metacrilato de etilo
ethyl methanoate metanoato de etilo
ethyl methyl ether éter etilmetílico
ethyl methyl ketone etilmetilcetona
ethyl monotartrate monotartrato de etilo
ethyl myristate miristato de etilo
ethyl nitrate nitrato de etilo
ethyl nitrite nitrito de etilo
ethyl nonanoate nonanoato de etilo
ethyl octanoate octanoato de etilo
ethyl octoate octoato de etilo
ethyl oleate oleato de etilo
ethyl orthophosphate ortofosfato de etilo
ethyl orthopropionate ortopropionato de
  etilo
ethyl orthosilicate ortosilicato de etilo
ethyl oxalate oxalato de etilo
ethyl oxide óxido de etilo
ethyl palmitate palmitato de etilo
ethyl perchlorate perclorato de etilo

ethyl pelargonate pelargonato de etilo
ethyl peroxide peróxido de etilo
ethyl phenyl ketone etilfenilcetona
ethyl phenylacetate fenilacetato de etilo
ethyl phosphate fosfato de etilo
ethyl phosphine fosfina de etilo
ethyl phthalate ftalato de etilo
ethyl propiolate propiolato de etilo
ethyl propionate propionato de etilo
ethyl pyrophosphate pirofosfato de etilo
ethyl pyruvate piruvato de etilo
ethyl salicylate salicilato de etilo
ethyl silicate silicato de etilo
ethyl succinate succinato de etilo
ethyl sulfate sulfato de etilo
ethyl sulfhydrate sulfhidrato de etilo
ethyl sulfide sulfuro de etilo
ethyl thiocyanate tiocianato de etilo
ethyl thioethane tioetano de etilo
ethyl thioethanol tioetanol de etilo
ethyl triacetylgallate triacetilgalato de etilo
ethyl valerate valerato de etilo
ethyl vanillate vanilato de etilo
ethyl vanillin vanilina de etilo
ethyl vinyl ether etilviniléter
ethylacetamide etilacetamida
ethylacetanilide etilacetanilida
ethylacetic acid ácido etilacético
ethylacetone etilacetona
ethylacetylene etilacetileno
ethylaluminum dichloride dicloruro de etilaluminio
ethylaluminum sesquichloride sesquicloruro de etilaluminio
ethylamine etilamina
ethylamine hydrobromide bromhidrato de etilamina
ethylamino etilamino
ethylaminoacetate etilaminoacetato
ethylaniline etilanilina
ethylanthraquinone etilantraquinona
ethylarsenious oxide óxido etilarsenioso
ethylate etilato
ethylation etilación
ethylaziridine etilaziridina
ethylbenzene etilbenceno
ethylbenzenesulfonic acid ácido etilbencenosulfónico

ethylbenzyl alcohol alcohol etilbencílico
ethylbenzyl chloride cloruro de etilbencilo
ethylbenzylaniline etilbencilanilina
ethylbutanol etilbutanol
ethylbutene etilbuteno
ethylbutyl etilbutilo
ethylbutyl acetate acetato de etilbutilo
ethylbutyl alcohol alcohol etilbutílico
ethylbutyl carbonate carbonato de etilbutilo
ethylbutyl ketone etilbutilcetona
ethylbutyl silicate silicato de etilbutilo
ethylbutylamine etilbutilamina
ethylbutyraldehyde etilbutiraldehído
ethylbutyric acid ácido etilbutírico
ethylcarbazole etilcarbazol
ethylcarbinol etilcarbinol
ethylcellulose etilcelulosa
ethylcyclohexane etilciclohexano
ethylcyclohexylamine etilciclohexilamina
ethylcyclopentane etilciclopentano
ethylcyclopentanone etilciclopentanona
ethyldichlorosilane etildiclorosilano
ethyldiethanolamine etildietanolamina
ethylene etileno
ethylene benzoate benzoato de etileno
ethylene bromide bromuro de etileno
ethylene bromohydrin bromohidrina de etileno
ethylene carbonate carbonato de etileno
ethylene chloride cloruro de etileno
ethylene chlorobromide clorobromuro de etileno
ethylene chlorohydrin clorohidrina de etileno
ethylene cyanide cianuro de etileno
ethylene cyanohydrin cianhidrina de etileno
ethylene diacetate diacetato de etileno
ethylene dibromide dibromuro de etileno
ethylene dichloride dicloruro de etileno
ethylene dicyanide dicianuro de etileno
ethylene diiodide diyoduro de etileno
ethylene dinitrate dinitrato de etileno
ethylene glycol etilenglicol
ethylene glycol diacetate diacetato de etilenglicol
ethylene glycol dibutyrate dibutirato de

etilenglicol

ethylene glycol diformate diformiato de etilenglicol

ethylene glycol dinitrate dinitrato de etilenglicol

ethylene glycol monoacetate monoacetato de etilenglicol

ethylene glycol monobutyl ether éter monobutílico de etilenglicol

ethylene glycol monoethyl ether éter monoetílico de etilenglicol

ethylene glycol monomethyl ether éter monometílico de etilenglicol

ethylene hydrate hidrato de etileno

ethylene hydride hidruro de etileno

ethylene nitrate nitrato de etileno

ethylene oxide óxido de etileno

ethylene-propylene etileno-propileno

ethylene resin resina etilénica

ethylene tetrabromide tetrabromuro de etileno

ethylene tetrachloride tetracloruro de etileno

ethylene thiourea tiourea de etileno

ethylene trichloride tricloruro de etileno

ethylene urea urea de etileno

ethyleneamine etilenamina

ethylenediamine etilendiamina

ethylenediaminetetraacetic acid ácido etilendiaminotetraacético

ethylenediaminetetraacetonitrile etilendiaminotetraacetonitrilo

ethylenedithiol etilenditiol

ethyleneimine etilenimina

ethylenenaphthalene etilenonaftaleno

ethylenesulfonic acid ácido etilensulfónico

ethylethanolamine etiletanolamina

ethylethylene etiletileno

ethylethyleneimine etiletilenimina

ethylfluorosulfonate etilfluorosulfonato

ethylfuran etilfurano

ethylheptane etilheptano

ethylhexaldehyde etilhexaldehído

ethylhexanal etilhexanal

ethylhexanediol etilhexanodiol

ethylhexanol etilhexanol

ethylhexenal etilhexenal

ethylhexoic acid ácido etilhexoico

ethylhexyl etilhexilo

ethylhexyl acetate acetato de etilhexilo

ethylhexyl acrylate acrilato de etilhexilo

ethylhexyl alcohol alcohol etilhexílico

ethylhexyl bromide bromuro de etilhexilo

ethylhexyl chloride cloruro de etilhexilo

ethylhexyl cyanoacetate cianoacetato de etilhexilo

ethylhexylamine etilhexilamina

ethylhexylaniline etilhexilanilina

ethylhydrazine etilhidrazina

ethylidene etilideno

ethylidene chloride cloruro de etilideno

ethylidene diacetate diacetato de etilideno

ethylidene dibromide dibromuro de etilideno

ethylidene diiodide diyoduro de etilideno

ethylidyne etilidino

ethylin etilina

ethylisobutylmethane etilisobutilmetano

ethylisohexanol etilisohexanol

ethyllithium etil-litio

ethylmagnesium bromide bromuro de etilmagnesio

ethylmagnesium chloride cloruro de etilmagnesio

ethylmethylacetylene etilmetilacetileno

ethylmorphine hydrochloride clorhidrato de etilmorfina

ethylmorpholine etilmorfolina

ethylnaphthylamine etilnaftilamina

ethyloic etiloico

ethylol etilol

ethylparaben etilparabeno

ethylpentane etilpentano

ethylphenol etilfenol

ethylphosphoric acid ácido etilfosfórico

ethylsulfuric acid ácido etisulfúrico

ethyltrichlorosilane etiltriclorosilano

ethylurethane etiluretano

ethyne etino

ethynyl etinilo

ethynylation etinilación

ethynylestradiol etinilestradiol

ettringite ettringita

eucairite eucairita

eucalyptol eucaliptol

eucarvone eucarvona
euchroite eucroíta
eucryptite eucriptita
eudalene eudaleno
eudialite eudialita
eudiometer eudiómetro
eugenic acid ácido eugénico
eugenol eugenol
eulytin eulitina
eupittonic acid ácido eupitónico
eupyrine eupirina
europium europio
europium chloride cloruro de europio
europium fluoride fluoruro de europio
europium nitrate nitrato de europio
europium oxalate oxalato de europio
europium oxide óxido de europio
europium sulfate sulfato de europio
eutectic eutéctico
eutectic alloy aleación eutéctica
eutectic mixture mezcla eutéctica
eutectic point punto eutéctico
eutectic system sistema eutéctico
eutectic temperature temperatura
    eutéctica
eutectoid eutectoide
eutrophication eutroficación
eutropic series serie eutrópica
eutropy eutropía
euxenite euxenita
evansite evansita
evaporate evaporar
evaporating dish plato de evaporación
evaporation evaporación
evaporation dish plato de evaporación
evaporator evaporador
evaporimeter evaporímetro
evaporite evaporita
even-even nucleus núcleo par-par
even-odd nucleus núcleo par-impar
evolution evolución
evolved gas analysis análisis de gases
    despedidos
exchange intercambio
exchange collision colisión de
    intercambio
exchange reaction reacción de
    intercambio
excimer excímero

exciplex complejo excitado
excitation excitación
excitation curve curva de excitación
excitation function función de excitación
excitation index índice de excitación
excitation spectrum espectro de
    excitación
excite excitar
excited excitado
excited atom átomo excitado
excited ion ion excitado
excited nucleus núcleo excitado
excited state estado excitado
exciter excitador
exciton excitón
exclusion exclusión
exclusion principle principio de
    exclusión
exo exo
exocarbon exocarburo
exocondensation exocondensación
exogenous exógeno
exopeptidase exopeptidasa
exosmosis exósmosis
exothermic exotérmico
exothermic compound compuesto
    exotérmico
exothermic reaction reacción exotérmica
exotic atom átomo exótico
expand expandirse
expander expansor
expansion expansión
experiment experimento
experimental experimental
experimentation experimentación
explode explotar
explosion explosión
explosion spectrum espectro de
    explosión
explosive explosivo
explosive reaction reacción explosiva
explosive substance substancia explosiva
exponential decay desintegración
    exponencial
exponential disintegration
    desintegración exponencial
exposure exposición
expression expresión
extender extensor, diluente

extinction extinción
extinction coefficient coeficiente de extinción
extract extracto
extractant extractante
extraction extracción
extraction apparatus aparato de extracción
extraction coefficient coeficiente de extracción

extractive distillation destilación extractiva
extractor extractor
extranuclear extranuclear
extrapolation extrapolación
extrusion extrusión
exude exudar
eye-wash lavador de ojos
Eyring equation ecuación de Eyring

F

fabianol fabianol
fabulite fabulita
fac fac
face-centered cubic cúbico centrado en las caras
face-centered cubic structure estructura cúbica centrada en las caras
face-centered lattice red centrada en las caras
face-centered structure estructura centrada en las caras
factor factor
factor quantity cantidad de factor
fagarine fagarina
Fahrenheit scale escala Fahrenheit
Fajans rules reglas de Fajans
Fajans-Soddy law ley de Fajans-Soddy
Fament process proceso de Fament
farad farad
faraday faraday
Faraday constant constante de Faraday
Faraday effect efecto de Faraday
Faraday law ley de Faraday
Faraday tube tubo de Faraday
faradiol faradiol
farinose farinosa
farnesene farneseno
farnesol farnesol
Farrar process proceso de Farrar
fast fijo, rápido
fast-atom bombardment bombardeo con átomos rápidos
fast chemical reaction reacción química rápida
fast neutron neutrón rápido
fast-neutron fission fisión por neutrones rápidos
fast-neutron spectrometry espectrometría de neutrones rápidos
fast reaction reacción rápida
fat grasa
fat hardening endurecimiento de grasa
fat soluble soluble en grasa

fat splitting división de grasas
fatty acid ácido graso
fatty alcohol alcohol graso
fatty amine amina grasa
fatty ester éster graso
fatty nitrile nitrilo graso
faujasite faujasita
Favorskii-Babayan synthesis síntesis de Favorskii-Babayan
Favorskii rearrangement reordenamiento de Favorskii
fayalite fayalita
feedback retroalimentación
feeder alimentador
feedstock materia prima para proceso
Fehling reagent reactivo de Fehling
Fehling solution solución de Fehling
Fehling test prueba de Fehling
Feiser solution solución de Feiser
Feist-Benary synthesis síntesis de Feist-Benary
feldspar feldespato
feldspathoid feldespatoide
felite felita
felspar feldespato
felvation felvación
femic fémico
femtochemistry femtoquímica
fen fen
fenchane fencano
fenchanol fencanol
fenchanone fencanona
fenchene fenqueno
fenchol fencol
fencholic acid ácido fencólico
fenchone fencona
fenchoxime fencoxima
fenchyl fenquilo
fenchyl alcohol alcohol fenquílico
fenitrothion fenitrotión
fennel oil aceite de hinojo
fensulfothion fensulfotiona
fentanyl citrate citrato de fentanilo

fenthion fentión
Fenton reaction reacción de Fenton
Fenton reagent reactivo de Fenton
ferberite ferberita
fergusonite fergusonita
ferment fermentar
fermentable fermentable
fermentation fermentación
fermentation tube tubo de fermentación
Fermi constant constante de Fermi
Fermi-Dirac statistics estadística de
    Fermi-Dirac
Fermi level nivel de Fermi
Fermi resonance resonancia de Fermi
Fermi surface superficie de Fermi
Fermi transition transición de Fermi
fermium fermio
Ferrario reaction reacción de Ferrario
ferrate ferrato
ferredoxin ferredoxina
ferri ferri
ferric férrico
ferric acetate acetato férrico
ferric acetylacetonate acetilacetonato
    férrico
ferric ammonium amonio férrico
ferric ammonium alum alumbre
    amónico férrico
ferric ammonium citrate citrato
    amónico férrico
ferric ammonium oxalate oxalato
    amónico férrico
ferric ammonium sulfate sulfato
    amónico férrico
ferric arsenate arseniato férrico
ferric arsenite arsenito férrico
ferric benzoate benzoato férrico
ferric bromide bromuro férrico
ferric chloride cloruro férrico
ferric chromate cromato férrico
ferric citrate citrato férrico
ferric dichromate dicromato férrico
ferric ferricyanide ferricianuro férrico
ferric ferrocyanide ferrocianuro férrico
ferric fluoride fluoruro férrico
ferric fluoroborate fluoroborato férrico
ferric fluorosilicate fluorosilicato férrico
ferric formate formiato férrico
ferric glycerophosphate glicerofosfato

    férrico
ferric hydrate hidrato férrico
ferric hydroxide hidróxido férrico
ferric hypophosphite hipofosfito férrico
ferric iodide yoduro férrico
ferric nitrate nitrato férrico
ferric octoate octoato férrico
ferric oleate oleato férrico
ferric oxalate oxalato férrico
ferric oxide óxido férrico
ferric oxide red rojo de óxido férrico
ferric perchlorate perclorato férrico
ferric perchloride percloruro férrico
ferric phenolate fenolato férrico
ferric phosphate fosfato férrico
ferric pyrophosphate pirofosfato férrico
ferric resinate resinato férrico
ferric sesquibromide sesquibromuro
    férrico
ferric sodium oxalate oxalato sódico
    férrico
ferric stearate estearato férrico
ferric subcarbonate subcarbonato férrico
ferric subsulfate subsulfato férrico
ferric sulfate sulfato férrico
ferric sulfide sulfuro férrico
ferric tallate talato férrico
ferric tannate tanato férrico
ferric thiocyanate tiocianato férrico
ferric tribromide tribromuro férrico
ferric trichloride tricloruro férrico
ferric trioxide trióxido férrico
ferric trisulfate trisulfato férrico
ferric vanadate vanadato férrico
ferrichrome ferricromo
ferricyanic acid ácido ferriciánico
ferricyanide ferricianuro
ferriferrous ferriferroso
ferrimagnetism ferrimagnetismo
ferrimanganese ferrimanganeso
ferrimanganic ferrimangánico
ferripotassium ferripotasio
ferripotassium cyanide cianuro
    ferripotásico
ferripyrine ferripirina
ferrisodium ferrisodio
ferrite ferrita
ferritic ferrítico
ferro ferro

ferroalloy ferroaleación
ferroaluminum ferroaluminio
ferroammonium ferroamonio
ferroboron ferroboro
ferrocene ferroceno
ferrocenoyl ferrocenoilo
ferrocenoyl chloride cloruro de ferrocenoilo
ferrocenoyl dichloride dicloruro de ferrocenoilo
ferrocenylborane ferrocenilborano
ferrocerium ferrocerio
ferrochrome ferrocromo
ferrochromium ferrocromo
ferrocyanic acid ácido ferrociánico
ferrocyanide ferrocianuro
ferrodolomite ferrodolomita
ferrodynamic ferrodinámico
ferroelectric ferroeléctrico
ferroelectricity ferroelectricidad
ferroferric ferroférrico
ferroferricyanide ferroferricianuro
ferroferrocyanide ferroferrocianuro
ferroin ferroína
ferromagnesite ferromagnesita
ferromagnesium ferromagnesio
ferromagnetic ferromagnético
ferromagnetic oxide óxido ferromagnético
ferromagnetism ferromagnetismo
ferromanganese ferromanganeso
ferromanganic ferromangánico
ferromanganous ferromanganoso
ferromolybdenum ferromolibdeno
ferron ferrón
ferronickel ferroníquel
ferroniobium ferroniobio
ferrophosphorus ferrofósforo
ferropotassium ferropotasio
ferropyrine ferropirina
ferrosilicon ferrosilicio
ferrosilite ferrosilita
ferrosoferric ferrosoférrico
ferrosoferric oxide óxido ferrosoférrico
ferrotitanium ferrotitanio
ferrotungsten ferrotungsteno
ferrous ferroso
ferrous acetate acetato ferroso
ferrous ammonium amonio ferroso

ferrous ammonium sulfate sulfato amónico ferroso
ferrous arsenate arseniato ferroso
ferrous bromide bromuro ferroso
ferrous carbonate carbonato ferroso
ferrous chloride cloruro ferroso
ferrous ferricyanide ferricianuro ferroso
ferrous ferrite ferrita ferrosa
ferrous ferrocyanide ferrocianuro ferroso
ferrous fluoride fluoruro ferroso
ferrous fluorosilicate fluorosilicato ferroso
ferrous fumarate fumarato ferroso
ferrous gluconate gluconato ferroso
ferrous hydroxide hidróxido ferroso
ferrous hypophosphite hipofosfito ferroso
ferrous hyposulfate hiposulfato ferroso
ferrous iodide yoduro ferroso
ferrous lactate lactato ferroso
ferrous naphthenate naftenato ferroso
ferrous nitrate nitrato ferroso
ferrous octoate octoato ferroso
ferrous oxalate oxalato ferroso
ferrous oxide óxido ferroso
ferrous perchlorate perclorato ferroso
ferrous phosphate fosfato ferroso
ferrous phosphide fosfuro ferroso
ferrous selenide seleniuro ferroso
ferrous sulfate sulfato ferroso
ferrous sulfide sulfuro ferroso
ferrous tantalate tantalato ferroso
ferrous tartrate tartrato ferroso
ferrous thiocyanate tiocianato ferroso
ferrous titanate titanato ferroso
ferrous tungstate tungstato ferroso
ferrovanadium ferrovanadio
ferroverdin ferroverdina
ferroxyl ferroxilo
ferrozirconium ferrozirconio
ferruginous ferruginoso
ferrum ferrum
fertile material material fértil
fertilizer fertilizante
ferulaldehyde ferulaldehído
ferulene feruleno
ferulic acid ácido ferúlico
fervanite fervanita

Feulgen reaction reacción de Feulgen
fiber fibra
fiber-optic fibroóptico
fiberglass fibra de vidrio
fibrid fibrido
fibril fibrilla
fibrin fibrina
fibrinogen fibrinógeno
fibroin fibroína
fibrolite fibrolita
fichtelite fictelita
ficin ficina
Fick diffusion law ley de difusión de Fick
fictile fíctil
field campo
field-emission microscope microscopio de emisión de campo
field intensity intensidad de campo
field-ion emission emisión iónica de campo
field-ion microscope microscopio iónico de campo
field ionization ionización de campo
field strength fuerza de campo
filament filamento
filicic acid ácido filícico
filixic acid ácido filíxico
filler relleno
film película
film badge dosímetro de película
filter filtro
filter aid facilitador de filtración
filter alum alumbre de filtro
filter bag bolsa de filtración
filter cone cono de filtración
filter crucible crisol de filtración
filter cylinder cilindro de filtración
filter discrimination discriminación de filtro
filter effectiveness efectividad de filtro
filter flask matraz de filtración
filter funnel embudo de filtración
filter medium medio para filtración
filter paper papel de filtro
filter photometry fotometría con filtro
filter pump bomba para filtración
filter sand arena de filtración
filter selectivity selectividad de filtro

filter spectrophotometer espectrofotómetro de filtro
filter tube tubo de filtro
filtered filtrado
filtering filtración
filtering flask matraz de filtración
filtrate filtrado
filtration filtración
filtration under pressure filtración bajo presión
fine chemical producto químico fino
fine structure estructura fina
fineness pureza, ley
finishing compound compuesto de acabado
finishing material material de acabado
Finkelstein reaction reacción de Finkelstein
fire alarm alarma de fuego
fire detector detector de fuego
fire extinguisher extintor de fuego
fire-extinguisher system sistema extintor de fuego
fire point punto de llama
fire protection protección contra fuego
fire resistance resistencia al fuego
fire resistant resistente al fuego
fireclay arcilla refractaria
fireproof a prueba de fuego
fireproof material material a prueba de fuego
first law of thermodynamics primera ley de termodinámica
first-order reaction reacción de primer orden
first-order spectrum espectro de primer orden
Fischer-Hepp rearrangement reordenamiento de Fischer-Hepp
Fischer indole synthesis síntesis de indoles de Fischer
Fischer oxazole synthesis síntesis de oxazoles de Fischer
Fischer phenylhydrazine synthesis síntesis de fenilhidrazinas de Fischer
Fischer polypeptide synthesis síntesis de polipéptidos de Fischer
Fischer projection proyección de Fischer
Fischer reagent reactivo de Fischer

Fischer salt sal de Fischer
Fischer solution solución de Fischer
Fischer-Tropsch process proceso de
  Fischer-Tropsch
Fischer-Tropsch reaction reacción de
  Fischer-Tropsch
Fischer-Tropsch synthesis síntesis de
  Fischer-Tropsch
fisetin fisetina
fish oil aceite de pescado
fissile físil
fissiochemistry fisioquímica
fission fisión
fission chamber cámara de fisión
fission cross-section sección transversal
  de fisión
fission energy energía de fisión
fission neutron neutrón de fisión
fission product producto de fisión
fission rate velocidad de fisión
fission spectrum espectro de fisión
fission threshold umbral de fisión
fission yield rendimiento de fisión
fissionable fisionable
Fittig reaction reacción de Fittig
Fittig synthesis síntesis de Fittig
fix fijar
fixation fijación
fixative fijativo
fixed carbon carbono fijo
fixed oil aceite fijo
fixed proportion proporción fija
fixing fijación
fixing agent agente fijador
flagstaffite flagstafita
flame llama
flame alarm alarma de llama
flame control control de llama
flame cutting oxicorte
flame detector detector de llama
flame-emission spectroscopy
  espectroscopia de emisión de llama
flame hardening endurecimiento por
  llama
flame-ionization detector detector de
  ionización por llama
flame photometer fotómetro de llama
flame photometry fotometría de llama
flame reaction reacción con llama

flame-resistant ignífugo
flame-retarding retardante de llama
flame spectrometer espectrómetro de
  llama
flame spectrometry espectrometría de
  llama
flame spectrophotometer
  espectrofotómetro de llama
flame spectrophotometry
  espectrofotometría de llama
flame spectrum espectro de llama
flame test prueba con llama
flameproof antideflagrante,
  incombustible
flameproof material material
  antideflagrante, material
  incombustible
flammability inflamabilidad
flammability limit límite de
  inflamabilidad
flammable inflamable
flammable material material inflamable
flammable substance substancia
  inflamable
flash destello, llama súbita luminosa,
  rebaba
flash distillation destilación instantánea
flash drying secado instantáneo
flash photolysis fotólisis instantánea
flash point punto de inflamabilidad
flash spectroscopy espectroscopia de
  destello
flask matraz
flat-bottom flask matraz de base plana
flavaniline flavanilina
flavanol flavanol
flavanone flavanona
flavazine flavazina
flavianic acid ácido flaviánico
flavin flavina
flavine flavina
flavoglaucin flavoglaucina
flavol flavol
flavone flavona
flavonoid flavonoide
flavonol flavonol
flavopannin flavopanina
flavophenine flavofenina
flavoprotein flavoproteína

flavopurpurin flavopurpurina
flavor sabor
flavoxanthin flavoxantina
flavylium flavilio
flaxseed oil aceite de linaza
Fleming tube tubo de Fleming
Fletcher furnace horno de Fletcher
flint pedernal
floatation separación por chorro de aire
floc masa flocosa
flocculant floculante
flocculate floculado
flocculating agent agente floculante
flocculation floculación
flocculent floculento
Flood equation ecuación de Flood
Flood reaction reacción de Flood
Flory temperature temperatura de Flory
flotation flotación
flotation activator activador de flotación
flotation agent agente de flotación
flow flujo
flow diagram diagrama de flujo
flower flor
flox flox
flucloxacillin flucloxacilina
fluctuation fluctuación
flue humero
fluellite fluellita
fluid fluido
fluidity fluidez
fluidization fluidización
fluo fluo
fluoboric acid ácido fluobórico
fluocerite fluocerita
fluoflavine fluoflavina
fluohydric acid ácido fluohídrico
fluor fluor
fluoracetamide fluoracetamida
fluoran fluorán
fluorandiol fluorandiol
fluoranthene fluoranteno
fluoranthraquinone fluorantraquinona
fluorapatite fluorapatita
fluoration fluoración
fluorene fluoreno
fluorenic acid ácido fluorénico
fluorenol fluorenol
fluorenone fluorenona

fluorenyl fluorenilo
fluorenylamine fluorenilamina
fluorenylidene fluorenilideno
fluorescein fluoresceína
fluorescence fluorescencia
fluorescence analysis análisis por fluorescencia
fluorescence efficiency eficiencia de fluorescencia
fluorescence indicator indicador de fluorescencia
fluorescence microscopy microscopia de fluorescencia
fluorescence spectroscopy espectroscopia de fluorescencia
fluorescence yield rendimiento de fluorescencia
fluorescent fluorescente
fluorescent dye colorante fluorescente
fluorescent pigment pigmento fluorescente
fluorescent radiation radiación fluorescente
fluorescin fluorescina
fluorgermanate fluorgermanato
fluorgermanic acid ácido fluorgermánico
fluoridation fluorización
fluoride fluoruro
fluorimeter fluorímetro
fluorimetry fluorimetría
fluorination fluorización
fluorine flúor
fluorine cyanide cianuro de flúor
fluorine halide haluro de flúor
fluorine hydride hidruro de flúor
fluorine nitrate nitrato de flúor
fluorine oxide óxido de flúor
fluorine polymer polímero con flúor
fluorinion fluorinión
fluoriodine fluoryodo
fluorion fluorión
fluorite fluorita
fluorization fluorización
fluoro fluoro
fluoroacetamide fluoroacetamida
fluoroacetanilide fluoroacetanilida
fluoroacetate fluoroacetato
fluoroacetic acid ácido fluoroacético
fluoroacetophenone fluoroacetofenona

fluoroalkane fluoroalcano
fluoroaniline fluoroanilina
fluoroapatite fluoroapatita
fluorobenzene fluorobenceno
fluoroborate fluoroborato
fluoroboric acid ácido fluorobórico
fluorocarbon fluorocarburo
fluorocarbon fiber fibra de
  fluorocarburo
fluorocarbon polymer polímero de
  fluorocarburo
fluorocarbon resin resina de
  fluorocarburo
fluorochemical fluoroquímico
fluorochromate fluorocromato
fluorodichloromethane
  fluorodiclorometano
fluoroelastomer fluoroelastómero
fluoroethanoic acid ácido fluoroetanoico
fluoroform fluoroformo
fluorogen fluorógeno
fluorohydrocarbon fluorohidrocarburo
fluoroiodate fluoroyodato
fluorometer fluorómetro
fluoromethane fluorometano
fluorometholone fluorometolona
fluorometric analysis análisis
  fluorométrico
fluorometry fluorometría
fluorone fluorona
fluoronium fluoronio
fluorophenol fluorofenol
fluorophore fluoróforo
fluorophosphonate fluorofosfonato
fluorophosphoric acid ácido
  fluorofosfórico
fluorophotometer fluorofotómetro
fluoroplumbic acid ácido fluoroplúmbico
fluoropolymer fluoropolímero
fluoroscope fluoroscopio
fluorosilicate fluorosilicato
fluorosilicic acid ácido fluorosilícico
fluorosulfonic acid ácido fluorosulfónico
fluorosulfuric acid ácido fluorosulfúrico
fluorothene fluoroteno
fluorotrichloromethane
  fluorotriclorometano
fluorspar espato flúor
fluoryl fluorilo

fluorylidine fluorilidina
fluosilicate fluosilicato
fluosilicic acid ácido fluosilícico
fluosulfonic acid ácido fluosulfónico
fluoxymesterone fluoximesterona
fluted filter paper papel de filtro
  plegado
flux flujo, fundente
fly ash ceniza de humero
foam espuma
fog niebla
foil hojuela, hoja
Fokker-Planck equation ecuación de
  Fokker-Planck
folacin folacina
folic acid ácido fólico
Folin apparatus aparato de Folin
Folin solution solución de Folin
folinic acid ácido folínico
fomometer fomómetro
fongisterol fongisterol
food additive aditivo de alimento
food color color de alimento
foots sedimentos de aceite
forbidden band banda prohibida
forbidden energy band banda de energía
  prohibida
forbidden line línea prohibida
forbidden transition transición prohibida
force constant constante de fuerza
forceps pinzas
forensic analysis análisis forense
forensic chemistry química forense
formal formal
formal charge carga formal
formaldehyde formaldehído
formaldehyde aniline anilina de
  formaldehído
formaldoxime formaldoxima
formalin formalina
formality formalidad
formamide formamida
formamidine formamidina
formamido formamido
formamine formamina
formamyl formamilo
Formanek indicator indicador de
  Formanek
formanilide formanilida

formaniline formanilina
formate formiato, formato
formation formación
formation constant constante de
    formación
formazan formazán
formazyl formacilo
formazyl hydride hidruro de formacilo
formiate formiato
formic acid ácido fórmico
formic aldehyde aldehído fórmico
formic ether éter fórmico
formic nitrile nitrilo fórmico
formimidoyl formimidoilo
formine formina
formohydrazide formohidrazida
formol formol
formolite formolita
formonitrile formonitrilo
formonitrolic acid ácido formonitrólico
formothion formotión
formoxime formoxima
formoxy formoxi
formoxyl formoxilo
formoxyl hydride hidruro de formoxilo
formoyl formoilo
formula fórmula
formula weight peso de fórmula, peso
    molecular
formulation formulación
formyl formilo
formyl bromide bromuro de formilo
formyl chloride cloruro de formilo
formyl fluoride fluoruro de formilo
formyl trichloride tricloruro de formilo
formyl triiodide triyoduro de formilo
formylacetic acid ácido formilacético
formylamide formilamida
formylamine formilamina
formylamino formilamino
formylation formilación
formylene formileno
formylhydrazine formilhidrazina
formylic acid ácido formílico
formyloxy formiloxi
formylpiperidine formilpiperidina
Forster reaction reacción de Forster
forsterite forsterita
fosfomycin fosfomicina

foshagite foshagita
fossil fósil
fossil fuel combustible fósil
fostion fostión
Foulger test prueba de Foulger
Fourier transform transformación de
    Fourier
Fourier transform infrared infrarrojo
    con transformación de Fourier
fowlerite fowlerita
foyaite foyaíta
fraction fracción
fractional fraccionario
fractional combustion combustión
    fraccionaria
fractional condensation condensación
    fraccionaria
fractional condensation tube tubo de
    condensación fraccionaria
fractional crystallization cristalización
    fraccionaria
fractional distillation destilación
    fraccionaria
fractional distillation tube tubo de
    destilación fraccionaria
fractional expression expresión
    fraccionaria
fractional filtration filtración
    fraccionaria
fractional precipitation precipitación
    fraccionaria
fractionating fraccionante
fractionating column columna
    fraccionante
fractionation fraccionación
fracture fractura
fragility fragilidad
fragrance fragancia
Franchimont reaction reacción de
    Franchimont
francium francio
Franck-Condon principle principio de
    Franck-Condon
franckeite franqueíta
francolite francolita
frangulic acid ácido frangúlico
frangulin frangulina
frangulinic acid ácido frangulínico
Frankland-Duppa reaction reacción de

Frankland-Duppa
Frankland method método de Frankland
Frankland notation notación de
   Frankland
Frankland reaction reacción de
   Frankland
Frankland synthesis síntesis de
   Frankland
franklinite franklinita
Frary metal metal de Frary
Frasch process proceso de Frasch
Fraude reagent reactivo de Fraude
fraunhofer fraunhofer
Fraunhofer lines líneas de Fraunhofer
Fraunhofer spectrum espectro de
   Fraunhofer
fraxetin fraxetina
fraxin fraxina
fraxinol fraxinol
free atom átomo libre
free charge carga libre
free electron electrón libre
free energy energía libre
free ion ion libre
free molecule molécula libre
free path recorrido libre
free radical radical libre
free sulfur azufre libre
freeze congelar
freeze drying liofilización
freezing congelación
freezing mixture mezcla para
   congelación
freezing point punto de congelación
freezing-point apparatus aparato de
   punto de congelación
freezing-point depression depresión de
   punto de congelación
freezing-point diagram diagrama de
   puntos de congelación
freezing-point elevation elevación de
   punto de congelación
freezing-point range intervalo de puntos
   de congelación
freezing-point reduction reducción de
   punto de congelación
freibergite freibergita
freierslebenite freierslebenita
Fremy salt sal de Fremy

Frenkel defect defecto de Frenkel
Frenkel-Kontorowa model modelo de
   Frenkel-Kontorowa
frequency frecuencia
Fresenius desiccator desecador de
   Fresenius
Freund acid ácido de Freund
Freund method método de Freund
Freund synthesis síntesis de Freund
Freundlich isotherm isoterma de
   Freundlich
freyalite freyalita
friction fricción
Friedel-Crafts condensation
   condensación de Friedel-Crafts
Friedel-Crafts reaction reacción de
   Friedel-Crafts
Friedlander synthesis síntesis de
   Friedlander
Friedrichs condenser condensador de
   Friedrichs
Fries rearrangement reordenamiento de
   Fries
Fries rule regla de Fries
frigorific frigorífico
frigorific mixture mezcla frigorífica
frit frita
fritting fritado
Froehde reagent reactivo de Froehde
frontier orbital orbital fronterizo
froth espuma
fructofuranosidase fructofuranosidasa
fructose fructosa
fructose diphosphate difosfato de
   fructosa
fructoside fructosida
fuberidazole fuberidazol
fuchsin fucsina
fuchsite fucsita
fucitol fucitol
fucose fucosa
fucosite fucosita
fucusine fucusina
fucusol fucusol
fuel combustible
fuel cell pila de combustible
fuel element elemento combustible
fuel oil aceite combustible
fugacity fugacidad

Fujimoto-Belleau reaction reacción de Fujimoto-Belleau
Fujiwara reaction reacción de Fujiwara
fulgenic acid ácido fulgénico
fulguration fulguración
fulgurator fulgurador
fulgurite fulgurita
fuller's earth tierra de batán
fullerene fulereno
fullerite fulerita
fulminate fulminato
fulminic acid ácido fulmínico
fulminuric acid ácido fulminúrico
fulvalene fulvaleno
fulvene fulveno
fumagillin fumagilina
fumaramic acid ácido fumarámico
fumaramide fumaramida
fumarate fumarato
fumarhydrazide fumarhidrazida
fumaric acid ácido fumárico
fumaroyl fumaroilo
fumaroyl chloride cloruro de fumaroilo
fumaryl fumarilo
fumaryl chloride cloruro de fumarilo
fume humo
fume cupboard alacena protectora contra humo, campana de humos
fume hood alacena protectora contra humo, campana de humos
fumigant fumigante
fuming humeante, fumante
fuming nitric acid ácido nítrico fumante
fuming sulfuric acid ácido sulfúrico fumante
functional group grupo funcional
fundamental fundamental
fundamental constant constante fundamental
fundamental particle partícula fundamental
fundamental research investigación fundamental
fundamental series serie fundamental
fundamental unit unidad fundamental
fungicide fungicida
fungus hongo
funnel embudo
funnel tube tubo con embudo

furacrylic acid ácido furacrílico
fural fural
furaldehyde furaldehído
furamide furamida
furan furano
furancarbinol furancarbinol
furancarboxylic acid ácido furancarboxílico
furandione furandiona
furanmethanethiol furanmetanotiol
furanmethylamine furanmetilamina
furano furano
furanose furanosa
furanoside furanosida
furanylmethyl furanilmetilo
furazan furazán
furazanyl furazanilo
furazolidone furazolidona
furfural furfural
furfuraldehyde furfuraldehído
furfuramide furfuramida
furfuran furfurano
furfurol furfurol
furfuryl furfurilo
furfuryl acetate acetato de furfurilo
furfuryl alcohol alcohol furfurílico
furfurylamide furfurilamida
furfurylamine furfurilamina
furfurylidene furfurilideno
furil furil
furildioxime furildioxima
furnace horno, crisol
furnace black negro de horno
furnace oil aceite de horno
furoamide furoamida
furoate furoato
furoic acid ácido furoico
furoin furoína
furol furol
furonic acid ácido furónico
furoyl furoilo
furoyl chloride cloruro de furoilo
furyl furilo
furylcarbinol furilcarbinol
furylidine furilidina
furylmethyl furilmetilo
fuse fundir
fused fundido
fused electrolyte electrólito fundido

fused quartz cuarzo fundido
fused ring anillo fundido
fused salt sal fundida
fused silica sílice fundida
fusel oil aceite de fusel
fusibility fusibilidad
fusible fusible
fusible alloy aleación fusible

fusidic acid ácido fusídico
fusion fusión
fusion energy energía de fusión
fusion heat calor de fusión
fusion point punto de fusion
fusion product producto de fusión
fusion tube tubo de fusión
fusionable fusionable

G

Gabbet solution solución de Gabbet
Gabriel-Colman rearrangement
    reordenamiento de Gabriel-Colman
Gabriel reaction reacción de Gabriel
Gabriel synthesis síntesis de Gabriel
gadinine gadinina
gadoleic acid ácido gadoleico
gadolinite gadolinita
gadolinium gadolinio
gadolinium bromide bromuro de
    gadolinio
gadolinium chloride cloruro de gadolinio
gadolinium fluoride fluoruro de
    gadolinio
gadolinium hydroxide hidróxido de
    gadolinio
gadolinium nitrate nitrato de gadolinio
gadolinium oxalate oxalato de gadolinio
gadolinium oxide óxido de gadolinio
gadolinium sulfate sulfato de gadolinio
gage medidor
gahnite gahnita
gaidic acid ácido gaídico
galactan galactano
galactaric acid ácido galactárico
galactosamine hydrochloride clorhidrato
    de galactosamina
galactose galactosa
galactosidase galactosidasa
galacturonic acid ácido galacturónico
galena galena
galenite galenita
gallacetophenone galacetofenona
gallamine galamina
gallate galato
gallein galeína
gallic gálico
gallic acid ácido gálico
gallic bromide bromuro gálico
gallic chloride cloruro gálico
gallic hydroxide hidróxido gálico
gallic iodide yoduro gálico

gallic oxide óxido gálico
gallin galina
gallium galio
gallium acetate acetato de galio
gallium antimonide antimoniuro de galio
gallium arsenide arseniuro de galio
gallium bromide bromuro de galio
gallium dibromide dibromuro de galio
gallium dichloride dicloruro de galio
gallium diiodide diyoduro de galio
gallium halide haluro de galio
gallium hydride hidruro de galio
gallium hydroxide hidróxido de galio
gallium iodide yoduro de galio
gallium monoxide monóxido de galio
gallium nitrate nitrato de galio
gallium oxide óxido de galio
gallium phosphate fosfato de galio
gallium phosphide fosfuro de galio
gallium sesquioxide sesquióxido de galio
gallium sulfate sulfato de galio
gallium sulfide sulfuro de galio
gallium trichloride tricloruro de galio
gallocyanine galocianina
gallogen galógeno
gallotannic acid ácido galotánico
gallotannin galotanino
gallous galoso
gallous bromide bromuro galoso
gallous oxide óxido galoso
galloyl galoilo
galvanic galvánico
galvanic action acción galvánica
galvanic anode ánodo galvánico
galvanic bath baño galvánico
galvanic cell pila galvánica
galvanic corrosion corrosión galvánica
galvanic current corriente galvánica
galvanic electricity electricidad
    galvánica
galvanic metallization metalización
    galvánica

galvanic series serie galvánica
galvanism galvanismo
galvanization galvanización
galvanize galvanizar
galvanized galvanizado
galvanized iron hierro galvanizado
galvanizing galvanización
galvanomagnetic galvanomagnético
galvanometer galvanómetro
galvanometric galvanométrico
galvanometry galvanometría
galvanoplasty galvanoplastia
gamma gamma
gamma acid ácido gamma
gamma-aminobutyric acid ácido gammaaminobutírico
gamma cross-section sección transversal gamma
gamma decay desintegración gamma
gamma disintegration desintegración gamma
gamma emission emisión gamma
gamma emitter emisor gamma
gamma ferric oxide óxido férrico gamma
gamma-globulin gammaglobulina
gamma iron hierro gamma
gamma radiation radiación gamma
gamma-ray source fuente de rayos gamma
gamma-ray spectrometer espectrómetro de rayos gamma
gamma-ray spectroscopy espectroscopia de rayos gamma
gamma-ray spectrum espectro de rayos gamma
gamma rays rayos gamma
gamma spectroscopy espectroscopia gamma
gamma transition transición gamma
gammexane gamexano
gammil gamilo
Gamow barrier barrera de Gamow
Gamow-Teller interaction interacción de Gamow-Teller
gangue ganga
ganomalite ganomalita
ganomatite ganomatita
ganophyllite ganofilita

garlic oil aceite de ajo
garnet granate
garnierite garnierita
gas gas
gas absorption absorción de gas
gas-admittance valve válvula de admisión de gas
gas admixture ratio razón de mezcla de gases
gas adsorption adsorción de gas
gas analysis análisis de gases
gas-analysis apparatus aparato de análisis de gases
gas black negro de gas
gas cell pila de gas
gas chromatography cromatografía de gases
gas chromatography infrared infrarrojo de cromatografía de gases
gas cleanup limpieza de gas
gas constant constante de gas
gas current corriente de gas
gas cylinder cilíndro de gas
gas density densidad de gas
gas detector detector de gases
gas discharge descarga gaseosa
gas electrode electrodo con gas
gas equation ecuación de gases
gas escape escape de gas
gas filter filtro de gas
gas flow flujo de gas
gas-flow control control de flujo de gas
gas generator generador de gas
gas hydrate hidrato de gas
gas inlet entrada de gas
gas-inlet valve válvula de entrada de gas
gas ionization ionización de gas
gas laws leyes de gases
gas leak fuga de gas
gas leakage fuga de gas
gas-liquid chromatography cromatografía gas-líquido
gas oil gasóleo
gas outlet salida de gas, toma de gas
gas pressure presión de gas
gas regulator regulador de gas
gas sniffer detector de gases
gas solubility solubilidad de gas
gas thermometer termómetro de gas

gaseous gaseoso
gaseous diffusion difusión gaseosa
gaseous discharge descarga gaseosa
gaseous mixture mezcla gaseosa
gaseous state estado gaseoso
gasification gasificación
gasohol gasohol
gasol gasol
gasoline gasolina
gasometric gasométrico
gasometric method método gasométrico
gasproof a prueba de gas
gassing desprendimiento de gas, absorción de gas, gaseamiento
Gastaldi synthesis síntesis de Gastaldi
gastric juice jugo gástrico
gastrin gastrina
Gattermann aldehyde synthesis síntesis de aldehídos de Gattermann
Gattermann-Koch reaction reacción de Gattermann-Koch
Gattermann-Koch synthesis síntesis de Gattermann-Koch
Gattermann reaction reacción de Gattermann
Gattermann synthesis síntesis de Gattermann
gauge medidor
gaultherin gaulterina
Gay-Lussac law ley de Gay-Lussac
Gay-Lussac tower torre de Gay-Lussac
gaylussite gailusita
gehlenite gehlenita
Geiger counter contador de Geiger
Geiger-Müller counter contador de Geiger-Müller
Geiger-Müller region región de Geiger-Müller
Geiger-Nutall rule regla de Geiger-Nutall
geissine geisina
gel gel
gel chromatography cromatografía de gel
gel electrophoresis electroforesis de gel
gel filtration filtración de gel
gel-permeation chromatography cromatografía de permeación en gel
gelatin gelatina

gelatin filter filtro de gelatina
gelatinization gelatinización
gelatinize gelatinizar
gelatinous gelatinoso
gelignite gelignita
gelling gelificación
gem gema, piedra preciosa
geminal geminal
geminal coupling acoplamiento geminal
gemstone piedra preciosa
genalkaloid genalcaloide
gene gen
general formula fórmula general
generate generar
generation generación
generation time tiempo de generación
generic genérico
generic formula fórmula genérica
generic name nombre genérico
genetic genético
Geneva system sistema de Ginebra
genistin genistina
gentamicin gentamicina
gentiobiose gentiobiosa
gentisic acid ácido gentísico
geochemistry geoquímica
geocronite geocronita
geometric conversion conversión geométrica
geometric isomer isómero geométrico
geometrical isomerism isomerismo geométrico
geranial geranial
geranialdehyde geranialdehído
geranic acid ácido geránico
geraniol geraniol
geranyl geranilo
geranyl acetate acetato de geranilo
geranyl butyrate butirato de geranilo
geranyl formate formiato de geranilo
geranyl propionate propionato de geranilo
Gerard reagent reactivo de Gerard
germanate germanato
germane germano
germanic germánico
germanic chloride cloruro germánico
germanide germaniuro
germanite germanita

germanium germanio
germanium dibromide dibromuro de
  germanio
germanium dichloride dicloruro de
  germanio
germanium diiodide diyoduro de
  germanio
germanium dioxide dióxido de germanio
germanium disulfide disulfuro de
  germanio
germanium halide haluro de germanio
germanium hydride hidruro de
  germanio
germanium hydroxide hidróxido de
  germanio
germanium monoxide monóxido de
  germanio
germanium oxide óxido de germanio
germanium potassium fluoride fluoruro
  potásico de germanio
germanium sulfide sulfuro de germanio
germanium telluride telururo de
  germanio
germanium tetrabromide tetrabromuro
  de germanio
germanium tetrachloride tetracloruro de
  germanio
germanium tetrafluoride tetrafluoruro
  de germanio
germanium tetrahydride tetrahidruro de
  germanio
germanium tetraiodide tetrayoduro de
  germanio
germanoformic acid ácido
  germanofórmico
germanous germanoso
germicidal germicida
germicide germicida
germination germinación
getter rarefactor, depurador de gas
gibberellic acid ácido giberélico
gibberellin giberelina
Gibbs adsorption equation ecuación de
  adsorción de Gibbs
Gibbs adsorption isotherm isoterma de
  adsorción de Gibbs
Gibbs-Donnan equilibrium equilibrio de
  Gibbs-Donnan
Gibbs-Duhem equation ecuación de

Gibbs-Duhem
Gibbs free energy energía libre de Gibbs
Gibbs function función de Gibbs
Gibbs-Helmholtz equation ecuación de
  Gibbs-Helmholtz
Gibbs phase rule regla de fases de Gibbs
Gibbs-Poynting equation ecuación de
  Gibbs-Poynting
Gibbs rule regla de Gibbs
gibbsite gibbsita
gibrel gibrel
gigantolite gigatolita
Giles flask matraz de Giles
gilpinite gilpinita
gilsonite gilsonita
ginger oil aceite de jengibre
Girard reagent reactivo de Girard
gismondite gismondita
gitonin gitonina
gitoxin gitoxina
glacial glacial
glacial acetic acid ácido acético glacial
glacial ethanoic acid ácido etanoico
  glacial
gladiolic acid ácido gladiólico
Glaser coupling acoplamiento de Glaser
glaserite glaserita
glass vidrio
glass beads perlas de vidrio
glass-ceramic vidrio cerámico
glass electrode electrodo de vidrio
glass enamel esmalte de vidrio
glass fiber fibra de vidrio
glass stirrer agitador de vidrio
glass stopper tapón de vidrio
glass-transition temperature
  temperatura de transición de vidrio
glass tube tubo de vidrio
glass tubing tubería de vidrio
glass wool lana de vidrio
Glauber salt sal de Glauber
glauberite glauberita
glaucochroite glaucocroíta
glaucodot glaucodot
glauconite glauconita
glaucophane glaucofana
glaucopicrine glaucopicrina
glaze barniz
gliadin gliadina

globulin globulina
glove box caja con guantes
glow luminiscencia
glucagon glucagón
glucan glucano
glucase glucasa
glucinium glucinio
glucogen glucógeno
gluconate gluconato
gluconic acid ácido glucónico
glucopyranose glucopiranosa
glucosamine glucosamina
glucosazone glucosazona
glucose glucosa
glucose isomerase isomerasa de glucosa
glucose oxidase oxidasa de glucosa
glucosidase glucosidasa
glucoside glucósido
glucosin glucosina
glucuronic acid ácido glucurónico
glucuronidase glucuronidasa
glucuronide glucuronida
glucuronolactone glucuronolactona
glue cola
gluside glusida
glutaconic acid ácido glutacónico
glutamic acid ácido glutámico
glutamine glutamina
glutamoyl glutamoilo
glutamyl glutamilo
glutamylcysteinylglycine
    glutamilcisteinilglicina
glutaraldehyde glutaraldehído
glutaric glutárico
glutaric acid ácido glutárico
glutaric anhydride anhídrido glutárico
glutaronitrile glutaronitrilo
glutathione glutationa
glutelin glutelina
gluten gluten
glutenin glutenina
glutethimide glutetimida
glutinic acid ácido glutínico
glutose glutosa
glycal glical
glyceral gliceral
glyceraldehyde gliceraldehído
glycerate glicerato
glyceric glicérico

glyceric acid ácido glicérico
glyceric aldehyde aldehído glicérico
glyceride glicérido
glycerin glicerina
glycerin carbonate carbonato de
    glicerina
glycerinate glicerinato
glycerine glicerina
glycero glicero
glycerol glicerol, glicerina
glycerol diacetate diacetato de glicerol,
    diacetato de glicerina
glycerol dichlorohydrin diclorohidrina
    de glicerol
glycerol distearate diesterato de glicerol
glycerol monoacetate monoacetato de
    glicerol
glycerol monochlorohydrin
    monoclorohidrina de glicerol
glycerol monolaurate monolaurato de
    glicerol
glycerol monooleate monooleato de
    glicerol
glycerol monostearate monoestearato de
    glicerol
glycerol trinitrate trinitrato de glicerol
glycerol tristearate triestearato de
    glicerol
glycerophosphate glicerofosfato
glycerophosphoric acid ácido
    glicerofosfórico
glycerose glicerosa
glyceroyl gliceroilo
glyceryl glicerilo
glyceryl abietate abietato de glicerilo
glyceryl chloride cloruro de glicerilo
glyceryl diacetate diacetato de glicerilo
glyceryl monoacetate monoacetato de
    glicerilo
glyceryl monolaurate monolaurato de
    glicerilo
glyceryl phthalate ftalato de glicerilo
glyceryl triacetate triacetato de glicerilo
glyceryl trinitrate trinitrato de glicerilo
glyceryl tripalmitate tripalmitato de
    glicerilo
glyceryl tristearate triestearato de
    glicerilo
glycide glicida

glycidic acid ácido glicídico
glycidol glicidol
glycidyl glicidilo
glycidyl ether éter glicidílico
glycine glicina
glycinium glicinio
glyco glico
glycocholic acid ácido glicocólico
glycogen glicógeno, glucógeno
glycogenase glicogenasa
glycogenesis glicogénesis
glycogenic acid ácido glicogénico
glycol glicol
glycol acetate acetato de glicol
glycol bromohydrin bromohidrina de
    glicol
glycol carbonate carbonato de glicol
glycol dichloride dicloruro de glicol
glycol diformate diformiato de glicol
glycol dimercaptoacetate
    dimercaptoacetato de glicol
glycol dimethyl ether éter dimetílico de
    glicol
glycol dinitrate dinitrato de glicol
glycol ester éster glicólico
glycol ether éter glicólico
glycol monoacetate monoacetato de
    glicol
glycol propionate propionato de glicol
glycol stearate estearato de glicol
glycol sulfhydrate sulfhidrato de glicol
glycolal glicolal
glycolaldehyde glicolaldehído
glycolic acid ácido glicólico
glycolic aldehyde aldehído glicólico
glycolide glicólido
glycolipid glicolípido
glycolonitrile glicolonitrilo
glycoloyl glicoloilo
glycolyl glicolilo
glycolysis glicólisis
glyconic acid ácido glicónico
glyconitrile gliconitrilo
glycoprotein glicoproteína
glycosaminoglycan glicosaminoglicano
glycosidase glicosidasa
glycoside glicósido, glucósido
glycosidic glicosídico
glycosidic bond enlace glicosídico

glycosylation glicosilación
glycyl glicilo
glycyl alcohol alcohol glicílico
glycylglycine glicilglicina
glyme glime
glyoxal glioxal
glyoxalase glioxalasa
glyoxalic acid ácido glioxálico
glyoxaline glioxalina
glyoxime glioxima
glyoxylase glioxilasa
glyoxylate cycle ciclo de glioxilato
glyoxylic acid ácido glioxílico
glyphtal gliftal
gmelinite gmelinita
gneiss gneis
gnoscopine gnoscopina
goethite goetita
goggles gafas de protección
Gogte synthesis síntesis de Gogte
gold oro
gold alloy aleación de oro
gold bromide bromuro de oro
gold bronze bronce de oro
gold chloride cloruro de oro
gold coating revestimiento de oro
gold cyanide cianuro de oro
gold dichloride dicloruro de oro
gold doping dopado con oro
gold foil hoja de oro
gold halide haluro de oro
gold hydroxide hidróxido de oro
gold iodide yoduro de oro
gold leaf hoja de oro, pan de oro
gold monobromide monobromuro de oro
gold monochloride monocloruro de oro
gold monoiodide monoyoduro de oro
gold number número de oro
gold oxide óxido de oro
gold pentafluoride pentafluoruro de oro
gold-plated dorado
gold-plating dorado
gold salt sal de oro
gold sulfide sulfuro de oro
gold tribromide tribromuro de oro
gold trichloride tricloruro de oro
gold tricyanide tricianuro de oro
gold triiodide triyoduro de oro
gold trioxide trióxido de oro

gold trisulfide trisulfuro de oro
Goldschmidt law ley de Goldschmidt
Goldschmidt process proceso de Goldschmidt
Goldschmidt reaction reacción de Goldschmidt
Gomberg-Bachmann reaction reacción de Gomberg-Bachmann
Gomberg reaction reacción de Gomberg
Gooch crucible crisol de Gooch
goslarite goslarita
gossypol gosipol
Göulard extract extracto de Göulard
Gould-Jacobs reaction reacción de Gould-Jacobs
Gouy-Chapman model modelo de Gouy-Chapman
Gouy layer capa de Gouy
grade grado
gradient gradiente
grading gradación
gradual gradual
graduate cilindro graduado
graduated graduado
graduated buret bureta graduada
graduated cylinder cilindro graduado
graduated pipet pipeta graduada
graduation graduación
Graebe-Ullman reaction reacción de Graebe-Ullman
Graebe-Ullman synthesis síntesis de Graebe-Ullman
graft copolymer copolímero con injerto
graft polymer polímero con injerto
grafting injerto
Graham law ley de Graham
Graham salt sal de Graham
grain alcohol alcohol de grano
gram gramo
gram-atom átomo-gramo
gram-atomic weight peso atómico-gramo
gram-equivalent equivalente-gramo
gram-equivalent weight peso equivalente-gramo
gram-ion ion-gramo
gram-molecular solution solución molecular-gramo
gram-molecular volume volumen molecular-gramo

gram-molecular weight peso molecular-gramo
gram-molecule molécula-gramo
gram-negative gramnegativo
gram-positive grampositivo
gramicidin gramicidina
gramine gramina
granite granito
granular granular
granularity granulosidad
granulation granulación
granule gránulo
granulose granulosa
grape sugar azúcar de uva
graphic formula fórmula gráfica
graphite grafito
graphite fiber fibra de grafito
graphitic acid ácido grafítico
graphitic oxide óxido grafítico
graphitization grafitización
grating rejilla, retículo
gravimetric gravimétrico
gravimetric analysis análisis gravimétrico
gravimetry gravimetría
gray gray
grease grasa
green salt sal verde
green vitriol vitriolo verde
greenhouse effect efecto de invernadero
greenockite greenockita
grid rejilla
grid spectrometer espectrómetro de rejilla
Griess reaction reacción de Griess
Griess reagent reactivo de Griess
Grignard degradation degradación de Grignard
Grignard reaction reacción de Grignard
Grignard reagent reactivo de Grignard
Grignard synthesis síntesis de Grignard
grind moler
grinding molido
griphite grifita
Grob fragmentation reaction reacción de fragmentación de Grob
grossularite grosularita
Grotrian diagram diagrama de Grotrian
Grottius-Draper law ley de Grottius-

Draper
ground molido, esmerilado, fundamental
ground glass vidrio esmerilado
ground-glass neck cuello de vidrio
  esmerilado
ground state estado fundamental
group grupo
group of elements grupo de elementos
group properties propiedades de grupo
Grove synthesis síntesis de Grove
grown crystal cristal cultivado
Grundmann synthesis síntesis de
  Grundmann
grunerite grunerita
gryolite griolita
guaiacol guayacol
guaiacol acetate acetato de guayacol
guaiol guaiol
guanamine guanamina
guanase guanasa
guanazyl guanacilo
guanidine guanidina
guanidine carbonate carbonato de
  guanidina
guanidine hydrochloride clorhidrato de
  guanidina
guanidine nitrate nitrato de guanidina
guanidine thiocyanate tiocianato de

guanidina
guanidinium guanidinio
guanidino guanidino
guanido guanido
guanine guanina
guano guano
guanosine guanosina
guanosine monophosphate monofosfato
  de guanosina
guanylic acid ácido guanílico
guanylurea sulfate sulfato de guanilurea
guar gum goma guar
Guareschi-Thorpe condensation
  condensación de Guareschi-Thorpe
Guerbet reaction reacción de Guerbet
Guldberg rule regla de Guldberg
gum acacia goma de acacia
gum arabic goma arábiga
gun metal bronce de cañón
guncotton algodón polvora, pólvora de
  algodón
gunpowder pólvora
Günzberg reagent reactivo de Günzberg
gutta-percha gutapercha
gynocardine ginocardina
gypsum yeso
gyrolite girolita
gyromagnetic ratio razón giromagnética

H

Haber process proceso de Haber
habit hábito
hafnium hafnio
hafnium boride boruro de hafnio
hafnium carbide carburo de hafnio
hafnium disulfide disulfuro de hafnio
hafnium hydroxide hidróxido de hafnio
hafnium nitride nitruro de hafnio
hafnium oxide óxido de hafnio
hafnium sulfate sulfato de hafnio
Haggenmacher equation ecuación de
    Haggenmacher
halazone halazona
half-cell semicelda, semipila
half-life media vida
half-reaction media reacción
halide haluro
halite halita
Hall effect efecto de Hall
Hall process proceso de Hall
Haller-Bauer reaction reacción de
    Haller-Bauer
halloysite halloysita
hallucinogen alucinógeno
halo halo
haloalkane haloalcano
haloamine haloamina
halocarbon halocarburo
halocarbon plastic plástico de
    halocarburo
halocarbon resin resina de halocarburo
halochromism halocromismo
haloform haloformo
halogen halógeno
halogen counter contador con halógeno
halogen lamp lámpara de halógeno
halogenate halogenar
halogenated halogenado
halogenating agent agente halogenante
halogenation halogenación
halogenide halogeniuro
halogenous halógeno
halohydrin halohidrina

haloid haloide
halonium halonio
haloperidol haloperidol
halothane halotano
Halphen reagent rectivo de Halphen
hamartite hamartita
Hammett equation ecuación de Hammett
Hammick-Illingworth rules reglas de
    Hammick-Illingworth
Hammick reaction reacción de
    Hammick
handedness asimetría, quiralidad
hanksite hanksita
Hantach pyrrole synthesis síntesis de
    pirroles de Hantach
Hantzsch pyridine synthesis síntesis de
    piridinas de Hantzsch
Hanus reagent reactivo de Hanus
Hanus solution solución de Hanus
hapto hapto
hard duro
hard acid ácido duro
hard base base dura
hard brass latón duro
hard bronze bronce duro
hard water agua dura
hardener endurecedor
hardening endurecimiento
hardness dureza
hardness of water dureza de agua
Hargreaves process proceso de
    Hargreaves
harmful radiation radiación dañina
harmful substance substancia dañina
harmine harmina
harmonic oscillator oscilador armónico
harmotome harmotoma
Harries ozonide reaction reacción de
    ozónidos de Harries
hartite hartita
hashish hachís
Hass chlorination rules reglas de
    clorinación de Hass

117

hausmannite hausmannita
hauyne hauyna
Haworth methylation metilación de Haworth
Haworth projection proyección de Haworth
Haworth synthesis síntesis de Haworth
Hayashi rearrangement reordenamiento de Hayasi
hazard warning advertencia de peligro
hazardous peligroso, perjudicial
hazardous area área peligrosa
hazardous chemical producto químico peligroso
hazardous environment ambiente peligroso
hazardous material material peligroso
hazardous substance substancia peligrosa
hazardous waste desperdicios peligrosos
heat calor
heat absorption absorción térmica, absorción de calor
heat balance equilibrio térmico
heat capacity capacidad térmica
heat conductivity conductividad térmica
heat content contenido térmico
heat cycle ciclo térmico
heat detector detector térmico
heat dissipation disipación térmica
heat effect efecto térmico
heat energy energía térmica
heat equivalent equivalente térmico
heat exchange intercambio térmico
heat exchanger intercambiador térmico
heat flow flujo térmico
heat index índice térmico
heat insulation termoaislamiento
heat of absorption calor de absorción
heat of activation calor de activación
heat of adhesion calor de adhesión
heat of adsorption calor de adsorción
heat of aggregation calor de agregación
heat of association calor de asociación
heat of atomization calor de atomización
heat of combination calor de combinación
heat of combustion calor de combustión
heat of compression calor de compresión

heat of condensation calor de condensación
heat of cooling calor de enfriamiento
heat of crystallization calor de cristalización
heat of decomposition calor de descomposición
heat of dilution calor de dilución
heat of dissociation calor de disociación
heat of dissolution calor de disolución
heat of emission calor de emisión
heat of evaporation calor de evaporación
heat of explosion calor de explosión
heat of formation calor de formación
heat of fuel calor de combustible
heat of fusion calor de fusión
heat of hydration calor de hidratación
heat of ionization calor de ionización
heat of isomerization calor de isomerización
heat of neutralization calor de neutralización
heat of oxidation calor de oxidación
heat of racemization calor de racemización
heat of reaction calor de reacción
heat of solidification calor de solidificación
heat of solution calor de solución
heat of sublimation calor de sublimación
heat of transition calor de transición
heat of vaporization calor de vaporización
heat transfer transferecia térmica
heating calentamiento, caldeo
heavy pesado
heavy alloy aleación pesada
heavy atom átomo pesado
heavy chemical producto químico pesado
heavy electron electrón pesado
heavy element elemento pesado
heavy hydrogen hidrógeno pesado
heavy isotope isótopo pesado
heavy metal metal pesado
heavy nucleus núcleo pesado
heavy oil aceite pesado
heavy oxygen oxígeno pesado
heavy spar espato pesado
heavy water agua pesada

hectorite hectorita
hedenbergite hedenbergita
hedonal hedonal
Hehner number número de Hehner
Heisenberg uncertainty principle
    principio de incertidumbre de
    Heisenberg
helcosol helcosol
helenine helenina
helenite helenita
Helferich method método de Helferich
helical helicoidal
helium helio
helium mass spectrometer
    espectrómetro de masa de helio
helium spectrometer espectrómetro de
    helio
helix hélice
Hell-Volhard-Zelinsky reaction reacción
    de Hell-Volhard-Zelinsky
Helmholtz equation ecuación de
    Helmholtz
Helmholtz free energy energía libre de
    Helmholtz
Helmholtz layer capa de Helmholtz
Helmholtz model modelo de Helmholtz
helvite helvita
hematein hemateína
hematin hematina
hematite hematita
hematoxylin hematoxilina
heme hem
hemel hemel
hemellitic acid ácido hemelítico
hemi hemi
hemiacetal hemiacetal
hemicellulose hemicelulosa
hemihedral hemihedral
hemimellitene hemimeliteno
hemin hemina
hemiquinonoid hemiquinonoide
hemlock cicuta
hemoglobin hemoglobina
hemp cáñamo
hempa hempa
hempseed oil aceite de cañamón
hendecane hendecano
Henderson-Hasselbach equation
    ecuación de Henderson-Hasselbach

heneicosane heneicosano
heneicosanoic acid ácido heneicosanoico
Henkel reaction reacción de Henkel
henna alheña
Henry law ley de Henry
Henry reaction reacción de Henry
hentriacontane hentriacontano
heparin heparina
hepta hepta
heptabarbital heptabarbital
heptachlor heptacloro
heptachlorepoxide heptaclorepóxido
heptacosane heptacosano
heptacosanoic acid ácido heptacosanoico
heptadecane heptadecano
heptadecanoic acid ácido heptadecanoico
heptadecanol heptadecanol
heptadecanone heptadecanona
heptadecylglyoxalidine
    heptadecilglioxalidina
heptadiene heptadieno
heptafluorobutyric acid ácido
    heptafluorobutírico
heptahydrate heptahidrato
heptaldehyde heptaldehído
heptalene heptaleno
heptamethyldisilazane
    heptametildisilazano
heptamethylene heptametileno
heptamethylnonane heptametilnonano
heptanal heptanal
heptane heptano
heptanedicarboxylic acid ácido
    heptanodicarboxílico
heptanedioic acid ácido heptanodioico
heptanedioyl heptanodioilo
heptanoate heptanoato
heptanoic acid ácido heptanoico
heptanol heptanol
heptanone heptanona
heptanoyl heptanoilo
heptanoyl chloride cloruro de heptanoilo
heptavalent heptavalente
heptene hepteno
heptenyl heptenilo
heptoic acid ácido heptoico
heptose heptosa
heptoxide heptóxido
heptyl heptilo

heptyl acetate acetato de heptilo
heptyl alcohol alcohol heptílico
heptyl formate formiato de heptilo
heptyl heptoate heptoato de heptilo
heptyl pelargonate pelargonato de
  heptilo
heptylamine heptilamina
heptylene heptileno
heptylic acid ácido heptílico
heptyne heptino
herapathite herapatita
herbicide herbicida
hercynite hercinita
herderite herderita
Hermann-Mauguin system sistema de
  Hermann-Mauguin
hermetic hermético
heroin heroína
herrerite herrerita
Herz compound compuesto de Herz
Herz reaction reacción de Herz
hesperidin hesperidina
Hess law ley de Hess
hessite hessita
hessonite hessonita
hetero hetero
heteroaromatic heteroaromático
heteroatomic heteroatómico
heteroatomic ring anillo heteroatómico
heteroauxin heteroauxina
heteroazeotrope heteroazeótropo
heterobaric heterobárico
heterocyclic heterocíclico
heterocyclic atom átomo heterocíclico
heterocyclic compound compuesto
  heterocíclico
heterogeneous heterogéneo
heterogeneous catalysis catálisis
  heterogénea
heterogeneous reaction reacción
  heterogénea
heterogeneous system sistema
  heterogéneo
heteroleptic heteroléptico
heterolysis heterólisis
heterolytic heterolítico
heterolytic reaction reacción heterolítica
heterometry heterometría
heteromolybdate heteromolibdato

heteronuclear heteronuclear
heteronuclear molecule molécula
  heteronuclear
heteropolar heteropolar
heteropolar bond enlace heteropolar
heteropoly heteropoli
heteropolymer heteropolímero
heterotope heterótopo
heterotopic heterotópico
heterotype heterotipo
heulandite heulandita
Heumann-Pfleger synthesis síntesis de
  Heumann-Pfleger
hexa hexa
hexaamine hexaamina
hexaborane hexaborano
hexabromo hexabromo
hexabromoethane hexabromoetano
hexacalcium phytate fitato hexacálcico
hexachloro hexacloro
hexachloroacetone hexacloroacetona
hexachlorobenzene hexaclorobenceno
hexachlorobutadiene hexaclorobutadieno
hexachlorocyclohexane
  hexaclorociclohexano
hexachlorocyclopentadiene
  hexaclorociclopentadieno
hexachlorodiphenyl oxide óxido de
  hexaclorodifenilo
hexachloroethane hexacloroetano
hexachloromethyl ether éter
  hexaclorometílico
hexachloromethylcarbonate
  hexaclorometilcarbonato
hexachloronaphthalene
  hexacloronaftaleno
hexachlorophene hexaclorofeno
hexachloroplatinate hexacloroplatinato
hexachloropropane hexacloropropano
hexachloropropylene hexacloropropileno
hexacontane hexacontano
hexacosane hexacosano
hexacosanoic acid ácido hexacosanoico
hexacosanol hexacosanol
hexacosyl hexacosilo
hexacosyl alcohol alcohol hexacosílico
hexacyano hexaciano
hexacyanogen hexacianógeno
hexadecanal hexadecanal

hexadecane hexadecano
hexadecanoate hexadecanoato
hexadecanoic acid ácido hexadecanoico
hexadecanol hexadecanol
hexadecanoyl hexadecanoilo
hexadecanoyl chloride cloruro de hexadecanoilo
hexadecene hexadeceno
hexadecenoic acid ácido hexadecenoico
hexadecenolide hexadecenolida
hexadecyl hexadecilo
hexadecyl alcohol alcohol hexadecílico
hexadecyl iodide yoduro de hexadecilo
hexadecyl mercaptan hexadecilmercaptano
hexadecyltrichlorosilane hexadeciltriclorosilano
hexadecyne hexadecino
hexadiene hexadieno
hexadienoic acid ácido hexadienoico
hexadiyne hexadiíno
hexaethyl tetraphosphate tetrafosfato de hexaetilo
hexafluoroacetone hexafluoroacetona
hexafluorobenzene hexafluorobenceno
hexafluorodisilane hexafluorodisilano
hexafluoroethane hexafluoroetano
hexafluorophosphoric acid ácido hexafluorofosfórico
hexafluoropropene hexafluoropropeno
hexafluoropropylene hexafluoropropileno
hexafluoropropylene epoxide epóxido de hexafluoropropileno
hexafluorosilicate hexafluorosilicato
hexafluorosilicic acid ácido hexafluorosilícico
hexaglycerol hexaglicerol
hexagon hexágono
hexagonal close packing estructura compacta hexagonal
hexagonal system sistema hexagonal
hexagonite hexagonita
hexahelicine hexahelicina
hexahydrate hexahidrato
hexahydric hexahídrico
hexahydric acid ácido hexahídrico
hexahydric alcohol alcohol hexahídrico
hexahydro hexahidro

hexahydroaniline hexahidroanilina
hexahydroanthracene hexahidroantraceno
hexahydrobenzene hexahidrobenceno
hexahydrobenzoic acid ácido hexahidrobenzoico
hexahydrocresol hexahidrocresol
hexahydromethyl hexahidrometilo
hexahydrophenol hexahidrofenol
hexahydrophthalic acid ácido hexahidroftálico
hexahydropyridine hexahidropiridina
hexahydrotoluene hexahidrotolueno
hexahydroxy hexahidroxi
hexahydroxycyclohexane hexahidroxiciclohexano
hexahydroxylene hexahidroxileno
hexaiodo hexayodo
hexaldehyde hexaldehído
hexalin hexalina
hexametaphosphate hexametafosfato
hexamethonium chloride cloruro de hexametonio
hexamethylbenzene hexametilbenceno
hexamethyldisilane hexametildisilano
hexamethyldisilazane hexametildisilazano
hexamethylene hexametileno
hexamethylene diisocyanate diisocianato de hexametileno
hexamethylene glycol hexametilenglicol
hexamethylenediamine hexametilendiamina
hexamethyleneimine hexametilenimina
hexamethylenetetramine hexametilentetramina
hexamethylmelamine hexametilmelamina
hexamine hexamina
hexanal hexanal
hexanaphthene hexanafteno
hexane hexano
hexanediamide hexanodiamida
hexanedinitrile hexanodinitrilo
hexanedioate hexanodioato
hexanedioic acid ácido hexanodioico
hexanediol hexanodiol
hexanedione hexanodiona
hexanelactam hexanolactama

hexanetriol hexanotriol
hexanitro hexanitro
hexanoate hexanoato
hexanoic acid ácido hexanoico
hexanol hexanol
hexanone hexanona
hexanoyl hexanoilo
hexanoyl chloride cloruro de hexanoilo
hexaphenyldisilane hexafenildisilano
hexaphenylethane hexafeniletano
hexaprismo hexaprismo
hexatriacontane hexatriacontano
hexenal hexenal
hexene hexeno
hexenic hexénico
hexenoic acid ácido hexenoico
hexenol hexenol
hexenyl hexenilo
hexestrol hexestrol
hexetidine hexetidina
hexobarbital hexobarbital
hexobarbitone hexobarbitona
hexoic hexoico
hexoic acid ácido hexoico
hexokinase hexoquinasa
hexone hexona
hexose hexosa
hexotriose hexotriosa
hexyl hexilo
hexyl acetate acetato de hexilo
hexyl alcohol alcohol hexílico
hexyl bromide bromuro de hexilo
hexyl chloride cloruro de hexilo
hexyl cinnamaldehyde cinamaldehído de
    hexilo
hexyl ether éter hexílico
hexyl formate formiato de hexilo
hexyl mercaptan hexilmercaptano
hexyl methacrylate metacrilato de hexilo
hexylacetic acid ácido hexilacético
hexylamine hexilamina
hexylene hexileno
hexylene glycol hexilenglicol
hexylic acid ácido hexílico
hexylphenol hexilfenol
hexylresorcinol hexilresorcina
hexyltrichlorosilane hexiltriclorosilano
hexyne hexino
hexynol hexinol

hiddenite hiddenita
hielmite hielmita
high definition alta definición
high density alta densidad
high energy alta energía
high-energy bond enlace de alta energía
high-energy neutron neutrón de alta
    energía
high-energy particle partícula de alta
    energía
high-performance liquid
    chromatography cromatografía
    líquida de alto rendimiento
high polymer alto polímero
high pressure alta presión
high temperature alta temperatura
high vacuum alto vacío
high-vacuum distillation destilación a
    alto vacío
higher level nivel superior
highest occupied molecular orbital
    orbital molecular de máxima energía
    ocupado
Hilbert-Johnson reaction reacción de
    Hilbert-Johnson
Hilbert space espacio de Hilbert
Hill reaction reacción de Hill
hindrance impedimento
Hinsberg reaction reacción de Hinsberg
Hinsberg synthesis síntesis de Hinsberg
Hinsberg test prueba de Hinsberg
hippurate hipurato
hippuric acid ácido hipúrico
hippuroyl hipuroilo
histaminase histaminasa
histamine histamina
histidine histidina
histochemistry histoquímica
histone histona
Hoch-Campbell synthesis síntesis de
    Hoch-Campbell
Hoesch synthesis síntesis de Hoesch
Hofmann amine separation separación
    de aminas de Hofmann
Hofmann degradation degradación de
    Hofmann
Hofmann isonitrile synthesis síntesis de
    isonitrilos de Hofmann
Hofmann-Martius rearrangement

reordenamiento de Hofmann-Martius
Hofmann reaction reacción de Hofmann
Hofmann rule regla de Hofmann
Hofmann-Sand reaction reacción de
Hofmann-Sand
Hofmann transformation
transformación de Hofmann
Hofmeister series serie de Hofmeister
hogbomite hogbomita
hole hueco, abertura
holmium holmio
holmium chloride cloruro de holmio
holmium fluoride fluoruro de holmio
holmium oxide óxido de holmio
holmium oxalate oxalato de holmio
holocellulose holocelulosa
holography holografía
holohedral holohedral
homatropine homatropina
homilite homilita
homo homo
homoatomic ring anillo homoatómico
homocentric homocéntrico
homochromic homocrómico
homocyclic homocíclico
homogeneous homogéneo
homogeneous catalysis catálisis
homogénea
homogeneous combustion combustión
homogénea
homogeneous reaction reacción
homogénea
homogeneous series serie homogénea
homogeneous system sistema homogéneo
homogenization homogenización
homogentisic acid ácido homogentísico
homoisohydric homoisohídrico
homoleptic homoléptico
homologous series serie homóloga
homolysis homólisis
homolytic fission fisión homolítica
homolytic reaction reacción homolítica
homomorph homomorfo
homonuclear homonuclear
homonuclear molecule molécula
homonuclear
homophthalic acid ácido homoftálico
homopolar homopolar
homopolar adsorption adsorción

homopolar
homopolar bond enlace homopolar
homopolar crystal cristal homopolar
homopolymer homopolímero
homosalate homosalato
homosalicylic acid ácido homosalicílico
homotope homótopo
honey miel
hood alacena protectora, campana
Hooker reaction reacción de Hooker
hopcalite hopcalita
Hopkins-Cole reaction reacción de
Hopkins-Cole
hops lúpulo
horizontal chromatography
cromatografía horizontal
hormone hormona
hornblende hornablenda
horsfordite horsfordita
Hortvet cryoscope crioscopio de Hortvet
Hortvet sublimator sublimador de
Hortvet
Houben-Fischer synthesis síntesis de
Houben-Fischer
Houben-Hoesch reaction reacción de
Houben-Hoesch
Houben-Hoesch synthesis síntesis de
Houben-Hoesch
Houdriflow process proceso de
Houdriflow
Houdry process proceso de Houdry
howlite howlita
Huber reagent reactivo de Huber
Hübl reagent reactivo de Hübl
Hübl solution solución de Hübl
hübnerite hubnerita
Hückel approximation aproximación de
Hückel
Hückel rule regla de Hückel
Hudson isorotation rules reglas de
isorrotación de Hudson
Hudson lactone rule regla de lactonas de
Hudson
Hume-Rothery rule regla de Hume-
Rothery
humectant humectante
humic acid ácido húmico
humidification humidificación
humidimeter humidímetro

humidity humedad
humidity detector detector de humedad
humidity indicator indicador de
    humedad
humidity meter humidímetro
humidity-proof a prueba de humedad
humidity protection protección contra
    humedad
humidity sensor sensor de humedad
humus humus
Hund rules reglas de Hund
Hunsdiecker reaction reacción de
    Hunsdiecker
Hunter process proceso de Hunter
hyalophane hialofano
hyalosiderite hialosiderita
hyaluronic acid ácido hialurónico
hyaluronidase hialuronidasa
hybrid orbital orbital híbrido
hybridization hibridización
hydantoic acid ácido hidantoico
hydantoin hidantoína
hydnocarpic acid ácido hidnocárpico
hydracrylic acid ácido hidracrílico
hydralazine hydrochloride clorhidrato
    de hidralazina
hydramine hidramina
hydrargillite hidrargillita
hydrase hidrasa
hydrastine hidrastina
hydrate hidrato
hydrated hidratado
hydrated alumina alúmina hidratada
hydrated aluminum oxide óxido
    alumínico hidratado
hydrated lime cal hidratada
hydrated silica sílice hidratada
hydration hidratación
hydratropic acid ácido hidratrópico
hydraulic fluid fluido hidráulico
hydraulic fracturing fracturación
    hidráulica
hydrazi hidrazi
hydrazide hidrazida
hydrazidine hidrazidina
hydrazido hidrazido
hydrazine hidrazina
hydrazine chloride cloruro de hidrazina
hydrazine dihydrochloride diclorhidrato

de hidrazina
hydrazine formate formiato de hidrazina
hydrazine hydrate hidrato de hidrazina
hydrazine monobromide monobromuro
    de hidrazina
hydrazine monochloride monocloruro de
    hidrazina
hydrazine nitrate nitrato de hidrazina
hydrazine perchlorate perclorato de
    hidrazina
hydrazine sulfate sulfato de hidrazina
hydrazine tartrate tartrato de hidrazina
hydrazinium hidrazinio
hydrazino hidrazino
hydrazinobenzene hidrazinobenceno
hydrazo hidrazo
hydrazobenzene hidrazobenceno
hydrazoic acid ácido hidrazoico
hydrazone hidrazona
hydrazono hidrazono
hydrazotoluene hidrazotolueno
hydrazyl hidracilo
hydride hidruro
hydridic bridge puente hidrídico
hydrido hidrido
hydrin hidrina
hydriodic acid ácido yodhídrico
hydro hidro
hydroabietyl alcohol alcohol
    hidroabietílico
hydrobenzoin hidrobenzoína
hydrobiotite hidrobiotita
hydroboration hidroboración
hydrobromic acid ácido bromhídrico
hydrobromide bromhidrato
hydrocarbon hidrocarburo
hydrocarbon gas gas de hidrocarburos
hydrocarbon resin resina de
    hidrocarburos
hydrocarbon series serie de
    hidrocarburos
hydrocellulose hidrocelulosa
hydrochloric clorhídrico
hydrochloric acid ácido clorhídrico
hydrochloric ether éter clorhídrico
hydrochloride clorhidrato
hydrocinnamic acid ácido hidrocinámico
hydrocinnamic alcohol alcohol
    hidrocinámico

hydrocinnamic aldehyde aldehído hidrocinámico
hydrocinnamyl acetate acetato de hidrocinamilo
hydrocolloid hidrocoloide
hydrocortisone hidrocortisona
hydrocracking hidrocraqueo
hydrocyanic acid ácido cianhídrico
hydrocyanide hidrocianuro
hydrodealkylation hidrodesalquilación
hydrodistillation hidrodestilación
hydrodynamic radius radio hidrodinámico
hydroextractor hidroextractor
hydrofining hidrorrefinación
hydroflumethiazide hidroflumetiazida
hydrofluoric acid ácido fluorhídrico
hydrofluorosilicic acid ácido hidrofluorosilícico
hydrofluosilicic acid ácido hidrofluosilícico
hydroforming hidroformación
hydroformylation hidroformilación
hydrofuramide hidrofuramida
hydrogasification hidrogasificación
hydrogel hidrogel
hydrogen hidrógeno
hydrogen acceptor aceptor de hidrógeno
hydrogen atmosphere atmósfera de hidrógeno
hydrogen atom átomo de hidrógeno
hydrogen azide azida de hidrógeno
hydrogen bond enlace de hidrógeno
hydrogen bromide bromuro de hidrógeno
hydrogen carbonate carbonato de hidrógeno
hydrogen carrier portadora de hidrógeno
hydrogen chloride cloruro de hidrógeno
hydrogen cyanide cianuro de hidrógeno
hydrogen dioxide dióxido de hidrógeno
hydrogen disulfide disulfuro de hidrógeno
hydrogen electrode electrodo de hidrógeno
hydrogen equivalent equivalente de hidrógeno
hydrogen fluoride fluoruro de hidrógeno
hydrogen hexafluorosilicate hexafluorosilicato de hidrógeno
hydrogen inlet entrada de hidrógeno
hydrogen iodide yoduro de hidrógeno
hydrogen ion ion hidrógeno
hydrogen-ion concentration concentración de ion hidrógeno
hydrogen-ion determination determinación de ion hidrógeno
hydrogen-ion indicator indicador de ion hidrógeno
hydrogen-ion meter medidor de ion hidrógeno
hydrogen line línea de hidrógeno
hydrogen oxide óxido de hidrógeno
hydrogen peroxide peróxido de hidrógeno
hydrogen persulfide persulfuro de hidrógeno
hydrogen phosphide fosfuro de hidrógeno
hydrogen selenide seleniuro de hidrógeno
hydrogen spectrum espectro de hidrógeno
hydrogen sulfate sulfato de hidrógeno
hydrogen sulfide sulfuro de hidrógeno
hydrogen tellurate telurato de hidrógeno
hydrogen telluride teluluro de hidrógeno
hydrogenate hidrogenar
hydrogenated hidrogenado
hydrogenated oil aceite hidrogenado
hydrogenation hidrogenación
hydrogenolysis hidrogenólisis
hydrohematite hidrohematita
hydrol hidrol
hydrolase hidrolasa
hydroliquefaction hidrolicuefacción
hydrolized hidrolizado
hydrolized protein proteína hidrolizada
hydrolysis hidrólisis
hydrolyte hidrólito
hydrolytic hidrolítico
hydrolytic dissociation disociación hidrolítica
hydrolyze hidrolizar
hydromagnesite hidromagnesita
hydrometer hidrómetro
hydrone hidrona
hydronitric acid ácido hidronítrico

hydronium ion ion hidronio
hydroperoxide hidroperóxido
hydrophile hidrófilo
hydrophilic hidrofílico
hydrophilic colloid coloide hidrofílico
hydrophilite hidrofilita
hydrophobic hidrofóbico
hydrophobic bonding enlazado
    hidrofóbico
hydrophobic colloid coloide hidrofóbico
hydropolysulfide hidropolisulfuro
hydroponics hidropónica
hydroquinol hidroquinol
hydroquinone hidroquinona
hydroquinone benzyl ether éter
    bencílico de hidroquinona
hydroquinone diethyl ether éter dietílico
    de hidroquinona
hydroquinone dimethyl ether éter
    dimetílico de hidroquinona
hydroquinone monoethyl ether éter
    monoetílico de hidroquinona
hydroseleno hidroseleno
hydrosol hidrosol
hydrosolvation hidrosolvatación
hydrostatic hidrostático
hydrosulfide hidrosulfuro
hydrosulfite hidrosulfito
hydrosulfite-formaldehyde hidrosulfito-
    formaldehído
hydrosulfuric acid ácido hidrosulfúrico
hydrosulfurous acid ácido
    hidrosulfuroso
hydrotetrazone hidrotetrazona
hydrotrope hidrótropo
hydrous hidratado
hydroxamic acid ácido hidroxámico
hydroxide hidróxido
hydroximic acid ácido hidroxímico
hydroxo hidroxo
hydroxonium hidroxonio
hydroxy hidroxi
hydroxyacetal hidroxiacetal
hydroxyacetanilide hidroxiacetanilida
hydroxyacetic acid ácido hidroxiacético
hydroxyacetone hidroxiacetona
hydroxyacetophenone
    hidroxiacetofenona
hydroxyadipaldehyde

hidroxiadipaldehído
hydroxyalanine hidroxialanina
hydroxyamino hidroxiamino
hydroxyaniline hidroxianilina
hydroxyanisole hidroxianisol
hydroxyanthracene hidroxiantraceno
hydroxyapatite hidroxiapatita
hydroxyazobenzenesulfonic acid ácido
    hidroxiazobencenosulfónico
hydroxybenzaldehyde
    hidroxibenzaldehído
hydroxybenzamide hidroxibenzamida
hydroxybenzene hidroxibenceno
hydroxybenzoic acid ácido
    hidroxibenzoico
hydroxybenzophenone
    hidroxibenzofenona
hydroxybenzyl alcohol alcohol
    hidroxibencílico
hydroxybutanal hidroxibutanal
hydroxybutyraldehyde
    hidroxibutiraldehído
hydroxybutyranilide hidroxibutiranilida
hydroxybutyric acid ácido
    hidroxibutírico
hydroxycamphane hidroxicanfano
hydroxycarotene hidroxicaroteno
hydroxycerussite hidroxicerusita
hydroxycholine hidroxicolina
hydroxycholestane hidroxicolestano
hydroxycitronellal hidroxicitronelal
hydroxycobalamin hidroxicobalamina
hydroxycorticosterone
    hidroxicorticosterona
hydroxydimethylbenzene
    hidroxidimetilbenceno
hydroxydiphenyl hidroxidifenilo
hydroxydiphenylamine
    hidroxidifenilamina
hydroxyethanesulfonic acid ácido
    hidroxietanosulfónico
hydroxyethyl carbamate carbamato de
    hidroxietilo
hydroxyethyl methacrylate metacrilato
    de hidroxietilo
hydroxyethyl piperazine piperazina de
    hidroxietilo
hydroxyethylacetamide
    hidroxietilacetamida

hydroxyethylamine hidroxietilamina
hydroxyethylcellulose hidroxietilcelulosa
hydroxyethylethylenediamine
   hidroxietiletilendiamina
hydroxyethylhydrazine
   hidroxietilhidrazina
hydroxyfenchane hidroxifencano
hydroxyhydrazide hidroxihidrazida
hydroxyimino hidroxiimino
hydroxyl hidroxilo
hydroxyl ion ion hidroxilo
hydroxylamine hidroxilamina
hydroxylamine acid sulfate sulfato ácido
   de hidroxilamina
hydroxylamine hydrochloride
   clorhidrato de hidroxilamina
hydroxylamine sulfate sulfato de
   hidroxilamina
hydroxylammonium chloride cloruro de
   hidroxilamonio
hydroxylation hidroxilación
hydroxymercuricresol
   hidroximercuricresol
hydroxymesitylene hidroximesitileno
hydroxymethane hidroximetano
hydroxymethyl hidroximetilo
hydroxymethylene hidroximetileno
hydroxymethylene carbonate carbonato
   de hidroximetileno
hydroxymethylfuraldehyde
   hidroximetilfuraldehído
hydroxynaphthalene hidroxinaftaleno
hydroxynaphthoic acid ácido
   hidroxinaftoico
hydroxynaphthoic anilide anilida
   hidroxinaftoica
hydroxynaphthoquinone
   hidroxinaftoquinona
hydroxyoleic acid ácido hidroxioleico
hydroxyphenol hidroxifenol
hydroxyphenyl hidroxifenilo
hydroxyphenylalanine
   hidroxifenilalanina
hydroxyphenylmercuric chloride
   cloruro hidroxifenilmercúrico
hydroxypiperidine hidroxipiperidina
hydroxyprogesterone
   hidroxiprogesterona
hydroxyproline hidroxiprolina

hydroxypropanone hidroxipropanona
hydroxypropionic acid ácido
   hidroxipropiónico
hydroxypropionitrile
   hidroxipropionitrilo
hydroxypropyl acrylate acrilato de
   hidroxipropilo
hydroxypropyl cellulose
   hidroxipropilcelulosa
hydroxypropyl methylcellulose
   hidroxipropilmetilcelulosa
hydroxypropylamine hidroxipropilamina
hydroxyquinoline hidroxiquinolina
hydroxyquinoline benzoate benzoato de
   hidroxiquinolina
hydroxyquinoline sulfate sulfato de
   hidroxiquinolina
hydroxysalicylic acid ácido
   hidroxisalicílico
hydroxystearic acid ácido
   hidroxiesterárico
hydroxystearyl alcohol alcohol
   hidroxiestearílico
hydroxysuccinic acid ácido
   hidroxisuccinico
hydroxytoluene hidroxitolueno
hydroxytoluic acid ácido hidroxitoluico
hydroxytriacontane hidroxitriacontano
hydroxytryptamine hidroxitriptamina
hydrozincite hidrocincita
hygric acid ácido hígrico
hygrometer higrómetro
hygrometric higrométrico
hygromycin higromicina
hygroscopic higroscópico
hyoscine hioscina
hyoscyamine hiosciamina
hyper hiper
hyperchromic hipercrómico
hyperconjugation hiperconjugación
hyperfine structure estructura hiperfina
hyperon hiperón
hyperoxide hiperóxido
hypersorption hipersorción
hypertonic hipertónico
hypertonic solution solución hipertónica
hypo hipo
hypoallergenic hipoalergénico
hypobromate hipobromato

hypobromite hipobromito
hypobromous acid ácido hipobromoso
hypochlorite hipoclorito
hypochlorous acid ácido hipocloroso
hypochromic hipocrómico
hypofluorite hipofluorita
hypoiodate hipoyodato
hyponitrile hiponitrilo
hyponitrous acid ácido hiponitroso
hypophamine hipofamina
hypophosphate hipofosfato
hypophosphite hipofosfito

hypophosphoric acid ácido hipofosfórico
hypophosphorous acid ácido hipofosforoso
hyposulfate hiposulfato
hyposulfite hiposulfito
hyposulfuric acid ácido hiposulfúrico
hyposulfurous acid ácido hiposulfuroso
hypothesis hipótesis
hypotonic hipotónico
hypotonic solution solución hipotónica
hypoxanthine hipoxantina
hysteresis histéresis

I

ibogaine ibogaína
ibuprofen ibuprofeno
ic ico
ice hielo
ice point punto de congelación
Iceland moss liquen islándico
Iceland spar espato de Islandia
ichthammol ictamol
ichthyocolla ictiocola
ichthyolate ictiolato
icosa icosa
icosahedron icosaedro
icosane icosano
icosanoic acid ácido icosanoico
icosanol icosanol
icosyl icosilo
ide uro, iuro, ido
ideal crystal cristal ideal
ideal gas gas ideal
ideal solution solución ideal
idene ideno
identification limit límite de
    identificación
identity period periodo de identidad
idiochromatic idiocromático
idose idosa
idyne idino
ignition ignición
ignition point punto de ignición
ignition temperature temperatura de
    ignición
Ilkovic equation ecuación de Ilkovic
illinium ilinio
ilmenite ilmenita
ilvaite ilvaíta
imazalil imazalil
imbibition imbibición
imidazole imidazol
imidazoletrione imidazoltriona
imidazolidone imidazolidona
imidazolium imidazolio
imidazolyl imidazolilo
imide imida

imidic acid ácido imídico
imidine imidina
imido imido
imidoyl imodoilo
imine imina
iminio iminio
iminium iminio
imino imino
imino acid iminoácido
iminobispropylamine
    iminobispropilamina
iminodiacetic acid ácido iminodiacético
iminodiacetonitrile iminodiacetonitrilo
iminourea iminourea
immersion inmersión
immiscible inmiscible
immunochemistry inmunoquímica
immunoglobulin inmunoglobulina
impact impacto
impalpable impalpable
impeller impulsor
imperial green verde imperial
implosion implosión
impregnate impregnar
impregnation impregnación
impurity impureza
in vitro in vitro
in vivo in vivo
inactive inactivo
incandescence incandescencia
incandescent incandescente
incendiary incendiario
incendiary gel gel incendiario
incidence incidencia
incineration incineración
inclusion complex complejo de inclusión
inclusion compound compuesto de
    inclusión
incompatible incompatible
incomplete equilibrium equilibrio
    incompleto
incomplete reaction reacción incompleta
incompressible incompresible

incompressible volume volumen
incompresible
incubation incubación
indamine indamina
indan indano
indandione indandiona
indanthrene indantreno
indanthrone indantrona
indanyl indanilo
indazole indazol
indene indeno
indenone indenona
indenyl indenilo
indeterminate indeterminado
index índice
Indian red rojo de la India
indican indicán
indicator indicador
indicator paper papel indicador
indicator range intervalo de indicador
indigo índigo, añil
indigo blue azul de índigo
indigo red rojo de índigo
indigotin indigotina
indirect dye colorante indirecto
indium indio
indium acetylacetonate acetilacetonato
de indio
indium antimonide antimoniuro de indio
indium arsenide arseniuro de indio
indium bromide bromuro de indio
indium chloride cloruro de indio
indium dichloride dicloruro de indio
indium hydrogensulfide hidrogenosufuro
de indio
indium hydroxide hidróxido de indio
indium nitrate nitrato de indio
indium oxide óxido de indio
indium phosphide fosfuro de indio
indium sulfate sulfato de indio
indium sulfide sulfuro de indio
indium telluride teluluro de indio
indium trichloride tricloruro de indio
indogen indógeno
indogenide indogenuro
indole indol
indoleacetic acid ácido indolacético
indolebutyric acid ácido indolbutírico
indolizine indolizina

indolol indolol
indolyl indolilo
indomethacin indometacina
indophenine reaction reacción de
indofenina
indophenol indofenol
indoxyl indoxilo
induced emission emisión inducida
induced fission fisión inducida
induced reaction reacción inducida
induction inducción
induction period periodo de inducción
inductive effect efecto inductivo
indurite indurita
industrial industrial
industrial alcohol alcohol industrial
industrial carbon carbón industrial
industrial chemistry química industrial
industrial diamond diamante industrial
industrial waste desperdicios industriales
indyl indilo
ine ina
inelastic inelástico
inelastic collision colisión inelástica
inert inerte
inert atmosphere box caja con atmósfera
inerte
inert complex complejo inerte
inert gas gas inerte
inert substance substancia inerte
infinite dilution dilución infinita
inflame inflamar
inflamer inflamador
inflammability inflamabilidad
inflammable inflamable
inflammableness inflamabilidad
infrared infrarrojo
infrared absorption spectrum espectro
de absorción de infrarrojo
infrared detector detector de infrarrojo
infrared microscope microscopio de
infrarrojo
infrared radiation radiación infrarroja
infrared source fuente de infrarrojo
infrared spectrometer espectrómetro de
infrarrojo
infrared spectroscopy espectroscopia de
infrarrojo
infrared spectrum espectro de infrarrojo

infusion infusión
ingot iron hierro de lingote
inhibition inhibición
inhibitor inhibidor
initial boiling point punto de ebullición inicial
initiating explosive explosivo iniciador
initiator iniciador
ink tinta
inlet entrada
ino ino
inorganic inorgánico
inorganic acid ácido inorgánico
inorganic chemistry química inorgánica
inorganic polymer polímero inorgánico
inosamine inosamina
inosine inosina
inosinic acid ácido inosínico
inositol inositol
insecticide insecticida
insertion reaction reacción de inserción
insolubility insolubilidad
insoluble insoluble
insolubleness insolubilidad
instrument instrumento
instrumentation instrumentación
insulator aislador
insulin insulina
intensity intensidad
intensive properties propiedades intensivas
interatomic interatómico
interatomic distance distancia interatómica
interatomic space espacio interatómico
intercalation compound compuesto de intercalación
interdiffusion interdifusión
interelectrode interelectródico
interface interfase
interfacial interfacial
interfacial film película interfacial
interfacial surface energy energía superficial interfacial
interference interferencia
interferon interferón
interhalogen interhalógeno
interionic interiónico
interionic attraction atracción

interiónica
intermediate producto intermedio
intermediate bond enlace intermedio
intermetallic intermetálico
intermetallic compound compuesto intermetálico
intermolecular intermolecular
intermolecular force fuerza intermolecular
internal interno
internal compensation compensación interna
internal conversion conversión interna
internal energy energía interna
internal phase fase interna
internal reaction reacción interna
internal symmetry simetría interna
international system sistema internacional
international unit unidad internacional
interstice intersticio
interstitial intersticial
intoxication intoxicación
intraannular intraanular
intraatomic intraatómico
intraatomic space espacio intraatómico
intramolecular intramolecular
intramolecular compensation compensación intramolecular
intramolecular condensation condensación intramolecular
intramolecular reaction reacción intramolecular
intranuclear intranuclear
intrinsic intrínseco
intrinsic viscosity viscosidad intrínseca
introfaction introfacción
intron intrón
inulin inulina
inulinase inulinasa
invariant point punto invariante
invention invención
inversion inversión
inversion point punto de inversión
inversion temperature temperatura de inversión
invert sugar azúcar invertido
invertase invertasa
iodate yodato

iodeosin yodeosina

iodic acid ácido yódico

iodic acid anhydride anhídrido de ácido yódico

iodide yoduro

iodimetry yodimetría

iodine yodo

iodine acetate acetato de yodo

iodine bisulfide bisulfuro de yodo

iodine bromide bromuro de yodo

iodine chloride cloruro de yodo

iodine cyanide cianuro de yodo

iodine dioxide dióxido de yodo

iodine disulfide disulfuro de yodo

iodine fluoride fluoruro de yodo

iodine monobromide monobromuro de yodo

iodine monochloride monocloruro de yodo

iodine number número de yodo

iodine oxide óxido de yodo

iodine pentabromide pentabromuro de yodo

iodine pentafluoride pentafluoruro de yodo

iodine pentoxide pentóxido de yodo

iodine sulfate sulfato de yodo

iodine test prueba de yodo

iodine tincture tintura de yodo

iodine trichloride tricloruro de yodo

iodine value valor de yodo

iodipamide yodipamida

iodisan yodisán

iodite yodita

iodized yodado

iodized oil aceite yodado

iodoacetic acid ácido yodoacético

iodoalkane yodoalcano

iodoaniline yodoanilina

iodobenzene yodobenceno

iodoethane yodoetano

iodoethylene yodoetileno

iodoform yodoformo

iodogorgoic acid ácido yodogorgoico

iodomethane yodometano

iodometric yodométrico

iodometry yodometría

iodonium yodonio

iodophor yodóforo

iodopropane yodopropano

iodoso yodoso

iodosobenzene yodosobenceno

iodosuccinimide yodosuccinimida

iodosyl yodosilo

iodoxy yodoxi

iodoxyl yodoxilo

iodyl yodilo

iolite iolita

ion ion

ion activity actividad iónica

ion avalanche avalancha iónica

ion charge carga iónica

ion cloud nube iónica

ion detector detector iónico

ion exchange intercambio iónico

ion-exchange chromatography cromatografía de intercambio iónico

ion-exchange column columna de intercambio iónico

ion-exchange resin resina de intercambio iónico

ion exclusion exclusión iónica

ion-exclusion chromatography cromatografía de exclusión iónica

ion migration migración iónica

ion pair par iónico

ion pump bomba iónica

ion retardation retardo iónico

ion source fuente iónica

ionic iónico

ionic atmosphere atmósfera iónica

ionic bond enlace iónico

ionic bonding enlazado iónico

ionic charge carga iónica

ionic conductance conductancia iónica

ionic crystal cristal iónico

ionic detergent detergente iónico

ionic equilibrium equilibrio iónico

ionic gel gel iónico

ionic polymerization polimerización iónica

ionic product producto iónico

ionic radius radio iónico

ionic solution solución iónica

ionic strength fuerza iónica

ionization ionización

ionization chamber cámara de ionización

ionization coefficient coeficiente de

ionización
ionization constant constante de
   ionización
ionization current corriente de
   ionización
ionization degree grado de ionización
ionization effect efecto de ionización
ionization energy energía de ionización
ionization potential potencial de
   ionización
ionize ionizar
ionized ionizado
ionized gas gas ionizado
ionizing ionizante
ionizing potential potencial de ionización
ionizing radiation radiación de
   ionización
ionizing solvent solvente de ionización
ionogen ionógeno
ionogenic ionogénico
ionography ionografía
ionol ionol
ionomer ionómero
ionomer resin resina de ionómero
ionone ionona
iprodione iprodiona
iridic chloride cloruro irídico
iridium iridio
iridium bromide bromuro de iridio
iridium chloride cloruro de iridio
iridium disulfide disulfuro de iridio
iridium oxide óxido de iridio
iridium potassium chloride cloruro
   potásico de iridio
iridium sesquioxide sesquióxido de
   iridio
iridium tetrabromide tetrabromuro de
   iridio
iridium tetrachloride tetracloruro de
   iridio
iridium tetraiodide tetrayoduro de iridio
iridium tribromide tribromuro de iridio
iridium trichloride tricloruro de iridio
iridium trioxide trióxido de iridio
irido irido
iridosmine iridosmina
Irish moss musgo de Irlanda
iron hierro
iron acetate acetato de hierro

iron ammonium sulfate sulfato amónico
   de hierro
iron arsenate arseniato de hierro
iron benzoate benzoato de hierro
iron black negro de hierro
iron blue azul de hierro
iron bromide bromuro de hierro
iron carbide carburo de hierro
iron carbonate carbonato de hierro
iron carbonyl carbonilo de hierro
iron chloride cloruro de hierro
iron chromate cromato de hierro
iron citrate citrato de hierro
iron dichloride dicloruro de hierro
iron disulfate disulfato de hierro
iron disulfide disulfuro de hierro
iron ethanoate etanoato de hierro
iron ferrocyanide ferrocianuro de hierro
iron fluoride fluoruro de hierro
iron formate formiato de hierro
iron hexacyanoferrate hexacianoferrato
   de hierro
iron hexafluorosilicate hexafluorosilicato
   de hierro
iron hydroxide hidróxido de hierro
iron iodide yoduro de hierro
iron monoxide monóxido de hierro
iron-nickel alloy aleación de hierro-
   níquel
iron nitrate nitrato de hierro
iron octoate octoato de hierro
iron ore mineral de hierro
iron oxalate oxalato de hierro
iron oxide óxido de hierro
iron oxide yellow amarillo de óxido de
   hierro
iron pentacarbonyl pentacarbonilo de
   hierro
iron perchlorate perclorato de hierro
iron phenolate fenolato de hierro
iron phosphate fosfato de hierro
iron phosphide fosfuro de hierro
iron powder polvo de hierro
iron protochloride protocloruro de
   hierro
iron protosulfide protosulfuro de hierro
iron pyrite pirita de hierro
iron pyrolignite pirolignita de hierro
iron pyrophosphate pirofosfato de hierro

iron resinate resinato de hierro
iron sponge esponja de hierro
iron sulfate sulfato de hierro
iron sulfide sulfuro de hierro
iron sulfite sulfito de hierro
iron thiocyanate tiocianato de hierro
iron thiosulfate tiosulfato de hierro
iron trichloride tricloruro de hierro
iron trioxide trióxido de hierro
irone irona
irradiating irradiante
irradiation irradiación
irreversible irreversible
irreversible colloid coloide irreversible
irreversible process proceso irreversible
irreversible reaction reacción
    irreversible
Irving-Williams order órden de Irving-
    Williams
isatin isatina
isatoic anhydride anhídrido isatoico
isazol isazol
isentropic process proceso isentrópico
isethionic acid ácido isetiónico
Ising model modelo de Ising
isinglass colapez, mica
iso iso
isoactivity line línea de isoactividad
isoamyl isoamilo
isoamyl acetate acetato de isoamilo
isoamyl alcohol alcohol isoamílico
isoamyl benzoate benzoato de isoamilo
isoamyl benzyl ether éter bencílico de
    isoamilo
isoamyl butyrate butirato de isoamilo
isoamyl chloride cloruro de isoamilo
isoamyl ether éter isoamílico
isoamyl formate formiato de isoamilo
isoamyl furoate furoato de isoamilo
isoamyl isovalerate isovalerato de
    isoamilo
isoamyl mercaptan isoamilmercaptano
isoamyl nitrite nitrito de isoamilo
isoamyl pelargonate pelargonato de
    isoamilo
isoamyl phthalate ftalato de isoamilo
isoamyl propionate propionato de
    isoamilo
isoamyl salicylate salicilato de isoamilo

isoamyl valerate valerato de isoamilo
isoamyldichloroarsine
    isoamildicloroarsina
isoamylene isoamileno
isoascorbic acid ácido isoascórbico
isobar isóbaro
isobaric spin espín isobárico
isobestic point punto isobéstico
isoborneol isoborneol
isobornyl isobornilo
isobornyl acetate acetato de isobornilo
isobornyl chloride cloruro de isobornilo
isobornyl salicylate salicilato de
    isobornilo
isobornyl thiocyanoacetate
    tiocianoacetato de isobornilo
isobutane isobutano
isobutanol isobutanol
isobutanolamine isobutanolamina
isobutene isobuteno
isobutoxy isobutoxi
isobutyl isobutilo
isobutyl acetate acetato de isobutilo
isobutyl acrylate acrilato de isobutilo
isobutyl alcohol alcohol isobutílico
isobutyl aldehyde aldehído isobutílico
isobutyl benzoate benzoato de isobutilo
isobutyl carbinol isobutilcarbinol
isobutyl chloroformate cloroformiato de
    isobutilo
isobutyl cinnamate cinamato de isobutilo
isobutyl cyanoacrylate cianoacrilato de
    isobutilo
isobutyl ether éter isobutílico
isobutyl furoate furoato de isobutilo
isobutyl isobutyrate isobutirato de
    isobutilo
isobutyl mercaptan isobutilmercaptano
isobutyl methacrylate metacrilato de
    isobutilo
isobutyl phenylacetate fenilacetato de
    isobutilo
isobutyl propionate propionato de
    isobutilo
isobutyl salicylate salicilato de isobutilo
isobutyl stearate estearato de isobutilo
isobutyl valerate valerato de isobutilo
isobutylamine isobutilamina
isobutylaminobenzoate

isobutilaminobenzoato
isobutylbenzene isobutilbenceno
isobutylene isobutileno
isobutyraldehyde isobutiraldehído
isobutyric isobutírico
isobutyric acid ácido isobutírico
isobutyric anhydride anhídrido
    isobutírico
isobutyronitrile isobutironitrilo
isobutyroyl isobutiroilo
isobutyroyl chloride cloruro de
    isobutiroilo
isobutyryl isobutirilo
isocetyl laurate laurato de isocetilo
isocetyl myristate miristato de isocetilo
isocetyl oleate oleato de isocetilo
isocetyl stearate estearato de isocetilo
isochore isócora
isocinchomeronic acid ácido
    isocincomerónico
isocrotonic acid ácido isocrotónico
isocyanate isocianato
isocyanate resin resina de isocianato
isocyanic acid ácido isociánico
isocyanide isocianuro
isocyano isociano
isocyanurate isocianurato
isocyanuric acid ácido isocianúrico
isocyclic isocíclico
isodecaldehyde isodecaldehído
isodecane isodecano
isodecanoic acid ácido isodecanoico
isodecanol isodecanol
isodecyl isodecilo
isodecyl chloride cloruro de isodecilo
isodispersion isodipersión
isodrin isodrina
isodurene isodureno
isodynamic isodinámico
isoelectric isoeléctrico
isoelectric point punto isoeléctrico
isoelectric precipitation precipitación
    isoeléctrica
isoelectronic isoelectrónico
isoelectronic sequence secuencia
    isoelectrónica
isoenzyme isoenzima
isoeugenol isoeugenol
isoheptane isoheptano

isohexane isohexano
isolate aislar
isolated aislado
isolated double bond enlace doble
    aislado
isolation aislamiento
isoleptic isoléptico
isoleucine isoleucina
isolog isólogo
isomer isómero
isomerase isomerasa
isomeric isomérico
isomeric shift deslizamiento isomérico
isomeric transition transición isomérica
isomerism isomerismo
isomerization isomerización
isometric isométrico
isomolecule isomolécula
isomorph isomorfo
isomorphic isomórfico
isomorphism isomorfismo
isonicotinic acid ácido isonicotínico
isonitrile isonitrilo
isonitro isonitro
isonitroso isonitroso
isonitrosoketone isonitrosocetona
isononyl alcohol alcohol isononílico
isooctane isooctano
isooctene isoocteno
isooctyl adipate adipato de isooctilo
isooctyl alcohol alcohol isooctílico
isooctyl palmitate palmitato de isooctilo
isooctyl thioglycolate tioglicolato de
    isooctilo
isoparaffin isoparafina
isopentaldehyde isopentaldehído
isopentane isopentano
isopentanoic acid ácido isopentanoico
isopentyl alcohol alcohol isopentílico
isophorone isoforona
isophthalic acid ácido isoftálico
isophthaloyl chloride cloruro de
    isoftaloilo
isopolyacid isopoliácido
isopolyester isopoliéster
isopolymolybdate isopolimolibdato
isopolymorphism isopolimorfismo
isopolytungstate isopolitungstato
isoprene isopreno

isoprene polymer polímero de isopreno
isoprenoid isoprenoide
isopropanol isopropanol
isopropanolamine isopropanolamina
isopropenyl acetate acetato de
  isopropenilo
isopropenyl chloride cloruro de
  isopropenilo
isopropoxyethanol isopropoxietanol
isopropyl isopropilo
isopropyl acetate acetato de isopropilo
isopropyl alcohol alcohol isopropílico
isopropyl bromide bromuro de
  isopropilo
isopropyl butyrate butirato de isopropilo
isopropyl chloride cloruro de isopropilo
isopropyl cyanide cianuro de isopropilo
isopropyl ether éter isopropílico
isopropyl iodide yoduro de isopropilo
isopropyl mercaptan
  isopropilmercaptano
isopropyl nitrate nitrato de isopropilo
isopropylacetone isopropilacetona
isopropylamine isopropilamina
isopropylaminoethanol
  isopropilaminoetanol
isopropylaniline isopropilanilina
isopropylbenzene isopropilbenceno
isopropylbenzyl isopropilbencilo
isopropylcarbinol isopropilcarbinol
isopropylcresol isopropilcresol
isopropylethylene isopropiletileno
isopropylidene isopropilideno
isopropylmethylbenzene
  isopropilmetilbenceno
isopropylnaphthalene isopropilnaftaleno
isopropylphenol isopropilfenol
isopropyltoluene isopropiltolueno
isopulegol isopulegol
isoquinoline isoquinolina
isosafrole isosafrol
isosteric isostérico

isosterism isosterismo
isostructural isoestructural
isotactic isotáctico
isotherm isoterma
isothermal process proceso isotérmico
isothiocyanate isotiocianato
isothiocyanato isotiocianato
isothiocyanic acid ácido isotiociánico
isotone isótono
isotonic isotónico
isotonic solution solución isotónica
isotope isótopo
isotope abundance abundancia isotópica
isotope effect efecto isotópico
isotope separation separación de
  isótopos
isotope separator separador de isótopos
isotope shift deslizamiento isotópico
isotopic isotópico
isotopic abundance abundancia isotópica
isotopic composition composición
  isotópica
isotopic indicator indicador isotópico
isotopic mass masa isotópica
isotopic number número isotópico
isotopic ratio razón isotópica
isotopic reaction reacción isotópica
isotopic spin espín isotópico
isotopic tracer trazador isotópico
isotopic weight peso isotópico
isotron isotrón
isotropic isotrópico
isovalent isovalente
isovalent hyperconjugation
  hiperconjugación isovalente
isovaleraldehyde isovaleraldehído
isovaleric acid ácido isovalérico
isoxazolyl isoxazolilo
itaconic acid ácido itacónico
ite ita, ito
ium io
Ivanov reagent reactivo de Ivanov

J

Jablonski diagram diagrama de Jablonski

jaborine jaborina

Jacobsen rearrangement reordenamiento de Jacobsen

Jacquemart reagent reactivo de Jacquemart

jade jade

jadeite jadeíta

Jahn-Teller effect efecto de Jahn-Teller

Janovsky reaction reacción de Janovsky

Japp-Klingemann reaction reacción de Japp-Klingemann

jasmine oil aceite de jazmín

jasmone jazmona

jasper jaspe

jelly jalea

johimbine johimbina

jojoba oil aceite de jojoba

Jones oxidation oxidación de Jones

Jones reageant reactivo de Jones

Jones reductor reductor de Jones

joseite joseíta

Joule effect efecto de Joule

Joule-Kelvin effect efecto de Joule-Kelvin

Joule law ley de Joule

Joule-Thomas coefficient coeficiente de Joule-Thomas

Joule-Thomson effect efecto de Joule-Thomson

juniper tar aceite de cada

juniperic acid ácido junipérico

K

kainite cainita
kalinite calinita
kanamycin kanamicina
kanamycin sulfate sulfato de kanamicina
kaolin caolin
kaolinite caolinita
kappa kappa
karat quilate
karaya gum goma karaya
Karl Fischer reagent reactivo de Karl
    Fischer
Karplus equation ecuación de Karplus
Keesom relationship relación de Keesom
Kekulé ring anillo de Kekulé
Kekulé structure estructura de Kekulé
Kelvin effect efecto de Kelvin
Kelvin scale escala Kelvin
Kendall-Mattox reaction reacción de
    Kendall-Mattox
keratin queratina
keratinase queratinasa
kernite kernita
kerogen kerógeno
kerosene queroseno
ketal cetal
ketene cetena
ketimide cetimida
ketimine cetimina
ketine cetina
keto ceto
keto acid cetoácido
ketoamine cetoamina
ketobenzotriazine cetobenzotriazina
ketoglutaric acid ácido cetoglutárico
ketohexose cetohexosa
ketol cetol
ketomalonic acid ácido cetomalónico
ketone cetona
ketone body cuerpo cetónico
ketopentose cetopentosa
ketopropionic acid ácido cetopropiónico
ketose cetosa
ketosis cetosis

ketovaleric acid ácido cetovalérico
ketoxime cetoxima
ketyl cetilo
key atom átomo clave
Keyes process proceso de Keyes
Kharasch abnormal addition adición
    anormal de Kharasch
kieselguhr kieselguhr
kieserite kieserita
Kiliani-Fischer synthesis síntesis de
    Kiliani-Fischer
kiln horno
kilogram kilogramo
kimberlite kimberlita
kinetic cinético
kinetic effect efecto cinético
kinetic energy energía cinética
kinetic theory teoría cinética
kinetics cinética
kinic acid ácido quínico
Kipp apparatus aparato de Kipp
Kishner cyclopropane synthesis síntesis
    de ciclopropanos de Kishner
Kjeldahl flask matraz de Kjeldahl
Kjeldahl method método de Kjeldahl
Kjeldahl test prueba de Kjeldahl
Klein reagent reactivo de Klein
kleinite kleinita
knock golpeteo
knocking golpeteo
Knoevenagel reaction reacción de
    Knoevenagel
Knoop hardness dureza de Knoop
Knoop scale escala de Knoop
Knorr pyrazole synthesis síntesis de
    pirazoles de Knorr
Knorr pyrrole synthesis síntesis de
    pirroles de Knorr
Knorr quinoline synthesis síntesis de
    quinolinas de Knorr
Knudsen flow flujo de Knudsen
Koch acid ácido de Koch
Koch-Haaf carboxylation carboxilación

de Koch-Haaf
Kochi reaction reacción de Kochi
Kohlrausch equation ecuación de
　Kohlrausch
kojic acid ácido cójico
Kolbe method método de Kolbe
Kolbe reaction reacción de Kolbe
Kolbe synthesis síntesis de Kolbe
Komarowsky reaction reacción de
　Komarowsky
Konowaloff rule regla de Konowaloff
Kopp law ley de Kopp
Kopp rule regla de Kopp
koppite koppita
Körner method método de Körner
kovar kovar
Krafft degradation degradación de
　Krafft

Kramers theorem teorema de Kramers
Krebs cycle ciclo de Krebs
kremersite kremersita
Kröhnke aldehyde synthesis síntesis de
　aldehídos de Kröhnke
Kroll process proceso de Kroll
krügite krugita
krypton criptón
Kucherov reaction reacción de
　Kucherov
Kuhn-Winterstein reaction reacción de
　Kuhn-Winterstein
Kundt effect efecto de Kundt
Kundt rule regla de Kundt
kunzite kunzita
kyanite quianita
kynurenine quinurenina

L

label   marcador, etiqueta
LaBel tube   tubo de LaBel
labeled   marcado, rotulado
labeled compound   compuesto marcado
labeling   marcado, rotulación
labile   lábil
labile complex   complejo lábil
laboratory   laboratorio
laboratory apparatus   aparato de laboratorio
laboratory conditions   condiciones de laboratorio
laboratory device   dispositivo de laboratorio
laboratory environment   ambiente de laboratorio
laboratory equipment   equipo de laboratorio
laboratory experiment   experimento de laboratorio
laboratory machinery   maquinaria de laboratorio
laboratory standard   patrón de laboratorio
labradorite   labradorita
laccase   lacasa
lachrymator   lacrimógeno
lachrymatory   lacrimógeno
lacmoid   lacmoide
lacquer   laca
lactalbumin   lactalbúmina
lactam   lactama
lactamic acid   ácido lactámico
lactamide   lactamida
lactase   lactasa
lactate   lactato
lactic acid   ácido láctico
lactic anhydride   anhídrido láctico
lactide   lactida
lactim   lactima
lactobiose   lactobiosa
lactogenic   lactogénico
lactoglobulin   lactoglobulina

lactol   lactol
lactolide   lactolida
lactone   lactona
lactonic acid   ácido lactónico
lactonitrile   lactonitrilo
lactoprene   lactopreno
lactose   lactosa
lactoyl   lactoilo
lactyl   lactilo
ladder polymer   polímero en escalera
Ladenburg rearrangement   reordenamiento de Ladenburg
lagoriolite   lagoriolita
lake   laca
lambda   lambda
lambda phenomenon   fenómeno lambda
lambda point   punto lambda
lambert   lambert
Lambert-Beer law   ley de Lambert-Beer
Lambert law   ley de Lambert
laminarin   laminarina
laminate   laminado
laminated   laminado
lampblack   negro de humo
lanarkite   lanarkita
Landau levels   niveles de Landau
Landé interval rule   regla de intervalos de Landé
landsbergite   landsbergita
langbeinite   langbeinita
Langmuir adsorption isotherm   isoterma de adsorción de Langmuir
Langmuir-Blodgett film   película de Langmuir-Blodgett
Langmuir theory   teoría de Langmuir
lanoceric acid   ácido lanocérico
lanolin   lanolina
lanosterol   lanosterol
lansfordite   lansfordita
lanthana   lantana
lanthanide   lantánido
lanthanide contraction   contracción lantánida

lanthanide series serie lantánida
lanthanum lantano
lanthanum acetate acetato de lantano
lanthanum ammonium nitrate nitrato amónico de lantano
lanthanum antimonide antimoniuro de lantano
lanthanum arsenide arseniuro de lantano
lanthanum carbonate carbonato de lantano
lanthanum chloride cloruro de lantano
lanthanum chloroanilate cloroanilato de lantano
lanthanum fluoride fluoruro de lantano
lanthanum nitrate nitrato de lantano
lanthanum oxalate oxalato de lantano
lanthanum oxide óxido de lantano
lanthanum phosphide fosfuro de lantano
lanthanum sesquioxide sesquióxido de lantano
lanthanum sulfate sulfato de lantano
lanthanum sulfide sulfuro de lantano
lanthanum trioxide trióxido de lantano
lanthionine lantionina
lapis lazuli lapislázuli
Laporte selection rule regla de selección de Laporte
lard manteca
lard oil aceite de manteca
larvicide larvicida
laser láser
laser spectroscopy espectroscopia de láser
Lassaigne test prueba de Lassaigne
latent energy energía latente
latent heat calor latente
latent solvent solvente latente
lateral chain cadena lateral
laterite laterita
latex látex
lattice red, celosía, retículo
lattice energy energía reticular
laudanidine laudanidina
laudanine laudanina
laudanosine laudanosina
laudanum láudano
Laue pattern patrón de Laue
laughing gas gas hilarante
laumonite laumonita

lauraldehyde lauraldehído
laurel leaves oil aceite de hojas de laurel
Laurent acid ácido de Laurent
lauric acid ácido láurico
lauric aldehyde aldehído láurico
laurite laurita
laurone laurona
lauroyl lauroilo
lauroyl chloride cloruro de lauroilo
lauroyl peroxide peróxido de lauroilo
lauroylsarcosine lauroilsarcosina
lauryl laurilo
lauryl acetate laurilacetato
lauryl alcohol alcohol laurílico
lauryl aldehyde aldehído laurílico
lauryl bromide laurilbromuro
lauryl chloride laurilcloruro
lauryl mercaptan laurilmercaptano
lauryl methacrylate laurilmetacrilato
lauryl sulfate laurilsulfato
laurylpyridinium chloride cloruro de laurilpiridinio
lautal lautal
lavender oil aceite de lavanda
law of alternation ley de alternación
law of conservation ley de conservación
law of definite composition ley de composición definida
law of definite proportions ley de proporciones definidas
law of diffusion ley de difusión
law of dilution ley de dilución
law of displacement ley de desplazamiento
law of distribution ley de distribución
law of equilibrium ley de equilibrio
law of mass action ley de acción de masas
law of mass conservation ley de conservación de masa
law of multiple proportions ley de proporciones múltiples
law of partition ley de partición
law of radioactive displacement ley de desplazamiento radiactivo
law of reciprocal proportions ley de proporciones recíprocas
lawrencium laurencio
laws of thermodynamics leyes de

termodinámica
layer capa
lazulite lazulita
lazurite lazurita
Le Chatelier principle principio de Le
    Chatelier
leaching lixiviación
lead plomo
lead accumulator acumulador de plomo
lead acetate acetato de plomo
lead antimonate antimoniato de plomo
lead arsenate arseniato de plomo
lead arsenite arsenito de plomo
lead azide azida de plomo
lead battery batería de plomo
lead benzoate benzoato de plomo
lead biorthophosphate biortofosfato de
    plomo
lead borate borato de plomo
lead borosilicate borosilicato de plomo
lead bromate bromato de plomo
lead bromide bromuro de plomo
lead butyrate butirato de plomo
lead carbonate carbonato de plomo
lead cell celda de plomo
lead chlorate clorato de plomo
lead chloride cloruro de plomo
lead chromate cromato de plomo
lead citrate citrato de plomo
lead-coated plomado
lead-coating plomado
lead cyanate cianato de plomo
lead cyanide cianuro de plomo
lead dichromate dicromato de plomo
lead dioxide dióxido de plomo
lead dithionate ditionato de plomo
lead ethanoate etanoato de plomo
lead fluoride fluoruro de plomo
lead fluoroborate fluoroborato de plomo
lead fluorosilicate fluorosilicato de
    plomo
lead formate formiato de plomo
lead fumarate fumarato de plomo
lead glass vidrio de plomo
lead halide haluro de plomo
lead hexacyanoferrate hexacianoferrato
    de plomo
lead hexafluorosilicate hexafluorosilicato
    de plomo

lead hydroxide hidróxido de plomo
lead hyposulfite hiposulfito de plomo
lead iodate yodato de plomo
lead iodide yoduro de plomo
lead lactate lactato de plomo
lead laurate laurato de plomo
lead linoleate linoleato de plomo
lead malate malato de plomo
lead maleate maleato de plomo
lead metasilicate metasilicato de plomo
lead molybdate molibdato de plomo
lead mononitroresorcinate
    mononitrorresorcinato de plomo
lead monosulfide monosulfuro de plomo
lead monoxide monóxido de plomo
lead naphthalenesulfonate
    naftalenosulfonato de plomo
lead naphthenate naftenato de plomo
lead nitrate nitrato de plomo
lead nitrite nitrito de plomo
lead octoate octoato de plomo
lead oleate oleato de plomo
lead orthophosphate ortofosfato de
    plomo
lead oxalate oxalato de plomo
lead oxide óxido de plomo
lead palmitate palmitato de plomo
lead pentasulfide pentasulfuro de plomo
lead perchlorate perclorato de plomo
lead peroxide peróxido de plomo
lead phosphate fosfato de plomo
lead phosphinate fosfinato de plomo
lead phosphite fosfito de plomo
lead phthalate ftalato de plomo
lead picrate picrato de plomo
lead propionate propionato de plomo
lead protoxide protóxido de plomo
lead pyrophosphate pirofosfato de
    plomo
lead resinate resinato de plomo
lead salicylate salicilato de plomo
lead selenate seleniato de plomo
lead selenide seleniuro de plomo
lead sesquioxide sesquióxido de plomo
lead silicate silicato de plomo
lead silicofluoride silicofluoruro de
    plomo
lead stannate estannato de plomo
lead stearate estearato de plomo

lead subacetate subacetato de plomo

lead subcarbonate subcarbonato de plomo

lead suboxide subóxido de plomo

lead sulfate sulfato de plomo

lead sulfide sulfuro de plomo

lead sulfite sulfito de plomo

lead sulfocyanide sulfocianuro de plomo

lead superoxide superóxido de plomo

lead tallate talato de plomo

lead tartrate tartrato de plomo

lead telluride telururo de plomo

lead tetraacetate tetraacetato de plomo

lead tetrachloride tetracloruro de plomo

lead tetraoxide tetróxido de plomo

lead thiocyanate tiocianato de plomo

lead thiosulfate tiosulfato de plomo

lead titanate titanato de plomo

lead trinitroresorcinate trinitrorresorcinato de plomo

lead tungstate tungstato de plomo

lead vanadate vanadato de plomo

lead wolframate wolframato de plomo

leaf hoja

leak fuga

leakage fuga

leakage radiation radiación de fuga

leaky con fugas

leather cuero

Lebedev process proceso de Lebedev

lecithin lecitina

Leclanché cell pila de Leclanché

lectin lectina

Leduc rule regla de Leduc

legal chemistry química legal

Lehmsted-Tanasescu reaction reacción de Lehmsted-Tanasescu

lemon oil aceite de limón

lemongrass oil aceite de luisa

lenacil lenacil

lene leno

leonardite leonardita

leonite leonita

lepargylic acid ácido lepargílico

lepidine lepidina

lepidolite lepidolita

lepidone lepidona

lepton leptón

lethal letal

lethal dose dosis letal

Letts synthesis síntesis de Letts

leucine leucina

leucite leucita

Leuckart reaction reacción de Leuckart

leuco leuco

leucoaniline leucoanilina

leucoline leucolina

leuconic acid ácido leucónico

leucovorin leucovorina

leucyl leucilo

leukonite leuconita

levo levo

levorotatory levorrotatorio

levulinic acid ácido levulínico

levulinic aldehyde aldehído levulínico

levulose levulosa

Lewis acid ácido de Lewis

Lewis atom átomo de Lewis

Lewis base base de Lewis

Lewis electron theory teoría electrónica de Lewis

Lewis metal metal de Lewis

Lewis structure estructura de Lewis

Lewis theory teoría de Lewis

lewisite lewisita

liberation liberación

licanic acid ácido licánico

lid tapa

lidocaine lidocaína

Lieben reaction reacción de Lieben

Liebermann reaction reacción de Liebermann

Liebig condenser condensador de Liebig

ligand ligando

ligand-field theory teoría de campo de ligandos

ligase ligasa

light emission emisión de luz

light flash destello de luz

light hydrocarbon hidrocarburo ligero

light hydrogen hidrógeno ligero

light metal metal ligero

light oil aceite ligero

light quantum cuanto de luz

light scattering dispersión de luz

light-sensitive sensible a la luz

light spectrum espectro de luz

light water agua ligera

lignin lignina
lignite lignita
lignocaine lignocaína
lignoceric acid ácido lignocérico
lignosulfonate lignosulfonato
ligroin ligroína
lime cal
lime oil aceite de lima
lime water agua de cal
limestone caliza
limited stability estabilidad limitada
limiting density densidad limitadora
limiting factor factor limitador
limonene limoneno
limonite limonita
linalool linalol
linalyl linalilo
linalyl acetate acetato de linalilo
linalyl formate formiato de linalilo
linalyl isobutyrate isobutirato de linalilo
linalyl propionate propionato de linalilo
linamarin linamarina
linarite linarita
lincomycin lincomicina
lindane lindano
Linde process proceso de Linde
lindgrenite lindgrenita
line spectrum espectro de líneas
linear lineal
linear molecule molécula lineal
linear polymer polímero lineal
linkage enlace, encadenamieto
linnaeite linnaeíta
linoleate linoleato
linoleic acid ácido linoleico
linolenic acid ácido linolénico
linolenyl alcohol alcohol linolenílico
linolic acid ácido linólico
linseed oil aceite de linaza
linseed oil meal harina de aceite de
    linaza
lionite lionita
lipase lipasa
lipid lípido
lipolysis lipólisis
lipophilic lipofílico
lipoprotein lipoproteína
lipotropic agent agente lipotrópico
lipoxidase lipoxidasa

lipoxygenase lipoxigenasa
liquation licuación
liquefaction licuefacción
liquefied licuado
liquefied gas gas licuado
liquefied natural gas gas natural licuado
liquefied petroleum gas gas de petróleo
    licuado
liqueur licor aromático
liquid líquido
liquid air aire líquido
liquid chromatography cromatografía
    líquida
liquid crystal cristal líquido
liquid dioxide dióxido líquido
liquid gas gas líquido
liquid glass vidrio líquido
liquid-liquid extraction extracción
    líquido-líquido
liquid nitrogen nitrógeno líquido
liquid oxygen oxígeno líquido
liquid-solid equilibrium equilibrio
    líquido-sólido
liquid state estado líquido
liquid-vapor equilibrium equilibrio
    líquido-vapor
liquor licor
liroconite liroconita
liter litro
litharge litarge
lithia litina
lithic acid ácido lítico
lithium litio
lithium acetate acetato de litio
lithium aluminate aluminato de litio
lithium aluminum deuteride deuteruro
    alumínico de litio
lithium aluminum hydride hidruro
    alumínico de litio
lithium amide amida de litio
lithium arsenate arseniato de litio
lithium benzoate benzoato de litio
lithium bicarbonate bicarbonato de litio
lithium borate borato de litio
lithium borohydride borohidruro de litio
lithium bromide bromuro de litio
lithium carbide carburo de litio
lithium carbonate carbonato de litio
lithium chlorate clorato de litio

lithium chloride cloruro de litio
lithium chromate cromato de litio
lithium citrate citrato de litio
lithium deuteride deuteruro de litio
lithium dichromate dicromato de litio
lithium fluoride fluoruro de litio
lithium fluorophosphate fluorofosfato de
    litio
lithium germanate germanato de litio
lithium halide haluro de litio
lithium hexachloroplatinate
    hexacloroplatinato de litio
lithium hexafluorosilicate
    hexafluorosilicato de litio
lithium hydride hidruro de litio
lithium hydrogen carbonate carbonato
    de hidrógeno de litio
lithium hydroxide hidróxido de litio
lithium hypochlorite hipoclorito de litio
lithium iodate yodato de litio
lithium iodide yoduro de litio
lithium lactate lactato de litio
lithium metaborate metaborato de litio
lithium metasilicate metasilicato de litio
lithium methoxide metóxido de litio
lithium methylate metilato de litio
lithium molybdate molibdato de litio
lithium myristate miristato de litio
lithium niobate niobato de litio
lithium nitrate nitrato de litio
lithium nitride nitruro de litio
lithium orthophosphate ortofosfato de
    litio
lithium oxalate oxalato de litio
lithium oxide óxido de litio
lithium palmitate palmitato de litio
lithium perchlorate perclorato de litio
lithium peroxide peróxido de litio
lithium phosphate fosfato de litio
lithium ricinoleate ricinoleato de litio
lithium salicylate salicilato de litio
lithium silicate silicato de litio
lithium stearate estearato de litio
lithium sulfate sulfato de litio
lithium sulfide sulfuro de litio
lithium sulfite sulfito de litio
lithium tetraborate tetraborato de litio
lithium tetrahydroaluminate
    tetrahidroaluminato de litio

lithium thiocyanate tiocianato de litio
lithium titanate titanato de litio
lithium tungstate tungstato de litio
lithium vanadate vanadato de litio
lithium zirconate zirconato de litio
lithocholic acid ácido litocólico
litidionite litidionita
litmus tornasol
litmus paper papel de tornasol
litre litro
livingstoneite livingstonita
lixiviation lixiviación
lobeline lobelina
localization localización
locant ubicador
logwood palo de campeche
London formula fórmula de London
lone pair par sin compartir
long-necked flask matraz de cuello largo
lorandite lorandita
lorazepam lorazepam
Loschmidt number número de
    Loschmidt
Lossen rearrangement reordenamiento
    de Lossen
low energy baja energía
low-energy electron diffraction
    difracción electrónica de baja energía
low-energy neutron neutrón de baja
    energía
low-energy particle partícula de baja
    energía
lowest occupied molecular orbital
    orbital molecular de mínima energía
    ocupado
Lowry-Brønsted theory teoría de
    Lowry-Brønsted
lubricant lubricante
lubricating oil aceite lubricante
lubrication lubricación
Lucas theory teoría de Lucas
luciferin luciferina
ludwigite ludwigita
Lugol solution solución de Lugol
luminance luminancia
luminescence luminiscencia
luminescent luminiscente
luminiferous luminífero
luminol luminol

luminometry luminometría
luminosity luminosidad
luminous luminoso
Lundergardh vaporizer vaporizador de Lundergardh
lupinidine lupinidina
luster brillo
lutein luteína
luteotropin luteotropina
lutetia lutecia
lutetium lutecio
lutetium chloride cloruro de lutecio
lutetium fluoride fluoruro de lutecio
lutetium nitrate nitrato de lutecio
lutetium oxide óxido de lutecio
lutetium sulfate sulfato de lutecio
lutidine lutidina
lycopene licopeno
lycopodium licopodio

lye lejía
Lyman continuum continuo de Lyman
Lyman series serie de Lyman
lyogel liogel
lyophilic liofílico
lyophilization liofilización
lyophobic liofóbico
lyotopic liotópico
lyotrope liótropo
lyotropic series serie liotrópica
Lyovac process proceso de Lyovac
lysergic acid ácido lisérgico
lysergic acid diethylamide dietilamida de ácido lisérgico
lysidine lisidina
lysine lisina
lysis lisis
lysozyme lisozima
lyxose lixosa

M

Macquer salt sal de Macquer
macro macro
macroanalysis macroanálisis
macrochemistry macroquímica
macrocyclic macrocíclico
macrolide macrólido
macromolecular macromolecular
macromolecular crystal cristal
    macromolecular
macromolecule macromolécula
macroscopic macroscópico
macrose macrosa
Maddrell salt sal de Maddrell
Madelung constant constante de
    Madelung
Madelung synthesis síntesis de
    Madelung
magenta magenta
magic numbers números mágicos
magnalite magnalita
magnalium magnalio
magnesia magnesia
magnesioferrite magnesioferrita
magnesite magnesita
magnesium magnesio
magnesium acetate acetato magnésico,
    acetato de magnesio
magnesium acetylacetonate
    acetilacetonato magnésico
magnesium alloy aleación magnésica
magnesium aluminate aluminato
    magnésico
magnesium amide amida magnésica
magnesium ammonium phosphate
    fosfato amónico magnésico
magnesium arsenate arseniato
    magnésico
magnesium benzoate benzoato
    magnésico
magnesium biphosphate bifosfato
    magnésico
magnesium borate borato magnésico
magnesium bromate bromato magnésico

magnesium bromide bromuro magnésico
magnesium carbonate carbonato
    magnésico
magnesium chlorate clorato magnésico
magnesium chloride cloruro magnésico
magnesium chromate cromato
    magnésico
magnesium citrate citrato magnésico
magnesium dichromate dicromato
    magnésico
magnesium dihydrogensulfate
    dihidrogenosulfato magnésico
magnesium dioxide dióxido magnésico
magnesium fluoride fluoruro magnésico
magnesium fluosilicate fluosilicato
    magnésico
magnesium flux fundente magnésico
magnesium formate formiato magnésico
magnesium gluconate gluconato
    magnésico
magnesium halide haluro magnésico
magnesium hexafluorosilicate
    hexafluorosilicato magnésico
magnesium hydride hidruro magnésico
magnesium hydrogen carbonate
    carbonato de hidrógeno magnésico
magnesium hydroxide hidróxido
    magnésico
magnesium hyposulfite hiposulfito
    magnésico
magnesium iodate yodato magnésico
magnesium iodide yoduro magnésico
magnesium lactate lactato magnésico
magnesium laurate laurato magnésico
magnesium lauryl sulfate laurilsulfato
    magnésico
magnesium limestone caliza magnésica
magnesium methoxide metóxido
    magnésico
magnesium methylate metilato
    magnésico
magnesium molybdate molibdato
    magnésico

magnesium myristate miristato magnésico

magnesium nitrate nitrato magnésico

magnesium nitride nitruro magnésico

magnesium nitrite nitrito magnésico

magnesium oleate oleato magnésico

magnesium orthophosphate ortofosfato magnésico

magnesium oxalate oxalato magnésico

magnesium oxide óxido magnésico

magnesium palmitate palmitato magnésico

magnesium perborate perborato magnésico

magnesium perchlorate perclorato magnésico

magnesium permanganate permanganato magnésico

magnesium peroxide peróxido magnésico

magnesium phosphate fosfato magnésico

magnesium phosphide fosfuro magnésico

magnesium phosphinate fosfinato magnésico

magnesium phosphonate fosfonato magnésico

magnesium propionate propionato magnésico

magnesium pyrophosphate pirofosfato magnésico

magnesium salicylate salicilato magnésico

magnesium silicate silicato magnésico

magnesium silicide siliciuro magnésico

magnesium silicofluoride silicofluoruro magnésico

magnesium stannate estannato magnésico

magnesium stannide estanniuro magnésico

magnesium stearate estearato magnésico

magnesium succinate succinato magnésico

magnesium sulfate sulfato magnésico

magnesium sulfide sulfuro magnésico

magnesium sulfite sulfito magnésico

magnesium tartrate tartrato magnésico

magnesium thiocyanate tiocianato magnésico

magnesium thiosulfate tiosulfato magnésico

magnesium trisilicate trisilicato magnésico

magnesium tungstate tungstato magnésico

magnesium urate urato magnésico

magnesium wolframate wolframato magnésico

magnesium zirconate zirconato magnésico

magnesyl megnesilo

magnetic magnético

magnetic activity actividad magnética

magnetic deflection deflexión magnética

magnetic element elemento magnético

magnetic moment momento magnético

magnetic quantum number número cuántico magnético

magnetic resonance resonancia magnética

magnetic resonance imaging formación de imágenes por resonancia magnética

magnetic separation separación magnética

magnetic spectrum espectro magnético

magnetic stirrer agitador magnético

magnetic susceptibility susceptibilidad magnética

magnetism magnetismo

magnetite magnetita

magnetization magnetización

magnetochemistry magnetoquímica

magnetohydrodynamics magnetohidrodinámica

magnetometric titration titulación magnetométrica, valoración magnetométrica

magneton magnetón

magnolite magnolita

magnolium magnolio

Magnus rule regla de Magnus

Magnus salt sal de Magnus

Maillard reaction reacción de Maillard

Majorana force fuerza de Majorana

malachite malaquita

malachite green verde malaquita

malamide malamida

Malaprade reaction reacción de Malaprade
malate malato
malathion malatión
malchite malquita
maldonite maldonita
maleamic acid ácido maleámico
maleate maleato
maleic acid ácido maleico
maleic anhydride anhídrido maleico
maleic hydrazide hidrazida maleica
malenoid malenoide
maleoyl maleoilo
malic acid ácido málico
malic amide amida málica
mallardite mallardita
malleability maleabilidad
malleable maleable
malonaldehydic acid ácido malonaldehídico
malonamide malonamida
malonate malonato
malonic acid ácido malónico
malonic amide amida malónica
malonic dinitrile dinitrilo malónico
malonic ester éster malónico
malonic ester synthesis síntesis de ésteres malónicos
malonic mononitrile mononitrilo malónico
malononitrile malononitrilo
malonyl malonilo
malonyl urea malonilurea
maloyl maloilo
malt malta
malt extract extracto de malta
maltase maltasa
maltha malta
malthene malteno
maltol maltol
maltose maltosa
mandarin oil aceite de mandarina
mandelic acid ácido mandélico
Mandelin reagent reactivo de Mandelin
mandelonitrile mandelonitrilo
maneb maneb
manganate manganato
manganese manganeso
manganese acetate acetato de manganeso

manganese arsenate arseniato de manganeso
manganese binoxide binóxido de manganeso
manganese black negro de manganeso
manganese borate borato de manganeso
manganese boride boruro de manganeso
manganese-boron manganeso-boro
manganese bromide bromuro de manganeso
manganese bronze bronce de manganeso
manganese carbide carburo de manganeso
manganese carbonate carbonato de manganeso
manganese carbonyl carbonilo de manganeso
manganese chloride cloruro de manganeso
manganese chromate cromato de manganeso
manganese citrate citrato de manganeso
manganese dioxide dióxido de manganeso
manganese dithionate ditionato de manganeso
manganese fluoride fluoruro de manganeso
manganese gluconate gluconato de manganeso
manganese glycerophosphate glicerofosfato de manganeso
manganese halide haluro de manganeso
manganese heptoxide heptóxido de manganeso
manganese hexacyanoferrate hexacianoferrato de manganeso
manganese hexafluorosilicate hexafluorosilicato de manganeso
manganese hydrate hidrato de manganeso
manganese hydroxide hidróxido de manganeso
manganese hypophosphite hipofosfito de manganeso
manganese iodide yoduro de manganeso
manganese lactate lactato de manganeso
manganese linoleate linoleato de manganeso

manganese monoxide monóxido de manganeso

manganese naphthenate naftenato de manganeso

manganese nitrate nitrato de manganeso

manganese octoate octoato de manganeso

manganese oleate oleato de manganeso

manganese oxalate oxalato de manganeso

manganese oxide óxido de manganeso

manganese peroxide peróxido de manganeso

manganese phosphate fosfato de manganeso

manganese phosphinate fosfinato de manganeso

manganese potassium sulfate sulfato potásico de manganeso

manganese protoxide protóxido de manganeso

manganese resinate resinato de manganeso

manganese sesquioxide sesquióxido de manganeso

manganese silicate silicato de manganeso

manganese sulfate sulfato de manganeso

manganese sulfide sulfuro de manganeso

manganese sulfite sulfito de manganeso

manganese tallate talato de manganeso

manganese tetrachloride tetracloruro de manganeso

manganese tetrafluoride tetrafluoruro de manganeso

manganese tetraoxide tetróxido de manganeso

manganese-titanium manganeso-titanio

manganese trioxide trióxido de manganeso

manganese tungstate tungstato de manganeso

manganic mangánico

manganic acetylacetonate acetilacetonato mangánico

manganic acid ácido mangánico

manganic fluoride fluoruro mangánico

manganic hydroxide hidróxido mangánico

manganic oxide óxido mangánico

manganin manganina

manganite manganita

manganosite manganosita

manganostilbite manganoestilbita

manganous manganoso

manganous acetate acetato manganoso

manganous arsenate arseniato manganoso

manganous bromide bromuro manganoso

manganous carbonate carbonato manganoso

manganous chloride cloruro manganoso

manganous chromate cromato manganoso

manganous citrate citrato manganoso

manganous fluoride fluoruro manganoso

manganous formate formiato manganoso

manganous hydroxide hidróxido manganoso

manganous iodide yoduro manganoso

manganous nitrate nitrato manganoso

manganous orthophosphate ortofosfato manganoso

manganous oxide óxido manganoso

manganous phosphate fosfato manganoso

manganous pyrophosphate pirofosfato manganoso

manganous silicate silicato manganoso

manganous sulfate sulfato manganoso

manganous sulfide sulfuro manganoso

manganous sulfite sulfito manganoso

manganous tartrate tartrato manganoso

Manila resin resina de Manila

manna maná

Mannich condensation reaction reacción de condensación de Mannich

Mannich reaction reacción de Mannich

mannite manita

mannitol manitol

mannitol hexanitrate hexanitrato de manitol

mannitose manitosa

mannose manosa

manometer manómetro

Marathon-Howard process proceso de Marathon-Howard

marble mármol

marcasite marcasita
margaric acid ácido margárico
margarine margarina
margarite margarita
marialite marialita
marihuana mariguana
marijuana mariguana
Mark-Houwink equation ecuación de
　　Mark-Houwink
Markovnikoff rule regla de
　　Markovnikoff
Markovnikov rule regla de Markovnikov
Marme reagent reactivo de Marme
marsh gas gas de pantano
Marsh test prueba de Marsh
martensite martensita
Martinet synthesis síntesis de Martinet
martite martita
martonite martonita
mascagnite mascagnita
maser máser
mash templa
masked element elemento enmascarado
masked radical radical enmascarado
masking enmascaramiento
mass masa
mass action law ley de acción de masas
mass conservation law ley de
　　conservación de masa
mass defect defecto de masa
mass fraction fracción de masa
mass number número de masa
mass spectrograph espectrógrafo de
　　masa
mass spectrometer espectrómetro de
　　masa
mass spectrometry espectrometría de
　　masa
mass spectroscope espectroscopio de
　　masa
mass spectroscopy espectroscopia de
　　masa
mass spectrum espectro de masa
mass-to-charge ratio razón de masa a
　　carga
massicot masicote
master batch lote matriz
master equation ecuación maestra
mastication masticación

material material
materialization materialización
materials handling manejo de materiales
matildite matildita
matlockite matlockita
matrix matriz
matrix spectrophotometry
　　espectrofotometría de matriz
matte mata
matter materia
maxivalence maxivalencia
Maxwell-Boltzmann distribution
　　distribución de Maxwell-Boltzmann
Mayer reagent reactivo de Mayer
McFayden-Stevens reaction reacción de
　　McFayden-Stevens
McLafferty rearrangement
　　reordenamiento de McLafferty
McMillan-Mayer theory teoría de
　　McMillan-Mayer
mean free path recorrido libre medio
mean free time tiempo libre medio
mean life vida media
measurement medida
measurement error error de medida
measuring beaker vaso de laboratorio de
　　medida
measuring flask matraz de medida
mechanical analysis análisis mecánico
mechanism mecanismo
mechanism of reaction mecanismo de
　　reacción
mechanochemistry mecanoquímica
meclizine hydrochloride clorhidrato de
　　meclizina
meconin meconina
medicinal chemistry química medicinal
medium medio
Meisenheimer complex complejo de
　　Meisenheimer
Meisenheimer rearrangement
　　reordenamiento de Meisenheimer
melamine melamina
melamine resin resina de melamina
melaniline melanilina
melanin melanina
melibiose melibiosa
melissic acid ácido melísico
melissyl alcohol alcohol melisílico

mellitate melitato
mellitic acid ácido melítico
melonite melonita
melphalan melfalano
melt fundir, derretir
melting point punto de fusión
melting-point depression depresión de
    punto de fusión
membrane membrana
membrane hydrolysis hidrólisis por
    membrana
memtetrahydrophthalic anhydride
    anhídrido memtetrahidroftálico
menachanite menacanita
menadione menadiona
Mendeleev law ley de Mendeleev
mendelevium mendelevio
Mendius reaction reacción de Mendius
mendozite mendocita
meneghinite meneghinita
menhaden oil aceite de menhaden
meniscus menisco
Menschutkin reaction reacción de
    Menschutkin
mensuration mensuración
menthacamphor mentalcanfor
menthadiene mentadieno
menthane mentano
menthanediamine mentanodiamina
menthanol mentanol
menthanone mentanona
menthene menteno
menthenol mentenol
menthenyl mentenilo
menthol mentol
menthol valerate valerato de mentol
menthonaphthene mentonafteno
menthone mentona
menthyl mentilo
menthyl acetate acetato de mentilo
menthyl salicylate salicilato de mentilo
meperidine hydrochloride clorhidrato de
    meperidina
meprobamate meprobamato
mercaptal mercaptal
mercaptamine mercaptamina
mercaptan mercaptano
mercapto mercapto
mercaptoacetic acid ácido

mercaptoacético
mercaptobenzoic acid ácido
    mercaptobenzoico
mercaptobenzothiazole
    mercaptobenzotiazol
mercaptoethanol mercaptoetanol
mercaptol mercaptol
mercaptophenyl mercaptofenilo
mercaptopropionic acid ácido
    mercaptopropiónico
mercaptosuccinic acid ácido
    mercaptosuccínico
mercaptothiazoline mercaptotiazolina
mercapturic acid ácido mercaptúrico
mercuration mercuración
mercurial mercurial
mercuric mercúrico
mercuric acetate acetato mercúrico
mercuric arsenate arseniato mercúrico
mercuric barium bromide bromuro
    bárico mercúrico
mercuric barium iodide yoduro bárico
    mercúrico
mercuric benzoate benzoato mercúrico
mercuric borate borato mercúrico
mercuric bromide bromuro mercúrico
mercuric carbonate carbonato mercúrico
mercuric chloride cloruro mercúrico
mercuric chromate cromato mercúrico
mercuric cyanate cianato mercúrico
mercuric cyanide cianuro mercúrico
mercuric dichromate dicromato
    mercúrico
mercuric fluoride fluoruro mercúrico
mercuric hydroxide hidróxido mercúrico
mercuric iodide yoduro mercúrico
mercuric lactate lactato mercúrico
mercuric nitrate nitrato mercúrico
mercuric oleate oleato mercúrico
mercuric oxalate oxalato mercúrico
mercuric oxide óxido mercúrico
mercuric oxycyanide oxicianuro
    mercúrico
mercuric phosphate fosfato mercúrico
mercuric potassium cyanide cianuro
    potásico mercúrico
mercuric salicylate salicilato mercúrico
mercuric sodium phenolsulfate
    fenolsulfato sódico mercúrico

mercuric stearate estearato mercúrico
mercuric succinate succinato mercúrico
mercuric sulfate sulfato mercúrico
mercuric sulfide sulfuro mercúrico
mercuric sulfocyanate sulfocianato
    mercúrico
mercuric thiocyanate tiocianato
    mercúrico
mercuride mercururo
mercurification mercurificación
mercurization mercurización
mercurous mercurioso
mercurous acetate acetato mercurioso
mercurous acetylide acetiluro
    mercurioso
mercurous bromide bromuro mercurioso
mercurous chlorate clorato mercurioso
mercurous chloride cloruro mercurioso
mercurous chromate cromato
    mercurioso
mercurous formate formiato mercurioso
mercurous iodide yoduro mercurioso
mercurous nitrate nitrato mercurioso
mercurous oxalate oxalato mercurioso
mercurous oxide óxido mercurioso
mercurous phosphate fosfato
    mercurioso
mercurous sulfate sulfato mercurioso
mercurous sulfide sulfuro mercurioso
mercurous tartrate tartrato mercurioso
mercury mercurio
mercury acetate acetato de mercurio
mercury alloy aleación de mercurio
mercury amalgam amalgama de
    mercurio
mercury arsenate arseniato de mercurio
mercury benzoate benzoato de mercurio
mercury bichloride bicloruro de
    mercurio
mercury bromide bromuro de mercurio
mercury cell pila de mercurio
mercury chloride cloruro de mercurio
mercury chromate cromato de mercurio
mercury cyanide cianuro de mercurio
mercury dichromate dicromato de
    mercurio
mercury dithiocarbonate ditiocarbonato
    de mercurio
mercury fulminate fulminato de

    mercurio
mercury hexacyanoferrate
    hexacianoferrato de mercurio
mercury hexafluorosilicate
    hexafluorosilicato de mercurio
mercury nitrate nitrato de mercurio
mercury oxide óxido de mercurio
mercury peroxide peróxido de mercurio
mercury phosphate fosfato de mercurio
mercury selenide seleniuro de mercurio
mercury telluride telururo de mercurio
mercury tetraborate tetraborato de
    mercurio
meridional meridional
merwinite merwinita
mesaconic acid ácido mesacónico
mescaline mescalina
mesitite mesitita
mesitol mesitol
mesityl mesitilo
mesityl oxide óxido de mesitilo
mesitylene mesitileno
mesitylenesulfonyl chloride cloruro de
    mesitilensulfonilo
meso meso
mesocolloid mesocoloide
mesoionic mesoiónico
mesomer mesómero
mesomeric mesomérico
mesomerism mesomería
mesomorphic mesomórfico
mesomorphic state estado mesomórfico
meson mesón
meson mass masa mesónica
mesonic mesónico
mesoscopic mesoscópico
mesothorium mesotorio
mesotomy mesotomía
mesoxalic acid ácido mesoxálico
mesoxalo mesoxalo
mesoxalyl mesoxalilo
mesyl mesilo
mesyl chloride cloruro de mesilo
meta meta
metabolic metabólico
metabolism metabolismo
metabolite metabolito
metaborate metaborato
metaboric acid ácido metabórico

metachromasia metacromasia
metachromatism metacromatismo
metaformaldehyde meaformaldehído
metal metal
metal-clad blindado
metal-coated metalizado
metal deposition deposición metálica
metal electrode electrodo metálico
metal film película metálica
metal foil hoja metálica
metal glass vidrio metálico
metal-metal bond enlace metal-metal
metalation metalación
metaldehyde metaldehído
metallic metálico
metallic binding forces fuerzas ligantes
  metálicas
metallic bond enlace metálico
metallic bonding enlazado metálico
metallic conduction conducción metálica
metallic crystal cristal metálico
metallic soap jabón metálico
metallify metalificar
metallization metalización
metallize metalizar
metallized metalizado
metallizing metalización
metallocene metaloceno
metallochromic metalocrómico
metallochromic indicator indicador
  metalocrómico
metallography metalografía
metalloid metaloide
metallurgy metalurgia
metalorganic metalorgánico
metamer metámero
metamerism metamerismo
metanilic acid ácido metalínico
metaphenylene metafenileno
metaphosphate metafosfato
metaphosphoric acid ácido
  metafosfórico
metasilicic acid ácido metasilícico
metasomatosis metasomatosis
metastable metaestable
metastable electron electrón metaestable
metastable ion ion metaestable
metastable phase fase metaestable
metastable state estado metaestable

metastasic electron electrón metastásico
metastasis metástasis
metastructure metaestructura
metathesis descomposición doble
metathetical reaction reacción
  metatética
metatitanic acid ácido metatitánico
meteoric meteórico
metepa metepa
meter metro, medidor
methacetin metacetina
methacrolein metacroleína
methacrylamide metacrilamida
methacrylate metacrilato
methacrylic acid ácido metacrílico
methacrylonitrile metacrilonitrilo
methacryloyl chloride cloruro de
  metacriloilo
methadone metadona
methadone hydrochloride clorhidrato de
  metadona
methallydine diacetate diacetato de
  metalidina
methallyl metalilo
methallyl alcohol alcohol metalílico
methanal metanal
methanamide metanamida
methanation metanación
methane metano
methanecarboxylic acid ácido
  metanocarboxílico
methanedicarboxylic acid ácido
  metanodicarboxílico
methanesulfonate metanosulfonato
methanesulfonic acid ácido
  metanosulfónico
methanesulfonyl chloride cloruro de
  metanosulfonilo
methanethiol metanotiol
methano metano
methanoate metanoato
methanoic acid ácido metanoico
methanol metanol
methanolate metanolato
methanoyl metanoilo
methenamine metenamina
methene meteno
methenyl metenilo
methicillin meticilina

methide metida
methine metina
methionic acid ácido metiónico
methionine metionina
methionyl metionilo
methomyl metomilo
methose metosa
methotrexate metotrexato
methoxalyl metoxalilo
methoxide metóxido
methoxy metoxi
methoxyacetaldehyde
  metoxiacetaldehído
methoxyacetanilide metoxiacetanilida
methoxyacetic acid ácido metoxiacético
methoxyacetophenone
  metoxiacetofenona
methoxyamine metoxiamina
methoxybenzaldehyde
  metoxibenzaldehído
methoxybenzene metoxibenceno
methoxybenzoic acid ácido
  metoxibenzoico
methoxybenzyl alcohol alcohol
  metoxibencílico
methoxybutanol metoxibutanol
methoxyethyl metoxietilo
methoxyl metoxilo
methoxynaphthalene metoxinaftaleno
methoxyphenol metoxifenol
methoxyphenyl metoxifenilo
methoxypropanol metoxipropanol
methoxypropylamine metoxipropilamina
methyl metilo
methyl abietate abietato de metilo
methyl acetate acetato de metilo
methyl acetoacetate acetoacetato de
  metilo
methyl acetone metilacetona
methyl acetylricinoleate acetilricinoleato
  de metilo
methyl acrylate acrilato de metilo
methyl alcohol alcohol metílico
methyl arachidate araquidato de metilo
methyl behenate behenato de metilo
methyl benzoate benzoato de metilo
methyl blue azul de metilo
methyl borate borato de metilo
methyl bromide bromuro de metilo

methyl butyrate butirato de metilo
methyl caprate caprato de metilo
methyl caproate caproato de metilo
methyl caprylate caprilato de metilo
methyl carbonate carbonato de metilo
methyl ceroate ceroato de metilo
methyl chloride cloruro de metilo
methyl chloroacetate cloroacetato de
  metilo
methyl chloroformate cloroformiato de
  metilo
methyl chlorosilane clorosilano de
  metilo
methyl chlorosulfonate clorosulfonato de
  metilo
methyl cinnamate cinamato de metilo
methyl cyanate cianato de metilo
methyl cyanide cianuro de metilo
methyl cyanoacetate cianoacetato de
  metilo
methyl cyanoformate cianoformiato de
  metilo
methyl dichloroacetate dicloroacetato de
  metilo
methyl dichlorostearate dicloroestearato
  de metilo
methyl elaidate elaidato de metilo
methyl ester éster metílico
methyl ether éter metílico
methyl fluoride fluoruro de metilo
methyl fluorosulfonate fluorosulfonato
  de metilo
methyl formate formiato de metilo
methyl gallate galato de metilo
methyl glycol metilglicol
methyl heneicosanoate heneicosanoato
  de metilo
methyl heptadecanoate heptadecanoato
  de metilo
methyl hydrazine hidrazina de metilo
methyl hydride hidruro de metilo
methyl iodide yoduro de metilo
methyl isocyanate isocianato de metilo
methyl isonicotinate isonicotinato de
  metilo
methyl isothiocyanate isotiocianato de
  metilo
methyl lactate lactato de metilo
methyl laurate laurato de metilo

methyl lauroleate lauroleato de metilo
methyl lignocerate lignocerato de metilo
methyl linoleate linoleato de metilo
methyl linolenate linolenato de metilo
methyl methacrylate metacrilato de metilo
methyl myristate miristato de metilo
methyl nitrate nitrato de metilo
methyl nitrite nitrito de metilo
methyl oleate oleato de metilo
methyl orange anaranjado de metilo
methyl oxide óxido de metilo
methyl palmitate palmitato de metilo
methyl pentadecanoate pentadecanoato de metilo
methyl propionate propionato de metilo
methyl pyruvate piruvato de metilo
methyl red rojo de metilo
methyl ricinoleate ricinoleato de metilo
methyl salicylate salicilato de metilo
methyl stearate estearato de metilo
methyl sulfate sulfato de metilo
methyl sulfide sulfuro de metilo
methyl tridecanoate tridecanoato de metilo
methyl violet violeta de metilo
methyl yellow amarillo de metilo
methylacetic acid ácido metilacético
methylacetophenone metilacetofenona
methylacetylene metilacetileno
methylacrylamide metilacrilamida
methylal metilal
methylalanine metilalanina
methylallyl acetate acetato de metilalilo
methylallyl alcohol alcohol metilalílico
methylallyl chloride cloruro de metilalilo
methylaluminum metilaluminio
methylamine metilamina
methylamino metilamino
methylaminoacetic acid ácido metilaminoacético
methylaminophenol metilaminofenol
methylamyl metilamilo
methylamyl acetate acetato de metilamilo
methylamyl alcohol alcohol metilamílico
methylamyl ketone metilamilcetona
methylaniline metilanilina

methylanisole metilanisol
methylanthracene metilantraceno
methylanthranilate metilantranilato
methylanthraquinone metilantraquinona
methylate metilar
methylated metilado
methylbenzene metilbenceno
methylbenzoic acid ácido metilbenzoico
methylbenzyl metilbencilo
methylbenzyl alcohol alcohol metilbencílico
methylbenzyl ether éter metilbencílico
methylbenzylamine metilbencilamina
methylbutane metilbutano
methylbutanol metilbutanol
methylbutene metilbuteno
methylbutyl ketone metilbutilcetona
methylbutylamine metilbutilamina
methylbutynol metilbutinol
methylcellulose metilcelulosa
methylcholanthrene metilcolantreno
methylcyclohexane metilciclohexano
methylcyclohexanol metilciclohexanol
methylcyclohexanone metilciclohexanona
methylcyclohexanyl metilciclohexanilo
methylcyclohexylamine metilciclohexilamina
methylcyclopentane metilciclopentano
methyldichloroarsine metildicloroarsina
methyldichlorosilane metildiclorosilano
methyldiethanolamine metildietanolamina
methyldioxolane metildioxolano
methyldiphenylamine metildifenilamina
methyldipropylmethane metildipropilmetano
methylene metileno
methylene blue azul de metileno
methylene bromide bromuro de metileno
methylene chloride cloruro de metileno
methylene dichloride dicloruro de metileno
methylene diiodide diyoduro de metileno
methylene iodide yoduro de metileno
methylene oxide óxido de metileno
methylenecyclopentadiene metilenociclopentadieno
methylenedisalicylic acid ácido

metilenodisalicílico
methylethyl ketone metiletilcetona
methylethylcellulose metiletilcelulosa
methylformanilide metilformanilida
methylfuran metilfurano
methylfurfurylamine metilfurfulrilamina
methylheptane metilheptano
methylheptenone metilheptenona
methylheptylamine metilheptilamina
methylhexane metilhexano
methylhexyl ketone metilhexilcetona
methylhydrazine metilhidrazina
methylhydrazone metilhidrazona
methylhydroxybutanone
     metilhidroxibutanona
methylic metílico
methylidyne metilidino
methylindole metilindol
methylisoamyl ketone
     metilisoamilcetona
methylisobutyl carbinol
     metilisobutilcarbinol
methylisobutyl ketone
     metilisobutilcetona
methylmagnesium bromide bromuro de
     metilmagnesio
methylmagnesium chloride cloruro de
     metilmagnesio
methylmagnesium iodide yoduro de
     metilmagnesio
methylmaleic acid ácido metilmaleico
methylmercury cyanide cianuro de
     metilmercurio
methylmethane metilmetano
methylnaphthalene metilnaftaleno
methylnitrobenzene metilnitrobenceno
methylol metilol
methylolurea metilolurea
methylparaben metilparabeno
methylpentadiene metilpentadieno
methylpentaldehyde metilpentaldehído
methylpentane metilpentano
methylpentanoic acid ácido
     metilpentanoico
methylpentanol metilpentanol
methylpentene metilpenteno
methylpentose metilpentosa
methylphenyl ether éter metilfenílico
methylphenyl ketone metilfenilcetona

methylphenyldichlorosilane
     metilfenildiclorosilano
methylphosphonic acid ácido
     metilfosfónico
methylphosphoric acid ácido
     metilfosfórico
methylpiperazine metilpiperazina
methylpropane metilpropano
methylpropene metilpropeno
methylpropyl ketone metilpropilcetona
methylpyrrole metilpirrol
methylpyrrolidine metilpirrolidina
methylquinoline metilquinolina
methylstyrene metilestireno
methylsulfonic acid ácido metilsulfónico
methylsulfonyl metilsulfonilo
methylsulfuric acid ácido metilsulfúrico
methyltaurine metiltaurina
methyltrichlorosilane metiltriclorosilano
methyltrinitrobenzene
     metiltrinitrobenceno
methylundecanoate metilundecanoato
methyne metino
metopon hydrochloride clorhidrato de
     metopón
metre metro
mevalonic acid ácido mevalónico
Meyer formula fórmula de Meyer
Meyer law ley de Meyer
Meyer reaction reacción de Meyer
Meyer synthesis síntesis de Meyer
meymacite meimacita
miargyrite miargirita
miazine miazina
mica mica
micelle micela
Michael condensation condensación de
     Michael
Michael reaction reacción de Michael
Michaelis-Arbuzov reaction reacción de
     Michaelis-Arbuzov
Michaelis constant constante de
     Michaelis
miconazole nitrate nitrato de miconazol
micril micrilo
micro micro
microanalysis microanálisis
microanalyzer microanalizador
microbalance microbalanza

microbe microbio
microbicide microbicida
microbody microcuerpo
microcapsule microcápsula
microchemical microquímico
microchemistry microquímica
microcrystal microcristal
microcrystalline microcristalino
microcrystalline wax cera
    microcristalina
microcurie microcurie
microdensitometer microdensitómetro
microdiffusion microdifusión
microelectrode microelectrodo
microelectrolysis microelectrólisis
microelectrophoresis microelectroforesis
microencapsulation microencapsulación
microgalvanometer microgalvanómetro
microgammil microgamilo
microgram microgramo
micrography micrografía
microincineration microincineración
microline microlina
micrometer micrómetro
micron micrón
micronutrient micronutriente
microorganism microorganismo
microphotometer microfotómetro
micropipet micropipeta
microporous microporoso
microprobe microsonda
microreaction microrreacción
microscope microscopio
microscopic microscópico
microscopic examination examen
    microscópico
microscopy microscopia
microsecond microsegundo
microspectrophotometer
    microespectrofotómetro
microspectroscope microespectroscopio
microwave microonda
microwave absorber absorbedor de
    microondas
microwave spectrometer espectrómetro
    de microondas
microwave spectroscope espectroscopio
    de microondas
microwave spectroscopy espectroscopia

de microondas
microwave spectrum espectro de
    microondas
miemite miemita
Miescher degradation degradación de
    Miescher
Mignonac reaction reacción de
    Mignonac
migration migración
migration current corriente de
    migración
migration velocity velocidad de
    migración
mildew moho
milk of magnesia leche de magnesia
milk sugar azúcar de leche
milky nublado
mill molino
milli mili
millibar milibarra
millicurie milicurie
milliequivalent miliequivalente
milligram miligramo
milliliter mililitro
millimeter milímetro
millimeter of mercury milímetro de
    mercurio
millimole milimol
millinormal milinormal
Millon base base de Millon
Millon reagent reactivo de Millon
Millon test prueba de Millon
mimetite mimetita
mineral mineral
mineral acid ácido mineral
mineral alkali álcali mineral
mineral carbon carbón mineral
mineral fat grasa mineral
mineral jelly jalea mineral
mineral oil aceite mineral
mineral pigment pigmento mineral
mineral pitch brea mineral
mineral tar alquitrán mineral
mineral wax cera mineral
mineralization mineralización
mineralogy mineralogía
minimum ionizing speed velocidad
    mínima de ionización
minium minio

minivalence minivalencia
minulite minulita
mirabilite mirabilita
mirror-image imagen de espejo
mirror nuclei núcleos espejos
mischmetal mischmetal
miscibility miscibilidad
miscible miscible
misfire falla de encendido
mispickel mispiquel
mitomycin mitomicina
Mitsunobu reaction reacción de
   Mitsunobu
mixed acid ácido mixto
mixed crystal cristal mixto
mixed indicator indicador mixto
mixed melting point punto de fusión
   mixto
mixed salt sal mixta
mixing mezcla
mixite mixita
mixo mixo
mixture mezcla
mobility movilidad
mode modo
model modelo
moderator moderador
modification modificación
modulation spectroscopy espectroscopia
   de modulación
modulus of elasticity módulo de
   elasticidad
Mohr method método de Mohr
Mohr salt sal de Mohr
Mohr titration titulación de Mohr,
   valoración de Mohr
Mohs scale escala de Mohs
moiety porción, mitad
moistness humedad
moisture humedad
moisture-proof a prueba de humedad
mol mol
molal molal
molal concentration concentración molal
molal conductivity conductividad molal
molal solution solución molal
molal specific heat calor específico
   molal
molal volume volumen molal

molality molalidad
molar molar
molar conductivity conductividad molar
molar heat calor molar
molar heat capacity capacidad térmica
   molar
molar latent heat calor latente molar
molar solution solución molar
molar specific heat calor específico
   molar
molar surface superficie molar
molar susceptibility susceptibilidad
   molar
molar volume volumen molar
molar weight peso molar
molarity molaridad
molasses melaza
mold moho
mole mol
mole fraction fracción molar
mole volume volumen molar
molecular molecular
molecular adhesion adhesión molecular
molecular amplitude amplitud molecular
molecular association asociación
   molecular
molecular attraction atracción molecular
molecular beam haz molecular
molecular biology biología molecular
molecular colloid coloide molecular
molecular combination combinación
   molecular
molecular concentration concentración
   molecular
molecular conductivity conductividad
   molecular
molecular density densidad molecular
molecular depression depresión
   molecular
molecular diagram diagrama molecular
molecular diamagnetism diamagnetismo
   molecular
molecular diameter diámetro molecular
molecular dipole dipolo molecular
molecular dispersion dispersión
   molecular
molecular dissociation disociación
   molecular
molecular distillation destilación

molecular

molecular elevation　elevación molecular

molecular energy level　nivel de energía molecular

molecular equation　ecuación molecular

molecular field　campo molecular

molecular film　película molecular

molecular flow　flujo molecular

molecular flux　flujo molecular

molecular formula　fórmula molecular

molecular free path　trayectoria libre molecular

molecular frequency　frecuencia molecular

molecular gas　gas molecular

molecular heat capacity　capacidad térmica molecular

molecular ion beam　haz iónico molecular

molecular ion collision　colisión iónica molecular

molecular lattice　red molecular

molecular level　nivel molecular

molecular mass　masa molecular

molecular number　número molecular

molecular orbital　orbital molecular

molecular paramagnetism paramagnetismo molecular

molecular polarizability　polarizabilidad molecular

molecular rays　rayos moleculares

molecular rearrangement reordenamiento molecular

molecular relaxation　relajación molecular

molecular repulsion　repulsión molecular

molecular rotation　rotación molecular

molecular sieve　cedazo molecular

molecular solution　solución molecular

molecular specific heat　calor específico molecular

molecular spectroscopy　espectroscopia molecular

molecular spectrum　espectro molecular

molecular still　destilador molecular

molecular structure　estructura molecular

molecular velocity　velocidad molecular

molecular vibration　vibración molecular

molecular volume　volumen molecular

molecular weight　peso molecular

molecularity　molecularidad

molecule　molécula

Molisch test　prueba de Molisch

molten　fundido

molten salt　sal fundida

molybdate　molibdato

molybdenite　molibdenita

molybdenum　molibdeno

molybdenum acetylacetonate acetilacetonato de molibdeno

molybdenum blue　azul de molibdeno

molybdenum boride　boruro de molibdeno

molybdenum chloride　cloruro de molibdeno

molybdenum dichloride　dicloruro de molibdeno

molybdenum dioxide　dióxido de molibdeno

molybdenum disilicide　disiliciuro de molibdeno

molybdenum disulfide　disulfuro de molibdeno

molybdenum hexacarbonyl hexacarbonilo de molibdeno

molybdenum hydroxide　hidróxido de molibdeno

molybdenum orange　anaranjado de molibdeno

molybdenum oxide　óxido de molibdeno

molybdenum pentachloride　pentacloruro de molibdeno

molybdenum sesquioxide　sesquióxido de molibdeno

molybdenum sulfide　sulfuro de molibdeno

molybdenum tetrasulfide　tetrasulfuro de molibdeno

molybdenum trioxide　trióxido de molibdeno

molybdenum trisulfide　trisulfuro de molibdeno

molybdenyl　molibdenilo

molybdic　molíbdico

molybdic acid　ácido molíbdico

molybdic anhydride　anhídrido molíbdico

molybdite　molibdita

molybdosilicic acid　ácido molibdosilícico

molysite molisita
monad mónada
monatomic monoatómico
monazite monacita
Mond process proceso de Mond
monetite monetita
monistic monístico
monistic compound compuesto
  monístico
mono mono
monoacetate monoacetato
monoacetin monoacetina
monoacid monoácido
monoamine monoamina
monoamino acid monoaminoácido
monoammonium monoamonio
monoatomic monoatómico
monoatomic gas gas monoatómico
monoatomic molecule molécula
  monoatómica
monobasic monobásico
monobasic calcium phosphate fosfato
  cálcico monobásico
monobromated monobromado
monobromo monobromo
monocalcium fluoride fluoruro
  monocálcico
monocalcium phosphate fosfato
  monocálcico
monochloro monocloro
monochloroacetone monocloroacetona
monochlorobenzene monoclorobenceno
monochloroethane monocloroetano
monochloromethane monoclorometano
monochlorophenol monoclorofenol
monochromatic monocromático
monodisperse monodisperso
monodisperse system sistema
  monodisperso
monoester monoéster
monoethanolamine monoetanolamina
monoethylamine monoetilamina
monofunctional monofuncional
monofunctional compound compuesto
  monofuncional
monoglyceride monoglicérido
monoglyme monoglime
monohydrate monohidrato
monohydric monohídrico

monohydric alcohol alcohol monohídrico
monolayer monocapa
monomer monómero
monomethylamine monometilamina
monometric monométrico
monomolecular monomolecular
monomolecular film película
  monomolecular
monomorphous monomorfo
mononuclear mononuclear
monophosphate monofosfato
monoprotic acid ácido monoprótico
monorefringent monorrefringente
monosaccharide monosacárido
monosodium glutamate glutamato
  monosódico
monosulfide monosulfuro
monoterpene monoterpeno
monotropic monotrópico
monotropy monotropía
monovalent monovalente
monoxide monóxido
monoxychlor monoxicloro
montan wax cera montana
montanite montanita
monticellite monticelita
montmorillonite montmorillonita
montroydite montroidita
mordant mordiente
morenosite morenosita
morin morina
morphine morfina
morpholine morfolina
morphology morfología
morphosan morfosán
Morse equation ecuación de Morse
Morse potential potencial de Morse
mortar mortero
mosaic gold oro musivo, oro de mosaico
Moseley diagram diagrama de Moseley
Moseley law ley de Moseley
Mössbauer effect efecto de Mössbauer
Mössbauer spectroscopy espectroscopia
  de Mössbauer
mu mu
mucic acid ácido múcico
mucopeptide mucopéptido
mucopolysaccharide mucopolisacárido
mullite mullita

multi multi
multiple bond enlace múltiple
multiple bonding enlazado múltiple
multiple decay desintegración múltiple
multiple disintegration desintegración múltiple
multiple proportions law ley de proporciones múltiples
multiplet multiplete
multivalent multivalente
Muntz metal metal Muntz
muon muon
murexide murexida
muriatic acid ácido muriático
muscarine muscarina
muscovite muscovita
musk almizcle
musk xylene xileno de almizcle
mustard gas gas mostaza
mustard oil aceite de mostaza
mutagenic agent agente mutagénico

mutamer mutámero
mutamerism mutamerismo
mutarotation mutarrotación
mutase mutasa
mutual exclusion rule regla de exclusión mutua
mutuality of phases mutualidad de fases
mycotoxin micotoxina
myoglobin mioglobina
myokinase mioquinasa
myosin miosina
myrcene mirceno
myricyl miricilo
myristic acid ácido mirístico
myristoyl miristoilo
myristoyl peroxide peróxido de miristoilo
myristyl alcohol alcohol miristílico
myristyl chloride cloruro de miristilo
myrrh mirra
myxin mixina

N

nabam nabam
nacre nácar
nacrite nacrita
nadorite nadorita
naftolen naftolén
nakrite nacrita
naled naled
nalidixic acid ácido nalidíxico
nalorphine nalorfina
naloxone naloxona
nanometer nanómetro
nanotechnology nanotecnología
nantokite nantoquita
napalin napalina
napalm napalm
naphtha nafta
naphthacene naftaceno
naphthacetol naftecetol
naphthacridine naftacridina
naphthal naftal
naphthaldehyde naftaldehído
naphthalene naftaleno
naphthaleneacetic acid ácido
    naftalenacético
naphthalenecarboxylic acid ácido
    naftalencarboxílico
naphthalenediamine naftalendiamina
naphthalenedicarboxylic acid ácido
    naftalendicarboxílico
naphthalenediol naftalendiol
naphthalenedisulfonic acid ácido
    naftalendisulfónico
naphthalenesulfinic acid ácido
    naftalensulfínico
naphthalenesulfonic acid ácido
    naftalensulfónico
naphthalenethiol naftalentiol
naphthalenyl naftalenilo
naphthalic acid ácido naftálico
naphthalide naftalida
naphthalimido naftalimido
naphthalin naftalina
naphthamide naftamida

naphthamine naftamina
naphthane naftano
naphthene nafteno
naphthenic acid ácido nafténico
naphthenyl naftenilo
naphthieno naftieno
naphthindene naftindeno
naphthionic acid ácido naftiónico
naphtho nafto
naphthodianthrene naftodiantreno
naphthoic acid ácido naftoico
naphthoic aldehyde aldehído naftoico
naphthol naftol
naphthol blue azul de naftol
naphthol green verde de naftol
naphthol orange anaranjado de naftol
naphthol yellow amarillo de naftol
naphtholate naftolato
naphtholsulfonic acid ácido
    naftolsulfónico
naphthonitrile naftonitrilo
naphthoquinone naftoquinona
naphthoresorcinol naftorresorcina
naphthothiazole naftotiazol
naphthoxy naftoxi
naphthoxyacetic acid ácido
    naftoxiacético
naphthoyl naftoilo
naphthyl naftilo
naphthyl alcohol alcohol naftílico
naphthyl aldehyde aldehído naftílico
naphthylacetic acid ácido naftilacético
naphthylamine naftilamina
naphthylamine hydrochloride
    clorhidrato de naftilamina
naphthylaminesulfonic acid ácido
    naftilaminosulfónico
naphthylbenzene naftilbenceno
naphthylbenzoate naftilbenzoato
naphthylene naftileno
naphthylenediamine naftilendiamina
naphthylethyl ether naftiletiléter
naphthylmercuric naftilmercúrico

163

naphthylmethanol naftilmetanol
naphthylmethyl naftilmetilo
naphthylmethylene naftilmetileno
naphthylnaphthyl naftilnaftilo
naphthyloxy naftiloxi
naphthylthiourea naftiltiourea
naphthyridine naftiridina
naproxen naproxén
naptalam naptalam
narceine narceína
narcotic narcótico
naringin naringina
nascent nasciente
nascent hydrogen hidrógeno nasciente
native compound compuesto nativo
native element elemento nativo
native metal metal nativo
natrolite natrolita
natron natrón
natural natural
natural abundance abundancia natural
natural base base natural
natural gas gas natural
Nazarov cyclization reaction reacción de
    ciclización de Nazarov
neatsfoot oil aceite de pata de vaca
Neber rearrangement reordenamiento
    de Neber
nectar nectar
Néel temperature temperatura de Néel
Nef reaction reacción de Nef
Nef synthesis síntesis de Nef
negative negativo
negative adsorption adsorción negativa
negative catalysis catálisis negativa
negative element elemento negativo
negative group grupo negativo
negative ion ion negativo
negative radical radical negativo
nematic nemático
nematic crystal cristal nemático
nematic phase fase nemática
Nencki reaction reacción de Nencki
Nenitzescu indole synthesis síntesis de
    indoles de Nenitzescu
neo neo
neocianite neocianita
neodecanoic acid ácido neodecanoico
neodymium neodimio

neodymium acetate acetato de neodimio
neodymium acetylacetate acetilacetato
    de neodimio
neodymium ammonium nitrate nitrato
    amónico de neodimio
neodymium carbonate carbonato de
    neodimio
neodymium chloride cloruro de
    neodimio
neodymium fluoride fluoruro de
    neodimio
neodymium hydroxide hidróxido de
    neodimio
neodymium iodide yoduro de neodimio
neodymium nitrate nitrato de neodimio
neodymium oxalate oxalato de neodimio
neodymium oxide óxido de neodimio
neodymium phosphate fosfato de
    neodimio
neodymium sulfate sulfato de neodimio
neodymium sulfide sulfuro de neodimio
neohexane neohexano
neolite neolita
neomycin neomicina
neon neón
neon lamp lámpara de neón
neopentane neopentano
neopentanoic acid ácido neopentanoico
neopentyl neopentilo
neopentyl glycol neopentilglicol
neoprene neopreno
neotridecanoic acid ácido
    neotridecanoico
nephelite nefelita
nephelometer nefelómtero
nephelometry nefelometría
nephrite nefrita
neptunium neptunio
neptunium dioxide dióxido de neptunio
neptunyl neptunilo
Nernst-Einstein equation ecuación de
    Nernst-Einstein
Nernst equation ecuación de Nernst
Nernst law ley de Nernst
Nernst potential potencial de Nernst
Nernst-Thomson rule regla de Nernst-
    Thomson
nerol nerol
nerolidol nerolidol

nerve gas gas neurotóxico
nervone nervona
nervonic acid ácido nervónico
neryl nerilo
nesquehonite nesquehonita
Nessler reagent reactivo de Nessler
Nessler tubes tubos de Nessler
neurine neurina
neutral neutro
neutral atom átomo neutro
neutral compound compuesto neutro
neutral element elemento neutro
neutral molecule molécula neutra
neutral oil aceite neutro
neutral reaction reacción neutra
neutral red rojo neutro
neutral salt sal neutra
neutral solution solución neutra
neutral violet violeta neutra
neutrality neutralidad
neutralization neutralización
neutralization equivalent equivalente de neutralización
neutralization number número de neutralización
neutralization ratio razón de neutralización
neutralize neutralizar
neutralized neutralizado
neutralizing neutralizante
neutrino neutrino
neutron neutrón
neutron absorber absorbedor neutrónico, absorbedor de neutrones
neutron absorption absorción neutrónica
neutron activation analysis análisis de activación neutrónica
neutron beam haz neutrónico
neutron binding energy energía de separación de neutrón
neutron bombardment bombardeo neutrónico
neutron capture captura neutrónica
neutron converter convertidor de neutrones
neutron current corriente neutrónica
neutron cycle ciclo neutrónico
neutron density densidad de neutrones
neutron diffraction difracción neutrónica

neutron energy energía neutrónica
neutron excess exceso de neutrones
neutron flux flujo neutrónico
neutron-induced inducido por neutrones
neutron-induced reaction reacción inducida por neutrones
neutron magnetic moment momento magnético neutrónico
neutron multiplication multiplicación neutrónica
neutron number número de neutrones
neutron producer productor de neutrones
neutron rest mass masa en reposo de neutrón
neutron source fuente de neutrones
neutron spectrometer espectrómetro neutrónico
neutron spectrometry espectrometría neutrónica
neutron spectroscopy espectroscopia neutrónica
neutron spectrum espectro neutrónico
neutronic neutrónico
neutronics neutrónica
Neville-Winter acid ácido de Neville-Winter
Newman projection proyección de Newman
Newtonian fluid fluido newtoniano
niacin niacina
niacinamide niacinamida
niacinamide ascorbate ascorbato de niacinamida
nialamide nialamida
niccolite niccolita
nichrome nicromo
nickel níquel
nickel acetate acetato de níquel
nickel acetylacetonate acetilacetonato de níquel
nickel ammonium chloride cloruro amónico de níquel
nickel ammonium sulfate sulfato amónico de níquel
nickel arsenate arseniato de níquel
nickel benzoate benzoato de níquel
nickel bromide bromuro de níquel
nickel-cadmium níquel-cadmio

nickel carbonate carbonato de níquel
nickel carbonyl carbonilo de níquel
nickel chloride cloruro de níquel
nickel cyanide cianuro de níquel
nickel dimethylglyoxime dimetilglioxima de níquel
nickel fluoride fluoruro de níquel
nickel formate formiato de níquel
nickel hexafluorosilicate hexafluorosilicato de níquel
nickel iodide yoduro de níquel
nickel-iron alloy aleación de níquel-hierro
nickel layer capa de níquel
nickel nitrate nitrato de níquel
nickel oxalate oxalato de níquel
nickel oxide óxido de níquel
nickel perchlorate perclorato de níquel
nickel peroxide peróxido de níquel
nickel phosphate fosfato de níquel
nickel potassium sulfate sulfato potásico de níquel
nickel protoxide protóxido de níquel
nickel-rhodium níquel-rodio
nickel sesquioxide sesquióxido de níquel
nickel-silver níquel-plata
nickel stannate estannato de níquel
nickel steel acero al níquel
nickel sulfate sulfato de níquel
nickel tetracarbonyl tetracarbonilo de níquel
nickel titanate titanato de níquel
nickelic niquélico
nickelic hydroxide hidróxido niquélico
nickelic oxide óxido niquélico
nickelic sulfide sulfuro niquélico
nickeline niquelina
nickelocene niqueloceno
nickelous niqueloso
nickelous acetate acetato niqueloso
nickelous arsenate arseniato niqueloso
nickelous bromide bromuro niqueloso
nickelous carbonate carbonato niqueloso
nickelous chloride cloruro niqueloso
nickelous formate formiato niqueloso
nickelous hydroxide hidróxido niqueloso
nickelous iodide yoduro niqueloso
nickelous nitrate nitrato niqueloso
nickelous oxalate oxalato niqueloso

nickelous oxide óxdo niqueloso
nickelous phosphate fosfato niqueloso
nickelous sulfate sulfato niqueloso
Nicol prism prisma de Nicol
nicotinamide nicotinamida
nicotine nicotina
nicotine hydrochloride clorhidrato de nicotina
nicotinic acid ácido nicotínico
nicotinic acid amide amida de ácido nicotínico
nicotinoyl nicotinoilo
nido nido
Niementowski quinazoline synthesis síntesis de quinazolinas de Niementowski
Niementowski quinoline synthesis síntesis de quinolinas de Niementowski
Nierenstein reaction reacción de Nierenstein
ninhydrin ninhidrina
niobate niobato
niobe oil aceite de niobe
niobic nióbico
niobic acid ácido nióbico
niobite niobita
niobium niobio
niobium bromide bromuro de niobio
niobium carbide carburo de niobio
niobium chloride cloruro de niobio
niobium diselenide diseleniuro de niobio
niobium fluoride fluoruro de niobio
niobium oxalate oxalato de niobio
niobium oxide óxido de niobio
niobium pentachloride pentacloruro de niobio
niobium pentoxide pentóxido de niobio
niobium potassium oxyfluoride oxifluoruro potásico de niobio
niobium silicide siliciuro de niobio
niobium-tin niobio-estaño
niobium-titanium niobio-titanio
niobium-uranium niobio-uranio
nioboxy nioboxi
niobus niobo
niobyl niobilo
nionel nionel
niranium niranio

nisin nisina
nisinic acid ácido nisínico
niter nitro
nitracidium nitracidio
nitralin nitralina
nitramide nitramida
nitramine nitramina
nitranilic acid ácido nitranílico
nitranilide nitranilida
nitraniline nitranilina
nitrate nitrato
nitrating acid ácido nitrante
nitration nitración
nitrene nitreno
nitric acid ácido nítrico
nitric ether éter nítrico
nitric oxide óxido nítrico
nitridation nitridación
nitride nitruro
nitrido nitrido
nitrification nitrificación
nitrilase nitrilasa
nitrile nitrilo
nitrile rubber caucho nitrilo
nitrilo nitrilo
nitrilotriacetic acid ácido
    nitrilotriacético
nitrilotriacetonitrile nitrilotriacetonitrilo
nitrite nitrito
nitrito nitrito
nitro nitro
nitro dye colorante nitro
nitroacetanilide nitroacetanilida
nitroamine nitroamina
nitroamino nitroamino
nitroaniline nitroanilina
nitroanisole nitroanisol
nitroanthracene nitroantraceno
nitroanthraquinone nitroantraquinona
nitroaromatic nitroaromático
nitrobarbituric acid ácido
    nitrobarbitúrico
nitrobarite nitrobarita
nitrobenzaldehyde nitrobenzaldehído
nitrobenzamide nitrobenzamida
nitrobenzene nitrobenceno
nitrobenzenesulfonic acid ácido
    nitrobencenosulfónico
nitrobenzoic acid ácido nitrobenzoico

nitrobenzonitrile nitrobenzonitrilo
nitrobenzotrifluoride
    nitrobenzotrifluoruro
nitrobenzoyl nitrobenzoilo
nitrobenzoyl chloride cloruro de
    nitrobenzoilo
nitrobenzyl nitrobencilo
nitrobenzyl alcohol alcohol
    nitrobencílico
nitrobenzyl bromide bromuro de
    nitrobencilo
nitrobenzyl cyanide cianuro de
    nitrobencilo
nitrobiphenyl nitobifenilo
nitrobromoform nitrobromoformo
nitrobutanol nitrobutanol
nitrocellulose nitrocelulosa
nitrochlorobenzene nitroclorobenceno
nitrochloroform nitrocloroformo
nitrocinnamic acid ácido nitrocinámico
nitrocobalamin nitrocobalamina
nitrocresol nitrocresol
nitrodiphenyl nitrodifenilo
nitrodiphenylamine nitrodifenilamina
nitroethane nitroetano
nitroethylpropanediol
    nitroetilpropanodiol
nitrofuran nitrofurano
nitrofurantoin nitrofurantoína
nitrogen nitrógeno
nitrogen chloride cloruro de nitrógeno
nitrogen cycle ciclo de nitrógeno
nitrogen dioxide dióxido de nitrógeno
nitrogen equivalent equivalente de
    nitrógeno
nitrogen fixation fijación de nitrógeno
nitrogen fluoride fluoruro de nitrógeno
nitrogen hydride hidruro de nitrógeno
nitrogen iodide yoduro de nitrógeno
nitrogen monoxide monóxido de
    nitrógeno
nitrogen mustard mostaza de nitrógeno
nitrogen oxide óxido de nitrógeno
nitrogen pentoxide pentóxido de
    nitrógeno
nitrogen peroxide peróxido de nitrógeno
nitrogen sulfide sulfuro de nitrógeno
nitrogen tetraoxide tetróxido de
    nitrógeno

nitrogen tribromide tribromuro de nitrógeno

nitrogen trichloride tricloruro de nitrógeno

nitrogen trifluoride trifluoruro de nitrógeno

nitrogen triiodide triyoduro de nitrógeno

nitrogen trioxide trióxido de nitrógeno

nitrogenase nitrogenasa

nitrogenated nitrogenado

nitrogenous base base nitrogenosa

nitroglycerin nitroglicerina

nitroguanidine nitroguanidina

nitrohydrochloric acid ácido nitroclorhídrico

nitrol nitrol

nitromagnesite nitromagnesita

nitromannite nitromanita

nitromannitol nitromanitol

nitromersol nitromersol

nitromethane nitrometano

nitromethoxyaniline nitrometoxianilina

nitromethylpropanediol nitrometilpropanodiol

nitron nitrón

nitronaphthalene nitronaftaleno

nitronaphthol nitronaftol

nitronaphthylamine nitronaftilamina

nitrone nitrona

nitronic acid ácido nitrónico

nitronium nitronio

nitronium perchlorate perclorato de nitronio

nitroparaffin nitroparafina

nitrophenetole nitrofenetol

nitrophenide nitrofenida

nitrophenol nitrofenol

nitrophenylacetic acid ácido nitrofenilacético

nitrophenylhydrazine nitrofenilhidrazina

nitrophenylpropiolic acid ácido nitrofenilpropiólico

nitrophosphate nitrofosfato

nitrophthalic acid ácido nitroftálico

nitropropane nitropropano

nitroquinoline nitroquinolina

nitrosalicylic acid ácido nitrosalicílico

nitroso nitroso

nitroso dye colorante nitroso

nitroso polymer polímero nitroso

nitrosoamine nitrosoamina

nitrosobenzene nitrosobenceno

nitrosobenzoic acid ácido nitrosobenzoico

nitrosodimethylamine nitrosodimetilamina

nitrosodimethylaniline nitrosodimetilanilina

nitrosoethane nitrosoetano

nitrosoguanidine nitrosoguanidina

nitrosomethylurea nitrosometilurea

nitrosonaphthol nitrosonaftol

nitrosonaphthylamine nitrosonaftilamina

nitrosonium nitrosonio

nitrosooxy nitrosooxi

nitrosophenol nitrosofenol

nitrosotoluene nitrosotolueno

nitrostarch nitroalmidón

nitrostyrene nitroestireno

nitrosulfathiazole nitrosulfatiazol

nitrosyl nitrosilo

nitrosyl bromide bromuro de nitrosilo

nitrosyl chloride cloruro de nitrosilo

nitrosyl fluoride fluoruro de nitrosilo

nitrosyl perchlorate perclorato de nitrosilo

nitrosylsulfuric acid ácido nitrosilsulfúrico

nitrothiophene nitrotiofeno

nitrotoluene nitrotolueno

nitrotoluidine nitrotoluidina

nitrotrichloromethane nitrotriclorometano

nitrotrifluoromethyl chlorobenzene nitrotrifluorometilclorobenceno

nitrourea nitrourea

nitrous acid ácido nitroso

nitrous oxide óxido nitroso

nitroxanthic acid ácido nitroxántico

nitroxyl nitroxilo

nitroxylene nitroxileno

nitroxylic acid ácido nitroxílico

nitryl nitrilo

nitryl chloride cloruro de nitrilo

nitryl fluoride fluoruro de nitrilo

nivenite nivenita

Nobel explosive explosivo de Nobel

nobelium nobelio

noble noble
noble gas gas noble
noble metal metal noble
noble potential potencial noble
nodule nódulo
nomenclature nomenclatura
non non
nona nona
nonacosane nonacosano
nonacosanol nonacosanol
nonacyclic nonacíclico
nonadecane nonadecano
nonadecanoic acid ácido nonadecanoico
nonadecanol nonadecanol
nonadecanone nonadecanona
nonahydrate nonahidrato
nonalactone nonalactona
nonanal nonanal
nonane nonano
nonanedioic acid ácido nonanodioico
nonanediol nonanodiol
nonanenitrile nonanonitrilo
nonanoate nonanoato
nonanoic acid ácido nonanoico
nonanol nonanol
nonanone nonanona
nonanoyl nonanoilo
nonanoyl chloride cloruro de nonanoilo
nonaqueous no acuoso
nonaqueous solution solución no acuosa
nonaqueous solvent solvente no acuosa
noncombustible no combustible
noncombustible material material no
    combustible
noncompetitive inhibition inhibición no
    competitiva
nondestructive no destructivo
nondestructive test prueba no destructiva
nonelectrolyte no electrólito
nonene noneno
nonferrous no ferroso
nonflammable no inflamable
nonideal no ideal
nonionic no iónico
nonionizing no ionizante
nonionizing solvent solvente no ionizante
nonlinear molecule molécula no lineal
nonmetal no metal
nonmetallic no metálico

nonnutritive sweetener endulcorante no
    nutritivo
nonoic acid ácido nonoico
nonpolar no polar
nonpolar compound compuesto no polar
nonpolar crystal cristal no polar
nonpolar molecule molécula no polar
nonpolar solvent solvente no polar
nonreducing sugar azúcar no reductor
nonstoichiometric no estequiométrico
nonyl nonilo
nonyl acetate acetato de nonilo
nonyl alcohol alcohol nonílico
nonyl aldehyde aldehído nonílico
nonyl bromide bromuro de nonilo
nonyl chloride cloruro de nonilo
nonyl hydride hidruro de nonilo
nonyl lactone lactona de nonilo
nonyl nonanoate nonanoato de nonilo
nonyl thiocyanate tiocianato de nonilo
nonylamine nonilamina
nonylbenzene nonilbenceno
nonylcarbinol nonilcarbinol
nonylene nonileno
nonylic acid ácido nonílico
nonylphenol nonilfenol
nonylphenoxyacetic acid ácido
    nonilfenoxiacético
nonyltrichlorosilane noniltriclorosilano
nonyne nonino
nopinene nopineno
nor nor
noradrenaline noradrenalina
norbornadiene norbornadieno
norbornene norborneno
norbornenemethanol norbornenometanol
norepinephrine norepinefrina
norgestrel norgestrel
norleucine norleucina
normal normal
normal conditions condiciones normales
normal hydrocarbon hidrocarburo
    normal
normal pressure presión normal
normal salt sal normal
normal solution solución normal
normal state estado normal
normal temperature temperatura normal
normality normalidad

Normant reagent reactivo de Normant
nornicotine nornicotina
noscapine noscapina
noselite noselita
notation notación
novobiocin novobiocina
novolak novolaca
noxious nocivo
noxious chemical producto químico
    nocivo
noxious gas gas nocivo
noxious substance substancia nociva
nuclear nuclear
nuclear absorption absorción nuclear
nuclear angular momentum momento
    angular nuclear
nuclear atom átomo nuclear
nuclear attraction atracción nuclear
nuclear binding energy energía de unión
    nuclear
nuclear bombardment bombardeo
    nuclear
nuclear capture captura nuclear
nuclear chain reaction reacción en
    cadena nuclear
nuclear charge carga nuclear
nuclear chemistry química nuclear
nuclear collision colisión nuclear
nuclear cross-section sección transversal
    nuclear
nuclear decay desintegración nuclear
nuclear-decay mode modo de
    desintegración nuclear
nuclear density densidad nuclear
nuclear disintegration desintegración
    nuclear
nuclear energy energía nuclear
nuclear energy level nivel de energía
    nuclear
nuclear equation ecuación nuclear
nuclear excitation excitación nuclear
nuclear fission fisión nuclear
nuclear force fuerza nuclear
nuclear fragment fragmento nuclear
nuclear fuel combustible nuclear
nuclear fusion fusión nuclear
nuclear ground state estado fundamental
    nuclear
nuclear heat calor nuclear

nuclear interaction interacción nuclear
nuclear isomer isómero nuclear
nuclear isomerism isomerismo nuclear
nuclear magnetic moment momento
    magnético nuclear
nuclear magnetic resonance resonancia
    magnética nuclear
nuclear magneton magnetón nuclear
nuclear mass masa nuclear
nuclear moment momento nuclear
nuclear neutron neutrón nuclear
nuclear number número nuclear
nuclear paramagnetism paramagnetismo
    nuclear
nuclear particle partícula nuclear
nuclear physics física nuclear
nuclear polarization polarización
    nuclear
nuclear potential energy energía
    potencial nuclear
nuclear power energía nuclear
nuclear process proceso nuclear
nuclear proton protón nuclear
nuclear quadripole moment momento
    cuadripolar nuclear
nuclear quadripole resonance
    resonancia cuadripolar nuclear
nuclear radiation radiación nuclear
nuclear radius radio nuclear
nuclear reaction reacción nuclear
nuclear reactor reactor nuclear
nuclear repulsion repulsión nuclear
nuclear resonance resonancia nuclear
nuclear scattering dispersión nuclear
nuclear species especie nuclear
nuclear spectroscopy espectroscopia
    nuclear
nuclear spectrum espectro nuclear
nuclear spin espín nuclear
nuclear stability estabilidad nuclear
nuclear structure estructura nuclear
nuclear theory teoría nuclear
nuclear transformation transformación
    nuclear
nuclear volume volumen nuclear
nuclear waste desperdicios nucleares
nuclease nucleasa
nucleate nucleado
nucleation nucleación

nucleic acid ácido nucleico
nucleogenesis nucleogénesis
nucleon nucleón
nucleonic nucleónico
nucleophile nucleófilo
nucleophilic nucleofílico
nucleophilic addition adición nucleofílica
nucleophilic substitution substitución nucleofílica
nucleoprotein nucleoproteína
nucleoside nucleósido

nucleotide nucleótido
nucleus núcleo
nuclide nucleido
nutrient nutriente
nutrient solution solución de nutrientes
nutrification nutrificación
nutrition nutrición
Nylander reagent reactivo de Nylander
nylidrin hydrochloride clorhidrato de nilidrina
nylon nilón

O

Obermayer reagent reactivo de
   Obermayer
occlusion oclusión
occupied molecular orbital orbital
   molecular ocupado
ocher ocre
ocimene ocimeno
octa octa
octabenzone octabenzona
octachloronaphthalene
   octacloronaftaleno
octacosane octacosano
octacosanoic acid ácido octacosanoico
octadeca octadeca
octadecanal octadecanal
octadecane octadecano
octadecanoic acid ácido octadecanoico
octadecanol octadecanol
octadecatrienoic acid ácido
   octadecatrienoico
octadecene octadeceno
octadecenoic acid ácido octadecenoico
octadecenol octadecenol
octadecenyl aldehyde aldehído
   octadecenílico
octadecyl octadecilo
octadecyl alcohol alcohol octadecílico
octadecyl isocyanate isocianato de
   octadecilo
octadecyltrichlorosilane
   octadeciltriclorosilano
octadiene octadieno
octafluorobutene octafluorobuteno
octafluorocyclobutane
   octafluorociclobutano
octafluoropropane octafluoropropano
octahedral octahedral
octahedro octahedro
octahedron octahedro
octahydrate octahidrato
octamethyltrisiloxane
   octametiltrisiloxano
octanal octanal

octane octano
octane number índice de octano
octane rating índice de octano
octanedicarboxylic acid ácido
   octanodicarboxílico
octanedioic acid ácido octanodioico
octanoate octanoato
octanoic acid ácido octanoico
octanol octanol
octanone octanona
octanoyl octanoilo
octanoyl chloride cloruro de octanoilo
octavalent octavalente
octene octeno
octet octeto
octet theory teoría de octeto
octoic acid ácido octoico
octomethylene octometileno
octyl octilo
octyl acetate acetato de octilo
octyl alcohol alcohol octílico
octyl aldehyde aldehído octílico
octyl bromide bromuro de octilo
octyl carbinol octilcarbinol
octyl chloride cloruro de octilo
octyl formate formiato de octilo
octyl gallate galato de octilo
octyl iodide yoduro de octilo
octyl mercaptan octilmercaptano
octyl peroxide peróxido de octilo
octyl phosphate fosfato de octilo
octylacetic acid ácido octilacético
octylamine octilamina
octylene octileno
octylene oxide óxido de octileno
octylmagnesium chloride cloruro de
   octilmagnesio
octylphenol octilfenol
octylphenyl octilfenilo
octylphenyl salicylate salicilato de
   octilfenilo
octyltrichlorosilane octiltriclorosilano
octyne octino

odd-even nucleus núcleo impar-par
odd-odd nucleus núcleo impar-impar
odor olor
odor intensity intensidad de olor
odor permanence permanencia de olor
odorant odorante
odoriferous odorífero
odorimetry odorimetría
odorometer odorómetro
oic oico
oid oide
oil aceite, petróleo
oil black negro de petróleo
oil gas gas de petróleo
oil of almond aceite de almendra
oil of anise aceite de anís
oil of apple aceite de manzana
oil of avocado aceite de aguacate
oil of banana aceite de plátano
oil of calamus aceite de cálamo
oil of camphor aceite de alcanfor
oil of cananga aceite de cananga
oil of caraway aceite de alcaravea
oil of cardamom aceite de cardamomo
oil of cascarilla aceite de cascarilla
oil of cashew aceite de anacardo
oil of cassia aceite de casia
oil of chamomile aceite de manzanilla
oil of chenopodium aceite de quenopodio
oil of cinnamon aceite de canela
oil of citronella aceite de citronela
oil of clove aceite de clavero
oil of coconut aceite de nuez de coco
oil of cod-liver aceite de hígado de bacalao
oil of copaiba aceite de copaiba
oil of cottonseed aceite de semilla de algodón
oil of croton aceite de crotón
oil of fennel aceite de hinojo
oil of flaxseed aceite de linaza
oil of garlic aceite de ajo
oil of ginger aceite de jengibre
oil of hempseed aceite de cañamón
oil of jasmine aceite de jazmín
oil of jojoba aceite de jojoba
oil of laurel leaves aceite de hojas de laurel
oil of lavender aceite de lavanda

oil of lemongrass aceite de luisa
oil of linseed aceite de linaza
oil of mustard aceite de mostaza
oil of neatsfoot aceite de pata de vaca
oil of orange-peel aceite de corteza de naranja
oil of palm aceite de palma
oil of palmarosa aceite de palmarosa
oil of parsley aceite de semilla de perejil
oil of patchouli aceite de pachulí
oil of pennyroyal aceite de poleo
oil of peppermint aceite de menta
oil of perilla aceite de perilla
oil of pine aceite de pino
oil of rapeseed aceite de colza
oil of ricinus aceite de ricino
oil of safflower aceite de cártamo
oil of sage aceite de salvia
oil of sandalwood aceite de sándalo
oil of sassafras aceite de sasafrás
oil of sesame aceite de sésamo
oil of shale aceite de pizarra
oil of spearmint aceite de menta verde
oil of sunflower aceite de girasol
oil of tangerine aceite de tangerina
oil of tarragon aceite de estragón
oil of thuja aceite de tuya
oil of tiglium aceite de tiglio
oil of verbena aceite de verbena
oil of vitriol aceite de vitriolo
oil of wintergreen aceite de pirola
oil of wormseed aceite de quenopondio
ointment ungüento
okonite okonita
ol ol
ole ol
oleamide oleamida
oleate oleato
olefin olefina
olefin copolymer copolímero olefínico
olefin fiber fibra olefínica
olefin resin resina olefínica
oleic acid ácido oleico
olein oleína
oleo oleo
oleoresin oleorresina
oleoyl oleoilo
oleoyl chloride cloruro de oleoilo
oleoylsarcosine oleoilsarcosina

oleyl oleilo
oleyl alcohol alcohol oleílico
oleyl aldehyde aldehído oleílico
oligo oligo
oligomer oligómero
oligonite oligonita
oligonucleotide oligonucleótido
oligopeptide oligopéptido
oligosaccharide oligosacárido
olive oil aceite de oliva
olivenite olivenita
olivetol olivetol
olivine olivina
omega omega
oncogen oncógeno
one ona
one-way tap espita unidireccional
onium onio
Onsager equation ecuación de Onsager
opacity opacidad
opal ópalo
opaque opaco
open chain cadena abierta
open shell capa abierta
operation operación
opiate opiato
opioid opioide
opium opio
Oppenauer oxidation oxidación de
  Oppenauer
optical activity actividad óptica
optical-activity index índice de actividad
  óptica
optical axis eje óptico
optical brightener abrillantador óptico
optical crystal cristal óptico
optical density densidad óptica
optical electrons electrones ópticos
optical fiber fibra óptica
optical glass vidrio óptico
optical isomer isómero óptico
optical microscope microscopio óptico
optical purity pureza óptica
optical rotation rotación óptica
optical rotatory dispersion dispersión
  rotatoria óptica
optical spectrograph espectrógrafo
  óptico
optical spectrometer espectrómetro
  óptico
optical spectroscope espectroscopio
  óptico
optical spectroscopy espectroscopia
  óptica
optical spectrum espectro óptico
optically ópticamente
optically active ópticamente activo
optimum conditions condiciones óptimas
orange cadmium cadmio anaranjado
orange mineral mineral anaranjado
orange oil aceite de corteza de naranja
orange-peel oil aceite de corteza de
  naranja
orbit órbita
orbital orbital
orbital angular momentum momento
  angular orbital
orbital electron electrón orbital
orbital moment momento orbital
orbital quantum number número
  cuántico orbital
orbital state estado orbital
orbital symmetry simetría orbital
orbital theory teoría orbital
orcin orcina
order orden
ordinal test prueba ordinal
ore mena, mineral
organic orgánico
organic acid ácido orgánico
organic analysis análisis orgánico
organic base base orgánica
organic chemistry química orgánica
organic combustion combustión orgánica
organic compound compuesto orgánico
organic pigment pigmento orgánico
organic radical radical orgánico
organic reaction mechanism mecanismo
  de reacción orgánica
organic salt sal orgánica
organic solvent solvente orgánico
organic synthesis síntesis orgánica
organoaluminum compound compuesto
  organoalumínico
organoborane organoborano
organoclay organoarcilla
organogel organogel
organolithium compound compuesto

organolítico
organomagnesium compound
compuesto organomagnesiano
organomercury compound compuesto
organomercúrico
organometallic organometálico
organometallic compound compuesto
organometálico
organometalloid organometaloide
organophosphorus compound
compuesto organofosforado
organosilane organosilano
organosilicon organosilíceo
organosol organosol
organozinc compound compuesto
organocíncico
Orgel diagram diagrama de Orgel
orientation effect efecto de orientación
ornithine ornitina
ornithuric acid ácido ornitúrico
orotic acid ácido orótico
orpiment oropimente
Orsat analyzer analizador de Orsat
Orsat apparatus aparato de Orsat
orthamine ortamina
orthite ortita
ortho orto
ortho acid ácido orto
orthoacetic acid ácido ortoacético
orthoarsenic acid ácido ortoarsénico
orthoboric acid ácido ortobórico
orthocarbonic acid ácido ortocarbónico
orthoclase ortoclasa
orthoformic ester éster ortofórmico
orthohydrogen ortohidrógeno
orthonitrogen ortonitrógeno
orthophosphate ortofosfato
orthophosphoric acid ácido
ortofosfórico
orthorhombic system sistema
ortorrómbico
orthosilicate ortosilicato
orthosilicic acid ácido ortosilícico
orthotungstic acid ácido ortotúngstico
osamine osamina
osazone osazona
oscillating crystal cristal oscilante
oscillating reaction reacción oscilante
oscillation oscilación

oscillometric titration titulación
oscilométrica, valoración
oscilométrica
ose osa
osmate osmiato
osmic acid ácido ósmico
osmic anhydride anhídrido ósmico
osmics ósmica
osmiridium osmiridio
osmium osmio
osmium chloride cloruro de osmio
osmium dichloride dicloruro de osmio
osmium dioxide dióxido de osmio
osmium disulfide disulfuro de osmio
osmium monoxide monóxido de osmio
osmium oxide óxido de osmio
osmium potassium chloride cloruro
potásico de osmio
osmium tetrachloride tetracloruro de
osmio
osmium tetraoxide tetróxido de osmio
osmocene osmoceno
osmolality osmolalidad
osmolarity osmolaridad
osmole osmol
osmometer osmómetro
osmometry osmometría
osmosis ósmosis
osmotic osmótico
osmotic cell célula osmótica
osmotic coefficient coeficiente osmótico
osmotic equivalent equivalente osmótico
osmotic gradient gradiente osmótico
osmotic pressure presión osmótica
osmous osmoso
osmous chloride cloruro osmoso
osone osono
osotriazole osotriazol
Ostwald dilution law ley de dilución de
Ostwald
Ostwald ripening maduración de
Ostwald
Ostwald rule regla de Ostwald
Oudeman law ley de Oudeman
ous oso, o
outer electron electrón exterior
outer orbital orbital exterior
outer shell capa exterior
outgassing desgaseamiento

ovalene ovaleno

Overhauser effect efecto de Overhauser

overlapping orbitals orbitales solapados

overpotential sobrepotencial

overvoltage sobretensión

ovicide ovicida

oxa oxa

oxalacetic acid ácido oxalacético

oxalate oxalato

oxaldehydic acid ácido oxaldehídico

oxalic acid ácido oxálico

oxalo oxalo

oxalyl oxalilo

oxalyl chloride cloruro de oxalilo

oxalyl dichloride dicloruro de oxalilo

oxalylurea oxalilurea

oxamic acid ácido oxámico

oxamide oxamida

oxamido oxamido

oxamoyl oxamoilo

oxanthrol oxantrol

oxatyl oxatilo

oxazole oxazol

oxetane oxetano

oxethyl oxetilo

oxetone oxetona

oxidant oxidante

oxidase oxidasa

oxidate oxidar

oxidation oxidación

oxidation number número de oxidación

oxidation potential potencial de
    oxidación

oxidation process proceso de oxidación

oxidation reaction reacción de oxidación

oxidation-reduction oxidación-reducción

oxidation-reduction cycle ciclo de
    oxidación-reducción

oxidation-reduction indicator indicador
    de oxidación-reducción

oxidation-reduction potential potencial
    de oxidación-reducción

oxidation-reduction reaction reacción de
    oxidación-reducción

oxidation state estado de oxidación

oxidative oxidativo

oxidative addition adición oxidativa

oxidative coupling acoplamiento
    oxidativo

oxide óxido

oxidimetry oxidimetría

oxidize oxidar

oxidizer oxidante

oxidizing oxidante

oxidizing acid ácido oxidante

oxidizing agent agente oxidante

oxidizing atmosphere atmósfera oxidante

oxidizing flame llama oxidante

oxidizing material material oxidante

oxidizing reaction reacción oxidante

oxidoreductase oxidorreductasa

oxime oxima

oximido oximido

oxine oxina

oxirane oxirano

oxirene oxireno

oxo oxo

oxo acid oxoácido

oxo process oxoproceso

oxo reaction oxorreacción

oxoethanoic acid ácido oxoetanoico

oxonium ion ion oxonio

oxosilane oxosilano

oxy oxi

oxyacetylene flame llama oxiacetilénica

oxyacid oxiácido

oxyamide oxiamida

oxyanion oxianión

oxyazo dye colorante oxiazoico

oxybenzoic acid ácido oxibezoico

oxycarbonyl oxicarbonilo

oxycyanogen oxicianógeno

oxydipropionitrile oxidipropionitrilo

oxygen oxígeno

oxygen absorbent absorbente de oxígeno

oxygen balance equilibrio de oxígeno

oxygen carrier portador de oxígeno

oxygen consumed oxígeno consumido

oxygen electrode electrodo de oxígeno

oxygen fluoride fluoruro de oxígeno

oxygen inlet entrada de oxígeno

oxygenate oxigenar

oxygenated oxigenado

oxygenation oxigenación

oxyhydrogen oxihidrógeno

oxyluminescence oxiluminiscencia

oxymethurea oximeturea

oxymethylene oximetileno

oxynaphthoic acid ácido oxinaftoico
oxyphosphorane oxifosforano
oxyquinoline oxiquinolina
oxyquinone oxiquinona
oxytetracycline oxitetraciclina
oxytoxin oxitoxina
ozocerite ozocerita
ozokerite ozoquerita
ozonation ozonación

ozonator ozonizador
ozone ozono
ozone layer capa de ozono
ozonidation ozonidación
ozonide ozónido
ozonization ozonización
ozonizer ozonizador
ozonolysis ozonólisis

P

Paal-Knorr synthesis síntesis de Paal-
Knorr
pachnolite pacnolita
packaging empaque, empaquetado
packed empaquetado
packed column columna empaquetada
packed tower torre empaquetada
packing empaquetamiento
packing effect efecto de
empaquetamiento
packing fraction fracción de
empaquetamiento
pagodite pagodita
paint pintura
paint remover quitapintura
paired electrons electrones emparejados
paligorskite paligorskita
palladate paladato
palladic paládico
palladium paladio
palladium amine amina de paladio
palladium bichloride bicloruro de
paladio
palladium black negro de paladio
palladium chloride cloruro de paladio
palladium diacetate diacetato de paladio
palladium dibromide dibromuro de
paladio
palladium dicyanide dicianuro de
paladio
palladium diiodide diyoduro de paladio
palladium dinitrate dinitrato de paladio
palladium dioxide dióxido de paladio
palladium disulfide disulfuro de paladio
palladium hydride hidruro de paladio
palladium hydroxide hidróxido de
paladio
palladium iodide yoduro de paladio
palladium monosulfite monosulfito de
paladio
palladium monoxide monóxido de
paladio
palladium nitrate nitrato de paladio

palladium oxide óxido de paladio
palladium potassium chloride cloruro
potásico de paladio
palladium sodium chloride cloruro
sódico de paladio
palladium subsulfate subsulfato de
paladio
palladium sulfate sulfato de paladio
palladous paladioso
palladous chloride cloruro paladioso
palladous iodide yoduro paladioso
palladous nitrate nitrato paladioso
palladous potassium chloride cloruro
potásico paladioso
palladous sodium chloride cloruro
sódico paladioso
pallamine palamina
pallas palas
palm oil aceite de palma
palmarosa oil aceite de palmarosa
palmic pálmico
palmierite palmierita
palmitate palmitato
palmitic acid ácido palmítico
palmitic cyanide cianuro palmítico
palmitin palmitina
palmitinic acid ácido palmitínico
palmitoleic acid ácido palmitoleico
palmitolic acid ácido palmitólico
palmitonitrile palmitonitrilo
palmitoyl palmitoilo
palmitoyl chloride cloruro de palmitoilo
palmityl alcohol alcohol palmitílico
palustric acid ácido palústrico
palygorskite paligorskita
pamoic acid ácido pamoico
pan pan
panchromatic pancromático
panclastite panclastita
pancreatin pancreatina
pandermite pandermita
Paneth rule regla de Paneth
Paneth technique técnica de Paneth

pannic acid ácido pánico
pantachromatic pantacromático
pantethine pantetina
panthenol pantenol
pantocaine pantocaína
pantolactone pantolactona
pantomorphism pantomorfismo
pantothenic acid ácido pantoténico
pantothenol pantotenol
papain papaína
papaverine papaverina
papaverine hydrochloride clorhidrato de papaverina
paper papel
paper chromatography cromatografía sobre papel
paper electrochromatography electrocromatografía sobre papel
paper electrophoresis electroforesis sobre papel
para para
para red rojo para
parabanic acid ácido parabánico
paraben parabeno
parabituminous parabituminoso
paracasein paracaseína
paracetaldehyde paracetaldehído
paracetamol paracetamol
paraconic acid ácido paracónico
paraconine paraconina
paracyanic acid ácido paraciánico
paracyanogen paracianógeno
paradiazine paradiazina
paraffin parafina
paraffin oil aceite de parafina
paraffin wax cera de parafina
paraformaldehyde paraformaldehído
paragonite paragonita
parahelium parahelio
parahydrogen parahidrógeno
paralactic acid ácido paraláctido
paraldehyde paraldehído
paraldol paraldol
parallax paralaje
parallax error error de paralaje
parallel spins espines paralelos
parallelosterism paralelosterismo
paralysol paralisol
paramagnetic paramagnético

paramagnetic resonance resonancia paramagnética
paramagnetic spectrum espectro paramagnético
paramagnetism paramagnetismo
parameter parámetro
paramorph paramorfo
paramorphism paramorfismo
paranitraniline paranitranilina
paranitraniline red rojo de paranitranilina
parapectic acid ácido parapéctico
paraquat paraquat
pararosaniline pararrosanilina
pararosolic acid ácido pararosólico
parasite parásito
parasiticide parasiticida
parasorbic acid ácido parasórbico
parathion paratión
parawolframate parawolframato
paraxanthine paraxantina
paraxylene paraxileno
parent compound compuesto base
parent molecule molécula base
parent name nombre base
parietic acid ácido pariético
parinaric acid ácido parinárico
Paris green verde de París
Paris red rojo de París
Paris violet violeta de París
parisite parisita
Parkes process proceso de Parkes
paromomycin paromomicina
paromomycin sulfate sulfato de paromomicina
paroxazine paroxazina
parsley oil aceite de semilla de perejil
parthenine partenina
partial parcial
partial condenser condensador parcial
partial molal quantity cantidad molal parcial
partial molar volume volumen molar parcial
partial pressure presión parcial
partial racemization racemización parcial
partially permeable membrane membrana parcialmente permeable

particle partícula
particle accelerator acelerador de
  partículas
particle counting conteo de partículas
particle distribution distribución de
  partículas
particle emission emisión de partículas
particle energy energía de partícula
particle mass masa de partícula
particle scattering dispersión de
  partículas
particle size tamaño de partícula
particulate particulado
partinium partinio
partition partición
partition chromatography cromatografía
  de partición
partition coeffcient coeficiente de
  partición
partition function función de partición
partition law ley de partición
parvoline parvolina
parvuline parvulina
parylene parileno
Pascal law ley de Pascal
Pascal rules reglas de Pascal
Paschen-Back effect efecto de Paschen-
  Back
Paschen series serie de Paschen
Passerini reaction reacción de Passerini
passivation pasivación
passivator pasivador
passive pasivo
passive iron hierro pasivo
passive metal metal pasivo
passivity pasividad
paste pasta
Pasteur effect efecto de Pasteur
Pasteur filter filtro de Pasteur
Pasteur reaction reacción de Pasteur
Pasteur salt solution solución de sal de
  Pasteur
pasteurization pasteurización
patchouli oil aceite de pachulí
patentability patentabilidad
Paterno-Buchi reaction reacción de
  Paterno-Buchi
pathochemistry patoquímica
pathogenic patogénico

pathway vía
patronite patronita
Patterson function función de Patterson
Patterson synthesis síntesis de Patterson
Pattinson process proceso de Pattinson
paucine paucina
Pauli exclusion principle principio de
  exclusión de Pauli
Pauli-Fermi exclusion principle
    principio de exclusión de Pauli-Fermi
Pauli principle principio de Pauli
Pauli rule regla de Pauli
Pavy solution solución de Pavy
peacock blue azul pavo real
peak pico, punta
peanut oil aceite de maní
peanut oil meal harina de aceite de maní
pearl ash cenizas de perla
pearl essence esencia de perla
pearl white blanco perla
Pearson solution solución de Pearson
peat turba
peat tar alquitrán de turba
pebulate pebulato
Pechmann pyrazole synthesis síntesis de
  pirazoles de Pechmann
pectate pectato
pectic péctico
pectic acid ácido péctico
pectin pectina
pectin sugar azúcar de pectina
pectinase pectinasa
pectization pectización
pectolite pectolita
pectose pectosa
pedesis pedesis
pegmatite pegmatita
pelargonic acid ácido pelargónico
pelargonic alcohol alcohol pelargónico
pelargonic aldehyde aldehído
  pelargónico
pelargonin pelargonina
pelargonitrile pelargonitrilo
pelargonyl pelargonilo
pelargonyl chloride cloruro de
  pelargonilo
pelargonyl peroxide peroxido de
  pelargonilo
Peligot salt sal de Pelilgot

pellet pastilla, comprimido
pelletierine peletierina
Pellizzari reaction reacción de Pellizzari
pellotine pelotina
Pelouze synthesis síntesis de Pelouze
penetrant penetrante
penetration penetración
penicillamine penicilamina
penicillanase penicilanasa
penicillanic acid ácido penicilánico
penicillic acid ácido penicílico
penicillin penicilina
Penning ionization ionización de Penning
pennyroyal oil aceite de poleo
Pensky-Martens apparatus aparato de Pensky-Martens
penta penta
pentaamino pentaamino
pentabasic pentabásico
pentaborane pentaborano
pentabromide pentabromuro
pentabromo pentabromo
pentabromobenzene pentabromobenceno
pentacarbonyl pentacarbonilo
pentacarboxylic pentacarboxílico
pentacene pentaceno
pentacetate pentacetato
pentachloride pentacloruro
pentachloro pentacloro
pentachloroaniline pentacloroanilina
pentachlorobenzene pentaclorobenceno
pentachloroethane pentacloroetano
pentachloronaphthalene pentacloronaftaleno
pentachloronitrobenzene pentacloronitrobenceno
pentachlorophenol pentaclorofenol
pentachlorothiophenol pentaclorotiofenol
pentacosamic acid ácido pentacosámico
pentacosane pentacosano
pentacosanoic acid ácido pentacosanoico
pentacyclic pentacíclico
pentad pentada
pentadecane pentadecano
pentadecanoic acid ácido pentadecanoico
pentadecanol pentadecanol
pentadecanolide pentadecanolida

pentadecanone pentadecanona
pentadecenyl pentadecenilo
pentadecyl pentadecilo
pentadecyl alcohol alcohol pentadecílico
pentadecylic acid ácido pentadecílico
pentadiene pentadieno
pentadienedioic acid ácido pentadienodioico
pentaerythritol pentaeritritol
pentaerythritol tetraacetate tetraacetato de pentaeritritol
pentaerythritol tetranitrate tetranitrato de pentaeritritol
pentaethylbenzene pentaetilbenceno
pentafluoride pentafluoruro
pentagalloyl pentagaloilo
pentaglucose pentaglucosa
pentaglycerin pentaglicerina
pentaglycol pentaglicol
pentahydrate pentahidrato
pentahydro pentahidro
pentahydroxy pentahidroxi
pentahydroxycyclohexane pentahidroxiciclohexano
pentaiodo pentayodo
pentalin pentalina
pentamethyl pentametilo
pentamethylbenzene pentametilbenceno
pentamethylbenzoic acid ácido pentametilbenzoico
pentamethylene pentametileno
pentamethylene bromide bromuro de pentametileno
pentamethylene dibromide dibromuro de pentametileno
pentamethylene glycol pentametilenglicol
pentamethylene oxide óxido de pentametileno
pentamethyleneamine pentametilenamina
pentamethylenediamine pentametilendiamina
pentamethylphenol pentametilfenol
pentamino pentaamino
pentanal pentanal
pentane pentano
pentane thermometer termómetro de pentano

pentanecarboxylic acid ácido
   pentanocarboxílico
pentanedicarboxylic acid ácido
   pentanodicarboxílico
pentanedioic acid ácido pentanodioico
pentanediol pentanodiol
pentanedione pentanodiona
pentanethiol pentanotiol
pentanoic acid ácido pentanoico
pentanol pentanol
pentanone pentanona
pentaoxide pentaóxido
pentaprismo pentaprismo
pentasodium triphosphate trifosfato
   pentasódico
pentasulfide pentasulfuro
pentatriacontane pentatriacontano
pentavalent pentavalente
pentazdiene pentazdieno
pentazocine pentazocina
pentazolyl pentazolilo
pentel pentel
pentene penteno
pentenedioic acid ácido pentenodioico
pentenic acid ácido penténico
pentenoic acid ácido pentenoico
pentenol pentenol
pentenyl pentenilo
penthrite pentrita
pentinic acid ácido pentínico
pentite pentita
pentitol pentitol
pentlandite pentlandita
pentobarbital pentobarbital
pentoic acid ácido pentoico
pentol pentol
pentolite pentolita
pentonic acid ácido pentónico
pentosan pentosana
pentose pentosa
pentoside pentósido
pentoxide pentóxido
pentoxyphenol pentoxifenol
pentrite pentrita
pentyl pentilo
pentyl acetate acetato de pentilo
pentyl alcohol alcohol pentílico
pentyl aldehyde aldehído pentílico
pentyl bromide bromuro de pentilo

pentyl butanoate butanoato de pentilo
pentyl carbamate carbamato de pentilo
pentyl chloride cloruro de pentilo
pentyl chlorocarbonate clorocarbonato
   de pentilo
pentyl cyanide cianuro de pentilo
pentyl ether éter pentílico
pentyl formate formiato de pentilo
pentyl furoate furoato de pentilo
pentyl hydrate hidrato de pentilo
pentyl hydride hidruro de pentilo
pentyl iodide yoduro de pentilo
pentyl isocyanide isocianuro de pentilo
pentyl ketone pentilcetona
pentyl nitrate nitrato de pentilo
pentyl oxide óxido de pentilo
pentyl phenyl ketone pentilfenilcetona
pentyl phenylhydrazine fenilhidrazina
   de pentilo
pentyl phthalate ftalato de pentilo
pentyl propionate propionato de pentilo
pentyl salicylate salicilato de pentilo
pentyl sulfate sulfato de pentilo
pentyl thiocyanate tiocianato de pentilo
pentyl valerate valerato de pentilo
pentyl xanthate xantato de pentilo
pentylacetic ether éter pentilacético
pentylamine pentilamina
pentylbenzene pentilbenceno
pentylbenzoate pentilbenzoato
pentylcinnamaldehyde
   pentilcinamaldehído
pentylene pentileno
pentylformic acid ácido pentilfórmico
pentylidene pentilideno
pentylidyne pentilidino
pentyloxy pentiloxi
pentyloxyhydrate pentiloxihidrato
pentyloxyphenol pentiloxifenol
pentylphenol pentilfenol
pentylurea pentilurea
pentylurethane pentiluretano
pentyne pentino
peppermint camphor alcanfor de menta
peppermint oil aceite de menta
pepsin pepsina
pepsinogen pepsinógeno
peptidase peptidasa
peptide péptido

peptide bond enlace de péptido
peptide linkage enlace de péptido
peptization peptización
peptone peptona
per per
peracetic acid ácido peracético
peracid perácido
peracidity peracidez
peralcohol peralcohol
perbenzoic acid ácido perbenzoico
perborate perborato
perbromate perbromato
perbromo perbromo
percarbamide percarbamida
percarbide percarburo
percarbonate percarbonato
percarbonic acid ácido percarbónico
percentage porcentaje
perchlorate perclorato
perchloric acid ácido perclórico
perchloric acid anhydride anhídrido de ácido perclórico
perchloride percloruro
perchloro percloro
perchlorobenzene perclorobenceno
perchlorocyclopentadiene perclorociclopentadieno
perchloroethane percloroetano
perchloroether percloroéter
perchloroethylene percloroetileno
perchloromethane perclorometano
perchloromethyl perclorometilo
perchloromethyl mercaptan perclorometilmercaptano
perchloropropylene percloropropileno
perchloryl perclorilo
perchloryl fluoride fluoruro de perclorilo
perchromate percromato
perchromic acid ácido percrómico
percolate percolar
percolation percolación
percolator percolador
percrystallization percristalización
perdistillation perdestilación
perfect gas gas perfecto
perfect solution solución perfecta
perfluoro perfluoro
perfluorobutene perfluorobuteno

perfluorobutyric acid ácido perfluorobutírico
perfluorocarbon perfluorocarburo
perfluorocyclobutane perfluorociclobutano
perfluoroethylene perfluoroetileno
perfluoropropane perfluoropropano
perfluoropropene perfluoropropeno
perfluorosulfonic acid ácido perfluorosulfónico
perforated plate placa perforada
performic acid ácido perfórmico
perfume perfume
perfusion perfusión
pergenol pergenol
perhydrate perhidrato
perhydro perhidro
perhydrol perhidrol
perhydronaphthalene perhidronaftaleno
peri peri
peri acid ácido peri
periclase periclasa
pericline periclina
pericyclic pericíclico
pericyclic reaction reacción pericíclica
pericyclo periciclo
perilla oil aceite de perilla
perilogic series serie perilógica
perimidine perimidina
perimidinyl perimidinilo
perimorph perimorfo
period periodo
periodate peryodato
periodic acid ácido peryódico
periodic chain cadena periódica
periodic classification clasificación periódica
periodic law ley periódica
periodic property propiedad periódica
periodic series serie periódica
periodic system sistema periódico
periodic table tabla periódica
periodicity periodicidad
peritectic peritéctico
peritectic point punto peritéctico
peritectic system sistema peritéctico
peritectic temperature temperatura peritéctica
peritectoid peritectoide

Perkin alicyclic synthesis síntesis alicíclica de Perkin
Perkin reaction reacción de Perkin
Perkin rearrangement reordenamiento de Perkin
perkinetic pericinético
Perkow reaction reacción de Perkow
perlite perlita
permalloy permaleación
permanent gas gas permanente
permanent hardness dureza permanente
permanent white blanco permanente
permanganate permanganato
permanganate titration titulación con permanganato, valoración con permanganato
permanganic acid ácido permangánico
permanganyl permanganilo
permeability permeabilidad
permeable permeable
permeable membrane membrana permeable
permeametry permeametría
permeation impregnación
permethrin permetrina
permissible explosive explosivo permisible
permonosulfuric acid ácido permonosulfúrico
permutation permutación
permutite permutita
pernitric acid ácido pernítrico
perovskite perovskita
peroxidase peroxidasa
peroxidation peroxidación
peroxide peróxido
peroxide number número de peróxido
peroxide value valor de peróxido
peroxo peroxo
peroxomonosulfuric acid ácido peroxomonosulfúrico
peroxonitric acid ácido peroxonítrico
peroxy peroxi
peroxyacetic acid ácido peroxiacético
peroxybenzoic acid ácido peroxibenzoico
peroxybenzoyl nitrate nitrato de peroxibenzoilo
peroxychromium peroxicromo
peroxydol peroxidol

peroxyethanoic acid ácido peroxietanoico
peroxyformic acid ácido peroxifórmico
peroxymonosulfuric acid ácido peroximonosulfúrico
peroxysulfuric acid ácido peroxisulfúrico
perphenazine perfenazina
perphosphoric acid ácido perfosfórico
perrhenate perrenato
perrhenic acid ácido perrénico
Perrin equation ecuación de Perrin
Perrin rule regla de Perrin
Persian red rojo de Persia
persorption persorción
persulfate persulfato
persulfide persulfuro
persulfuric acid ácido persulfúrico
perthio pertio
perthiocarbonate pertiocarbonato
Peru balsam bálsamo de Perú
pervaporation pervaporación
perylene perileno
pesticide pesticida
pestle mano de mortero
petalite petalita
Peterson reaction reacción de Peterson
petri dish caja de Petri
petrochemical petroquímico
Petroff reagent reactivo de Petroff
petrolatum petrolato
petrolatum wax cera de petrolato
petroleum petróleo
petroleum asphalt asfalto de petróleo
petroleum benzin bencina de petróleo
petroleum coke coque de petróleo
petroleum ether éter de petróleo
petroleum gas gas de petróleo
petroleum jelly petrolato
petroleum naphtha nafta de petróleo
petroleum refining refinación de petróleo
petroleum wax cera de petróleo
petroxolin petroxolina
petzite petzita
pewter peltre
Pfeiffer effect efecto de Pfeiffer
Pfitzinger reaction reacción de Pfitzinger

Pfitzner-Moffatt oxidation oxidación de
    Pfitzner-Moffatt
pH pH
pH indicator indicador de pH
pH meter medidor de pH
pH recorder registrador de pH
pH scale escala de pH
pH value valor de pH
pharmaceutical farmacéutico
pharmacokinetics farmacocinética
pharmacolite farmacolita
pharmacology farmacología
phase fase
phase diagram diagrama de fases
phase equilibrium equilibrio de fases
phase reversal inversión de fase
phase rule regla de fases
phase transfer transferencia de fase
phase transition transición de fase
phasotropy fasotropía
phellandrene felandreno
phen fen
phenacaine hydrochloride clorhidrato de
    fenacaína
phenacetin fenacetina
phenacetol fenacetol
phenacyl fenacilo
phenacyl alcohol alcohol fenacílico
phenacyl chloride cloruro de fenacilo
phenacylidene fenacilideno
phenanthraquinone fenantraquinona
phenanthrene fenantreno
phenanthrenequinone fenantrenoquinona
phenanthrenyl fenantrenilo
phenanthridine fenantridina
phenanthrol fenantrol
phenanthroline fenantrolina
phenanthroline indicator indicador de
    fenantrolina
phenanthrone fenantrona
phenanthrophenazine fenantrofenacina
phenanthryl fenantrilo
phenanthrylene fenantrileno
phenarsazine chloride cloruro de
    fenarsazina
phenate fenato
phenazine fenazina
phenazone fenazona
phenazonium fenazonio

phenenyl fenenilo
phenethyl fenetilo
phenethyl acetate acetato de fenetilo
phenethyl alcohol alcohol fenetílico
phenethyl isobutyrate isobutirato de
    fenetilo
phenethylamine fenetilamina
phenethylene fenetileno
phenetidine fenetidina
phenetidino fenetidino
phenetole fenetol
phenetyl fenetilo
phenic acid ácido fénico
phenidone fenidona
phenindione fenindiona
pheniramine maleate maleato de
    feniramina
phenixin fenixina
phenmethyl fenmetilo
phenobarbital fenobarbital
phenobarbitone fenobarbitona
phenodiazine fenodiazina
phenol fenol
phenol acid ácido fenólico
phenol aldehyde aldehído fenólico
phenol bismuth bismuto de fenol
phenol coefficient coeficiente de fenol
phenol ether éter fenólico
phenol-formaldehyde resin resina de
    fenol-formaldehído
phenol-furfural resin resina de fenol-
    furfural
phenol red rojo fenol
phenol trinitrate trinitrato de fenol
phenolate fenolato
phenolate process proceso de fenolato
phenoldisulfonic acid ácido
    fenoldisulfónico
phenolic fenólico
phenolic acid ácido fenólico
phenolic resin resina fenólica
phenolphthalein fenolftaleína
phenolphthalide fenolftalida
phenolsafranin fenolsafranina
phenolsulfonate fenolsulfonato
phenolsulfonephthalein
    fenolsulfonaftaleína
phenolsulfonic acid ácido fenolsulfónico
phenone fenona

phenothiazine fenotiazina
phenothrin fenotrina
phenotole fenotol
phenoxazine fenoxazina
phenoxide fenóxido
phenoxy fenoxi
phenoxy resin resina fenoxi
phenoxyacetic acid ácido fenoxiacético
phenoxybenzene fenoxibenceno
phenoxydihydroxypropane
    fenoxidihidroxipropano
phenoxyethanol fenoxietanol
phenoxyethylpenicillin
    fenoxietilpenicilina
phenoxypropanediol fenoxipropanodiol
phenoxypropylene oxide óxido de
    fenoxipropileno
phenyl fenilo
phenyl acetate acetato de fenilo
phenyl azide azida de fenilo
phenyl bromide bromuro de fenilo
phenyl carbamate carbamato de fenilo
phenyl carbonate carbonato de fenilo
phenyl chloride cloruro de fenilo
phenyl cyanide cianuro de fenilo
phenyl ether éter fenílico
phenyl fluoride fluoruro de fenilo
phenyl hydrate hidrato de fenilo
phenyl hydroxide hidróxido de fenilo
phenyl isocyanate isocianato de fenilo
phenyl isocyanide isocianuro de fenilo
phenyl isothiocyanate isotiocianato de
    fenilo
phenyl ketone fenilcetona
phenyl mercaptan fenilmercaptano
phenyl methyl ketone fenilmetilcetona
phenyl mustard oil aceite de
    fenilmostasa
phenyl phthalate ftalato de fenilo
phenyl salicylate salicilato de fenilo
phenyl sulfate sulfato de fenilo
phenyl valerate valerato de fenilo
phenyl vinyl ketone fenilvinilcetona
phenylacetaldehyde fenilacetaldehído
phenylacetaldehyde dimethylacetal
    dimetilacetal de fenilacetaldehído
phenylacetamide fenilacetamida
phenylacetic acid ácido fenilacético
phenylacetonitrile fenilacetonitrilo

phenylacetyl fenilacetilo
phenylacetyl chloride cloruro de
    fenilacetilo
phenylalanine fenilalanina
phenylamine fenilamina
phenylaniline fenilanilina
phenylarsonic acid ácido fenilarsónico
phenylation fenilación
phenylazo fenilazo
phenylbarbital fenilbarbital
phenylbenzamide fenilbenzamida
phenylbenzene fenilbenceno
phenylbenzoate fenilbenzoato
phenylbenzoyl fenilbenzoilo
phenylbenzyl fenilbencilo
phenylbutane fenilbutano
phenylbutanoic acid ácido fenilbutanoico
phenylbutazone fenilbutazona
phenylbutene fenilbuteno
phenylbutynol fenilbutinol
phenylbutyric acid ácido fenilbutírico
phenylcarbamoyl fenilcarbamoilo
phenylcarbethoxypyrazolone
    fenilcarbetoxipirazolona
phenylcarbimide fenilcarbimida
phenylcarbinol fenilcarbinol
phenylcarbylamine chloride cloruro de
    fenilcarbilamina
phenylchloroform fenilcloroformo
phenylcyclidene hydrochloride
    clorhidrato de fenilciclideno
phenylcyclohexane fenilciclohexano
phenylcyclohexanol fenilciclohexanol
phenyldichloroarsine fenildicloroarsina
phenyldidecyl phosphite fosfito de
    fenildidecilo
phenyldiethanolamine
    fenildietanolamina
phenyldimethylurea fenildimetilurea
phenylene fenileno
phenylene blue azul fenileno
phenylenediamine fenilendiamina
phenylethane feniletano
phenylethanoic acid ácido feniletanoico
phenylethyl feniletilo
phenylethyl acetate acetato de feniletilo
phenylethyl alcohol alcohol feniletílico
phenylethyl mercaptan
    feniletilmercaptano

phenylethyl salicylate salicilato de feniletilo

phenylethylacetic acid ácido feniletilacético

phenylethylamine feniletilamina

phenylethylethanolamine feniletiletanolamina

phenylethylene feniletileno

phenylethylene glycol feniletilenglicol

phenylethylether feniletiléter

phenylformamide fenilformamida

phenylformic acid ácido fenilfórmico

phenylglycine fenilglicina

phenylglycol fenilglicol

phenylglycolic acid ácido fenilglicólico

phenylhydrazine fenilhidrazina

phenylhydrazone fenilhidrazona

phenylhydroxylamine fenilhidroxilamina

phenylic fenílico

phenylic acid ácido fenílico

phenylidene fenilideno

phenyllithium fenil-litio

phenylmagnesium fenilmagnesio

phenylmagnesium bromide bromuro de fenilmagnesio

phenylmagnesium chloride cloruro de fenilmagnesio

phenylmercuric fenilmercúrico

phenylmercuric acetate acetato fenilmercúrico

phenylmercuric benzoate benzoato fenilmercúrico

phenylmercuric borate borato fenilmercúrico

phenylmercuric chloride cloruro fenilmercúrico

phenylmercuric ethanoate etanoato fenilmercúrico

phenylmercuric hydroxide hidroxido fenilmercúrico

phenylmercuric lactate lactato fenilmercúrico

phenylmercuric naphthenate naftenato fenilmercúrico

phenylmercuric nitrate nitrato fenilmercúrico

phenylmercuric oleate oleato fenilmercúrico

phenylmercuric propionate propionato fenilmercúrico

phenylmercuric salicylate salicilato fenilmercúrico

phenylmethane fenilmetano

phenylmethanol fenilmetanol

phenylmethyl fenilmetilo

phenylmethyl acetate acetato de fenilmetilo

phenylmethylcarbinol fenilmetilcarbinol

phenylmethylene fenilmetileno

phenylmethylethanolamine fenilmetiletanolamina

phenylmorpholine fenilmorfolina

phenylnaphthalene fenilnaftaleno

phenylnaphthylamine fenilnaftilamina

phenylneopentyl phosphite fosfito de fenilneopentilo

phenylnitroamine fenilnitroamina

phenylnonane fenilnonano

phenylpentane fenilpentano

phenylphenol fenilfenol

phenylphosphine fenilfosfina

phenylphosphinic acid ácido fenilfosfínico

phenylphosphonic acid ácido fenilfosfónico

phenylpiperazine fenilpiperazina

phenylpropane fenilpropano

phenylpropanol fenilpropanol

phenylpropanone fenilpropanona

phenylpropenal fenilpropenal

phenylpropenoic acid ácido fenilpropenoico

phenylpropenol fenilpropenol

phenylpropionic acid ácido fenilpropiónico

phenylpropyl fenilpropilo

phenylpropyl acetate acetato de fenilpropilo

phenylpropyl alcohol alcohol fenilpropílico

phenylpropyl aldehyde aldehído fenilpropílico

phenylpropyl chloride cloruro de fenilpropilo

phenylpropyl ketone fenilpropilcetona

phenylpropylpyridine fenilpropilpiridina

phenylpyrazolide fenilpirazolida

phenylpyrazolidone fenilpirazolidona

phenylpyridine fenilpiridina
phenylsalicylic acid ácido fenilsalicílico
phenylstearic acid ácido fenilesteárico
phenylsulfanilic acid ácido
　　fenilsulfanílico
phenylsulfonic acid ácido fenilsulfónico
phenylsulfonyl fenilsulfonilo
phenylthiourea feniltiourea
phenyltoluene feniltolueno
phenyltrichlorosilane feniltriclorosilano
phenyltridecane feniltridecano
phenylurethane feniluretano
phenytoin fenitoína
pheromone feromona
phi phi
phillipite filipita
phloroglucinol floroglucinol
phonochemistry fonoquímica
phorate forato
phorbol forbol
phorone forona
phoryl forilo
phosalone fosalona
phosgene fosgeno
phosmet fosmet
phospha fosfa
phosphamic acid ácido fosfámico
phosphamide fosfamida
phosphamidon fosfamidón
phosphane fosfano
phosphatase fosfatasa
phosphate fosfato
phosphate buffer amortiguador de
　　fosfato
phosphate glass vidrio de fosfato
phosphate rock roca de fosfato
phosphate slag escoria de fosfato
phosphatide fosfátido
phosphazene fosfaceno
phosphenyl fosfenilo
phosphide fosfuro
phosphinate fosfinato
phosphine fosfina
phosphinic acid ácido fosfínico
phosphinico fosfinico
phosphinite fosfinita
phosphino fosfino
phosphinoso fosfinoso
phosphinous acid ácido fosfinoso

phosphinoyl fosfinoilo
phosphite fosfito
phospho fosfo
phosphobenzene fosfobenceno
phosphocreatine fosfocreatina
phosphodiester fosfodiéster
phosphodiesterase fosfodiesterasa
phosphoglyceride fosfoglicérido
phospholipase fosfolipasa
phospholipid fosfolípido
phosphomolybdate fosfomolibdato
phosphomolybdic acid ácido
　　fosfomolíbdico
phosphonate fosfonato
phosphonic acid ácido fosfónico
phosphonite fosfonita
phosphonitrile fosfonitrilo
phosphonium fosfonio
phosphonium bromide bromuro de
　　fosfonio
phosphonium chloride cloruro de
　　fosfonio
phosphonium hydroxide hidróxido de
　　fosfonio
phosphonium iodide yoduro de fosfonio
phosphonium salt sal de fosfonio
phosphono fosfono
phosphonoso fosfonoso
phosphonous acid ácido fosfonoso
phosphonoyl fosfonoilo
phosphoprotein fosfoproteína
phosphor fósforo
phosphor bronze bronce fosforado
phosphorane fosforano
phosphorate fosforato
phosphorated fosforado
phosphorescence fosforescencia
phosphorescent fosforescente
phosphoric fosfórico
phosphoric acid ácido fosfórico
phosphoric anhydride anhídrido
　　fosfórico
phosphoric bromide bromuro fosfórico
phosphoric chloride cloruro fosfórico
phosphoric oxide óxido fosfórico
phosphoric perbromide perbromuro
　　fosfórico
phosphoric perchloride percloruro
　　fosfórico

phosphoric sulfide sulfuro fosfórico
phosphorimetry fosforimetría
phosphorite fosforita
phosphoroso fosforoso
phosphorous acid ácido fosforoso
phosphorus fósforo
phosphorus bromide bromuro de fósforo
phosphorus chloride cloruro de fósforo
phosphorus cycle ciclo de fósforo
phosphorus heptasulfide heptasulfuro de fósforo
phosphorus hydride hidruro de fósforo
phosphorus nitride nitruro de fósforo
phosphorus oxide óxido de fósforo
phosphorus oxyacid oxiácido de fósforo
phosphorus oxybromide oxibromuro de fósforo
phosphorus oxychloride oxicloruro de fósforo
phosphorus pentabromide pentabromuro de fósforo
phosphorus pentachloride pentacloruro de fósforo
phosphorus pentafluoride pentafluoruro de fósforo
phosphorus pentaselenide pentaseleniuro de fósforo
phosphorus pentasulfide pentasulfuro de fósforo
phosphorus pentoxide pentóxido de fósforo
phosphorus persulfide persulfuro de fósforo
phosphorus sesquisulfide sesquisulfuro de fósforo
phosphorus sulfide sulfuro de fósforo
phosphorus thiochloride tiocloruro de fósforo
phosphorus tribromide tribromuro de fósforo
phosphorus trichloride tricloruro de fósforo
phosphorus triiodide triyoduro de fósforo
phosphorus trisulfide trisulfuro de fósforo
phosphoryl fosforilo
phosphoryl bromide bromuro de fosforilo

phosphoryl chloride cloruro de fosforilo
phosphoryl fluoride fluoruro de fosforilo
phosphoryl nitride nitruro de fosforilo
phosphorylase fosforilasa
phosphorylation fosforilación
phosphotungstate fosfotungstato
phosphotungstic acid ácido fosfotúngstico
photoacoustic spectroscopy espectroscopia fotoacústica
photocatalysis fotocatálisis
photocatalyst fotocatalizador
photocathode fotocátodo
photochemical fotoquímico
photochemical activation activación fotoquímica
photochemical catalysis catálisis fotoquímica
photochemical dissociation disociación fotoquímica
photochemical effect efecto fotoquímico
photochemical equivalent equivalente fotoquímico
photochemical equivalent law ley de equivalente fotoquímico
photochemical excitation excitación fotoquímica
photochemical induction inducción fotoquímica
photochemical oxidant oxidante fotoquímico
photochemical process proceso fotoquímico
photochemical reaction reacción fotoquímica
photochemical yield rendimiento fotoquímico
photochemistry fotoquímica
photochromic fotocrómico
photochromic compound compuesto fotocrómico
photochromism fotocromismo
photoconduction fotoconducción
photodecomposition fotodescomposición
photodegradation fotodegradación
photodeposition fotodeposición
photodimerization fotodimerización
photodisintegration fotodesintegración
photodissociation fotodisociación

photoelectric fotoeléctrico
photoelectric absorption absorción fotoeléctrica
photoelectric cell célula fotoeléctrica
photoelectric colorimeter colorímetro fotoeléctrico
photoelectron fotoelectrón
photoelectron spectromicroscopy espectromicroscopia de fotoelectrones
photoelectron spectroscopy espectroscopia de fotoelectrones
photoemission fotoemisión
photoemissive fotoemisivo
photoemitter fotoemisor
photographic fotográfico
photoionization fotoionización
photoluminescence fotoluminiscencia
photolysis fotólisis
photolyte fotólito
photolytic fotolítico
photometer fotómetro
photometric fotométrico
photometric analysis análisis fotométrico
photometric titration titulación fotométrica, valoración fotométrica
photon fotón
photon absorption absorción de fotones
photon emission emisión de fotones
photoneutron fotoneutrón
photonuclear fotonuclear
photonuclear reaction reacción fotonuclear
photooxidation fotooxidación
photopolymer fotopolímero
photopolymerization fotopolimerización
photoproton fotoprotón
photoreaction fotorreacción
photoreduction fotorreducción
photosensitive fotosensible
photosensitive substance substancia fotosensible
photosensitivity fotosensibilidad
photosensitization fotosensibilización
photosensitizer fotosensibilizador
photostabilization fotoestabilización
photosynthesis fotosíntesis
phototropism fototropismo
phototropy fototropia

photovoltaic fotovoltaico
photovoltaic cell célula fotovoltaica
phthalamic acid ácido ftalámico
phthalamide ftalamida
phthalate ftalato
phthalazine ftalazina
phthaldiamide ftaldiamida
phthalein ftaleína
phthalic acid ácido ftálico
phthalic anhydride anhídrido ftálico
phthalic ester éster ftálico
phthalide ftalida
phthalidyl ftalidilo
phthalidylidene ftalidilideno
phthalimide ftalimida
phthalimido ftalimido
phthalocyanine ftalocianina
phthalodinitrile ftalodinitrilo
phthalonitrile ftalonitrilo
phthaloyl ftaloilo
phthaloyl chloride cloruro de ftaloilo
phthalyl ftalilo
phthioic acid ácido ftioico
phycocolloid ficocoloide
physical analysis análisis físico
physical chemistry química física, fisicoquímica
physical constant constante física
physical phenomenon fenómeno físico
physical property propiedad física
physical quantity cantidad física
physical solution solución física
physical solvent solvente físico
physical testing pruebas físicas
physicochemical fisicoquímico
physostigmine fisostigmina
physostigmine sulfate sulfato de fisostigmina
phytane fitano
phytanic acid ácido fitánico
phytic acid ácido fítico
phytin fitina
phytochemistry fitoquímica
phytol fitol
phytotoxic fitotóxico
phytyl fitilo
pi pi
pi bond enlace pi
pi bonding enlazado pi

pi complex complejo pi
pi electron electrón pi
picene piceno
pickling decapado
picocurie picocurie
picoline picolina
picolinic acid ácido picolínico
picolyl picolilo
picolylamine picolilamina
picotite picotita
picramic acid ácido picrámico
picramide picramida
picraminic acid ácido picramínico
picrate picrato
picric acid ácido pícrico
picro picro
picrolonic acid ácido picrolónico
picromerite picromerita
picronitric acid ácido picronítrico
picrotoxin picrotoxina
picryl picrilo
picryl chloride cloruro de picrilo
Pictet-Hubert reaction reacción de
  Pictet-Hubert
piedmontite piedmontita
piezochemistry piezoquímica
piezocrystallization piezocristalización
piezoelectric effect efecto piezoeléctrico
piezoelectricity piezoelectricidad
pig iron hierro en lingotes
pigment pigmento
pigment volume concentration
  concentración de volmen de pigmento
pilocarpine pilocarpina
pilot plant planta piloto
Piloty-Robinson synthesis síntesis de
  Piloty-Robinson
pimaric acid ácido pimárico
pimaricin pimaricina
pimelic acid ácido pimélico
pimelite pimelita
pinacoid pinacoide
pinacol pinacol
pinacol condensation condensación de
  pinacol
pinacol rearrangement reordenamiento
  de pinacol
pinacolone pinacolona
pinacone pinacona

pinane pinano
pinanyl pinanilo
pine oil aceite de pino
pine tar alquitrán de pino
pine-tar oil aceite de alquitrán de pino
pine-tar pitch brea de alquitrán de pino
pinene pineno
pinene hydrochloride clorhidrato de
  pineno
pinic acid ácido pínico
Pinner reaction reacción de Pinner
Pinner synthesis síntesis de Pinner
pinocarveol pinocarveol
pipe clay tierra de pipas
pipecoline pipecolina
piperaline piperalina
piperazine piperazina
piperazine dihydrochloride diclorhidrato
  de piperazina
piperazine hexahydrate hexahidrato de
  piperazina
piperidine piperidina
piperidinium piperidinio
piperidino piperidino
piperidinoethanol piperidinoetanol
piperidyl piperidilo
piperine piperina
piperitol piperitol
piperitone piperitona
piperonal piperonal
piperonyl piperonilo
piperonyl butoxide butóxido de
  piperonilo
piperonylic acid ácido piperonílico
piperonylidene piperonilideno
piperonyloyl piperoniloilo
piperyl piperilo
piperylene piperileno
pipet pipeta
pipette pipeta
Piria reaction reacción de Piria
pitch pez, brea
pitchblende pecblenda
Pitzer equation ecuación de Pitzer
pivaldehyde pivaldehído
pivalic acid ácido piválico
Planck constant constante de Planck
plane of polarization plano de
  polarización

plane of symmetry plano de simetría
plane-polarized polarizado en un plano
plane-polarized light luz polarizada en un plano
planetary electrons electrones planetarios
plant planta
plasma plasma
plasmid plásmido
plasmin plasmina
plasmogen plasmógeno
plaster of Paris yeso de París
plastic plástico
plastic film película de plástico
plastic flow flujo plástico
plastic foam espuma de plástico
plastic pipe tubo de plástico
plasticity plasticidad
plasticizer plasticizador
plastisol plastisol
plastoquinone plastoquinona
plate placa
plate column columna de placas
plate efficiency eficiencia de placas
plated depositado, electrodepositado, metalizado, galvanoplasteado
plating depósito, electrodeposición, metalización, galvanoplastia
platinic platínico
platinic ammonium chloride cloruro amónico platínico
platinic chloride cloruro platínico
platinic oxide óxido platínico
platinic sodium chloride cloruro sódico platínico
platinic sulfate sulfato platínico
platinization platinado
platinochloride platinocloruro
platinocyanide platinocianuro
platinous platinoso
platinous chloride cloruro platinoso
platinous iodide yoduro platinoso
platinum platino
platinum amine amina de platino
platinum ammonium chloride cloruro amónico de platino
platinum bichloride bicloruro de platino
platinum black negro de platino
platinum chloride cloruro de platino

platinum-cobalt alloy aleación de platino-cobalto
platinum dibromide dibromuro de platino
platinum dichloride dicloruro de platino
platinum diiodide diyoduro de platino
platinum dioxide dióxido de platino
platinum iodide yoduro de platino
platinum-iridium alloy aleación de platino-iridio
platinum-lithium platino-litio
platinum monosulfide monosulfuro de platino
platinum oxide óxido de platino
platinum potassium chloride cloruro potásico de platino
platinum-rhodium alloy aleación de platino-rodio
platinum sodium chloride cloruro sódico de platino
platinum sponge esponja de platino
platinum sulfate sulfato de platino
platinum tetrachloride tetracloruro de platino
plazolite plazolita
pleochroism pleocroismo
plumbago plumbago
plumbate plumbato
plumbic plúmbico
plumbic acid ácido plúmbico
plumbous plumboso
plumbous oxide óxido plumboso
plumbous sulfide sulfuro plumboso
plutonium plutonio
plutonium oxide óxido de plutonio
Podbielniak analysis análisis de Podbielniak
poison veneno
poison gas gas venenoso
poison vapor vapor venenoso
poisonous venenoso
polar polar
polar compound compuesto polar
polar covalent bond enlace covalente polar
polar liquid líquido polar
polar molecule molécula polar
polar solvent solvente polar
polarimeter polarímetro

polarimetric analysis análisis polarimétrico
polarimetry polarimetría
polarity polaridad
polarizability polarizabilidad
polarizable polarizable
polarization polarización
polarization current corriente de polarización
polarization potential potencial de polarización
polarize polarizar
polarized polarizado
polarized electrode electrodo polarizado
polarized light luz polarizada
polarizer polarizador
polarizing polarizante
polarogram polarograma
polarographic analysis análisis polarográfico
polarography polarografía
Polenske number número de Polenske
pollucite polucita
pollution contaminación
polonium polonio
Polonovski reaction reacción de Polonovski
poly poli
polyacetal poliacetal
polyacetaldehyde poliacetaldehído
polyacetylene poliacetileno
polyacrylamide poliacrilamida
polyacrylate poliacrilato
polyacrylic acid ácido poliacrílico
polyacrylic fiber fibra poliacrílica
polyacrylonitrile poliacrilonitrilo
polyalcohol polialcohol
polyalkane polialcano
polyalkene polialqueno
polyalkylidene polialquilideno
polyallomer polialómero
polyamide poliamida
polyamide resin resina de poliamida
polyamine poliamina
polyamine-methylene resin resina de poliamina-metileno
polyaminotriazole poliaminotriazol
polyatomic poliatómico
polyatomic gas gas poliatómico

polyatomic molecule molécula poliatómica
polybasic polibásico
polybasic acid ácido polibásico
polybenzimidazole polibencimidazol
polyblend polimezcla
polybutadiene polibutadieno
polybutene polibuteno
polybutylene polibutileno
polycarbonate policarbonato
polycarboxylic policarboxílico
polycarboxylic acid ácido policarboxílico
polychlor policloro
polychloral policloral
polychlorinated biphenyl bifenilo policlorado
polychloroprene policloropreno
polychlorotrifluoroethane policlorotrifluoroetano
polychromatic policromático
polycondensation policondensación
polycyclic policíclico
polycyclic hydrocarbon hidrocarburo policíclico
polydimethylsiloxane polidimetilsiloxano
polydispersion polidispersión
polyelectrolyte polielectrólito
polyene polieno
polyester poliéster
polyester fiber fibra poliéster
polyester film película poliéster
polyester resin resina poliéster
polyester rubber caucho poliéster
polyethenoid polietenoide
polyether poliéter
polyether foam espuma poliéter
polyether glycol polieterglicól
polyether resin resina poliéter
polyethylene polietileno
polyethylene glycol polietilenglicol
polyethylene glycol chloride cloruro de polietilenglicol
polyethylene oxide óxido de polietileno
polyethylene resin resina de polietileno
polyethylene tetraphthalate tetraftalato de polietileno
polyethyleneimine polietilenimina
polyformaldehyde poliformaldehído
polyforming poliformación

polyfunctional polifuncional
polyfunctional system sistema polifuncional
polyfurfuryl alcohol alcohol polifurfurílico
polygen polígeno
polyglycerol poliglicerol
polyglycerol ester éster de poliglicerol
polyglycol poliglicol
polyglycol distearate diestearato de poliglicol
polyhalite polihalita
polyhexafluoropropene polihexafluoropropeno
polyhexamethylene polixematileno
polyhydrate polihidrato
polyhydric polihídrico
polyhydric alcohol alcohol polihídrico
polyhydric phenol fenol polihídrico
polyimide poliimida
polyimide resin resina de poliimida
polyiodide poliyoduro
polyisobutene poliisobuteno
polyisobutylene poliisobutileno
polyisocyanurate poliisocianurato
polyisoprene poliisopreno
polymer polímero
polymeric polimérico
polymeric reagent reactivo polimérico
polymerism polimerismo
polymerization polimerización
polymerize polimerizar
polymethacrylate resin resina de polimetacrilato
polymethanal polimetanal
polymethylbenzene polimetilbenceno
polymethylene glycol polimetilenglicol
polymethylpentene polimetilpenteno
polymorphism polimorfismo
polymyxin polimixina
polynuclear polinuclear
polynuclear hydrocarbon hidrocarburo polinuclear
polynuclidic polinuclídico
polyol poliol
polyolefin poliolefina
polyoxadiazole polioxadiazol
polyoxamide polioxamida
polyoxy polioxi

polyoxyalkylene resin resina de polioxialquileno
polyoxyethylene polioxietileno
polyoxyethylene monostearate monoesterato de polioxietileno
polyoxyethyleneoxypropylene polioxietilenoxipropileno
polyoxymethylene polioximetileno
polyoxypropylenediamine polioxipropilendiamina
polypeptide polipéptido
polyphase polifásico
polyphenyl polifenilo
polyphenylene oxide óxido de polifenileno
polyphosphate polifosfato
polyphosphoric acid ácido polifosfórico
polyprenol poliprenol
polypropylene polipropileno
polypropylene glycol polipropilenglicol
polypropylene oxide óxido de polipropileno
polypropylenebenzene polipropilenbenceno
polypyrrolidine polipirrolidina
polyrotaxame polirrotaxama
polysaccharide polisacárido
polysiloxane polisiloxano
polysorbate polisorbato
polyspiro poliespiro
polystyrene poliestireno
polysulfide polisulfuro
polysulfone polisulfona
polyterpene resin resina de politerpeno
polytetrafluoroethene politetrafluoroeteno
polytetrafluoroethylene politetrafluoroetileno
polytetrafluoroethylene resin resina de politetrafluoroetileno
polythene politeno
polythiadazole politiadazol
polythiazyl politiacilo
polythionate politionato
polythionic acid ácido politiónico
polyunsaturated poliinsaturado
polyunsaturated acid ácido poliinsaturado
polyunsaturated fat grasa poliinsaturada

polyurethane poliuretano
polyurethane resin resina de poliuretano
polyurethane rubber caucho de poliuretano
polyvalence polivalencia
polyvalent polivalente
polyvinyl polivinilo
polyvinyl acetal polivinilacetal
polyvinyl acetate acetato de polivinilo
polyvinyl alcohol alcohol polivinílico
polyvinyl butyral polivinilbutiral
polyvinyl carbazole polivinilcarbazol
polyvinyl chloride cloruro de polivinilo
polyvinyl chloride acetate acetato de cloruro de polivinilo
polyvinyl dichloride dicloruro de polivinilo
polyvinyl ether éter polivinílico
polyvinyl ethyl ether éter etílico polivinílico
polyvinyl fluoride fluoruro de polivinilo
polyvinyl isobutyl ether éter isobutílico polivinílico
polyvinyl methyl ether éter metílico polivinílico
polyvinyl resin resina de polivinilo
polyvinylidene polivinilideno
polyvinylidene chloride cloruro de polivinilideno
polyvinylidene fluoride fluoruro de polivinilideno
polyvinylidene resin resina de polivinilideno
polyvinylpyrrolidine polivinilpirrolidina
Pomeranz-Fritsch reaction reacción de Pomeranz-Fritsch
Ponzio reaction reacción de Ponzio
porcelain porcelana
porcelain enamel esmalte de porcelana
pore poro
porometric porométrico
porosimeter porosímetro
porosity porosidad
porous poroso
porphin porfina
porphyrin porfirina
porphyrinogen porfirinógeno
portland cement cemento portland
positive positivo

positive crystal cristal positivo
positive electron electrón positivo
positive element elemento positivo
positive group grupo positivo
positive ion ion positivo
positive radical radical positivo
positive-ray analysis análisis de rayos positivos
positron positrón
positron emission emisión de positrón
positronium positronio
postprecipitation posprecipitación
potasan potasán
potash potasa
potash alum alumbre potásico
potassic potásico
potassium potasio
potassium abietate abietato potásico
potassium acetate acetato potásico, acetato de potasio
potassium acetylsalicylate acetilsalicilato potásico
potassium acid carbonate carbonato ácido potásico
potassium acid fluoride fluoruro ácido potásico
potassium acid oxalate oxalato ácido potásico
potassium acid phosphate fosfato ácido potásico
potassium acid saccharate sacarato ácido potásico
potassium acid sulfate sulfato ácido potásico
potassium acid sulfite sulfito ácido potásico
potassium acid tartrate tartrato ácido potásico
potassium alginate alginato potásico
potassium alum alumbre potásico
potassium aluminate aluminato potásico
potassium aluminosilicate aluminosilicato potásico
potassium aluminum fluoride fluoruro alumínico potásico
potassium aluminum sulfate sulfato alumínico potásico
potassium aminobenzoate aminobenzoato potásico

potassium anthranilate antranilato potásico

potassium antimonate antimoniato potásico

potassium argentocyanide argentocianuro potásico

potassium arsenate arseniato potásico

potassium arsenite arsenito potásico

potassium aurate aurato potásico

potassium benzenedisulfonate bencenodisulfonato potásico

potassium benzoate benzoato potásico

potassium bicarbonate bicarbonato potásico

potassium bichromate bicromato potásico

potassium bifluoride bifluoruro potásico

potassium binoxalate binoxalato potásico

potassium biphthalate biftalato potásico

potassium bisulfate bisulfato potásico

potassium bisulfite bisulfito potásico

potassium bitartrate bitartrato potásico

potassium borate borato potásico

potassium borofluoride borofluoruro potásico

potassium borohydride borohidruro potásico

potassium bromate bromato potásico

potassium bromide bromuro potásico

potassium butanoate butanoato potásico

potassium butoxide butóxido potásico

potassium carbonate carbonato potásico

potassium chlorate clorato potásico

potassium chloride cloruro potásico

potassium chlorochromate clorocromato potásico

potassium chloroplatinate cloroplatinato potásico

potassium chromate cromato potásico

potassium cinnamate cinamato potásico

potassium citrate citrato potásico

potassium cuprocyanide cuprocianuro potásico

potassium cyanate cianato potásico

potassium cyanide cianuro potásico

potassium cyclamate ciclamato potásico

potassium dichloroisocyanurate dicloroisocianurato potásico

potassium dichromate dicromato potásico

potassium dicyanoargenate dicianoargenato potásico

potassium dicyanoaurate dicianoaurato potásico

potassium diphosphate difosfato potásico

potassium disulfate disulfato potásico

potassium disulfite disulfito potásico

potassium dithionate ditionato potásico

potassium ethanoate etanoato potásico

potassium feldspar feldespato potásico

potassium ferrate ferrato potásico

potassium ferricyanide ferricianuro potásico

potassium ferrite ferrita potásica

potassium ferrocyanide ferrocianuro potásico

potassium fluoride fluoruro potásico

potassium fluoroborate fluoroborato potásico

potassium fluosilicate fluosilicato potásico

potassium fluozirconate fluozirconato potásico

potassium formate formiato potásico

potassium gluconate gluconato potásico

potassium glutamate glutamato potásico

potassium glycerophosphate glicerofosfato potásico

potassium hexacarbonyl hexacarbonilo potásico

potassium hexachloroplatinate hexacloroplatinato potásico

potassium hexacyanoferrate hexacianoferrato potásico

potassium hexafluorophosphate hexafluorofosfato potásico

potassium hexanitrocobaltate hexanitrocobaltato potásico

potassium hippurate hipurato potásico

potassium hydrate hidrato potásico

potassium hydride hidruro potásico

potassium hydrogen carbonate carbonato de hidrógeno potásico

potassium hydrogen fluoride fluoruro de hidrógeno potásico

potassium hydrogen phosphate fosfato de hidrógeno potásico

potassium hydrogen phthalate ftalato de hidrógeno potásico

potassium hydrogen sulfate sulfato de hidrógeno potásico

potassium hydrogen tartrate tartrato de hidrógeno potásico

potassium hydrosulfide hidrosulfuro potásico

potassium hydroxide hidróxido potásico

potassium hypochlorite hipoclorito potásico

potassium hypophosphite hipofosfito potásico

potassium hyposulfate hiposulfato potásico

potassium hyposulfite hiposulfito potásico

potassium iodate yodato potásico

potassium iodide yoduro potásico

potassium laurate laurato potásico

potassium linoleate linoleato potásico

potassium manganate manganato potásico

potassium mercuric iodide yoduro mercúrico potásico

potassium metaarsenite metaarsenito potásico

potassium metabisulfite metabisulfito potásico

potassium metaphosphate metafosfato potásico

potassium methoxide metóxido potásico

potassium molybdate molibdato potásico

potassium monophosphate monofosfato potásico

potassium monosulfide monosulfuro potásico

potassium naphthenate naftenato potásico

potassium nitrate nitrato potásico

potassium nitrite nitrito potásico

potassium oleate oleato potásico

potassium orthophosphate ortofosfato potásico

potassium osmate osmiato potásico

potassium oxalate oxalato potásico

potassium oxide óxido potásico

potassium penicillin penicilina potásica

potassium pentasulfide pentasulfuro potásico

potassium percarbonate percarbonato potásico

potassium perchlorate perclorato potásico

potassium periodate peryodato potásico

potassium permanganate permanganato potásico

potassium peroxide peróxido potásico

potassium peroxoborate peroxoborato potásico

potassium peroxosulfate peroxosulfato potásico

potassium peroxydisulfate peroxidisulfato potásico

potassium persulfate persulfato potásico

potassium phosphate fosfato potásico

potassium phosphide fosfuro potásico

potassium phosphinate fosfinato potásico

potassium phosphite fosfito potásico

potassium phosphonate fosfonato potásico

potassium picrate picrato potásico

potassium polymetaphosphate polimetafosfato potásico

potassium polysulfide polisulfuro potásico

potassium prussiate prusiato potásico

potassium pyroborate piroborato potásico

potassium pyrophosphate pirofosfato potásico

potassium pyrosulfate pirosulfato potásico

potassium pyrosulfite pirosulfito potásico

potassium ricinoleate ricinoleato potásico

potassium salicylate salicilato potásico

potassium silicate silicato potásico

potassium silicofluoride silicofluoruro potásico

potassium sodium carbonate carbonato sódico potásico

potassium sodium tartrate tartrato sódico potásico

potassium sorbate sorbato potásico

potassium stannate estannato potásico

potassium stearate estearato potásico
potassium sulfate sulfato potásico
potassium sulfide sulfuro potásico
potassium sulfite sulfito potásico
potassium sulfocarbonate
    sulfocarbonato potásico
potassium sulfocyanate sulfocianato
    potásico
potassium sulfocyanide sulfocianuro
    potásico
potassium superoxide superóxido
    potásico
potassium tartrate tartrato potásico
potassium tellurite telurita potásica
potassium tetraborate tetraborato
    potásico
potassium tetrachloropalladate
    tetracloropaladato potásico
potassium tetracyanoaurate
    tetracianoaurato potásico
potassium tetrafluoroborate
    tetrafluoroborato potásico
potassium tetrahydroborate
    tetrahidroborato potásico
potassium tetraiodoaurate
    tetrayodoaurato potásico
potassium tetroxalate tetroxalato
    potásico
potassium thiocyanate tiocianato
    potásico
potassium thiosulfate tiosulfato potásico
potassium titanate titanato potásico
potassium titanium fluoride fluoruro de
    titanio potásico
potassium trichlorophenate
    triclorofenato potásico
potassium triiodide triyoduro potásico
potassium tripolyphosphate
    tripolifosfato potásico
potassium tungstate tungstato potásico
potassium undecylenate undecilenato
    potásico
potassium wolframate wolframato
    potásico
potassium xanthate xantato potásico
potassium zinc iodide yoduro de cinc
    potásico
potassium zirconium chloride cloruro
    de zirconio potásico

potential potencial
potentiometric potenciométrico
potentiometric titration titulación
    potenciométrica, valoración
    potenciométrica
pour point punto de flujo, punto de
    fluidez
powder polvo
powdered en polvo
powdered resin resina en polvo
Prandtl number número de Prandtl
praseodymium praseodimio
praseodymium acetate acetato de
    praseodimio
praseodymium chloride cloruro de
    praseodimio
praseodymium oxalate oxalato de
    praseodimio
praseodymium oxide óxido de
    praseodimio
praseodymium sulfate sulfato de
    praseodimio
precious metal metal precioso
precious stone piedra preciosa
precipitability precipitabilidad
precipitable precipitable
precipitant precipitante
precipitate precipitado
precipitated precipitado
precipitation precipitación
precipitation indicator indicador de
    precipitación
precipitation number número de
    precipitación
precipitator precipitador
precursor precursor
predissociation predisociación
prednisolone prednisolona
prednisone prednisona
preferential preferencial
Pregl procedure procedimento de Pregl
pregnanediol pregananodiol
pregnenedione pregnenodiona
pregnenolone pregnenolona
prehnite prehnita
prenyl prenilo
preparation preparación
prephenic acid ácido prefénico
prepolymer prepolímero

preservative preservativo
pressure presión
pressure effect efecto de presión
pressure filter filtro de presión
Prevost reaction reacción de Prevost
primary primario
primary alcohol alcohol primario
primary amine amina primaria
primary azo dye colorante azoico
    primario
primary electron electrón primario
primary emission emisión primaria
primary ion ion primario
primary ionization ionización primaria
primary nucleus núcleo primario
primary radiation radiación primaria
primary reaction reacción primaria
principal principal
principal axis eje principal
principal quantum number número
    cuántico principal
principal series serie principal
principal valency valencia principal
Prins reaction reacción de Prins
prism prisma
pristane pristano
probe sonda
procaine hydrochloride clorhidrato de
    procaína
procedure procedimiento
process proceso
processing procesamiento
prochiral proquiral
producer gas gas pobre
product producto
production producción
production reactor reactor de
    producción
progesterone progesterona
proguanil proguanil
projection proyección
prolactin prolactina
prolamine prolamina
proline prolina
prolyl prolilo
promazine promazina
promazine hydrochloride clorhidrato de
    promazina
promethazine prometazina

promethium prometio
promoter promotor
proof prueba, graduación normal
propadiene propadieno
propagation propagación
propagation rate velocidad de
    propagación
propanal propanal
propanamide propanamida
propane propano
propane hydrate hidrato de propano
propane sultone propanosultona
propanecarboxylic acid ácido
    propanocarboxílico
propanediamine propanodiamina
propanediol propanodiol
propanenitrile propanonitrilo
propanethiol propanotiol
propanil propanil
propano propano
propanoic acid ácido propanoico
propanol propanol
propanol nitrate nitrato de propanol
propanolamine propanolamina
propanolpyridine propanolpiridina
propanone propanona
propanoyl propanoilo
propanoyl chloride cloruro de
    propanoilo
propargite propargita
propargyl propargilo
propargyl alcohol alcohol propargílico
propargyl bromide bromuro de
    propargilo
propargyl chloride cloruro de propargilo
propellant propelente
propenal propenal
propene propeno
propene polymer polímero de propeno
propenenitrile propenonitrilo
propenethiol propenotiol
propenetricarboxylic acid ácido
    propenotricarboxílico
propenoic acid ácido propenoico
propenol propenol
propenyl propenilo
propenyl alcohol alcohol propenílico
propenyl guaethol propenilguaetol
propenyl hexanoate hexanoato de

propenilo
propenyl isothiocyanate isotiocianato de
propenilo
propenylamine propenilamina
propenylanisole propenilanisol
propenylene propenileno
propenylidene propenilideno
propenylthiourea propeniltiourea
properties propiedades
propinyl propinilo
propiolactone propiolactona
propiolic acid ácido propiólico
propioloyl propioloilo
propionaldehyde propionaldehído
propionate propionato
propione propiona
propionic acid ácido propiónico
propionic aldehyde aldehído propiónico
propionic anhydride anhídrido
propiónico
propionic ether éter propiónico
propionitrile propionitrilo
propiono propiono
propionyl propionilo
propionyl chloride cloruro de propionilo
propionyl peroxide peróxido de
propionilo
propionylbenzene propionilbenceno
propiophenone propiofenona
proportion proporción
proportional proporcional
proportional ionization ionización
proporcional
propoxy propoxi
propoxyphene propoxifeno
propoxypropanol propoxipropanol
propranonol propranonol
propyl propilo
propyl acetate acetato de propilo
propyl alcohol alcohol propílico
propyl aldehyde aldehído propílico
propyl benzoate benzoato de propilo
propyl bromide bromuro de propilo
propyl butanoate butanoato de propilo
propyl butyrate butirato de propilo
propyl carbamate carbamato de propilo
propyl chloride cloruro de propilo
propyl chlorosulfonate clorosulfonato de
propilo

propyl cyanide cianuro de propilo
propyl formate formiato de propilo
propyl furoate furoato de propilo
propyl gallate galato de propilo
propyl iodide yoduro de propilo
propyl mercaptan propilmercaptano
propyl nitrate nitrato de propilo
propyl nitrite nitrito de propilo
propyl pelargonate pelargonato de
propilo
propyl propionate propionato de propilo
propyl xanthate xantato de propilo
propylacetone propilacetona
propylamine propilamina
propylaniline propilanilina
propylbenzene propilbenceno
propylene propileno
propylene carbonate carbonato de
propileno
propylene chloride cloruro de propileno
propylene chlorohydrin clorohidrina de
propileno
propylene dichloride dicloruro de
propileno
propylene glycol propilenglicol
propylene glycol dinitrate dinitrato de
propilenglicol
propylene glycol distearate diestearato
de propilenglicol
propylene glycol monoricinoleate
monorricinoleato de propilenglicol
propylene glycol monostearate
monoestearato de propilenglicol
propylene glycol phenyl ether éter
fenílico de propilenglicol
propylene oxide óxido de propileno
propylenediamine propilendiamina
propyleneimine propilenimina
propylformic acid ácido propilfórmico
propylhydroxylamine
propilhidroxilamina
propylidene propilideno
propylmagnesium bromide bromuro de
propilmagnesio
propylmalonic acid ácido propilmalónico
propylparaben propilparabeno
propylpiperidine propilpiperidina
propylpyridine propilpiridina
propyltrichlorosilane propiltriclorosilano

propylurea propilurea
propynal propinal
propyne propino
propynol propinol
propynyl propinilo
prostaglandin prostaglandina
prosthetic group grupo prostético
protactinium protactinio
protamine protamina
protease proteasa
protecting group grupo protector
protective coating revestimiento protector
protective glass vidrio protector
protective gloves guantes protectores
protein proteína
proteolysis proteólisis
proteolytic enzyme enzima proteolítica
prothrombin protrombina
protium protio
proto proto
protogenic protogénico
protolysis protólisis
protolytic catalysis catálisis protolítica
proton protón
proton accelerator acelerador protónico, acelerador de protones
proton acceptor aceptor protónico
proton bombardment bombardeo protónico
proton capture captura protónica
proton cycle ciclo protónico
proton donor donador protónico
proton microscope microscopio protónico
proton number número protónico
protonation protonación
protonic acid ácido protónico
protonic bridge puente protónico
protophilic protofílico
protoplasm protoplasma
protosulfate protosulfato
prototropic prototrópico
protyl protilo
provitamin provitamina
Prussian blue azul de Prusia
prussic acid ácido prúsico
Pschorr reaction reacción de Pschorr
pseudo seudo, pseudo

pseudo acid seudoácido
pseudo base seudobase
pseudoaromatic seudoaromático
pseudoasymmetric seudoasimétrico
pseudocritical properties propiedades seudocríticas
pseudocumene seudocumeno
pseudocumidine seudocumidina
pseudocumol seudocumol
pseudocumyl seudocumilo
pseudohalogen seudohalógeno
pseudohexyl alcohol alcohol seudohexílico
pseudoisomerism seudoisomerismo
pseudoisotope seudoisótopo
pseudomorphic seudomórfico
pseudoreduced properties propiedades seudorreducidas
psi psi
psilocin psilocina
psilocybin psilocibina
psilomelane psilomelano
psychotropic drug droga psicotrópica
pteridine pteridina
pterin pterina
ptomaine ptomaína
ptyalin ptialina
pulegone pulegona
pullulanase pululanasa
pulp pulpa
pulse impulso
pulverization pulverización
pumice pómez
Pummerer rearrangement reordenamiento de Pummerer
pump bomba
pungent acre
Purdie methylation metilación de Purdie
pure puro
purification purificación
purified water agua purificada
purine purina
purity pureza
puromycin puromicina
purpurin purpurina
putrefaction putrefacción
putrescine putrescina
putty masilla
pyran pirano

pyranone piranona
pyranose piranosa
pyranoside piranosida
pyranthrene pirantreno
pyranyl piranilo
pyrazine pirazina
pyrazole pirazol
pyrazolidine pirazolidina
pyrazoline pirazolina
pyrazolone pirazolona
pyrazolyl pirazolilo
pyrene pireno
pyreno pireno
pyrenyl pirenilo
pyrethrin piretrina
pyrethroid piretroide
pyrethrolone piretrolona
pyrethrum piretro
pyridazine piridazina
pyridine piridina
pyridine oxide óxido de piridina
pyridine polymer polímero de piridina
pyridinecarboxylic acid ácido
    piridinocarboxílico
pyridinedicarboxylic acid ácido
    piridinodicarboxílico
pyridinepentacarboxylic acid ácido
    piridinopentacarboxílico
pyridinesulfonic acid ácido
    piridinosulfónico
pyridinium piridinio
pyridinyl piridinilo
pyridone piridona
pyridoxal piridoxal
pyridoxal hydrochloride clorhidrato de
    piridoxal
pyridoxal phosphate fosfato de piridoxal
pyridoxine piridoxina
pyridyl piridilo
pyridylamine piridilamina
pyridylcarbinol piridilcarbinol
pyrimidine pirimidina
pyrimidinyl pirimidinilo
pyrimithate pirimitato
pyrite pirita
pyrithiamine piritiamina
pyro piro
pyroacemic acid ácido piroacémico
pyroboric acid ácido pirobórico

pyrocatechol pirocatecol
pyrocellulose pirocelulosa
pyrochlore pirocloro
pyrochroite pirocroíta
pyrogallate pirogalato
pyrogallol pirogalol
pyrolan pirolán
pyroligneous acid ácido pirolígneo
pyrolusite pirolusita
pyrolysis pirólisis
pyromellitic acid ácido piromelítico
pyromellitic dianhydride dianhídrido
    piromelítico
pyrometer pirómetro
pyrometric cone cono pirométrico
pyrometry pirometría
pyromorphite piromorfita
pyromucic acid ácido piromúcico
pyrone pirona
pyrophoric pirofórico
pyrophoric material material pirofórico
pyrophosphate pirofosfato
pyrophosphoric acid ácido pirofosfórico
pyrophosphoryl pirofosforilo
pyrophyllite pirofilita
pyrosin pirosina
pyrosulfate pirosulfato
pyrosulfite pirosulfito
pyrosulfuric acid ácido pirosulfúrico
pyrosulfurous acid ácido pirosulfuroso
pyrosulfuryl pirosulfurilo
pyrosulfuryl chloride cloruro de
    pirosulfurilo
pyrotartaric acid ácido pirotartárico
pyrotechnics pirotecnia
pyrovanadic acid ácido pirovanádico
pyroxene piroxeno
pyroxylin piroxilina
pyrrhotite pirrotita
pyrrocoline pirrocolina
pyrrole pirrol
pyrrolidine pirrolidina
pyrrolidinyl pirrolidinilo
pyrrolidone pirrolidona
pyrroline pirrolina
pyrrolyl pirrolilo
pyrrone pirrona
pyrroyl pirroilo
pyrryl pirrilo

pyruvaldehyde piruvaldehído
pyruvic acid ácido pirúvico
pyruvic alcohol alcohol pirúvico

pyruvic aldehyde aldehído pirúvico
pyruvonitrile piruvonitrilo

Q

quadri cuadri
quadridentate ligand ligando
  cuadridentado
quadrimolecular cuadrimolecular
quadrivalence cuadrivalencia
quadrivalent cuadrivalente
quadruple point punto cuádruple
quadruple resonance resonancia
  cuádruple
quadrupole cuadripolo
qualitative cualitativo
qualitative analysis análisis cualitativo
qualitative reaction reacción cualitativa
qualitative test prueba cualitativa
quality assurance comprobación de
  calidad
quality control control de calidad
quanta cuantos
quantitative cuantitativo
quantitative analysis análisis cuantitativo
quantitative measurement medida
  cuantitativa
quantitative microanalysis microanálisis
  cuantitativo
quantitative reaction reacción
  cuantitativa
quantitative test prueba cuantitativa
quantity cantidad
quantivalence cuantivalencia
quantization cuantificación
quantize cuantificar
quantized cuantificado
quantized system sistema cuantificado
quantum cuanto
quantum chemistry química cuántica
quantum efficiency eficiencia cuántica
quantum equivalence equivalencia
  cuántica
quantum group grupo cuántico
quantum jump salto cuántico
quantum level nivel cuántico
quantum mechanics mecánica cuántica
quantum number número cuántico

quantum of energy cuanto de energía
quantum of light cuanto de luz
quantum of radiation cuanto de
  radiación
quantum state estado cuántico
quantum statistics estadística cuántica
quantum theory teoría cuántica
quantum transition transición cuántica
quantum unit unidad cuántica
quantum yield rendimiento cuántico
quark quark
quartet cuarteto
quartz cuarzo
quartz-controlled controlado por cuarzo
quartz crystal cristal de cuarzo
quartz rock cuarcita
quartzite cuarcita
quasicrystal cuasicristal
quassia cuasia
quaternary ammonium salt sal amónica
  cuaternaria
quaternary carbon carbono cuaternario
quaternary carbon atom átomo de
  carbono cuaternario
quaternary salt sal cuaternaria
quaternary system sistema cuaternario
quebracho quebracho
Quelet reaction reacción de Quelet
quench extinguir, apagar, enfriar
  rápidamente
quenching extinción, apagado,
  enfriamiento rápido
quercinic acid ácido quercínico
quercitol quercitol
quercitrin quercitrina
quicklime cal viva
quicksilver azogue
quinacrine quinacrina
quinaldic acid ácido quináldico
quinaldine quinaldina
quinamine quinamina
quinaphthol quinaftol
quinazoline quinazolina

quinhydrone quinhidrona
quinhydrone electrode electrodo de quinhidrona
quinic acid ácido quínico
quinidine quinidina
quinine quinina
quinine sulfate sulfato de quinina
quininic acid ácido quinínico
quininone quininona
quinitol quinitol
quinizarin quinizarina
quinogen quinógeno
quinol quinol
quinoline quinolina
quinoline blue azul de quinolina
quinoline dye colorante de quinolina
quinoline red rojo de quinolina
quinolinic acid ácido quinolínico

quinolinium quinolinio
quinolinol quinolinol
quinolinone quinolinona
quinolizine quinolizina
quinolone quinolona
quinolyl quinolilo
quinone quinona
quinonedioxime quinonadioxima
quinonyl quinonilo
quinosol quinosol
quinoxaline quinoxalina
quinoxalinyl quinoxalinilo
quinoyl quinoilo
quinquevalent quinquevalente
quintet quinteto
quintozene quintoceno
quinuclidine quinuclidina

R

Racah parameters parámetros de Racah
racemase racemasa
racemate racemato
racemation racemación
raceme racema
racemic racémico
racemic acid ácido racémico
racemic mixture mezcla racémica
racemic substance substancia racémica
racemization racemización
rad rad
radial chromatography cromatografía
   radial
radiance radiancia
radiant radiante
radiate radiar
radiated radiado
radiating radiante
radiation radiación
radiation absorber absorbedor de
   radiación
radiation biochemistry bioquímica de
   radiación
radiation catalysis catálisis de radiación
radiation chemistry química de
   radiación
radiation counter contador de radiación
radiation danger peligro de radiación
radiation detector detector de radiación
radiation dose dosis de radiación
radiation effect efecto de radiación
radiation-induced inducido por radiación
radiation intensity intensidad de
   radiación
radiation ionization ionización por
   radiación
radiation-sensitive sensible a radiación
radiation source fuente de radiación
radiation unit unidad de radiación
radiative radiativo
radiative capture captura radiativa
radiative correction corrección radiativa
radiative neutron capture captura

neutrónica radiativa
radiative transfer transferencia radiativa
radiator radiador
radical radical
radicofunctional nomenclature
   nomenclatura radicofuncional
radio radio
radioactinium radioactinio
radioactive radiactivo
radioactive atom átomo radiactivo
radioactive chain cadena radiactiva
radioactive clock reloj radiactivo
radioactive constant constante radiactiva
radioactive contamination
   contaminación radiactiva
radioactive dating datación radiactiva
radioactive decay desintegración
   radiactiva
radioactive decay constant constante de
   desintegración radiactiva
radioactive decay product producto de
   desintegración radiactiva
radioactive decay series serie de
   desintegración radiactiva
radioactive deposit depósito radiactivo
radioactive disintegration desintegración
   radiactiva
radioactive displacement law ley de
   desplazamiento radiactivo
radioactive element elemento radiactivo
radioactive emanation emanación
   radiactiva
radioactive emission emisión radiactiva
radioactive equilibrium equilibrio
   radiactivo
radioactive family familia radiactiva
radioactive fission fisión radiactiva
radioactive half-life media vida
   radiactiva
radioactive indicator indicador
   radiactivo
radioactive isotope isótopo radiactivo
radioactive material material radiactivo

radioactive metal metal radiactivo
radioactive mineral mineral radiactivo
radioactive nucleus núcleo radiactivo
radioactive nuclide nucleido radiactivo
radioactive product producto radiactivo
radioactive series serie radiactiva
radioactive source fuente radiactiva
radioactive tracer trazador radiactivo
radioactive transformation
    transformación radiactiva
radioactive waste desperdicios
    radiactivos
radioactivity radiactividad
radioactivity standard patrón de
    radiactividad
radioassay radioanálisis
radiobarium radiobario
radiobiology radiobiología
radiocarbon radiocarbono
radiocarbon dating datación por
    radiocarbono
radiocesium radiocesio
radiochemical radioquímico
radiochemical process proceso
    radioquímico
radiochemistry radioquímica
radiochromatography
    radiocromatografía
radiocrystallography radiocristalografía
radioelement radioelemento
radiofrequency spectrometer
    espectrómetro de radiofrecuencia
radiofrequency spectroscopy
    espectroscopia de radiofrecuencia
radiogenic radiogénico
radiography radiografía
radioimmunoassay radioinmunoanálisis
radioiodine radioyodo
radioisotope radioisótopo
radiolead radioplomo
radiolite radiolita
radiological radiológico
radiology radiología
radioluminescence radioluminiscencia
radiolysis radiólisis
radiometallography radiometalografía
radiometer radiómetro
radiometric radiométrico
radiometric analysis análisis

radiométrico
radiometric dating datación radiométrica
radiometric titration titulación
    radiométrica, valoración radiométrica
radiometry radiometría
radiomimetic radiomimético
radiomimetic substance substancia
    radiomimética
radion radión
radionitrogen radionitrógeno
radionuclide radionucleido
radiosodium radiosodio
radiotellurium radiotelurio
radiothor radiotor
radiothorium radiotorio
radium radio
radium bromide bromuro de radio
radium carbonate carbonato de radio
radium chloride cloruro de radio
radium sulfate sulfato de radio
radium unit unidad de radio
radius radio
radius ratio razón de radios
radon radón
raffinate rafinato
raffinose rafinosa
ralstonite ralstonita
Raman effect efecto de Raman
Raman lines líneas de Raman
Raman spectrophotometry
    espectrofotometría de Raman
Raman spectroscopy espectroscopia de
    Raman
Raman spectrum espectro de Raman
Ramberg-Backlund reaction reacción de
    Ramberg-Backlund
ramie ramio
Ramsauer effect efecto de Ramsauer
Ramsay-Shields equation ecuación de
    Ramsay-Shields
Ramsay-Young equation ecuación de
    Ramsay-Young
Ramsay-Young law ley de Ramsay-
    Young
rancid rancio
random copolymer copolímero aleatorio
Raney catalyst catalizador de Raney
Raney nickel níquel de Raney
Rankine scale escala Rankine

rankinite rankinita
ranksite ranksita
Raoult law ley de Raoult
rapeseed oil aceite de colza
rapid reaction reacción rápida
rare earth tierra rara
rare-earth element elemento de tierras raras
rare-earth metal metal de tierras raras
rare-earth salts sales de tierras raras
rare gas gas raro
rare metal metal raro
rarefaction rarefacción
rarefy rareficar
Raschig process proceso de Raschig
Raschig rings anillos de Raschig
rasorite rasorita
Rast method método de Rast
rate velocidad, tasa
rate constant constante de velocidad de reacción
rate-determining step paso determinante de velocidad
rate of decay velocidad de desintegración
rate of disintegration velocidad de desintegración
rate of formation velocidad de formación
rate of reaction velocidad de reacción
rate of sedimentation velocidad de sedimentación
ratio razón, relación
ratio of specific heats razón de calores específicos
rational formula fórmula racional
rauwolfia rauwolfia
raw material materia prima
ray rayo
Rayleigh line línea de Rayleigh
rayon rayón
reactant reactante
reaction reacción
reaction constant constante de reacción
reaction control control de reacción
reaction energy energía de reacción
reaction enthalpy number número de entalpía de reacción
reaction equation ecuación de reacción
reaction formula fórmula de reacción

reaction heat calor de reacción
reaction indicator indicador de reacción
reaction inhibitor inhibidor de reacción
reaction intermediate producto intermedio de reacción
reaction kinetics cinética de reacciones
reaction law ley de reacción
reaction mechanism mecanismo de reacción
reaction of first order reacción de primer orden
reaction of second order reacción de segundo orden
reaction order orden de reacción
reaction product producto de reacción
reaction promotor promotor de reacción
reaction rate velocidad de reacción
reaction step paso de reacción
reaction time tiempo de reacción
reaction velocity velocidad de reacción
reactivation reactivación
reactive reactivo
reactive anode ánodo reactivo
reactive atmosphere atmósfera reactiva
reactive bond enlace reactivo
reactive dye colorante reactivo
reactivity reactividad
reactor reactor
reagent reactivo
reagent solution solución de reactivo
real gas gas real
realgar rejalgar
rearrangement reordenamiento, redisposición, transposición
rearrangement reaction reacción de reordenamiento
recalescence recalescencia
receiver receptor
receptor receptor
reciprocal proportions law ley de proporciones recíprocas
recombinant deoxyribonucleic acid ácido desoxirribonucleico recombinante
recombination process proceso de recombinación
reconstitution reconstitución
recording balance balanza de registro
recovery recuperación

recrystallization recristalización
recrystallization nucleus núcleo de recristalización
recrystallize recristalizar
rectification rectificación
rectifying column columna rectificante
rectorite rectorita
recycling reciclado
red acetate acetato rojo
red arsenic arsénico rojo
red brass latón rojo
red glass vidrio rojo
red lead plomo rojo
red manganese manganeso rojo
red mercury oxide óxido de mercurio rojo
red metal metal rojo
red ocher ocre rojo
red oil aceite rojo
red oxide óxido rojo
red phosphorus fósforo rojo
red potassium chromate cromato potásico rojo
red precipitate precipitado rojo
reddingite redingita
redistillation redestilación
redox redox
redox catalyst catalizador redox
redox cell célula de redox
redox equilibrium equilibrio de redox
redox indicator indicador redox
redox potential potencial de redox
redox potentiometry potenciometría redox
redox reaction reacción de redox
redox series serie de redox
redox system sistema de redox
redox titration titulación redox, valoración redox
reduce reducir
reduced reducido
reduced mass masa reducida
reduced state estado reducido
reducer reductor
reducing reductor
reducing agent agente reductor
reducing atmosphere atmósfera reductora
reducing flame llama reductora

reducing material material reductor
reducing reaction reacción reductora
reducing sugar azúcar reductor
reductant reductor
reductase reductasa
reduction reducción
reduction potential potencial de reducción
reductone reductona
reductor reductor
Reed reaction reacción de Reed
reference referencia
reference electrode electrodo de referencia
refine refinar
refined refinado
refinement refinamiento
refinery refinería
refinery gas gas de refinería
refining refinación
reflux reflujo
reflux condenser condensador de reflujo
refluxing reflujo
Reformatsky reaction reacción de Reformatsky
reforming reformación
refraction refracción
refractive refractivo
refractive index índice de refracción
refractivity refractividad
refractometer refractómetro
refractory refractario
refrigerant refrigerante
refrigeration refrigeración
regelation regelación
regeneration regeneración
regenerative regenerativo
regioselectivity regioselectividad
Regnault method método de Regnault
regular polymer polímero regular
regular system sistema regular
regulator regulador
rehydration rehidratación
Reich process proceso de Reich
Reimer reaction reacción de Reimer
Reimer-Tiemann reaction reacción de Reimer-Tiemann
Reinecke acid ácido de Reinecke
Reinecke salt sal de Reinecke

reinforced plastic plástico reforzado
reinite reinita
Reinsch test prueba de Reinsch
Reissert indole synthesis síntesis de
　indoles de Reissert
Reissert reaction reacción de Reissert
relative relativo
relative atomic mass masa atómica
　relativa
relative density densidad relativa
relative error error relativo
relative humidity humedad relativa
relative molecular mass masa molecular
　relativa
relative stereochemistry estereoquímica
　relativa
relative weight peso relativo
relativistic relativista
relativistic particle partícula relativista
relativistic quantum theory teoría
　cuántica relativista
relativity relatividad
relaxation relajación
relaxation effect efecto de relajación
relaxation time tiempo de relajación
release liberación
rem rem
remote handling manejo remoto
remote-handling equipment equipo de
　manejo remoto
renaturation renaturación
renin renina
rennet cuajo
rennin renina
repeatable repetible
repeating unit unidad repetitiva
repellency repelencia
repellent repelente
repercolation repercolación
replacement reemplazo
replicate duplicar
replication duplicación
Reppe process proceso de Reppe
reprocessing reprocesamiento
reproducibility reproducibilidad
reproducible reproducible
reptation reptación
repulsion repulsión
resbenzophenone resbenzofenona

research investigación
research and development investigación
　y desarrollo
research laboratory laboratorio de
　investigación
resene reseno
reserpine reserpina
reserve reserva
residual residual
residual oil aceite residual
residue residuo
resin resina
resinamine resinamina
resinate resinato
resinification resinificación
resinography resinografía
resinoid resinoide
resinol resinol
resinous resinoso
resist reserva
resite resita
resitol resitol
resocyanin resocianina
resol resol
resolution resolución
resolving power poder de resolución
resonance resonancia
resonance energy energía de resonancia
resonance fluorescence fluorescencia de
　resonancia
resonance ionization spectroscopy
　espectroscopia de ionización en
　resonancia
resonance luminescence luminiscencia
　de resonancia
resonance neutron neutrón de resonancia
resonance radiation radiación de
　resonancia
resonance spectrum espectro de
　resonancia
resonant resonante
resorcin resorcina
resorcinol resorcina, resorcinol
resorcinol acetate acetato de resorcina
resorcinol blue azul de resorcina
resorcinol dimethyl ether éter dimetílico
　de resorcina
resorcinol-formaldehyde resin resina de
　resorcina-formaldehído

resorcinol monoacetate monoacetato de resorcina

resorcinol monobenzoate monobenzoato de resorcina

resorcinol phenolate fenolato de resorcina

resorcyl resorcilo

resorcylic acid ácido resorcílico

restitution restitución

restricted internal rotation rotación interna restringida

resultant resultante

retardation retardo

retarder retardador

retene reteno

retention retención

retention index índice de retención

retention time tiempo de retención

retention volume volumen de retención

Retger law ley de Retger

reticulated reticulado

retinal retinal

retinene retineno

retinite retinita

retinol retinol

retort retorta

retro retro

retrogradation retrogradación

retrograde condensation condensación retrógrada

retrograde evaporation evaporación retrógrada

retropinacol rearrangement reordenamiento de retropinacol

retrorsine retrorcina

retrosynthesis retrosíntesis

Reverdin reaction reacción de Reverdin

reversal inversión

reverse osmosis ósmosis inversa

reversible reversible

reversible action acción reversible

reversible chemical reaction reacción química reversible

reversible process proceso reversible

reversible reaction reacción reversible

reversion reversión

revertose revertosa

revive revivir

rhamnitol ramnitol

rhamnose ramnosa

rhenate renato

rhenic acid ácido rénico

rhenium renio

rhenium black negro de renio

rhenium chloride cloruro de renio

rhenium fluoride fluoruro de renio

rhenium heptasulfide heptasulfuro de renio

rhenium heptoxide heptóxido de renio

rhenium oxide óxido de renio

rhenium pentachloride pentacloruro de renio

rhenium sulfide sulfuro de renio

rhenium trichloride tricloruro de renio

rhenium trioxychloride trioxicloruro de renio

rheology reología

rheometer reómetro

rhigolene rigoleno

rhizonic acid ácido rizónico

rho rho

rho acid ácido rho

rhodalline rodalina

rhodamine rodamina

rhodanate rodanato

rhodanic acid ácido rodánico

rhodanide rodanida

rhodanine rodanina

rhodanometry rodanometría

rhodinol rodinol

rhodinyl rodinilo

rhodinyl acetate acetato de rodinilo

rhodite rodita

rhodium rodio

rhodium black negro de rodio

rhodium carbonyl chloride cloruro de rodiocarbonilo

rhodium chloride cloruro de rodio

rhodium hydroxide hidróxido de rodio

rhodium nitrate nitrato de rodio

rhodium oxide óxido de rodio

rhodium sulfate sulfato de rodio

rhodium sulfide sulfuro de rodio

rhodium trichloride tricloruro de rodio

rhodizite rodizita

rhodizonic acid ácido rodizónico

rhodochrosite rodocrosita

rhodolite rodolita

rhodonite rodonita
rhodopsin rodopsina
rhodoxanthin rodoxantina
rhombic rómbico
rhombic dodecahedron dodecahedro rómbico
rhombic sulfur azufre rómbico
rhombic system sistema rómbico
rhombohedral rombohedral
rhombohedral system sistema rombohedral
rhombohedron rombohedro
rhotanium rotanio
rhythmic deposition deposición rítmica
rhythmic precipitation precipitación rítmica
riboflavin riboflavina
riboflavin phosphate fosfato de riboflavina
ribofuranosyladenine ribofuranosiladenina
ribonic acid ácido ribónico
ribonuclease ribonucleasa
ribonucleic acid ácido ribonucleico
ribonucleoprotein ribonucleoproteína
ribose ribosa
ribosephosphoric acid ácido ribosfosfórico
riboside ribosida
ribosome ribosoma
ribosyl ribosilo
ribulose ribulosa
rich gas gas rico
rich mixture mezcla rica
richmondite richmondita
Richter law ley de Richter
ricin ricina
ricinic acid ácido ricínico
ricinine ricinina
ricinoleate ricinoleato
ricinoleic acid ácido ricinoleico
ricinolein ricinoleína
ricinoleyl alcohol alcohol ricinoleílico
ricinus oil aceite de ricino
rickardite rickardita
Riegler test prueba de Riegler
Riehm quinoline synthesis síntesis de quinolinas de Riehm
Riemschneider synthesis síntesis de Riemschneider

Riley oxidation oxidación de Riley
ring anillo
ring breakage rompimiento de anillo
ring closure cierre de anillo
ring compound compuesto en anillo
ring formation formación de anillo
ring isomerism isomerismo de anillo
ring reaction reacción de anillo
ring structure estructura de anillo
ring system sistema de anillo
ring test prueba de anillo
Ringer solution solución de Ringer
ripening maduración
ripidolite ripidolita
rissic acid ácido rísico
ristocetin ristocetina
Ritter reaction reacción de Ritter
roast cocer, calcinar
roasting cocido, calcinación
Robinson reaction reacción de Robinson
Rochelle salt sal de Rochelle
rock crystal cristal de roca
rock salt sal de roca
rocket fuel cobustible de cohete
rodenticide rodenticida
rodinol rodinol
roentgen roentgen
roentgenogram roentgenograma
Roesler process proceso de Roesler
Rohrbach solution solución de Rohrbach
röntgen roentgen
room temperature temperatura de sala, temperatura ambiente
rosaniline rosanilina
roscoelite roscoelita
Rose Bengal rosa de Bengala
roselite roselita
Rosenheim color test prueba de color de Rosenheim
Rosenmund reaction reacción de Rosenmund
Rosenmund reduction reducción de Rosenmund
Rosenmund-von Braun synthesis síntesis de Rosenmund-von Braun
Rosenstein process proceso de Rosenstein
rosin colofonia, resina

rosin oil aceite de colofonia
rosinol rosinol
rosolic acid ácido rosólico
rotamer rotámero
rotameter rotámetro
rotation rotación
rotational rotacional
rotational constant constante rotacional
rotational energy energía rotacional
rotational spectrum espectro rotacional
rotatory rotatorio
rotatory dispersion dispersión rotatoria
rotatory power poder rotatorio
rotaversion rotaversión
rotenone rotenona
Rothemund reaction reacción de Rothemund
Rowe rearrangement reordenamiento de Rowe
Rowland ghost fantasma de Rowland
rowlandite rowlandita
rubber caucho
rubber accelerator acelerador de caucho
rubber hydrochloride clorhidrato de caucho
rubber sponge esponja de caucho
rubber sulfide sulfuro de caucho
rubene rubeno
ruberite ruberita
ruberythric acid ácido ruberítrico
rubicene rubiceno
rubidine rubidina
rubidium rubidio
rubidium acetate acetato de rubidio
rubidium alum alumbre de rubidio
rubidium bromide bromuro de rubidio
rubidium carbonate carbonato de rubidio
rubidium chloride cloruro de rubidio
rubidium chromate cromato de rubidio
rubidium dichromate dicromato de rubidio
rubidium fluoride fluoruro de rubidio
rubidium hexachloroplatinate hexacloroplatinato de rubidio
rubidium hexafluorosilicate hexafluorosilicato de rubidio
rubidium hydride hidruro de rubidio
rubidium hydroxide hidróxido de rubidio

rubidium iodate yodato de rubidio
rubidium iodide yoduro de rubidio
rubidium nitrate nitrato de rubidio
rubidium oxide óxido de rubidio
rubidium perchlorate perclorato de rubidio
rubidium peroxide peróxido de rubidio
rubidium sulfate sulfato de rubidio
rubidium sulfide sulfuro de rubidio
rubidium sulfite sulfito de rubidio
rubidium tartrate tartrato de rubidio
rubitannic acid ácido rubitánico
rubrene rubreno
ruby rubí
ruby spinel rubí espinela
Ruff-Fenton degradation degradación de Ruff-Fenton
rust herrumbre, moho
ruthenate rutenato
ruthenic ruténico
ruthenic chloride cloruro ruténico
ruthenious rutenioso
ruthenium rutenio
ruthenium bromide bromuro de rutenio
ruthenium carbonyl carbonilo de rutenio
ruthenium chloride cloruro de rutenio
ruthenium dichloride dicloruro de rutenio
ruthenium fluoride fluoruro de rutenio
ruthenium hydroxide hidróxido de rutenio
ruthenium nitrosonitrate nitrosonitrato de rutenio
ruthenium oxide óxido de rutenio
ruthenium oxychloride oxicloruro de rutenio
ruthenium red rojo de rutenio
ruthenium sesquichloride sesquicloruro de rutenio
ruthenium sesquioxide sesquióxido de rutenio
ruthenium silicide siliciuro de rutenio
ruthenium sulfide sulfuro de rutenio
ruthenium tetraoxide tetróxido de rutenio
ruthenium trichloride tricloruro de rutenio
ruthenium trifluoride trifluoruro de

rutenio
ruthenocene rutenoceno
Rutherford nuclear atom átomo nuclear
  de Rutherford
Rutherford scattering dispersión de
  Rutherford
rutherfordite rutefordita
rutile rutilo

rutin rutina
rydberg rydberg
Rydberg constant constante de Rydberg
Rydberg correction corrección de
  Rydberg
Rydberg formula fórmula de Rydberg
Rydberg spectrum espectro de Rydberg

S

sabadilla cebadilla

Sabatier-Senderens reduction reducción de Sabatier-Senderens

sabinene sabineno

sabinic acid ácido sabínico

sabinol sabinol

saccharase sacarasa

saccharate sacarato

saccharic sacárico

saccharic acid ácido sacárico

saccharide sacárido

saccharification sacarificación

saccharify sacarificar

saccharimeter sacarímetro

saccharin sacarina

saccharinic acid ácido sacarínico

saccharonic acid ácido sacarónico

saccharose sacarosa

saccharose unit unidad de sacarosa

Sachse reaction reacción de Sachse

Sackur-Tetrode equation ecuación de Sackur-Tetrode

sacrificial protection protección sacrificial

safe storage almacenamiento seguro

safety seguridad

safety device dispositivo de seguridad

safety engineering ingeniería de seguridad

safety glass vidrio de seguridad

safety goggles anteojos de seguridad, gafas protectoras

safflower oil aceite de cártamo

safranine safranina

safrole safrol

sage oil aceite de salvia

sahlite sahlita

sal ammoniac sal amoníaco

sal soda sal sosa

salamide salamida

salic sálico

salicin salicina

salicoyl salicoilo

salicyl salicilo

salicyl alcohol alcohol salicílico

salicylal salicilal

salicylaldehyde salicilaldehído

salicylamide salicilamida

salicylanilide salicilanilida

salicylate salicilato

salicylated salicilado

salicylic acid ácido salicílico

salicylic aldehyde aldehído salicílico

salicylic amide amida salicílica

salicylidene salicilideno

salicylonitrile salicilonitrilo

salicyloyl saliciloilo

saligenin saligenina

saligenol saligenol

salimeter salímetro

saline salino

saline water agua salina

salinity salinidad

salinometer salinómetro

salmine salmina

salol salol

salt sal

salt bridge puente salino

salt cake torta de sal

salt of Lemery sal de Lemery

salt of tartar sal de tártaro

salting out precipitación asistida por sal

saltpeter salitre

samaric samárico

samaric bromide bromuro samárico

samaric chloride cloruro samárico

samaric hydroxide hidróxido samárico

samaric nitrate nitrato samárico

samaric oxalate oxalato samárico

samaric sulfate sulfato samárico

samaric sulfide sulfuro samárico

samarium samario

samarium chloride cloruro de samario

samarium oxide óxido de samario

samarium trioxide trióxido de samario

samarous samaroso

samarous chloride cloruro samaroso
samarous sulfate sulfato samaroso
samarskite samarskita
sample muestra
sampling muestreo, muestra
sand arena
sand bath baño de arena
sandalwood oil aceite de sándalo
sandarac sandaraca
Sandmeyer reaction reacción de
    Sandmeyer
Sandmeyer synthesis síntesis de
    Sandmeyer
sanguinaria sanguinaria
santalol santalol
santalyl santalilo
santalyl acetate acetato de santalilo
santalyl chloride cloruro de santalilo
saponification saponificación
saponification equivalent equivalente de
    saponificación
saponification number número de
    saponificación
saponification value valor de
    saponificación
saponin saponina
saponite saponita
sapphire zafiro
sarcolactate sarcolactato
sarcolactic acid ácido sarcoláctico
sarcolite sarcolita
sarcolysin sarcolisina
sarcosine sarcosina
sardinianite sardinianita
sardonyx sardónica
Sarett oxidation oxidación de Sarett
Sargent curve curva de Sargent
sarin sarina
sassafras oil aceite de sasafrás
satin spar espato satinado
saturable saturable
saturant saturante
saturate saturar
saturated saturado
saturated ammonia amoníaco saturado
saturated compound compuesto saturado
saturated fat grasa saturada
saturated hydrocarbon hidrocarburo
    saturado

saturated liquid líquido saturado
saturated solution solución saturada
saturated vapor vapor saturado
saturating saturante
saturation saturación
saturation effect efecto de saturación
saturation isomerism isomerismo de
    saturación
saturation level nivel de saturación
saturation point punto de saturación
saturation spectroscopy espectroscopia
    de saturación
saturator saturador
sawhorse projection proyección en
    cabrilla
saxitoxin saxitoxina
scale escala, balanza, costra, capa de
    óxido
scan exploración, barrido
scandium escandio
scandium chloride cloruro de escandio
scandium fluoride fluoruro de escandio
scandium hydroxide hidróxido de
    escandio
scandium oxalate oxalato de escandio
scandium oxide óxido de escandio
scandium sulfate sulfato de escandio
scandium sulfite sulfito de escandio
scanning exploración, barrido
scanning electron microscope
    microscopio electrónico de barrido
scanning electron microscopy
    microscopia electrónica de barrido
scapolite escapolita
scatter dispersión, esparcimiento
scattered disperso, difuso
scattered electrons electrones dispersos
scattering dispersión, esparcimiento
scavenger depurador
Schäffer acid ácido de Schäffer
Schäffer salt sal de Schäffer
Scheele green verde Scheele
scheelite scheelita
Scheibler reagent reactivo de Scheibler
Schiemann reaction reacción de
    Schiemann
Schiff base base de Schiff
Schiff reagent reactivo de Schiff
Schiff solution solución de Schiff

Schiff test prueba de Schiff
Schmidlin ketene synthesis síntesis de cetenas de Schmidlin
Schmidt lines líneas de Schmidt
Schmidt reaction reacción de Schmidt
Schoenflies system sistema de Schoenflies
Scholl reaction reacción de Scholl
Schomaker-Stevenson equation ecuación de Schomaker-Stevenson
schönite schonita
Schorigin reaction reacción de Schorigin
Schotten-Baumann reaction reacción de Schotten-Baumann
Schottky defect defecto de Schottky
Schrödinger equation ecuación de Schrödinger
Schweitzer reagent reactivo de Schweitzer
scientific method método científico
scintillating centelleante
scintillation centelleo
scintillation counter contador de centelleos
scintillation spectrometer espectrómetro de centelleos
scintillator centelleador
scission escisión
scleroscope escleroscopio
scolecite escolecita
scopolamine escopolamina
scopoline escopolina
scorodite escorodita
screen cedazo, tamiz, criba, pantalla
screening cribado, tamizado, apantallamiento
screening constant constante de apantallamiento
screening effect efecto de apantallamiento
scrubber depurador
scrubbing depuración
sealant sellador
sealed sellado
sebacic acid ácido sebácico
sebaconitrile sebaconitrilo
sebacoyl sebacoilo
sebacoyl chloride cloruro de sebacoilo
sebacylic acid ácido sebacílico

sec sec
seco seco
secobarbital secobarbital
second segundo
second law of thermodynamics segunda ley de termodinámica
second-order reaction reacción de segundo orden
secondary secundario
secondary alcohol alcohol secundario
secondary amine amina secundaria
secondary carbon carbono secundario
secondary carbon atom átomo de carbono secundario
secondary cell pila secundaria
secondary emission emisión secundaria
secondary-ion mass spectrometry espectrometría de masa de iones secundarios
secondary radiation radiación secundaria
secondary reaction reacción secundaria
sedanolic acid ácido sedanólico
sedative sedante
sediment sedimento
sedimentation sedimentación
sedimentation coefficient coeficiente de sedimentación
sedimentation constant constante de sedimentación
sedimentation equilibrium equilibrio de sedimentación
sedimentation potential potencial de sedimentación
sedimentation rate velocidad de sedimentación
sedimentation velocity velocidad de sedimentación
seed semilla
seeding siembra
Seger cone cono de Seger
segregation segregación
Seidlitz salt sal de Seidlitz
selection rules reglas de selección
selective selectivo
selectivity selectividad
selectivity coefficient coeficiente de selectividad
selena selena

selenate seleniato
selenic selénico
selenic acid ácido selénico
selenide seleniuro
seleninic acid ácido selenínico
selenino selenino
seleninyl seleninilo
selenious selenioso
selenious acid ácido selenioso
selenious oxide óxido selenioso
selenite selenita, selenito
selenium selenio
selenium bromide bromuro de selenio
selenium chloride cloruro de selenio
selenium dibromide dibromuro de selenio
selenium dichloride dicloruro de selenio
selenium diethyldithiocarbamate dietilditiocarbamato de selenio
selenium dioxide dióxido de selenio
selenium monobromide monobromuro de selenio
selenium monochloride monocloruro de selenio
selenium nitride nitruro de selenio
selenium oxide óxido de selenio
selenium oxyacid oxiácido de selenio
selenium oxychloride oxicloruro de selenio
selenium sulfide sulfuro de selenio
selenium tetrabromide tetrabromuro de selenio
selenium tetrafluoride tetrafluoruro de selenio
seleno seleno
selenocyanate selenocianato
selenofuran selenofurano
selenol selenol
selenomium selenomio
selenone selenona
selenonic acid ácido selenónico
selenonium selenonio
selenono selenono
selenonyl selenonilo
selenourea selenourea
selenous selenioso
selenous acid ácido selenioso
selenous anhydride anhídrido selenioso
selenoxide selenóxido

selenyl selenilo
self-absorption autoabsorción
self-ionization autoionización
self-organization autoorganización
self-reduction autorreducción
sellaite sellaíta
semi semi
semicarbazide semicarbazida
semicarbazide hydrochloride clorhidrato de semicarbazida
semicarbazido semicarbazido
semicarbazino semicarbazino
semicarbazone semicarbazona
semicarbazono semicarbazono
semiconductor semiconductor
semidine semidina
semiempirical semiempírico
semimetal semimetal
semimetallic semimetálico
semimicroanalysis semimicroanálisis
semimicrochemistry semimicroquímica
semipermeable semipermeable
semipermeable membrane membrana semipermeable
semiprecious semiprecioso
semisolid semisólido
semisynthetic semisintético
semivalence semivalencia
Semmler-Wolff reaction reacción de Semmler-Wolff
senarmontite senarmontita
seniority prioridad
sensitive sensible
sensitizer sensibilizador
sensor sensor
separating funnel embudo de separación
separation separación
separation energy energía de separación
separator separador
separatory funnel embudo de separación
sepia sepia
sepiolite sepiolita
septanose ring anillo de septanosa
septiphene septifeno
sequence rules reglas de secuencia
sequestering secuestrante
sequestering agent agente secuestrante
sequestration secuestro
serendipity serendipismo

series serie
series of lines serie de líneas
serine serina
Serini reaction reacción de Serini
serotonin serotonina
serpentine serpentina
serum suero
sesame oil aceite de sésamo
sesamin sesamina
sesamol sesamol
sesamolin sesamolina
sesone sesona
sesqui sesqui
sesquicarbonate sesquicarbonato
sesquichloride sesquicloruro
sesquioxide sesquióxido
sesquisalt sesquisal
sesquisoda sesquisosa
sesquiterpene sesquiterpeno
setting solidificación, endurecimiento
setting point punto de solidificación,
    punto de endurecimiento
settling asentamiento, estabilización
sexivalent sexivalente
sextet sexteto
shale oil aceite de pizarra
shape-selective catalyst catalizador
    selectivo de tamaño
Sharpless reaction reacción de Sharpless
shell capa, concha, cáscara
shell structure estructura de capas
shellac goma laca
shielding blindaje
shift deslizamiento, desplazamiento
shifting deslizamiento, desplazamiento
Shoelkopf acid ácido de Shoelkopf
sialic acid ácido siálico
side chain cadena lateral
side reaction reacción secundaria
siderite siderita
siderophile element elemento siderófilo
siderotilate siderotilato
Siemens-Halske process proceso de
    Siemens-Halske
Siemens process proceso de Siemens
sienna siena
sieve cedazo
sievert sievert
Sievert law ley de Sievert

sigma sigma
sigma bond enlace sigma
sigma electron electrón sigma
sigma function función sigma
sigma particle partícula sigma
sigma phenomenon fenómeno sigma
sigma value valor sigma
sigmatropic sigmatrópico
sigmatropic reaction reacción
    sigmatrópica
silane silano
silanediyl silanodiilo
silanetriyl silanotriilo
silanol silanol
silatrane silatrano
silazane silazano
silica sílice
silica gel gel de sílice
silica sand arena de sílice
silicane silicano
silicate silicato
silicate of soda silicato de sosa
siliceous silíceo
silicic silícico
silicic acid ácido silícico
silicide siliciuro
silicious silíceo
silico silico
silicobenzoic acid ácido silicobenzoico
silicobromoform silicobromoformo
silicochloroform silicocloroformo
silicoethane silicoetano
silicofluoride silicofluoruro
silicoheptane silicoheptano
silicoiodoform silicoyodoformo
silicol silicol
silicomanganese silicomanganeso
silicomethane silicometano
silicon silicio
silicon alloy aleación de silicio
silicon boride boruro de silicio
silicon bromide bromuro de silicio
silicon bronze bronce de silicio
silicon carbide carburo de silicio
silicon chloride cloruro de silicio
silicon-copper silicio-cobre
silicon-copper alloy aleación de silicio-
    cobre
silicon dioxide dióxido de silicio

silicon disulfide disulfuro de silicio
silicon fluoride fluoruro de silicio
silicon-gold alloy aleación de silicio-oro
silicon hydride hidruro de silicio
silicon iodide yoduro de silicio
silicon monoxide monóxido de silicio
silicon nitride nitruro de silicio
silicon octachloride octacloruro de silicio
silicon oxide óxido de silicio
silicon oxychloride oxicloruro de silicio
silicon rubber caucho de silicio
silicon sulfide sulfuro de silicio
silicon tetrabromide tetrabromuro de silicio
silicon tetrachloride tetracloruro de silicio
silicon tetrafluoride tetrafluoruro de silicio
silicon tetrahydride tetrahidruro de silicio
silicon tetraiodide tetrayoduro de silicio
silicone silicona
silicone oil aceite de silicona
silicone rubber caucho de silicona
silicotungstic acid ácido silicotúngstico
silicyl silicilo
silicylene silicileno
silk seda
sillimanite silimanita
siloxane siloxano
siloxy siloxi
silumin silumina
silver plata
silver acetate acetato de plata
silver acetylide acetiluro de plata
silver alloy aleación de plata
silver amalgam amalgama de plata
silver antimonide antimoniuro de plata
silver arsenate arseniato de plata
silver arsenide arseniuro de plata
silver arsenite arsenito de plata
silver azide azida de plata
silver benzamide benzamida de plata
silver benzoate benzoato de plata
silver borate borato de plata
silver bromate bromato de plata
silver bromide bromuro de plata
silver bronze bronce de plata

silver carbonate carbonato de plata
silver chlorate clorato de plata
silver chloride cloruro de plata
silver chlorite clorito de plata
silver chromate cromato de plata
silver citrate citrato de plata
silver cyanate cianato de plata
silver cyanide cianuro de plata
silver dichromate dicromato de plata
silver dithionate ditionato de plata
silver ferricyanide ferricianuro de plata
silver fluoride fluoruro de plata
silver fluoroborate fluoroborato de plata
silver fluosilicate fluosilicato de plata
silver fulminate fulminato de plata
silver hexacyanoferrate hexacianoferrato de plata
silver hexafluorosilicate hexafluorosilicato de plata
silver hypochlorite hipoclorito de plata
silver hypophosphate hipofosfato de plata
silver iodate yodato de plata
silver iodide yoduro de plata
silver lactate lacatato de plata
silver laurate laurato de plata
silver mercury iodide yoduro mercúrico de plata
silver migration migración de plata
silver myristate miristato de plata
silver nitrate nitrato de plata
silver nitride nitruro de plata
silver nitrite nitrito de plata
silver ore mineral de plata
silver orthophosphate ortofosfato de plata
silver oxalate oxalato de plata
silver oxide óxido de plata
silver palmitate palmitato de plata
silver perchlorate perclorato de plata
silver permanganate permanganato de plata
silver peroxide peróxido de plata
silver phosphate fosfato de plata
silver phosphide fosfuro de plata
silver picrate picrato de plata
silver-plated plateado
silver-plating plateado
silver potassium cyanide cianuro

potásico de plata

silver salicylate salicilato de plata

silver salt sal de plata

silver selenide seleniuro de plata

silver sodium chloride cloruro sódico de plata

silver sodium cyanide cianuro sódico de plata

silver sodium thiosulfate tiosulfato sódico de plata

silver stearate estearato de plata

silver suboxide subóxido de plata

silver sulfate sulfato de plata

silver sulfide sulfuro de plata

silver sulfite sulfito de plata

silver tartrate tartrato de plata

silver telluride telururo de plata

silver thiocyanate tiocianato de plata

silver thiosulfate tiosulfato de plata

silver vanadate vanadato de plata

silvestrene silvestreno

silvex silvex

silvichemical silviquímico

silylation sililación

silylene silileno

simazine simazina

Simmons-Smith reagent reactivo de Simmons-Smith

Simmons-Smith reaction reacción de Simmons-Smith

Simonini reaction reacción de Simonini

Simons process proceso de Simons

simple distillation destilación simple

simple salt sal simple

simultaneous reaction reacción simultánea

single bond enlace sencillo

singlet singulete

sinigrin sinigrina

sinter sínter, sinterizado, aglutinación

sintered sinterizado, aglutinado

sintering sinterización, aglutinación

sinterization sinterización, aglutinación

sinterize sinterizar, aglutinar

sisal sisal

sitosterol sitosterol

skatole escatol

sklodowskite sklodowskita

Skraup synthesis síntesis de Skraup

skutterudite skutterudita

slack apagado, flojo, lento

slag escoria

slaked lime cal apagada

slate pizarra

slate flour harina de pizarra

slow chemical reaction reacción química lenta

slow combustion combustión lenta

slow neutron neutrón lento

slow-neutron spectrometry espectrometría de neutrones lentos

slow reaction reacción lenta

sludge lodo, sedimento

slurry suspensión, suspensión acuosa, lechada

smalt esmaltín

smectic esméctico

smectic phase fase esméctica

smelt fundir

smelting fundido

Smiles rearrangement reordenamiento de Smiles

smog smog, niebla contaminante

smoke humo

smokeless fuel combustible sin humo

smokeless powder pólvora sin humo

snake venom veneno de serpiente

soap jabón

soapstone esteatita

soda sosa, soda

soda alum alumbre de sosa

soda ash ceniza de sosa

soda lime cal sodada

soda niter nitro de sosa

soda pulp pulpa de sosa

sodalite sodalita

sodamide sodamida

Soddy displacement law ley de desplazamiento de Soddy

sodium sodio

sodium abietate abietato sódico

sodium acetate acetato sódico, acetato de sodio

sodium acetone bisulfate bisulfato de acetona sódico

sodium acetylformate acetilformiato sódico

sodium acetylsalicylate acetilsalicilato

sódico

sodium acid carbonate carbonato ácido sódico

sodium acid chromate cromato ácido sódico

sodium acid fluoride fluoruro ácido sódico

sodium acid phosphate fosfato ácido sódico

sodium acid pyrophosphate pirofosfato ácido sódico

sodium acid sulfate sulfato ácido sódico

sodium acid sulfite sulfito ácido sódico

sodium acid tartrate tartrato ácido sódico

sodium alginate alginato sódico

sodium alum alumbre sódico

sodium aluminate aluminato sódico

sodium aluminosilicate aluminosilicato sódico

sodium aluminum hydride hidruro alumínico sódico

sodium aluminum phosphate fosfato alumínico sódico

sodium aluminum silicofluoride silicofluoruro alumínico sódico

sodium aluminum sulfate sulfato alumínico sódico

sodium amalgam amalgama sódica

sodium amide amida sódica

sodium aminohippurate aminohipurato sódico

sodium aminosalicylate aminosalicilato sódico

sodium ammonium phosphate fosfato amónico sódico

sodium ammonium sulfate sulfato amónico sódico

sodium amytal amital sódico

sodium anoxynaphthonate anoxinaftonato sódico

sodium antimonate antimoniato sódico

sodium antimonyl nitrate nitrato de antimonilo sódico

sodium arsanilate arsanilato sódico

sodium arsenate arseniato sódico

sodium arsenide arseniuro sódico

sodium arsenite arsenito sódico

sodium ascorbate ascorbato sódico

sodium aspartate aspartato sódico

sodium auribromide auribromuro sódico

sodium aurichloride auricloruro sódico

sodium aurothiomalate aurotiomalato sódico

sodium azide azida sódica

sodium barbiturate barbiturato sódico

sodium benzenesulfonate bencenosulfonato sódico

sodium benzoate benzoato sódico

sodium benzosulfimide benzosulfimida sódica

sodium benzyl succinate succinato de bencilo sódico

sodium beryllium fluoride fluoruro de berilio sódico

sodium bicarbonate bicarbonato sódico

sodium bichromate bicromato sódico

sodium bifluoride bifluoruro sódico

sodium binoxalate binoxalato sódico

sodium binoxide binóxido sódico

sodium biphosphate bifosfato sódico

sodium bismuthate bismutato sódico

sodium bisulfate bisulfato sódico

sodium bisulfide bisulfuro sódico

sodium bisulfite bisulfito sódico

sodium bitartrate bitartrato sódico

sodium bithionolate bitionolato sódico

sodium borate borato sódico

sodium boroformate boroformiato sódico

sodium borohydride bohidruro sódico

sodium bromate bromato sódico

sodium bromide bromuro sódico

sodium bromite bromito sódico

sodium butyrate butirato sódico

sodium cacodylate cacodilato sódico

sodium caprylate caprilato sódico

sodium carbide carburo sódico

sodium carbolate carbolato sódico

sodium carbonate carbonato sódico

sodium carbonate decahydrate decahidrato de carbonato sódico

sodium carbonate monohydrate monohidrato de carbonato sódico

sodium carbonate peroxide peróxido de carbonato sódico

sodium carbonate peroxohydrate peroxohidrato de carbonato sódico

sodium carboxymethylcellulose
carboximetilcelulosa sódica

sodium carminate carminato sódico

sodium caseinate caseinato sódico

sodium cellulosate celulosato sódico

sodium chlorate clorato sódico

sodium chloride cloruro sódico

sodium chlorite clorito sódico

sodium chloroacetate cloroacetato
sódico

sodium chloroaluminate cloroaluminato
sódico

sodium chloroaurate cloroaurato sódico

sodium chlorophthalate cloroftalato
sódico

sodium chloroplatinate cloroplatinato
sódico

sodium chloroplatinite cloroplatinita
sódica

sodium chlorosulfonate clorosulfonato
sódico

sodium chromate cromato sódico

sodium chromite cromita sódica

sodium cinnamate cinamato sódico

sodium citrate citrato sódico

sodium copper chloride cloruro de cobre
sódico

sodium copper cyanide cianuro de cobre
sódico

sodium cyanamide cianamida sódica

sodium cyanate cianato sódico

sodium cyanide cianuro sódico

sodium cyanoaurite cianoaurita sódica

sodium cyanoborohydride
cianoborohidruro sódico

sodium cyanocuprate cianocuprato
sódico

sodium cyanoethanesulfonate
cianoetanosulfonato sódico

sodium cyclamate ciclamato sódico

sodium cyclopentadienide
ciclopentadienida sódica

sodium dehydroacetate deshidroacetato
sódico

sodium dextran sulfate sulfato de
dextrano sódico

sodium diacetate diacetato sódico

sodium diatrizoate diatrizoato sódico

sodium dibutyldithiocarbamate

dibutilditiocarbamato sódico

sodium dichloroisocyanurate
dicloroisocianurato sódico

sodium dichloropropionate
dicloropropionato sódico

sodium dichromate dicromato sódico

sodium dicyanoaurate dicianoaurato
sódico

sodium dihydrogenorthoperiodate
dihidrogenoortoperyodato sódico

sodium dihydrogenphosphate
dihidrogenofosfato sódico

sodium dihydroxyethylglycine
dihidroxietilglicina sódica

sodium dimethylarsenate
dimetilarseniato sódico

sodium dimethyldithiocarbamate
dimetilditiocarbamato sódico

sodium dinitrocresolate dinitrocresolato
sódico

sodium dioxide dióxido sódico

sodium diphenylhydantoin
difenilhidantoína sódica

sodium dispersion dispersión sódica

sodium disulfate disulfato sódico

sodium disulfite disulfito sódico

sodium dithionate ditionato sódico

sodium dithionite ditionito sódico

sodium dithiosalicylate ditiosalicilato
sódico

sodium diuranate diuranato sódico

sodium divanadate divanadato sódico

sodium dodecylbenzenesulfonate
dodecilbencenosulfonato sódico

sodium edetate edetato sódico

sodium erythorbate eritorbato sódico

sodium ethanoate etanoato sódico

sodium ethanolate etanolato sódico

sodium ethoxide etoxido sódico

sodium ethylate etilato sódico

sodium ethylxanthate exilxantato sódico

sodium ferricyanide ferricianuro sódico

sodium ferrocyanide ferrocianuro sódico

sodium fluorescein fluoresceína sódica

sodium fluoride fluoruro sódico

sodium fluoroacetate fluoroacetato
sódico

sodium fluoroborate fluoroborato sódico

sodium fluorophosphate fluorofosfato

sódico

sodium fluorosilicate fluorosilicato sódico

sodium folate folato sódico

sodium formaldehyde bisulfite bisulfito de formaldehído sódico

sodium formate formiato sódico

sodium fusidate fusidato sódico

sodium germanate germanato sódico

sodium glucoheptonate glucoheptonato sódico

sodium gluconate gluconato sódico

sodium glutamate glutamato sódico

sodium glycolate glicolato sódico

sodium gold chloride cloruro de oro sódico

sodium gold cyanide cianuro de oro sódico

sodium guanylate guanilato sódico

sodium halide haluro sódico

sodium hexachloroosmate hexacloroosmiato sódico

sodium hexachloroplatinate hexacloroplatinato sódico

sodium hexacyanoferrate hexacianoferrato sódico

sodium hexafluoroaluminate hexafluoroaluminato sódico

sodium hexafluorosilicate hexafluorosilicato sódico

sodium hexametaphosphate hexametafosfato sódico

sodium hexanitrocobaltate hexanitrocobaltato sódico

sodium hydrate hidrato sódico

sodium hydride hidruro sódico

sodium hydrogen carbonate carbonato de hidrógeno sódico

sodium hydrogen difluoride difluoruro de hidrógeno sódico

sodium hydrogen fluoride fluoruro de hidrógeno sódico

sodium hydrogen peroxide peróxido de hidrógeno sódico

sodium hydrogen phosphate fosfato de hidrógeno sódico

sodium hydrogen sulfate sulfato de hidrógeno sódico

sodium hydrogen sulfide sulfuro de hidrógeno sódico

sodium hydrogen sulfite sulfito de hidrógeno sódico

sodium hydrogen tartrate tartrato de hidrógeno sódico

sodium hydrosulfide hidrosulfuro sódico

sodium hydrosulfite hidrosulfito sódico

sodium hydroxide hidróxido sódico

sodium hypobromite hipobromito sódico

sodium hypochlorite hipoclorito sódico

sodium hyponitrite hiponitrito sódico

sodium hypophosphite hipofosfito sódico

sodium hyposulfate hiposulfato sódico

sodium hyposulfite hiposulfito sódico

sodium inosinate inosinato sódico

sodium iodate yodato sódico

sodium iodide yoduro sódico

sodium iodohippurate yodohipurato sódico

sodium iothalamate iotalamato sódico

sodium iron pyrophosphate pirofosfato de hierro sódico

sodium isoascorbate isoascorbato sódico

sodium isobutylxanthate isobutilxantato sódico

sodium isovalerate isovalerato sódico

sodium lactate lactato sódico

sodium lamp lámpara de sodio

sodium lauryl sulfate laurilsulfato sódico

sodium-lead alloy aleación de sodio-plomo

sodium lead hyposulfite hiposulfito de plomo sódico

sodium lead thiosulfate tiosulfato de plomo sódico

sodium malonic ester éster malónico sódico

sodium manganate manganato sódico

sodium mercaptoacetate mercaptoacetato sódico

sodium mercaptobenzothiazole mercaptobenzotiazol sódico

sodium metabisulfite metabisulfito sódico

sodium metaborate metaborato sódico

sodium metanilate metanilato sódico

sodium metaperiodate metaperyodato sódico

sodium metaphosphate metafosfato sódico

sodium metasilicate metasilicato sódico

sodium metavanadate metavanadato sódico

sodium methacrylate metacrilato sódico

sodium methanoate metanoato sódico

sodium methanolate metanloato sódico

sodium methiodal metiodal sódico

sodium methoxide metóxido sódico

sodium methyl carbonate carbonato de metilo sódico

sodium methyl siliconate siliconato de metilo sódico

sodium methylate metilato sódico

sodium molybdate molibdato sódico

sodium molybdophosphate molibdofosfato sódico

sodium molybdosilicate molibdosilicato sódico

sodium monosulfide monosulfuro sódico

sodium monoxide monóxido sódico

sodium naphthalenesulfonate naftalenosulfonato sódico

sodium naphthenate naftenato sódico

sodium naphthionate naftionato sódico

sodium niobate niobato sódico

sodium nitranilate nitranilato sódico

sodium nitrate nitrato sódico

sodium nitride nitruro sódico

sodium nitrilotriacetate nitrilotriacetato sódico

sodium nitrite nitrito sódico

sodium nitroferricyanide nitroferricianuro sódico

sodium nitrophenolate nitrofenolato sódico

sodium novobiocin novobiocina sódica

sodium octyl sulfate octilsulfato sódico

sodium oleate oleato sódico

sodium orthophosphate ortofosfato sódico

sodium orthosilicate ortosilicato sódico

sodium orthovanadate ortovanadato sódico

sodium oxalate oxalato sódico

sodium oxide óxido sódico

sodium palconate palconato sódico

sodium palladium chloride cloruro de paladio sódico

sodium palmitate palmitato sódico

sodium paraperiodate paraperyodato sódico

sodium pentaborate pentaborato sódico

sodium pentaborate decahydrate decahidrato de pentaborato sódico

sodium pentachlorophenate pentaclorofenato sódico

sodium pentobarbital pentobarbital sódico

sodium perborate perborato sódico

sodium perborate monohydrate monohidrato de perborato sódico

sodium perborate tetrahydrate tetrahidrato de perborato sódico

sodium percarbonate percarbonato sódico

sodium perchlorate perclorato sódico

sodium periodate peryodato sódico

sodium permanganate permanganato sódico

sodium peroxide peróxido sódico

sodium peroxodicarbonate peroxodicarbonato sódico

sodium peroxydisulfate peroxidisulfato sódico

sodium persulfate persulfato sódico

sodium phenate fenato sódico

sodium phenoacetate fenoacetato sódico

sodium phenobarbital fenobarbital sódico

sodium phenolate fenolato sódico

sodium phenolsulfonate fenolsulfonato sódico

sodium phenylacetate fenilacetato sódico

sodium phenylphenate fenilfenato sódico

sodium phenylphenolate fenilfenolato sódico

sodium phenylphosphinate fenilfosfinato sódico

sodium phosphate fosfato sódico

sodium phosphide fosfuro sódico

sodium phosphinate fosfinato sódico

sodium phosphite fosfito sódico

sodium phosphoaluminate fosfoaluminato sódico

sodium phosphomolybdate fosfomolibdato sódico

sodium phosphonate fosfonato sódico
sodium phosphotungstate fosfotungstato sódico
sodium phosphovanadate fosfovanadato sódico
sodium phthalate ftalato sódico
sodium phytate fitato sódico
sodium picramate picramato sódico
sodium platinichloride platinicloruro sódico
sodium platinochloride platinocloruro sódico
sodium plumbate plumbato sódico
sodium plumbite plumbito sódico
sodium polyphosphate polifosfato sódico
sodium polysulfide polisulfuro sódico
sodium-potassium alloy aleación de sodio-potasio
sodium potassium tartrate tartrato potásico sódico
sodium propionate propionato sódico
sodium prussiate prusiato sódico
sodium pyroantimonate piroantimoniato sódico
sodium pyroborate piroborato sódico
sodium pyrophosphate pirofosfato sódico
sodium pyrophosphate peroxide peróxido de pirofosfato sódico
sodium pyroracemate pirorracemato sódico
sodium pyrosulfate pirosulfato sódico
sodium pyrosulfite pirosulfito sódico
sodium pyrovanadate pirovanadato sódico
sodium pyruvate piruvato sódico
sodium resinate resinato sódico
sodium rhodanate rodanato sódico
sodium ricinoleate ricinoleato sódico
sodium saccharin saccarina sódica
sodium salicylate salicilato sódico
sodium sarcosinate sarcosinato sódico
sodium secobarbital secobarbital sódico
sodium selenate seleniato sódico
sodium selenite selenito sódico
sodium sesquicarbonate sesquicarbonato sódico
sodium sesquisilicate sesquisilicato sódico

sodium silicate silicato sódico
sodium silicoaluminate silicoaluminato sódico
sodium silicofluoride silicofluoruro sódico
sodium silicomolybdate silicomolibdato sódico
sodium silicotungstate silicotungstato sódico
sodium silver chloride cloruro de plata sódico
sodium silver cyanide cianuro de plata sódico
sodium silver thiosulfate tiosulfato de plata sódico
sodium sodioacetate sodioacetato sódico
sodium sorbate sorbato sódico
sodium stannate estannato sódico
sodium stannite estannita sódica
sodium stearate estearato sódico
sodium stearoyl lactylate estearoiloilactilato sódico
sodium styrenesulfonate estirenosulfonato sódico
sodium subsulfite subsulfito sódico
sodium succinate succinato sódico
sodium sulfalizarate sulfalizarato sódico
sodium sulfanilate sulfanilato sódico
sodium sulfantimonate sulfantimoniato sódico
sodium sulfate sulfato sódico
sodium sulfate decahydrate decahidrato de sulfato sódico
sodium sulfate monohydrate monohidrato de sulfato sódico
sodium sulfhydrate sulfhidrato sódico
sodium sulfide sulfuro sódico
sodium sulfite sulfito sódico
sodium sulfobromophthalein sulfobromoftaleína sódica
sodium sulfocarbolate sulfocarbolato sódico
sodium sulfocarbonate sulfocarbonato sódico
sodium sulfocyanate sulfocianato sódico
sodium sulfocyanide sulfocianuro sódico
sodium sulfonate sulfonato sódico
sodium sulfopropionitrile sulfopropionitrilo sódico

sodium sulforicinoleate sulforricinoleato
sódico
sodium sulfoxylate sulfoxilato sódico
sodium superoxide superóxido sódico
sodium tartrate tartrato sódico
sodium tellurate telurato sódico
sodium tellurite telurita sódica
sodium tetraborate tetraborato sódico
sodium tetrabromoaurate
tetrabromoaurato sódico
sodium tetrachloroaluminate
tetracloroaluminato sódico
sodium tetrachloroaurate
tetracloroaurato sódico
sodium tetrachloropalladate
tetracloropaladato sódico
sodium tetrachlorophenate
tetraclorofenato sódico
sodium tetrachloroplatinate
tetracloroplatinato sódico
sodium tetradecyl sulfate
tetradecilsulfato sódico
sodium tetrahydroborate
tetrahidroborato sódico
sodium tetraphenylborate
tetrafenilborato sódico
sodium tetraphosphate tetrafosfato
sódico
sodium tetrasulfide tetrasulfuro sódico
sodium thioantimonate tioantimoniato
sódico
sodium thioaurate tioaurato sódico
sodium thiocarbonate tiocarbonato
sódico
sodium thiocyanate tiocianato sódico
sodium thioglycolate tioglicolato sódico
sodium thiosulfate tiosulfato sódico
sodium titanate titanato sódico
sodium toluate toluato sódico
sodium toluenesulfonate
toluenosulfonato sódico
sodium trichloroacetate tricloroacetato
sódico
sodium trichlorophenate triclorofenato
sódico
sodium triphosphate trifosfato sódico
sodium tripolyphosphate tripolifosfato
sódico
sodium trithiocarbonate tritiocarbonato

sódico
sodium trititanate trititanato sódico
sodium tungstate tungstato sódico
sodium tungstophosphate tungstofosfato
sódico
sodium tungstosilicate tungstosilicato
sódico
sodium undecylenate undecilenato
sódico
sodium uranate uranato sódico
sodium valproate valproato sódico
sodium vanadate vanadato sódico
sodium vanadophosphate vanadofosfato
sódico
sodium warfarin warfarina sódica
sodium wolframate wolframato sódico
sodium xanthate xantato sódico
sodium xanthogenate xantogenato sódico
sodium xylenesulfonate xilenosulfonato
sódico
sodium zirconium glycolate glicolato de
zirconio sódico
sodium zirconium lactate lactato de
zirconio sódico
sodium zirconium sulfate sulfato de
zirconio sódico
sodyl sodilo
soft acid ácido blando
soft base base blanda
soft water agua blanda
softener ablandador
soil suelo, tierra, tierra negra
soil conditioner acondicionador de suelo
sol sol
solan solana
solanine solanina
solar cell célula solar
solar energy energía solar
solar furnace horno solar
solar pond tanque solar
solate solato
solation solación
solder soldante
solid sólido
solid-liquid equilibrium equilibrio
sólido-líquido
solid solution solución sólida
solid state estado sólido
solid-state chemistry química de estado

sólido
solidification solidificación
solidify solidificar
solidus solidus
solubility solubilidad
solubility coefficient coeficiente de solubilidad
solubility curve curva de solubilidad
solubility product producto de solubilidad
solubility-product constant constante de producto de solubilidad
solubility test prueba de solubilidad
solubilize solubilizar
soluble soluble
soluble glass vidrio soluble
soluble oil aceite soluble
soluble starch almidón soluble
solute soluto
solution solución, disolución
solution conductivity conductividad de solución
solution pressure presión de disolución
solutrope solutropo
solvate solvato
solvation solvatación
solvatochromism solvatocromismo
solvent solvente, disolvente
solvent drying secado por solvente
solvent extraction extracción por solvente
solvent refining refinación por solvente
solventless sin solvente
solvolysis solvólisis
solvolytic solvolítico
solvus solvus
Sommelet-Hauser rearrangement reordenamiento de Sommelet-Hauser
Sommelet reaction reacción de Sommelet
Sonn-Muller method método de Sonn-Muller
sonochemistry sonoquímica
sonolysis sonólisis
soot hollín
sorbent sorbente
sorbic acid ácido sórbico
sorbide sórbido
sorbitan sorbitán

sorbitol sorbitol
sorbitol anhydride anhídrido de sorbitol
sorbose sorbosa
Sörensen indicator indicador de Sörensen
Soret effect efecto de Soret
sorghum zahína
sorption sorción
sosoloid sosoloide
sour agrio, rancio
source fuente
Soxhlet extractor extractor de Soxhlet
soybean soja
soybean oil aceite de soja
space espacio
spallation espalación
spandex espandex
Spanish white blanco de España
spar espato
spark chispa
sparking chisporroteo
spearmint menta verde
spearmint oil aceite de menta verde
species especie
specific específico
specific activity actividad específica
specific charge carga específica
specific conductance conductancia específica
specific conductivity conductividad específica
specific energy energía específica
specific gravity gravedad específica
specific heat calor específico
specific heat capacity capacidad térmica específica
specific reaction rate velocidad de reacción específica
specific refractivity refractividad específica
specific rotation rotación específica
specific volume volumen específico
specific weight peso específico
spectra espectros
spectral espectral
spectral analysis análisis espectral
spectral line línea espectral
spectral region región espectral
spectral series serie espectral

spectral shift deslizamiento espectral
spectral tube tubo espectral
spectro espectro
spectroanalysis espectroanálisis
spectroanalytical espectroanalítco
spectrochemical espectroquímico
spectrochemical analysis análsis
    espectroquímico
spectrochemical series serie
    espectroquímica
spectrochemistry espectroquímica
spectrofluorometer espectrofluorómetro
spectrofluorometric
    espectrofluorométrico
spectrofluorometry espectrofluorometría
spectrogram espectrograma
spectrograph espectrógrafo
spectrographic espectrográfico
spectrographically espectrográficamente
spectrography espectrografía
spectrometer espectrómetro
spectrometric espectrométrico
spectrometric analysis análisis
    espectrométrico
spectrometric instrument instrumento
    espectrométrico
spectrometrically espectrométricamente
spectrometry espectrometría
spectromicroscope espectromicroscopio
spectromicroscopic
    espectromicroscópico
spectrophotometer espectrofotómetro
spectrophotometric espectrofotométrico
spectrophotometric analysis análisis
    espectrofotométrico
spectrophotometry espectrofotometría
spectropolarimeter espetropolarímetro
spectropolarimetric espetropolarimétrico
spectroradiometer espectrorradiómetro
spectroscope espectroscopio
spectroscopic espectroscópico
spectroscopic analysis análisis
    espectroscópico
spectroscopically espectroscópicamente
spectroscopy espectroscopia
spectrum espectro
spectrum analysis análisis espectral
spectrum analyzer analizador espectral
spectrum line línea espectral

spectrum series serie espectral
specular especular
spelter cinc industrial
Sperry process proceso de Sperry
sperrylite esperrilita
sphalerite esfalerita
sphingomyelin esfingomielina
sphingosine esfingosina
spin espín
spin magnetic moment momento
    magnético de espín
spin of electron espín de electrón
spin-orbit coupling acoplamiento espín-
    orbital
spin-orbit interaction interacción espín-
    orbital
spin quantum number número cuántico
    de espín
spin resonance resonancia de espín
spin-spin coupling acoplamiento espín-
    espín
spin-spin interaction interacción espín-
    espín
spinel espinela
spinning electron electrón giratorio
spinodal curve curva espinodal
spiran espirano
spirit espíritu
spiro atom átomo espiro
spiro system sistema espiro
spiro union unión espiro
spirochete espiroqueta
spironolactone espironolactona
spiropentane espiropentano
splitting división, partición,
    desdoblamiento
spodumene espodumena
sponge esponja
spontaneous espontáneo
spontaneous combustion combustión
    espontánea
spontaneous decay desintegración
    espontánea
spontaneous disintegration
    desintegración espontánea
spontaneous emission emisión
    espontánea
spontaneous fission fisión espontánea
spontaneous heating calentamiento

espontáneo
spontaneous ignition ignición espontánea
spontaneous reaction reacción
    espontánea
sputter evaporación catódica,
    chisporroteo, deposición electrónica
sputtering evaporación catódica,
    chisporroteo, deposición electrónica
squalane escualano
squalene escualeno
stability estabilidad
stability constant constante de
    estabilidad
stabilization estabilización
stabilization energy energía de
    estabilización
stabilizer estabilizador
stable estable
stable equilibrium equilibrio estable
stable isobar isóbaro estable
stable isotope isótopo estable
stable nucleus núcleo estable
stable system sistema estable
stacking apilamiento
Staedel-Rugheimer pyrazine synthesis
    síntesis de pirazinas de Staedel-
    Rugheimer
staggered conformation conformación
    alternada
stain tintura, colorante
stainless steel acero inoxidable
stalagmometer estalagmómetro
stalagmometry estalagmometría
standard patrón, estándar, norma
standard cell pila patrón
standard conditions condiciones
    normales
standard pressure presión normal
standard solution solución normal
standard temperature temperatura
    normal
standard temperature and pressure
    temperatura y presión normales
standardization estandarización,
    normalización
standardize estandarizar, normalizar
standardized estandarizado, normalizado
stannane estannano
stannate estannato

stannic estánnico
stannic acid ácido estánnico
stannic anhydride anhídrido estánnico
stannic bromide bromuro estánnico
stannic chloride cloruro estánnico
stannic chromate cromato estánnico
stannic fluoride fluoruro estánnico
stannic hydroxide hidróxido estánnico
stannic iodide yoduro estánnico
stannic oxide óxido estánnico
stannic phosphide fosfuro estánnico
stannic sulfate sulfato estánnico
stannic sulfide sulfuro estánnico
stannite estannita
stannous estannoso
stannous acetate acetato estannoso
stannous bromide bromuro estannoso
stannous chloride cloruro estannoso
stannous chromate cromato estannoso
stannous citrate citrato estannoso
stannous ethylhexoate etilhexoato
    estannoso
stannous fluoride fluoruro estannoso
stannous hydroxide hidróxido estannoso
stannous iodide yoduro estannoso
stannous octoate octoato estannoso
stannous oleate oleato estannoso
stannous oxalate oxalato estannoso
stannous oxide óxido estannoso
stannous pyrophosphate pirofosfato
    estannoso
stannous sulfate sulfato estannoso
stannous sulfide sulfuro estannoso
stannous tartrate tartrato estannoso
stannum estaño
stannyl estannilo
starch almidón
starch dialdehyde dialdehído de almidón
starch phosphate fosfato de almidón
starch syrup jarabe de almidón
starch xanthate xantato de almidón
Stark effect efecto de Stark
Stark-Einstein law ley de Stark-Einstein
Stark-Luneland effect efecto de Stark-
    Luneland
state estado
static estático
stationary phase fase estacionaria
stationary state estado estacionario

statistical mechanics mecánica
 estadística
Staudinger reaction reacción de
 Staudinger
Stead reagent reactivo de Stead
steady state estado estacionario
steam vapor de agua
steam bath baño de vapor
steam distillation destilación al vapor
steam point punto de vapor
steam reforming reformación al vapor
stearaldehyde estearaldehído
stearamide estearamida
stearate estearato
stearic acid ácido esteárico
stearin estearina
stearone estearona
stearonitrile estearonitrilo
stearoyl estearoilo
stearoyl chloride cloruro de estearoilo
stearoylaminophenol
 estearoilaminofenol
stearyl estearilo
stearyl alcohol alcohol estearílico
stearyl mercaptan estearilmercaptano
stearyl methacrylate metacrilato de
 estearilo
steatite esteatita
steel acero
Steffen process proceso de Steffen
Stengel process proceso de Stengel
step paso, etapa
stereoblock polymer polímero de
 estereobloques
stereochemistry estereoquímica
stereoisomer estereoisómero
stereoisomerism estereoisomerismo
stereoregular estereorregular
stereoregular polymer polímero
 estereorregular
stereoscope estereoscopio
stereoselectivity estereoselectividad
stereospecific estereoespecífico
stereospecific catalyst catalizador
 estereoespecífico
stereospecific polymer polímero
 estereoespecífico
stereospecific reaction reacción
 estereoespecífica

stereospecific synthesis síntesis
 estereoespecífica
steric estérico
steric effect efecto estérico
steric geometry geometría estérica
steric hindrance impedimento estérico
sterile esteril
sterilization esterilización
sterling silver plata esterlina
steroid esteriode
sterol esterol
Stevens rearrangement reordenamiento
 de Stevens
stibic anhydride anhídrido estíbico
stibine estibina
stibino estibino
stibinoso estibinoso
stibnite estibnita
stibo estibo
stibonium estibonio
stibono estibono
stibonoso estibonoso
stiboso estiboso
Stieglitz rearrangement reordenamiento
 de Stieglitz
stigmasterol estigmasterol
stilbene estilbeno
stilbene dye colorante de estilbeno
stilbestrol estilbestrol
stilbite estilbita
still destilador, alambique
stimulation estimulación
stirrer agitador
stirring rod varilla agitadora
Stobbe condensation condensación de
 Stobbe
Stobbe reaction reacción de Stobbe
stochastic estocástico
stochastic process proceso estocástico
Stock system sistema de Stock
Stoddard solvent solvente de Stoddard
stoichiometric estequiométrico
stoichiometric coefficient coeficiente
 estequiométrico
stoichiometric compound compuesto
 estequiométrico
stoichiometric mixture mezcla
 estequiométrica
stoichiometry estequiometría

Stokes law ley de Stokes
Stokes line línea de Stokes
Stokes shift deslizamiento de Stokes
Stolle synthesis síntesis de Stolle
stopper tapón
stopper with tap tapón con llave
storage almacenamiento
storage battery batería de acumuladores
storage guidelines normas de
    almacenamiento
storage precautions precauciones de
    almacenamiento
Stork enamine reaction reacción de
    enaminas de Stork
straight chain cadena recta
straight-chain hydrocarbon
    hidrocarburo de cadena recta
strain deformación, tensión
strain theory teoría de deformación
streaming potential potencial de
    circulación
streamline filtration filtración por
    bordes
Strecker degradation degradación de
    Strecker
Strecker reaction reacción de Strecker
Strecker sulfite alkylation alquilación de
    sulfitos de Strecker
Strecker synthesis síntesis de Strecker
strength fuerza, intensidad
streptolin estreptolina
streptomycin estreptomicina
stress estrés, tensión, deformación
stripped atom átomo despojado de
    electrones
stripping retroextracción, extracción,
    eliminación, lavado
strong fuerte
strong acid ácido fuerte
strong base base fuerte
strong electrolyte electrólito fuerte
strontia estroncia
strontianite estroncianita
strontium estroncio
strontium acetate acetato de estroncio
strontium arsenite arsenito de estroncio
strontium bicarbonate bicarbonato de
    estroncio
strontium bromate bromato de estroncio

strontium bromide bromuro de estroncio
strontium carbonate carbonato de
    estroncio
strontium chlorate clorato de estroncio
strontium chloride cloruro de estroncio
strontium chromate cromato de
    estroncio
strontium dioxide dióxido de estroncio
strontium dithionate ditionato de
    estroncio
strontium fluoride fluoruro de estroncio
strontium formate formiato de estroncio
strontium hexafluorosilicate
    hexafluorosilicato de estroncio
strontium hydrogen carbonate
    carbonato de hidrógeno de estroncio
strontium hydroxide hidróxido de
    estroncio
strontium hyposulfite hiposulfito de
    estroncio
strontium iodide yoduro de estroncio
strontium lactate lactato de estroncio
strontium molybdate molibdato de
    estroncio
strontium monosulfide monosulfuro de
    estroncio
strontium monoxide monóxido de
    estroncio
strontium nitrate nitrato de estroncio
strontium nitrite nitrito de estroncio
strontium oxalate oxalato de estroncio
strontium oxide óxido de estroncio
strontium perchlorate perclorato de
    estroncio
strontium peroxide peróxido de
    estroncio
strontium phosphate fosfato de estroncio
strontium-potassium chlorate clorato de
    estroncio-potasio
strontium saccharate sacarato de
    estroncio
strontium salicylate salicilato de
    estroncio
strontium sulfate sulfato de estroncio
strontium sulfide sulfuro de estroncio
strontium sulfite sulfito de estroncio
strontium tartrate tartrato de estroncio
strontium thiosulfate tiosulfato de
    estroncio

strontium titanate titanato de estroncio

strontium zirconate zirconato de estroncio

structural estructural

structural antagonist antagonista estructural

structural formula fórmula estructural

structural isomerism isomerismo estructural

structure estructura

strychnidine estricnidina

strychnine estricnina

Stuffer rule regla de Stuffer

styphnic acid ácido estífnico

styracin estiracina

styralyl acetate acetato estiralílico

styralyl alcohol alcohol estiralílico

styrenated oil aceite estirenado

styrene estireno

styrene-acrylonitrile resin resina de estireno-acrilonitrilo

styrene-butadiene rubber caucho de estireno-butadieno

styrene glycol estirenglicol

styrene monomer monómero de estireno

styrene nitrosite nitrosita de estireno

styrene oxide óxido de estireno

styrene plastic plástico de estireno

styrene polymer polímero de estireno

styryl estirilo

sub sub

subacetate subacetato

subatomic subatómico

subatomic decomposition descomposición subatómica

subatomic particle partícula subatómica

subatomic reaction reacción subatómica

subatomics subatómica

subcarbonate subcarbonato

suberane suberano

suberic acid ácido subérico

suberone suberona

suberyl suberilo

sublation sublación

sublethal subletal

sublevel subnivel

sublimate sublimar

sublimation sublimación

sublimation point punto de sublimación

sublimator sublimador

subnitrate subnitrato

subnuclear subnuclear

subnuclear particle partícula subnuclear

suboxide subóxido

subshell subcapa

substance substancia

substance concentration concentración de substancia

substantive dye colorante substantivo

substituent átomo substituyente, grupo substituyente

substitute substituir

substitution substitución

substitution reaction reacción de substitución

substrate substrato

substructure subestructura

subtilin subtilina

succinaldehyde succinaldehído

succinamic acid ácido succinámico

succinamoyl succinamoilo

succinamyl succinamilo

succinate succinato

succinic acid ácido succínico

succinic acid peroxide peróxido de ácido succínico

succinic anhydride anhídrido succínico

succinimide succinimida

succinimido succinimido

succiniodimide succiniodimida

succinonitrile succinonitrilo

succinyl succinilo

succinyl chloride cloruro de succinilo

succinyl oxide óxido de succinilo

sucrase sucrasa

sucrate sucrato

sucrol sucrol

sucrose sucrosa, sacarosa

sucrose acetate isobutyrate acetatoisobutirato de sucrosa

sucrose monostearate monoestearato de sucrosa

sucrose octaacetate octaacetato de sucrosa

sucrose polyester poliéster de sucrosa

suction pump bomba de succión

sugar azúcar

sugar of lead azúcar de plomo

sugar of milk azúcar de leche
sugar substitute substituto de azúcar
sugarcane wax cera de caña de azúcar
sulfa drug droga sulfa, sulfonamida
sulfacetamide sulfacetamida
sulfadiazine sulfadiazina
sulfadimidine sulfadimidina
sulfaguanidine sulfaguanidina
sulfaldehyde sulfaldehído
sulfamate sulfamato
sulfamerazine sulfamerazina
sulfamethizole sulfametizol
sulfamic sulfámico
sulfamic acid ácido sulfámico
sulfamide sulfamida
sulfamidic acid ácido sulfamídico
sulfaminic acid ácido sulfamínico
sulfamoyl sulfamoilo
sulfamyl sulfamilo
sulfane sulfano
sulfanilamide sulfanilamida
sulfanilate sulfanilato
sulfanilic acid ácido sulfanílico
sulfarsenate sulfarseniato
sulfatase sulfatasa
sulfate sulfato
sulfate mineral mineral de sulfato
sulfathiazole sulfatiazol
sulfation sulfatación
sulfazide sulfazida
sulfenamide sulfenamida
sulfenic sulfénico
sulfeno sulfeno
sulfenone sulfenona
sulfhydryl sulfhidrilo
sulfide sulfuro
sulfide dye colorante al sulfuro
sulfime sulfima
sulfimide sulfimida
sulfinate sulfinato
sulfine sulfina
sulfino sulfino
sulfinoxide sulfonóxido
sulfinyl sulfinilo
sulfinyl bromide bromuro de sulfinilo
sulfinyl chloride cloruro de sulfinilo
sulfinyl fluoride fluoruro de sulfinilo
sulfinylimide sulfinilimida
sulfite sulfito

sulfite pulp pulpa de sulfito
sulfo sulfo
sulfoacetic acid ácido sulfoacético
sulfoamino sulfoamino
sulfobenzoic acid ácido sulfobenzoico
sulfobenzoic anhydride anhídrido
  sulfobenzoico
sulfocarbanilide sulfocarbanilida
sulfocarbimide sulfocarbimida
sulfocarbolic acid ácido sulfocarbólico
sulfocarbolide sulfocarbolida
sulfocarbonic acid ácido sulfocarbónico
sulfocyanate sulfocianato
sulfocyanic acid ácido sulfociánico
sulfocyanide sulfocianuro
sulfohydrate sulfohidrato
sulfolane sulfolano
sulfoleic acid ácido sulfoleico
sulfonamide sulfonamida
sulfonate sulfonato
sulfonation sulfonación
sulfone sulfona
sulfonic acid ácido sulfónico
sulfonic anhydride anhídrido sulfónico
sulfonium sulfonio
sulfonohydrazonic acid ácido
  sulfonohidrazónico
sulfonyl sulfonilo
sulfonyl chloride cloruro de sulfonilo
sulfonyl fluoride fluoruro de sulfonilo
sulfonyldiphenol sulfonildifenol
sulfophthalic anhydride anhídrido
  sulfoftálico
sulfosalicylic acid ácido sulfosalicílico
sulfoselenide sulfoseleniuro
sulfoxide sulfóxido
sulfoxylate sulfoxilato
sulfoxylic acid ácido sulfoxílico
sulfur azufre
sulfur bichloride bicloruro de azufre
sulfur bromide bromuro de azufre
sulfur chloride cloruro de azufre
sulfur cycle ciclo de azufre
sulfur dichloride dicloruro de azufre
sulfur dioxide dióxido de azufre
sulfur dye colorante al azufre
sulfur fluoride fluoruro de azufre
sulfur hexafluoride hexafluoruro de
  azufre

sulfur hexaiodide hexayoduro de azufre
sulfur iodide yoduro de azufre
sulfur monobromide bromuro de azufre
sulfur monochloride monocloruro de azufre
sulfur monofluoride monofluoruro de azufre
sulfur monoxide monóxido de azufre
sulfur oxide óxido de azufre
sulfur oxyacid oxiácido de azufre
sulfur oxychloride oxicloruro de azufre
sulfur pentachloride pentacloruro de azufre
sulfur subchloride subcloruro de azufre
sulfur test prueba de azufre
sulfur tetrachloride tetracloruro de azufre
sulfur tetrafluoride tetrafluoruro de azufre
sulfur trioxide trióxido de azufre
sulfurated sulfurado
sulfurated lime cal sulfurada
sulfuration sulfuración
sulfuretted sulfurado
sulfuric sulfúrico
sulfuric acid ácido sulfúrico
sulfuric anhydride anhídrido sulfúrico
sulfuric chloride cloruro sulfúrico
sulfuric monobromide monobromuro sulfúrico
sulfuric monochloride monocloruro sulfúrico
sulfuric oxychloride oxicloruro sulfúrico
sulfurous sulfuroso
sulfurous acid ácido sulfuroso
sulfurous oxychloride oxicloruro sulfuroso
sulfuryl sulfurilo
sulfuryl chloride cloruro de sulfurilo
sulfuryl fluoride fluoruro de sulfurilo
Sullivan reaction reacción de Sullivan
sulphur azufre
sultam sultama
sultone sultona
sulvanite sulvanita
sunflower oil aceite de girasol
superacid superácido
superactinide superactínido
superalloy superaleación

supercarbonate supercarbonato
superconductivity superconductividad
superconductor superconductor
supercooled subenfriado
supercooled liquid líquido subenfriado
supercooling subenfriado
supercritical supercrítico
supercritical fluid fluido supercrítico
superfluid superfluido
superfluidity superfluidez
superheating supercalentamiento
superheavy superpesado
superheavy element elemento superpesado
supernatant sobrenadante
supernormal supernormal
superoxide superóxido
superoxide dismutase dismutasa superóxida
superpalite superpalita
superphosphate superfosfato
superphosphoric acid ácido superfosfórico
superplasticity superplasticidad
superpolymer superpolímero
supersaturated supersaturado
supersaturated solution solución supersaturada
supersaturation supersaturación
supersolubility supersolubilidad
supra supra
supramolecular chemistry química supramolecular
surface superficie
surface-active agent agente tensoactivo
surface activity actividad superficial
surface area área superficial
surface chemistry química superficial
surface combustion combustión superficial
surface compound compuesto superficial
surface energy energía superficial
surface orientation orientación superficial
surface pressure presión superficial
surface reaction reacción superficial
surface tension tensión superficial
surface viscocity viscocidad superficial
surfactant surfactante

surfusion surfusión
surrosion surrosión
suspended solids sólidos suspendidos
suspending agent agente de suspensión
suspension suspensión
suspensoid suspensoide
suxamethonium chloride cloruro de suxametonio
Swarts reaction reacción de Swarts
sweet oil aceite dulce
sweet water agua dulce
sweeten endulzar
sweetener endulcorante, endulzador
sydnone sidnona
sylvan silvano
sylvanite silvanita
sylvic acid ácido sílvico
sylvine silvina
sylvinite silvinita
sylvite silvita
sylyl sililo
symbol símbolo
symmetric simétrico
symmetrical simétrico
symmetrical carbon atom átomo de carbono simétrico
symmetrical compound compuesto simétrico
symmetry simetría
syndiotactic sindiotáctico
syneresis sinéresis
synergism sinergismo
synergist sinergista
syntactic sintáctico
synthesis síntesis
synthesis gas gas de síntesis
synthesize sintetizar
synthetic sintético
synthetic detergent detergente sintético
synthetic fiber fibra sintética
synthetic fuel combustible sintético
synthetic gas gas sintético
synthetic natural gas gas natural sintético
synthetic oil aceite sintético
synthetic resin resina sintética
synthetic rubber caucho sintético
synthol sintol
syrosingopine sirosingopina
syrup jarabe
systematic name nombre sistemático
systemic sistémico
Szilard-Chalmers effect efecto de Szilard-Chalmers

T

table salt sal de mesa
tabun tabún
tachiol taquiol
taconite taconita
tactic polymer polímero táctico
tacticity tacticidad
tactosol tactosol
taenite taenita
Tafel rearrangement reordenamiento de
    Tafel
tag marcador
tagged atom átomo marcado
tagged molecule molécula marcada
tailings residuos, desechos, colas
talc talco
tall oil resina líquida
tallow sebo
tallow oil aceite de sebo
talose talosa
Tanabe-Sugano diagram diagrama de
    Tanabe-Sugano
tanacetyl alcohol alcohol tanacetílico
tangerine oil aceite de tangerina
tannase tanasa
tannic acid ácido tánico
tannin tanino
tanning curtido
tannyl tanilo
tantalate tantalato
tantalic acid ácido tantálico
tantalic chloride cloruro tantálico
tantalite tantalita
tantalous tantaloso
tantalous chloride cloruro tantaloso
tantalum tántalo, tantalio
tantalum alcoholate alcoholato de
    tántalo
tantalum bromide bromuro de tántalo
tantalum carbide carburo de tántalo
tantalum chloride cloruro de tántalo
tantalum disulfide disulfuro de tántalo
tantalum fluoride fluoruro de tántalo
tantalum nitride nitruro de tántalo

tantalum oxide óxido de tántalo
tantalum pentabromide pentabromuro
    de tántalo
tantalum pentachloride pentacloruro de
    tántalo
tantalum pentafluoride pentafluoruro de
    tántalo
tantalum pentoxide pentóxido de tántalo
tantalum potassium fluoride fluoruro
    potásico de tántalo
tantiron tantirón
tapioca tapioca
tapiolite tapolita
tar alquitrán, brea
tar acid ácido de alquitrán
tar base base de alquitrán
tar camphor alcanfor de alquitrán
tar oil aceite de alquitrán
tar soap jabón de alquitrán
tare tara
target blanco
tarnish deslustre, descoloramiento
tarragon oil aceite de estragón
tartar emetic tártaro emético
tartaric acid ácido tartárico
tartaroyl tartaroilo
tartrate tartrato
tartrazine tartrazina
tartronic acid ácido tartrónico
tartronoyl tartronoilo
taste sabor, gusto
tau tau
tau particle partícula tau
tau value valor tau
Tauber test prueba de Tauber
taurine taurina
taurocholic acid ácido taurocólico
tauryl taurilo
tautomer tautómero
tautomerase tautomerasa
tautomeric tautomérico
tautomerism tautomería
tautourea tautourea

taxol taxol
technetium tecnecio
technical técnico
Teclu burner mechero de Teclu
teichoic acid ácido teicoico
tektite tectita
Teller-Redlich rule regla de Teller-Redlich
tellurate telurato
telluric acid ácido telúrico
telluric bromide bromuro telúrico
telluric lead plomo telúrico
telluride telururo
tellurinic acid ácido telurínico
tellurinyl telurinilo
tellurite telurita
tellurium telurio
tellurium bromide bromuro de telurio
tellurium chloride cloruro de telurio
tellurium dibromide dibromuro de telurio
tellurium dichloride dicloruro de telurio
tellurium dioxide dióxido de telurio
tellurium disulfide disulfuro de telurio
tellurium fluoride fluoruro de telurio
tellurium hexafluoride hexafluoruro de telurio
tellurium hydroxide hidróxido de telurio
tellurium iodide yoduro de telurio
tellurium monoxide monóxido de telurio
tellurium nitrate nitrato de telurio
tellurium oxide óxido de telurio
tellurium oxychloride oxicloruro de telurio
tellurium sulfate sulfato de telurio
tellurium sulfide sulfuro de telurio
tellurium tetrabromide tetrabromuro de telurio
tellurium trioxide trióxido de telurio
telluroketone telurocetona
telluronium teluronio
tellurous teluroso
tellurous acid ácido teluroso
tellurous bromide bromuro teluroso
tellurous chloride cloruro teluroso
telluryl telurilo
telodrin telodrina
telomerization telomerización
telomerization reaction reacción de telomerización

temper templar
temperature temperatura
temperature coefficient coeficiente de temperatura
temperature control control de temperatura
temperature degree grado de temperatura
temperature regulator regulador de temperatura
temperature scale escala de temperatura
temperature-sensitive sensible a la temperatura
tenderization ablandamiento
tenderize ablandar
tennantite tennantita
tenorite tenorita
tenth-normal solution solución decinormal
tephroite tefroíta
ter ter
teratogen teratógeno
terbia terbia
terbium terbio
terbium chloride cloruro de terbio
terbium hydroxide hidróxido de terbio
terbium nitrate nitrato de terbio
terbium oxide óxido de terbio
terbium sulfate sulfato de terbio
terebene terebeno
terephthalaldehyde tereftalaldehído
terephthalic acid ácido tereftálico
terephthalonitrile tereftalonitrilo
terephthaloyl chloride cloruro de tereftaloilo
terephthaloylbenzoic acid ácido tereftaloilbenzoico
terminology terminología
ternary ternario
ternary compound compuesto ternario
ternary system sistema ternario
teroxide teróxido
terpadiene terpadieno
terpene terpeno
terpenoid terpenoide
terpenol terpenol
terphenyl terfenilo
terpilenol terpilenol

terpin hydrate hidrato de terpina
terpinene terpineno
terpineol terpineol
terpinolene terpinoleno
terpinyl acetate acetato de terpinilo
terpolymer terpolímero
tert terc
tertiary terciario
tertiary alcohol alcohol terciario
tertiary amine amina terciaria
tertiary carbon carbono terciario
tertiary carbon atom átomo de carbono terciario
tervalent tervalente
test prueba
test paper papel de prueba
test tube probeta, tubo de ensayo
testosterone testosterona
tetra tetra
tetrabasic tetrabásico
tetrabenzyl tetrabencilo
tetraborane tetraborano
tetrabromo tetrabromo
tetrabromobenzene tetrabromobenceno
tetrabromoethane tetrabromoetano
tetrabromoethylene tetrabromoetileno
tetrabromofluorescein tetrabromofluoresceína
tetrabromomethane tetrabromometano
tetrabromosilane tetrabromosilano
tetrabutyl titanate titanato de tetrabutilo
tetrabutyl zirconate zirconato de tetrabutilo
tetrabutylammonium chloride cloruro de tetrabutilamonio
tetrabutyltin tetrabutilestaño
tetracaine tetracaína
tetracaine hydrochloride clorhidrato de tetracaína
tetracarboxybutane tetracarboxibutano
tetracarboxylic acid ácido tetracarboxílico
tetracene tetraceno
tetrachloride tetracloruro
tetrachloro tetracloro
tetrachloroaniline tetracloroanilina
tetrachlorobenzene tetraclorobenceno
tetrachlorodiphenylethane tetraclorodifeniletano

tetrachloroethane tetracloroetano
tetrachloroethylene tetracloroetileno
tetrachloromethane tetraclorometano
tetrachloronaphthalene tetracloronaftaleno
tetrachlorophenol tetraclorofenol
tetrachlorophthalic acid ácido tetracloroftálico
tetrachlorophthalic anhydride anhídrido tetracloroftálico
tetrachlorosalicylanilide tetraclorosalicilanilida
tetrachlorosilane tetraclorosilano
tetracosane tetracosano
tetracosanoic acid ácido tetracosanoico
tetracosyl tetracosilo
tetracyano tetraciano
tetracyanoethylene tetracianoetileno
tetracyanoquinonedimethane tetracianoquinonadimetano
tetracycline tetraciclina
tetracyclone tetraciclona
tetrad tetrada
tetradecane tetradecano
tetradecanoic acid ácido tetradecanoico
tetradecanol tetradecanol
tetradecene tetradeceno
tetradecyl tetradecilo
tetradecyl chloride cloruro de tetradecilo
tetradecyl thiol tetradeciltiol
tetradecylamine tetradecilamina
tetradecylene tetradecileno
tetradifon tetradifón
tetraethanolammonium hydroxide hidróxido de tetraetanolamonio
tetraethoxypropane tetraetoxipropano
tetraethyl tetraetilo
tetraethyl phosphate fosfato de tetraetilo
tetraethylbenzene tetraetilbenceno
tetraethylene glycol tetraetilenglicol
tetraethylene glycol dimethacrylate dimetacrilato de tetraetilenglicol
tetraethylene glycol monostearate monoestearato de tetraetilenglicol
tetraethylenepentamine tetraetilenopentamina
tetraethylgermanium tetraetilgermanio
tetraethylhexyl titanate titanato de tetraetilhexilo

tetraethyllead tetraetilplomo
tetraethyltin tetraetilestaño
tetrafluoroborate tetrafluoroborato
tetrafluorodichloroethane
tetrafluorodicloroetano
tetrafluoroethylene tetrafluoroetileno
tetrafluoroethylene epoxide epóxido de
tetrafluoroetileno
tetrafluorohydrazine
tetrafluorohidrazina
tetrafluoromethane tetrafluorometano
tetrafluorosilane tetrafluorosilano
tetragalloyl tetragaloilo
tetraglycol dichloride dicloruro de
tetraglicol
tetragonal tetragonal
tetragonal system sistema tetragonal
tetrahedral tetraédrico
tetrahedral angle ángulo tetraédrico
tetrahedral atom átomo tetraédrico
tetrahedral carbon carbono tetraédrico
tetrahedral compound compuesto
tetraédrico
tetrahedral coordination coordinación
tetraédrica
tetrahedron tetraedro
tetrahedronal tetraédrico
tetrahydrate tetrahidrato
tetrahydro tetrahidro
tetrahydrobenzene tetrahidrobenceno
tetrahydrocannibol tetrahidrocannibol
tetrahydrofuran tetrahidrofurano
tetrahydrofuran polymer polímero de
tetrahidrofurano
tetrahydrofurandimethanol
tetrahidrofuranodimetanol
tetrahydrofurfuryl alcohol alcohol
tetrahidrofurfurílico
tetrahydrofurfuryl benzoate benzoato
de tetrahidrofurfurilo
tetrahydrofurfuryl laurate laurato de
tetrahidrofurfurilo
tetrahydrofurfurylamine
tetrahidrofurfurilamina
tetrahydronaphthalene
tetrahidronaftaleno
tetrahydrophenol tetrahidrofenol
tetrahydrophthalic anhydride anhídrido
tetrahidroftálico

tetrahydropyranmethanol
tetrahidropiranometanol
tetrahydropyridine tetrahidropiridina
tetrahydrothiophene tetrahidrotiofeno
tetrahydroxy tetrahidroxi
tetrahydroxybenzene
tetrahidroxibenceno
tetrahydroxybenzoic acid ácido
tetrahidroxibenzoico
tetrahydroxybutane tetrahidroxibutano
tetraiodo tetrayodo
tetraisopropyl titanate titanato de
tetraisopropilo
tetraisopropyl zirconate zirconato de
tetraisopropilo
tetraketone tetracetona
tetrakis tetrakis
tetralin tetralina
tetralite tetralita
tetralone tetralona
tetramer tetrámero
tetramethoxypropane
tetrametoxipropano
tetramethyl tetrametilo
tetramethylammonium chloride cloruro
de tetrametilamonio
tetramethylbenzene tetrametilbenceno
tetramethylbenzidine tetrametilbencidina
tetramethylcyclobutanediol
tetrametilciclobutanodiol
tetramethyldiaminobenzhydrol
tetrametildiaminobenzhidrol
tetramethyldiaminobenzophenone
tetrametildiaminobenzofenona
tetramethyldiaminodiphenylsulfone
tetrametildiaminodifenilsulfona
tetramethylene tetrametileno
tetramethylenediamine
tetrametilendiamina
tetramethylethylenediamine
tetrametiletilendiamina
tetramethyllead tetrametilplomo
tetramethylmethane tetrametilmetano
tetramethylsilane tetrametilsilano
tetramethyltin tetrametilestaño
tetramethylurea tetrametilurea
tetramine tetramina
tetramolecular tetramolecular
tetramorphism tetramorfismo

tetranitrate tetranitrato
tetranitro tetranitro
tetranitroaniline tetranitroanilina
tetranitrobiphenyl tetranitrobifenilo
tetranitromethane tetranitrometano
tetranitrophenol tetranitrofenol
tetraoxide tetróxido, tetraóxido
tetraoxophosphoric acid ácido tetraoxofosfórico
tetraoxosulfuric acid ácido tetraoxosulfúrico
tetraphenyl tetrafenilo
tetraphenylbutadiene tetrafenilbutadieno
tetraphenylcyclopentadienone tetrafenilciclopentadienona
tetraphenylene tetrafenileno
tetraphenylethane tetrafeniletano
tetraphenylmethane tetrafenilmetano
tetraphenylsilane tetrafenilsilano
tetraphenyltin tetrafenilestaño
tetraphenylurea tetrafenilurea
tetraphosphoric acid ácido tetrafosfórico
tetraphosphorus selenide seleniuro de tetrafósforo
tetrapotassium pyrophosphate pirofosfato tetrapotásico
tetrapropylene tetrapropileno
tetrasilane tetrasilano
tetrasodium diphosphate difosfato tetrasódico
tetrasodium ethylenediaminetetraacetic acid ácido etilendiaminotetraacético tetrasódico
tetrasodium pyrophosphate pirofosfato tetrasódico
tetrathioarsenate tetratioarseniato
tetrathioarsenite tetratioarsenito
tetrathionate tetrationato
tetrathionic acid ácido tetratiónico
tetravalent tetravalente
tetrazene tetraceno
tetrazole tetrazol
tetrazolium tetrazolio
tetrazolyl tetrazolilo
tetrazone tetrazona
tetrel tetrel
tetritol tetritol
tetrol tetrol
tetrolic acid ácido tetrólico

tetrone tetrona
tetrose tetrosa
tetroxide tetróxido
tetryl tetrilo
texaphyrin texafirina
texture textura
thalenite thalenita
thalidomide talidomida
thallate talato
thallic tálico
thallium talio
thallium acetate acetato de talio
thallium alum alumbre de talio
thallium amalgam amalgama de talio
thallium bromide bromuro de talio
thallium carbonate carbonato de talio
thallium chloride cloruro de talio
thallium chromate cromato de talio
thallium ethanolate etanolato de talio
thallium fluoride fluoruro de talio
thallium formate formiato de talio
thallium hexafluorosilicate hexafluorosilicato de talio
thallium hydroxide hidróxido de talio
thallium iodide yoduro de talio
thallium monoxide monóxido de talio
thallium nitrate nitrato de talio
thallium oxide óxido de talio
thallium oxysulfide oxisulfuro de talio
thallium sesquichloride sesquicloruro de talio
thallium sulfate sulfato de talio
thallium sulfide sulfuro de talio
thallium sulfoxide sulfóxido de talio
thallium trifluoroacetate trifluoroacetato de talio
thallous talioso
thallous bromide bromuro talioso
thallous carbonate carbonato talioso
thallous chloride cloruro talioso
thallous hydroxide hidróxido talioso
thallous iodide yoduro talioso
thallous monoxide monóxido talioso
thallous nitrate nitrato talioso
thallous oxide óxido talioso
thallous sesquichloride sesquicloruro talioso
thallous sulfate sulfato talioso
thallous sulfide sulfuro talioso

thebaine tebaína
thebenidine tebenidina
theine teína
Thenard blue azul de Thenard
thenoic acid ácido tenoico
thenyl tenilo
thenyl alcohol alcohol tenílico
thenyldiamine tenildiamina
theobroma oil aceite de teobroma
theobromine teobromina
theophylline teofilina
theoretical teorético
theoretical plate placa teorética
thermal térmico, termal
thermal activation activación térmica
thermal analysis análisis térmico
thermal capacity capacidad térmica
thermal collision colisión térmica
thermal column columna térmica
thermal conductance conductancia
térmica
thermal conductivity conductividad
térmica
thermal constant constante térmica
thermal cracking craqueo térmico
thermal decomposition descomposición
térmica
thermal degradation degradación
térmica
thermal diffusion difusión térmica
thermal effect efecto térmico
thermal energy energía térmica
thermal equilibrium equilibrio térmico
thermal excitation excitación térmica
thermal expansion expansión térmica
thermal expansion coefficient
coeficiente de expansión térmica
thermal fission fisión térmica
thermal fragmentation fragmentación
térmica
thermal insulation aislamiento térmico
thermal intensity intensidad térmica
thermal ionization ionización térmica
thermal neutron neutrón térmico
thermal pollution contaminación térmica
thermal reactor reactor térmico
thermal reforming reformación térmica
thermal stability estabilidad térmica
thermal titration titulación térmica,

valoración térmica
thermalize termalizar
thermatomic process proceso
termatómico
thermic térmico, termal
thermionic effect efecto termiónico
thermite termita
thermo termo
thermoanalysis termoanálisis
thermobalance termobalanza
thermocatalytic termocatalítico
thermochemical termoquímico
thermochemical standard patrón
termoquímico
thermochemistry termoquímica
thermocouple termopar
thermocroic termocroico
thermodiffusion termodifusión
thermodynamic termodinámico
thermodynamic data datos
termodinámicos
thermodynamic potential potencial
termodinámico
thermodynamic stability estabilidad
termodinámica
thermodynamics termodinámica
thermodynamics law ley de
termodinámica
thermoelectric termoeléctrico
thermoelectricity termoelectricidad
thermofission termofisión
Thermofor process proceso de
Thermofor
thermofusion termofusión
thermogenic termogénico
thermograph termógrafo
thermogravimetric analysis análisis
termogravimétrico
thermokinetic analysis análisis
termocinético
thermoluminescence termoluminiscencia
thermoluminescent termoluminiscente
thermolysis termólisis
thermolytic termolítico
thermomagnetic effect efecto
termomagnético
thermometamorphism
termometamorfismo
thermometer termómetro

thermometer scale escala termométrica
thermometric termométrico
thermometric analysis análisis termométrico
thermometric titration titulación termométrica, valoración termométrica
thermonatrite termonatrita
thermoneutrality termoneutralidad
thermonuclear termonuclear
thermonuclear energy energía termonuclear
thermonuclear fusion fusión termonuclear
thermonuclear neutron neutron termonuclear
thermonuclear reaction reacción termonuclear
thermonuclear transformation transformación termonuclear
thermoplastic termoplástico
thermoplastic resin resina termoplástica
thermoregulator termorregulador
thermostable termoestable
thermostat termostato
thermostatic termostático
thermotropic termotrópico
theta theta
thetine tetina
thexyl texilo
thia tia
thiabendazole tiabendazol
thial tial
thiamine tiamina
thiamine hydrochloride clorhidrato de tiamina
thiamine pyrophosphate pirofosfato de tiamina
thianthrene tiantreno
thiazine tiazina
thiazole tiazol
thiazole dye colorante de tiazol
thiazolyl tiazolilo
thiazyl tiacilo
thick film película gruesa
thickener espesador
thickening espesamiento
thickening agent agente de espesamiento
Thiele apparatus aparato de Thiele

Thiele reaction reacción de Thiele
thienyl tienilo
thimerosal timerosal
thin film película fina
thin-film chromatography cromatografía de película fina
thin layer capa fina
thin-layer chromatography cromatografía de capa fina
thinner diluente
thio tio
thioacetal tioacetal
thioacetaldehyde tioacetaldehído
thioacetamide tioacetamida
thioacetanilide tioacetanilida
thioacetic acid ácido tioacético
thioacid tioácido
thioalcohol tioalcohol
thioaldehyde tioaldehído
thioanhydride tioanhídrido
thioarsenate tioarseniato
thioate tioato
thiobarbituric acid ácido tiobarbitúrico
thiobenzaldehyde tiobenzaldehído
thiobenzamide tiobenzamida
thiobenzoic acid ácido tiobenzoico
thiobenzophenone tiobenzofenona
thiocarbamoyl tiocarbamoilo
thiocarbanilide tiocarbanilida
thiocarbonate tiocarbonato
thiocarbonic acid ácido tiocarbónico
thiocarbonyl tiocarbonilo
thiocarboxy tiocarboxi
thiocarboxylic acid ácido tiocarboxílico
thiochrome tiocromo
thiocyanate tiocianato
thiocyanato tiocianato
thiocyanic acid ácido tiociánico
thiocyanide tiocianuro
thiocyanogen tiocianógeno
thiodiglycol tiodiglicol
thiodiglycolic acid ácido tiodiglicólico
thiodiphenol tiodifenol
thiodipropionic acid ácido tiodipropiónico
thiodipropionitrile tiodipropionitrilo
thioester tioéster
thioether tioéter
thioethyl alcohol alcohol tioetílico

thioflavine tioflavina
thiofuran tiofurano
thioglycerol tioglicerol
thioglycolic acid ácido tioglicólico
thiohydantoin tiohidantoína
thiohydroxy tiohidroxi
thioic acid ácido tioico
thioketone tiocetona
thiol tiol
thiolactic acid ácido tioláctico
thiolate tiolato
thiomalic acid ácido tiomálico
thiomolybdate tiomolibdato
thionate tionato
thionic acid ácido tiónico
thionyl tionilo
thionyl bromide bromuro de tionilo
thionyl chloride cloruro de tionilo
thiopental sodium tiopental sódico
thiophane tiofano
thiophene tiofeno
thiophenealdehyde tiofenaldehído
thiophenol tiofenol
thiophenyl tiofenilo
thiophosgene tiofosgeno
thiophosphoric acid ácido tiofosfórico
thiophosphoryl tiofosforilo
thioridazine tioridazina
thiosalicylic acid ácido tiosalicílico
thiosemicarbazide tiosemicarbazida
thiosorbitol tiosorbitol
thiosulfate tiosulfato
thiosulfonic acid ácido tiosulfónico
thiosulfuric acid ácido tiosulfúrico
thiotungstate tiotungstato
thiouracil tiouracil
thiourea tiourea
thioxanthene tioxanteno
thioxo tioxo
thioxylenol tioxilenol
thiram tiram
third law of thermodynamics tercera ley
  de termodinámica
third-order reaction reacción de tercer
  orden
thixotropy tixotropía
Thomas-Fermi model modelo de
  Thomas-Fermi
thomsonite thomsonita

thoria toria
thorianite torianita
thoriated toriado
thorin torina
thorite torita
thorium torio
thorium anhydride anhídrido de torio
thorium carbide carburo de torio
thorium chloride cloruro de torio
thorium dioxide dióxido de torio
thorium disulfide disulfuro de torio
thorium fluoride fluoruro de torio
thorium hydroxide hidróxido de torio
thorium nitrate nitrato de torio
thorium oxalate oxalato de torio
thorium oxide óxido de torio
thorium picrate picrato de torio
thorium reactor reactor de torio
thorium series serie de torio
thorium sulfate sulfato de torio
thorium tetrachloride tetracloruro de
  torio
thoron torón
Thorpe reaction reacción de Thorpe
thortveitite thortveitita
three-necked flask matraz de tres cuellos
threnardite trenardita
threo treo
threonine treonina
threose treosa
threshold umbral
threshold detector detector de umbral
threshold energy energía umbral
threshold limit value valor umbral límite
threshold of detection umbral de
  detección
threshold value valor umbral
thrombin trombina
thuja oil aceite de tuya
thujene tuyeno
thujone tuyona
thulia tulia
thulium tulio
thulium chloride cloruro de tulio
thulium oxalate oxalato de tulio
thulium oxide óxido de tulio
thymic acid ácido tímico
thymidine timidina
thymidylic acid ácido timidílico

thymine timina
thymol timol
thymol blue azul de timol
thymol iodide yoduro de timol
thymolphthalein timolftaleína
thymolsulfonephthalein
   timolsulfonaftaleína
thymyl timilo
thyronine tironina
thyrothricin tirotricina
thyrotropic hormone hormona
   tirotrópica
thyroxine tiroxina
Tiemann rearrangement reordenamiento
   de Tiemann
tiglic acid ácido tíglico
tiglium oil aceite de tiglio
timolol timolol
tin estaño, hojalata
tin anhydride anhídrido de estaño
tin ash ceniza de estaño
tin bisulfide bisulfuro de estaño
tin bromide bromuro de estaño
tin bronze bronce de estaño
tin chloride cloruro de estaño
tin chromate cromato de estaño
tin dichloride dicloruro de estaño
tin difluoride difluoruro de estaño
tin dioxide dióxido de estaño
tin fluoride fluoruro de estaño
tin hexafluorosilicate hexafluorosilicato
   de estaño
tin hydride hidruro de estaño
tin hydroxide hidróxido de estaño
tin iodide yoduro de estaño
tin monosulfide monosulfuro de estaño
tin oxide óxido de estaño
tin perchloride percloruro de estaño
tin peroxide peróxido de estaño
tin-plate estañar
tin-plated estañado
tin-plating estañado
tin protochloride protocloruro de estaño
tin protosulfide protosulfuro de estaño
tin protoxide protóxido de estaño
tin resinate resinato de estaño
tin salt sal de estaño
tin sulfate sulfato de estaño
tin sulfide sulfuro de estaño

tin tetrabromide tetrabromuro de estaño
tin tetrachloride tetracloruro de estaño
tin tetraiodide tetrayoduro de estaño
tincal tincal
tincture tintura
tinned estañado
tinning estañado
tinplate hojalata
tinstone casiterita
tintometer tintómetro
Tishchenko reaction reacción de
   Tishchenko
titanate titanato
titanellow amarillo de titanio
titania titania
titanic titánico
titanic acid ácido titánico
titanic chloride cloruro titánico
titanic hydroxide hidróxido titánico
titanic sulfate sulfato titánico
titanium titanio
titanium acetylacetonate acetilacetonato
   de titanio
titanium alloy aleación de titanio
titanium ammonium oxalate oxalato
   amónico de titanio
titanium boride boruro de titanio
titanium bromide bromuro de titanio
titanium carbide carburo de titanio
titanium chelate quelato de titanio
titanium chloride cloruro de titanio
titanium diboride diboruro de titanio
titanium dichloride dicloruro de titanio
titanium dioxide dióxido de titanio
titanium disilicide disiliciuro de titanio
titanium disulfide disulfuro de titanio
titanium ester éster de titanio
titanium ferrocene ferroceno de titanio
titanium fluoride fluoruro de titanio
titanium hydride hidruro de titanio
titanium hydroxide hidróxido de titanio
titanium iodide yoduro de titanio
titanium isopropylate isopropilato de
   titanio
titanium monoxide monóxido de titanio
titanium nitrate nitrato de titanio
titanium nitride nitruro de titanio
titanium ore mineral de titanio
titanium oxalate oxalato de titanio

titanium oxide óxido de titanio
titanium peroxide peróxido de titanio
titanium potassium fluoride fluoruro
  potásico de titanio
titanium sesquisulfate sesquisulfato de
  titanio
titanium sponge esponja de titanio
titanium sulfate sulfato de titanio
titanium tetrachloride tetracloruro de
  titanio
titanium trichloride tricloruro de titanio
titanium trioxide trióxido de titanio
titanium white blanco de titanio
titanocene dichloride dicloruro de
  titanoceno
titanous titanoso
titanous chloride cloruro titanoso
titanous oxalate oxalato titanoso
titanous sulfate sulfato titanoso
titanyl titanilo
titanyl sulfate sulfato de titanilo
titer título
titrant solución de titulación, solución de
  valoración
titration titulación, valoración,
  volumetría
tobacco tabaco
Tobias acid ácido de Tobias
tocopherol tocoferol
tocophersolan tocofersolán
tolan tolano
tolidine tolidina
Tollens reagent reactivo de Tollens
Tollens test prueba de Tollens
tolnaftate tolnaftato
tolualdehyde tolualdehído
toluene tolueno
toluene trichloride tricloruro de tolueno
toluene trifluoride trifluoruro de tolueno
toluenediamine toluendiamina
toluenediisocyanate toluendiisocianato
toluenesulfamine toluensulfamina
toluenesulfanilide toluensulfanilida
toluenesulfonamide toluensulfonamida
toluenesulfonic acid ácido
  toluensulfónico
toluenesulfonyl toluensulfonilo
toluenesulfonyl chloride cloruro de
  toluensulfonilo

toluenethiol toluentiol
toluenyl toluenilo
toluic acid ácido toluico
toluic aldehyde aldehído toluico
toluidine toluidina
toluidine red rojo de toluidina
toluidino toluidino
toluoyl toluoilo
toluquinone toluquinona
toluyl toluilo
toluylene toluileno
toluylenediamine toluilendiamina
toluylic acid ácido toluílico
tolyl tolilo
tolyl chloride cloruro de tolilo
tolyl isobutyrate isobutirato de tolilo
tolylacetic acid ácido tolilacético
tolylaldehyde tolilaldehído
tolyldiethanolamine tolildietanolamina
tolylene tolileno
tolylenediamine tolilendiamina
tolylidene tolilideno
tomatine tomatina
tongs tenazas
tonka tonca
topaz topacio
topochemical reaction reacción
  topoquímica
topochemistry topoquímica
topotactic topotáctico
toroid toriode
torr torr
torsion balance balanza de torsión
torula yeast levadura de tórula
tosyl tosilo
tosylation tosilación
total heat calor total
total reflux reflujo total
Toth process proceso de Toth
touchstone piedra de toque
tourmaline turmalina
tower torre
toxaphene toxafeno
toxic tóxico
toxic properties propiedades tóxicas
toxic substance substancia tóxica
toxication intoxicación
toxicity toxicidad
toxicology toxicología

toxin toxina
trace traza
trace analysis análisis de trazas
trace element microelemento, oligoelemento
tracer trazador, indicador radiactivo
tracer atom átomo trazador
tracer compound compuesto trazador
tracer element elemento trazador
tragacanth gum goma de tragacanto
tranexamic acid ácido tranexámico
trans trans
trans effect efecto trans
trans-isomer isómero trans
transactinide transactínido
transactinide element elemento transactínido
transalkylation transalquilación
transaminase transaminasa
transamination transaminación
transducer transductor
transesterification transesterificación
transfer coefficient coeficiente de transferencia
transfer ribonucleic acid ácido ribonucleico de transferencia
transferase transferasa
transference number número de transferencia
transformation transformación
transformation series serie de transformaciones
transformer oil aceite de transformador
transition transición
transition effect efecto de transición
transition element elemento de transición
transition interval intervalo de transición
transition point punto de transición
transition state estado de transición
transition temperature temperatura de transición
transition time tiempo de transición
transitional transicional
transmission diffraction difracción por transmisión
transmittance transmitancia
transmutation transmutación

transoid transoide
transparency transparencia
transport transporte
transport coefficient coeficiente de transporte
transport number número de transporte
transposition transposición
transuranic transuránico
transuranic element elemento transuránico
Traube purine synthesis síntesis de purinas de Traube
traumatic acid ácido traumático
travertine travertino
tremolite tremolita
tretamine tretamina
tri tri
triacetate triacetato
triacetin triacetina
triacetoneamine triacetonamina
triacontane triacontano
triacontanoic acid ácido triacontanoico
triacontanol triacontanol
triacontyl triacontilo
triad tríada
trialkylsilanol trialquilsilanol
triallyl trialilo
triallyl cyanurate cianurato de trialilo
triallyl phosphate fosfato de trialilo
triallylamine trialilamina
triamcinolone triamcinolona
triamine triamina
triaminobenzene triaminobenceno
triaminotoluene triaminotolueno
triamylamine triamilamina
triamylbenzene triamilbenceno
triangular diagram diagrama triangular
triangulo triangulo
triarylmethane dye colorante de triarilmetano
triatomic triatómico
triatomic molecule molécula triatómica
triazane triazano
triazano triazano
triazene triazeno
triazeno triazeno
triazine triazina
triazinyl triazinilo
triazole triazol

triazone resin resina de triazona
tribasic tribásico
tribasic acid ácido tribásico
tribasic calcium phosphate fosfato cálcico tribásico
tribenzyl tribencilo
tribenzylamine tribencilamina
triboluminescence triboluminiscencia
tribromide tribromuro
tribromo tribromo
tribromoacetaldehyde tribromoacetaldehído
tribromoacetic acid ácido tribromoacético
tribromoaniline tribromoanilina
tribromobenzene tribromobenceno
tribromoethanal tribromoetanal
tribromoethane tribromoetano
tribromoethanol tribromoetanol
tribromomethane tribromometano
tribromophenol tribromofenol
tribromopropane tribromopropano
tribromosalicylanilide tribromosalicilanilida
tributyl tributilo
tributyl borate borato de tributilo
tributyl citrate citrato de tributilo
tributyl phosphate fosfato de tributilo
tributyl phosphine fosfina de tributilo
tributyl phosphite fosfito de tributilo
tributyl phosphorothiolate fosforotiolato de tributilo
tributylaluminum tributilaluminio
tributylamine tributilamina
tributylborane tributilborano
tributyltin tributilestaño
tributyltin chloride cloruro de tributilestaño
tributyltin oxide óxido de tributilestaño
tricalcium citrate citrato tricálcico
tricalcium orthophosphate ortofosfato tricálcico
tricalcium phosphate fosfato tricálcico
tricalcium silicate silicato tricálcico
tricarbimide tricarbimida
tricarbon dioxide dióxido de tricarbono
tricarboxylic acid cycle ciclo de ácido tricarboxílico
trichlorfon triclorfón

trichloride tricloruro
trichloro tricloro
trichloroacetaldehyde tricloroacetaldehído
trichloroacetamide tricloroacetamida
trichloroacetic acid ácido tricloroacético
trichloroanisole tricloroanisol
trichlorobenzene triclorobenceno
trichloroborazole tricloroborazol
trichlorobromomethane triclorobromometano
trichlorobutylene oxide óxido de triclorobutileno
trichloroethanal tricloroetanal
trichloroethane tricloroetano
trichloroethanoic acid ácido tricloroetanoico
trichloroethanol tricloroetanol
trichloroethene tricloroeteno
trichloroethylene tricloroetileno
trichlorofluoromethane triclorofluorometano
trichloroisocyanuric acid ácido tricloroisocianúrico
trichloromethane triclorometano
trichloromethyl triclorometilo
trichloromethyl chloroformate cloroformiato de triclorometilo
trichloromethylphosphonic acid ácido triclorometilfosfónico
trichloromethylsulfenyl chloride cloruro de triclorometilsulfenilo
trichloronaphthalene tricloronaftaleno
trichloronitrosomethane tricloronitrosometano
trichlorophenol triclorofenol
trichloropropane tricloropropano
trichlorosilane triclorosilano
trichlorotoluene triclorotolueno
trichlorotrifluoroacetone triclorotrifluoroacetona
trichlorotrifluoroethane triclorotrifluoroetano
triclinic system sistema triclínico
tricosane tricosano
tricosanoic acid ácido tricosanoico
tricosanol tricosanol
tricresyl tricresilo
tricresyl phosphate fosfato de tricresilo

tricresyl phosphite fosfito de tricresilo
tricyanic acid ácido triciánico
tricyano triciano
tricyclic tricíclico
tricyclodecane triciclodecano
tridecane tridecano
tridecanoic acid ácido tridecanoico
tridecanol tridecanol
tridecene tridecano
tridecyl tridecilo
tridecyl alcohol alcohol tridecílico
tridecylbenzene tridecilbenceno
tridecylic acid ácido tridecílico
tridodecyl amine tridodecilamina
tridymite tridimita
triethanolamine trietanolamina
triethanolamine lauryl sulfate
    laurilsulfato de trietanolamina
triethoxy trietoxi
triethoxyhexane trietoxihexano
triethoxymethane trietoximetano
triethoxymethoxypropane
    trietoximetoxipropano
triethyl trietilo
triethyl borate borato de trietilo
triethyl citrate citrato de trietilo
triethyl phosphate fosfato de trietilo
triethyl phosphite fosfito de trietilo
triethyl phosphorothioate fosforotioato
    de trietilo
triethylaluminum trietilaluminio
triethylamine trietilamina
triethylbenzene trietilbenceno
triethylborane trietilborano
triethylene trietileno
triethylene glycol trietilenglicol
triethylene glycol dichloride dicloruro
    de trietilenglicol
triethylene glycol didecanoate
    didecanoato de trietilenglicol
triethylene glycol dihydroabietate
    dihidoabietato de trietilenglicol
triethylene glycol dimethyl ether éter
    dimetílico de trietilenglicol
triethylene glycol dioctoate dioctoato de
    trietilenglicol
triethylene glycol dipelargonate
    dipelargonato de trietilenglicol
triethylene glycol dipropionate

dipropionato de trietilenglicol
triethylenediamine trietilendiamina
triethylenemelamine trietilenmelamina
triethylenephosphoramide
    trietilenfosforamida
triethylenetetramine trietilentetramina
triethylenetriamine trietilentriamina
triethylmethane trietilmetano
triethylorthoformate trietilortoformiato
triflate triflato
trifluoro trifluoro
trifluoroacetic acid ácido
    trifluoroacético
trifluoroamine oxide óxido de
    trifluoroamina
trifluorobromomethane
    trifluorobromometano
trifluorochloromethane
    trifluoroclorometano
trifluoroethanoic acid ácido
    trifluoroetanoico
trifluoroiodomethane
    trifluoroyodometano
trifluoromethyl trifluorometilo
trifluoronitrosomethane
    trifluoronitrosometano
trifluorostyrene trifluoroestireno
trifluorotrichloroethane
    trifluorotricloroetano
trigalloyl trigaloilo
triglyceride triglicérido
triglycerol triglicerol
triglycine triglicina
triglycol dichloride dicloruro de triglicol
triglyme triglime
trigol trigol
trigonal trigonal
trigonal bipyramid bipirámide trigonal
trigonal bipyramidal coordination
    coordinación bipiramidal trigonal
trigonal system sistema trigonal
trigonelline trigonelina
trihexyl trihexilo
trihexylene glycol biborate biborato de
    trihexilenglicol
trihydrate trihidrato
trihydric trihídrico
trihydric alcohol alcohol trihídrico
trihydroxy trihidroxi

trihydroxyanthraquinone trihidroxiantraquinona

trihydroxybenzene trihidroxibenceno

trihydroxybenzoic acid ácido trihidroxibenzoico

trihydroxyethylamine stearate estearato de trihidroxietilamina

trihydroxypropane trihidroxipropano

triiodo triyodo

triiodomethane triyodometano

triiodothyronine triyodotironina

triisobutylaluminum triisobutilaluminio

triisobutylene triisobutileno

triisooctyl phosphite fosfito de triisooctilo

triisopropanolamine triisopropanolamina

triisopropyl phosphite fosfito de triisopropilo

triketone tricetona

trilauryl trilaurilo

trilauryl phosphite trilaurilfosfito

trilauryl trithiophosphite trilauriltritiofosfito

trilaurylamine trilaurilamina

trimagnesium phosphate fosfato de trimagnesio

trimellitic acid ácido trimelítico

trimellitic anhydride anhídrido trimelítico

trimer trímero

trimercuric orthophosphate ortofosfato trimercúrico

trimercurous orthophosphate ortofosfato trimercurioso

trimesoyl trichloride tricloruro de trimesoilo

trimetallic trimetálico

trimethadione trimetadiona

trimethano trimetano

trimethoxy trimetoxi

trimethoxyboroxine trimetoxiboroxina

trimethoxymethane trimetoximetano

trimethyl trimetilo

trimethyl borate borato de trimetilo

trimethyl phosphate fosfato de trimetilo

trimethyl phosphite fosfito de trimetilo

trimethylacetic acid ácido trimetilacético

trimethyladipic acid ácido trimetiladípico

trimethylaluminum trimetilaluminio

trimethylamine trimetilamina

trimethylamine oxide óxido de trimetilamina

trimethylaniline trimetilanilina

trimethylbenzene trimetilbenceno

trimethylbenzoic acid ácido trimetilbenzoico

trimethylbutane trimetilbutano

trimethylchlorosilane trimetilclorosilano

trimethylcyclododecatriene trimetilciclododecatrieno

trimethylcyclohexane trimetilciclohexano

trimethylcyclohexanol trimetilciclohexanol

trimethylene trimetileno

trimethylene bromide bromuro de trimetileno

trimethylene chlorohydrin clorohidrina de trimetileno

trimethylene glycol trimetilenglicol

trimethylene oxide óxido de trimetileno

trimethylhexane trimetilhexano

trimethylhexanol trimetilhexanol

trimethylmethane trimetilmetano

trimethylnonanone trimetilnonanona

trimethylolethane trimetiloletano

trimethylolpropane trimetilolpropano

trimethylpentane trimetilpentano

trimethylpentanediol trimetilpentanodiol

trimethylpentene trimetilpenteno

trimethylpyridine trimetilpiridina

trimethylsilyl trimetilsililo

trimethyltin trimetilestaño

trimolecular trimolecular

trimorphism trimorfismo

trinitrate trinitrato

trinitro trinitro

trinitroaniline trinitroanilina

trinitroanisole trinitroanisol

trinitrobenzene trinitrobenceno

trinitrobenzoic acid ácido trinitrobenzoico

trinitroglycerin trinitroglicerina

trinitromethane trinitrometano

trinitronaphthalene trinitronaftaleno

trinitrophenol trinitrofenol

trinitroresorcinol trinitrorresorcina

trinitrotoluene trinitrotolueno
trinor trinor
trioctadecyl phosphite fosfito de trioctadecilo
trioctyl phosphate fosfato de trioctilo
trioctylphosphinic oxide óxido trioctilfosfínico
triol triol
trione triona
triose triosa
trioxa trioxa
trioxane trioxano
trioxide trióxido
trioxime trioxima
trioxin trioxina
trioxoboric acid ácido trioxobórico
trioxosulfuric acid ácido trioxosulfúrico
trioxygen trioxígeno
trioxymethylene trioximetileno
tripalmitin tripalmitina
triphasic trifásico
triphenol trifenol
triphenyl trifenilo
triphenyl phosphate fosfato de trifenilo
triphenyl phosphite fosfito de trifenilo
triphenylacetic acid ácido trifenilacético
triphenylantimony trifenilantimonio
triphenylbenzene trifenilbenceno
triphenylboron trifenilboro
triphenylcarbinol trifenilcarbinol
triphenylene trifenileno
triphenylguanidine trifenilguanidina
triphenylmethane dye colorante de trifenilmetano
triphenylmethyl trifenilmetilo
triphenylphosphine trifenilfosfina
triphenylphosphorus trifenilfósforo
triphenyltin trifenilestaño
triphenyltin chloride cloruro de trifenilestaño
triphenyltin hydroxide hidróxido de trifenilestaño
triphosgene trifosgeno
triphosphoric acid ácido trifosfórico
triple bond enlace triple
triple point punto triple
triple superphosphate superfosfato triple
triplet triplete
triplite triplita

tripod trípode
tripoli trípoli
tripolite tripolita
tripotassium orthophosphate ortofosfato tripotásico
tripotassium phosphate fosfato tripotásico
tripropyl tripropilo
tripropylaluminum tripropilaluminio
tripropylamine tripropilamina
tripropylene tripropileno
tripropylene glycol tripropilenglicol
triptane triptano
triptycene tripticeno
tris tris
trischloroethyl phosphite fosfito de triscloroetilo
trisethylhexyl phosphate fosfato de trisetilhexilo
trisethylhexyl phosphite fosfito de trisetilhexilo
trishydroxyphenylpropane trishidroxifenilpropano
trisilane trisilano
trisodium citrate citrato trisódico
trisodium ethylenediaminetetraacetic acid ácido etilendiaminotetraacético trisódico
trisodium orthophosphate ortofosfato trisódico
trisodium phosphate fosfato trisódico
trisulfide trisulfuro
triterpene triterpeno
trithioacetaldehyde tritioacetaldehído
trithiocarbonate tritiocarbonato
trithionic acid ácido tritiónico
tritio tritio
tritium tritio
tritolyl phosphate fosfato de tritolilo
triton tritón
triturate triturar
trityl tritilo
triuranium octoxide octóxido de triuranio
trivalent trivalente
trivial name nombre trivial
trona trona
Trona process proceso de Trona
tropanol tropanol

tropic acid ácido trópico
tropilidene tropilideno
tropine tropina
tropolone tropolona
tropoyl tropoilo
tropylium tropilio
truth serum suero de la verdad
trypsin tripsina
trypsinogen tripsinógeno
tryptophan triptófano
tryptophyl triptofila
Tscherniac-Einhorn reaction reacción
   de Tscherniac-Einhorn
tung oil aceite de tung
tungstate tungstato
tungsten tungsteno
tungsten alloy aleación de tungsteno
tungsten boride boruro de tungsteno
tungsten bronze bronce de tungsteno
tungsten carbide carburo de tungsteno
tungsten carbonyl carbonilo de
   tungsteno
tungsten chloride cloruro de tungsteno
tungsten dichloride dicloruro de
   tungsteno
tungsten diselenide diseleniuro de
   tungsteno
tungsten disulfide disulfuro de tungsteno
tungsten fluoride fluoruro de tungsteno
tungsten hexacarbonyl hexacarbonilo de
   tungsteno
tungsten hexachloride hexacloruro de
   tungsteno
tungsten lake laca de tungsteno
tungsten oxide óxido de tungsteno
tungsten oxychloride oxicloruro de
   tungsteno
tungsten pentachloride pentacloruro de
   tungsteno
tungsten pentoxide pentóxido de

   tungsteno
tungsten silicide siliciuro de tungsteno
tungsten steel acero al tungsteno
tungsten trioxide trióxido de tungsteno
tungstic túngstico
tungstic acid ácido túngstico
tungstic anhydride anhídrido túngstico
tungstic oxide óxido túngstico
tungstite tungstita
tungstophosphate tungstofosfato
tungstosilicate tungstosilicato
tunnel effect efecto túnel
turbidimetric analysis análisis
   turbidimétrico
turbidimetry turbidimetría
turbidity turbidez
turbidity point punto de turbidez
Turkey red rojo turco
Turnbull blue azul de Turnbull
turpentine trementina
turpentine camphor alcanfor de
   trementina
turpentine oil aceite de trementina
turquoise turquesa
twinning maclado
Twitchell process proceso de Twitchell
Twitchell reagent reactivo de Twitchell
two-dimensional chromatography
   cromatografía bidimensional
Tyndall effect efecto de Tyndall
type tipo
typical típico
tyramine tiramina
Tyrian purple púrpura de Tiro
tyrocidine tirocidina
tyrosinase tirosinasa
tyrosine tirosina
tyrosyl tirosilo
tysonite tisonita

U

U-tube tubo en U
ubiquinone ubiquinona
ulexine ulexina
ulexite ulexita
Ullman reaction reacción de Ullman
ulmin ulmina
ultimate analysis análisis elemental
ultra ultra
ultracentrifuge ultracentrífuga
ultrafiltration ultrafiltración
ultramarine blue azul ultramarino
ultramarine green verde ultramarino
ultramicroscope ultramicroscopio
ultramicroscopic ultramicroscópico
ultrapure ultrapuro
ultrarapid ultrarrápido
ultrasensitive ultrasensible
ultrasonic ultrasónico
ultrasonics ultrasónica
ultraviolet ultravioleta
ultraviolet absorber absorbedor
    ultravioleta
ultraviolet absorption absorción
    ultravioleta
ultraviolet-absorption
    spectrophotometry
    espectrofotometría de absorción
    ultravioleta
ultraviolet densitometry densitometría
    ultravioleta
ultraviolet light luz ultravioleta
ultraviolet radiation radiación
    ultravioleta
ultraviolet spectrometer espectrómetro
    ultravioleta
ultraviolet spectrophotometry
    espectrofotometría ultravioleta
ultraviolet spectroscopy espectroscopia
    ultravioleta
ultraviolet spectrum espectro
    ultravioleta
ultraviolet stabilizer estabilizador
    ultravioleta

ultraviolet-visible spectroscopy
    espectroscopia ultravioleta-visible
umangite umangita
umbellic acid ácido umbélico
umbelliferone umbeliferona
umbellulone umbelulona
umber tierra de sombra
unary unario
uncertainty principle principio de
    incertidumbre
uncoupling phenomena fenómenos de
    desacoplamiento
undeca undeca
undecalactone undecalactona
undecanal undecanal
undecane undecano
undecanoic acid ácido undecanoico
undecanol undecanol
undecanone undecanona
undecene undeceno
undecenoic acid ácido undecenoico
undecenol undecenol
undecenyl undecenilo
undecyl undecilo
undecyl alcohol alcohol undecílico
undecylamine undecilamina
undecylene undecileno
undecylenic acid ácido undecilénico
undecylenic alcohol alcohol undecilénico
undecylenyl acetate acetato de
    undecilenilo
undecylic acid ácido undecílico
undercooling subenfriado
uni uni
uniaxial uniaxial
unified atomic mass constant constante
    de masa atómica unificada
unimolecular unimolecular
unimolecular film película unimolecular
unimolecular reaction reacción
    unimolecular
union unión
unit unidad

univalent univalente
universal indicator indicador universal
unoccupied molecular orbital orbital
    molecular desocupado
unpaired electrons electrones no
    emparejados
unpolarized light luz no polarizada
unsaturated insaturado, no saturado
unsaturated compound compuesto
    insaturado
unsaturated hydrocarbon hidrocarburo
    insaturado
unsaturated solution solución no
    saturada
unsaturation insaturación, no saturación
unstable inestable
unstable element elemento inestable
unstable isotope isótopo inestable
unstable nucleus núcleo inestable
unsym asim
unsymmetrical asimétrico
ur ur
uracil uracil
uralite uralita
uramido uramido
uramine uramina
uramino uramino
uranate uranato
urania urania
uranic uránico
uranic acid ácido uránico
uranic chloride cloruro uránico
uranic oxide óxido uránico
uranin uranina
uranine uranina
uranine yellow amarillo de uranina
uraninite uraninita
uranite uranita
uranium uranio
uranium acetate acetato de uranio
uranium bromide bromuro de uranio
uranium carbide carburo de uranio
uranium chloride cloruro de uranio
uranium content contenido de uranio
uranium decay series serie de
    desintegración de uranio
uranium dicarbide dicarburo de uranio
uranium dioxide dióxido de uranio
uranium enrichment enriquecimiento de

uranio
uranium fission fisión de uranio
uranium fluoride fluoruro de uranio
uranium glass vidrio de uranio
uranium hexafluoride hexafluoruro de
    uranio
uranium hydride hidruro de uranio
uranium hydroxide hidróxido de uranio
uranium iodide yoduro de uranio
uranium monocarbide monocarburo de
    uranio
uranium nitrate nitrato de uranio
uranium ore mineral de uranio
uranium oxide óxido de uranio
uranium oxychloride oxicloruro de
    uranio
uranium-radium series serie de uranio-
    radio
uranium reactor reactor de uranio
uranium series serie de uranio
uranium sulfate sulfato de uranio
uranium tetrabromide tetrabromuro de
    uranio
uranium tetrachloride tetracloruro de
    uranio
uranium tetrafluoride tetrafluoruro de
    uranio
uranium trioxide trióxido de uranio
uranocene uranoceno
uranocircite uranocircita
uranophane uranofano
uranospherite uranosferita
uranothallite uranotalita
uranous uranoso
uranous chloride cloruro uranoso
uranous oxide óxido uranoso
uranyl uranilo
uranyl acetate acetato de uranilo
uranyl benzoate benzoato de uranilo
uranyl chloride cloruro de uranilo
uranyl fluoride fluoruro de uranilo
uranyl formate formiato de uranilo
uranyl hexacyanoferrate
    hexacianoferrato de uranilo
uranyl hydroxide hidróxido de uranilo
uranyl nitrate nitrato de uranilo
uranyl oxide óxido de uranilo
uranyl sulfate sulfato de uranilo
uranyl sulfide sulfuro de uranilo

urea urea
urea acetate acetato de urea
urea adduct aducto de urea
urea-ammonium orthophosphate
    ortofosfato de urea-amonio
urea-ammonium phosphate fosfato de
    urea-amonio
urea anhydride anhídrido de urea
urea cycle ciclo de urea
urea-formaldehyde resin resina de urea-
    formaldehído
urea nitrate nitrato de urea
urea oxalate oxalato de urea
urea peroxide peróxido de urea
urease ureasa
Urech synthesis síntesis de Urech
ureide ureido
ureido ureido

urethane uretano
urethane resin resina de uretano
urethano uretano
uric acid ácido úrico
uridine uridina
uridine monophosphate monofosfato de
    uridina
uridine phosphate fosfato de uridina
uridylic acid ácido uridílico
urobenzoic acid ácido urobenzoico
uronic acid ácido urónico
uronium uronio
urotropin urotropina
ursin ursina
urylon urilón
usnic acid ácido úsnico
uwarowite uvarovita

V

vacancy vacante
vacant vacante
vaccenic acid ácido vaccénico
vacuum vacío
vacuum condensing point punto de
  condensación al vacío
vacuum crystallizer cristalizador al
  vacío
vacuum deposition deposición al vacío
vacuum distillation destilación al vacío
vacuum filtration filtración al vacío
vacuum forming formación al vacío
vacuum metalizing metalización al vacío
vacuum pump bomba de vacío
valacidin valacidina
valence valencia
valence angle ángulo de valencia
valence band banda de valencia
valence bond enlace de valencia
valence electrons electrones de valencia
valence isomerization isomerización de
  valencia
valence number número de valencia
valence shell capa de valencia
valence tautomerism tautomerismo de
  valencia
valence theory teoría de valencia
valency valencia
valentinite valentinita
valeral valeral
valeraldehyde valeraldehído
valeramide valeramida
valerate valerato
valerian oil aceite de valeriana
valerianic acid ácido valeriánico
valeric acid ácido valérico
valeric aldehyde aldehído valérico
valeric anhydride anhídrido valérico
valerolactone valerolactona
valeryl valerilo
valine valina
valyl valilo
van der Waals adsorption adsorción de

van der Waals
van der Waals attraction atracción de
  van der Waals
van der Waals constant constante de van
  der Waals
van der Waals covolume covolumen de
  van der Waals
van der Waals equation ecuación de van
  der Waals
van der Waals forces fuerzas de van der
  Waals
van der Waals molecule molécula de
  van der Waals
van der Waals radius radio de van der
  Waals
van der Waals structure estructura de
  van der Waals
Van Slyke determination determinación
  de Van Slyke
Van Slyke method método de Van Slyke
van't Hoff equation ecuación de van't
  Hoff
van't Hoff factor factor de van't Hoff
van't Hoff formula fórmula de van't
  Hoff
van't Hoff isochore isócora de van't
  Hoff
van't Hoff isotherm isoterma de van't
  Hoff
van't Hoff law ley de van't Hoff
van't Hoff theory teoría de van't Hoff
vanadate vanadato
vanadic acid ácido vanádico
vanadic acid anhydride anhídrido de
  ácido vanádico
vanadic anhydride anhídrido vanádico
vanadic sulfate sulfato vanádico
vanadic sulfide sulfuro vanádico
vanadinite vanadinita
vanadite vanadita
vanadium vanadio
vanadium acetylacetonate
  acetilacetonato de vanadio

vanadium bromide bromuro de vanadio
vanadium carbide carburo de vanadio
vanadium chloride cloruro de vanadio
vanadium dichloride dicloruro de vanadio
vanadium difluoride difluoruro de vanadio
vanadium dioxide dióxido de vanadio
vanadium disulfide disulfuro de vanadio
vanadium ethylate etilato de vanadio
vanadium fluoride fluoruro de vanadio
vanadium hexacarbonyl hexacarbonilo de vanadio
vanadium hydroxide hidróxido de vanadio
vanadium monoxide monóxido de vanadio
vanadium nitride nitruro de vanadio
vanadium oxide óxido de vanadio
vanadium oxychloride oxicloruro de vanadio
vanadium oxydichloride oxidicloruro de vanadio
vanadium oxytrichloride oxitricloruro de vanadio
vanadium pentasulfide pentasulfuro de vanadio
vanadium pentoxide pentóxido de vanadio
vanadium sesquioxide sesquióxido de vanadio
vanadium suboxide subóxido de vanadio
vanadium sulfate sulfato de vanadio
vanadium sulfide sulfuro de vanadio
vanadium tetrachloride tetracloruro de vanadio
vanadium tetraoxide tetróxido de vanadio
vanadium trichloride tricloruro de vanadio
vanadium trioxide trióxido de vanadio
vanadium trisulfide trisulfuro de vanadio
vanadol vanadol
vanadous vanadioso
vanadous acid ácido vanadioso
vanadous chloride cloruro vanadioso
vanadous hydroxide hidróxido vanadioso
vanadyl vanadilo

vanadyl bromide bromuro de vanadilo
vanadyl chloride cloruro de vanadilo
vanadyl sulfate sulfato de vanadilo
vanadylic vanadílico
vanadylic chloride cloruro vanadílico
vanadylous vanadiloso
vancomycin hydrochloride clorhidrato de vancomicina
vanillic acid ácido vainíllico
vanillic aldehyde aldehído vainíllico
vanillin vainillina
vanilloyl vainilloilo
vanillyl vainillilo
vanillylidene vainillilideno
vapor vapor
vapor bath baño de vapor
vapor density densidad de vapor
vapor-liquid equilibrium equilibrio vapor-líquido
vapor-phase chromatography cromatografía en fase vapor
vapor pressure presión de vapor
vapor-pressure osmometer osmómetro de presión de vapor
vapor tension tensión de vapor
vaporization vaporización
vaporize vaporizar
variscite variscita
varnish barniz
Varrentrapp reaction reacción de Varrentrapp
Vaska compound compuesto de Vaska
vasopressin vasopresina
vat dye colorante de tina
Vauquelin salt sal de Vauquelin
vauquelinite vauquelinita
vector model of atomic structure modelo vectorial de estructura atómica
vegetable dye colorante vegetal
vegetable gum goma vegetal
vegetable oil aceite vegetal
vehicle vehículo
velocity velocidad
velocity constant constante de velocidad
velocity of reaction velocidad de reacción
Venetian red rojo veneciano
venturi venturi
veratraldehyde veratraldehído

veratric acid ácido verátrico
veratrole veratrol
veratroyl veratroilo
veratryl veratrilo
veratrylidene veratrilideno
verbena oil aceite de verbena
verbenone verbenona
verdigris cardenillo
vermiculite vermiculita
vermillion bermellón
vernolepin vernolepina
verxite verxita
vessel recipiente
vetivone vetivona
vibration vibración
vibrational energy energía vibracional
vibrational level nivel vibracional
vibrational spectrum espectro
    vibracional
vic vec
vicinal vecinal
Victoria blue azul Victoria
Victoria green verde Victoria
villiaumite villiaumita
Vilsmeier-Haack reaction reacción de
    Vilsmeier-Haack
Vilsmeier reagent reactivo de Vilsmeier
vinegar vinagre
vinic acid ácido vínico
vinyl vinilo
vinyl acetal resin resina de vinilacetal
vinyl acetate acetato de vinilo
vinyl acetate resin resina de acetato de
    vinilo
vinyl alcohol alcohol vinílico
vinyl bromide bromuro de vinilo
vinyl butyl ether vinilbutiléter
vinyl butyrate butirato de vinilo
vinyl chloride cloruro de vinilo
vinyl chloride resin resina de cloruro de
    vinilo
vinyl cyanide cianuro de vinilo
vinyl ether viniléter
vinyl ether resin resina de viniléter
vinyl ethyl ether viniletiléter
vinyl fluoride fluoruro de vinilo
vinyl iodide yoduro de vinilo
vinyl isobutyl ether vinilisobutiléter
vinyl ketone vinilcetona

vinyl methyl ether vinilmetiléter
vinyl methyl ketone vinilmetilcetona
vinyl oxide óxido de vinilo
vinyl polymer polímero de vinilo
vinyl polymerization polimerización de
    vinilo
vinyl propionate propionato de vinilo
vinyl resin resina de vinilo
vinyl stearate estearato de vinilo
vinyl sulfide sulfuro de vinilo
vinyl trichloride tricloruro de vinilo
vinylacetonitrile vinilacetonitrilo
vinylacetylene vinilacetileno
vinylamine vinilamina
vinylation vinilación
vinylbenzene vinilbenceno
vinylcarbazole vinilcarbazol
vinylcyclohexene vinilciclohexeno
vinylcyclohexene dioxide dióxido de
    vinilciclohexeno
vinylcyclohexene monoxide monóxido
    de vinilciclohexeno
vinylene vinileno
vinylethoxyethyl sulfide sulfuro de
    viniletoxietilo
vinylethylene viniletileno
vinylethylhexyl ether viniletilhexiléter
vinylethylpyridine viniletilpiridina
vinylidene vinilideno
vinylidene chloride cloruro de vinilideno
vinylidene chloride polymer polímero de
    cloruro de vinilideno
vinylidene fluoride fluoruro de
    vinilideno
vinylidene resin resina de vinilideno
vinylimine vinilimina
vinylmagnesium chloride cloruro de
    vinilmagnesio
vinylog vinílogo
vinylpyridine vinilpiridina
vinylpyrrolidone vinilpirrolidona
vinylstyrene vinilestireno
vinyltoluene viniltolueno
vinyltrichlorosilane viniltriclorosilano
vioform vioformo
violanthrone violantrona
violuric acid ácido violúrico
viomycin viomicina
virtual level nivel virtual

virtual particle partícula virtual
virtual quantum cuanto virtual
virtual state estado virtual
virus virus
viscid viscoso
viscometer viscosímetro
viscoplastic viscoplástico
viscose viscosa
viscosimeter viscosímetro
viscosimetry viscosimetría
viscosity viscosidad
viscosity index índice de viscosidad
viscous viscoso
visible absorption spectrophotometry
    espectrofotometría de absorción
    visible
visible spectrophotometry
    espectrofotometría visible
visible spectrum espectro visible
visual colorimetry colometría visual
vitamin vitamina
vitamin complex complejo vitamínico
vitellin vitelina
vitreous vítreo
vitrification vitrificación
vitriol vitriolo
vivianite vivianita
Voight amination aminación de Voight
volatile volátil
volatile compound compuesto volátil
volatile fluid fluido volátil
volatile organic compound compuesto
    orgánico volátil
volatile solvent solvente volátil

volatility volatilidad
volatility product producto de volatilidad
volatilization volatilización
volatilize volatilizar
Volhard-Erdmann cyclization
    ciclización de Volhard-Erdmann
Volhard method método de Volhard
Volhard solution solución de Volhard
Volhard titration titulación de Volhard,
    valoración de Volhard
Volta series serie de Volta
voltaic cell pila voltaica
voltaite voltaíta
voltzite voltzina
volume volumen
volumenometer volumenómetro
volumeter volúmetro
volumetric volumétrico
volumetric analysis análisis volumétrico
volumetric factor factor volumétrico
volumetric flask matraz volumétrico
volumetric pipet pipeta volumétrica
volumetric solution solución volumétrica
von Braun reaction reacción de von
    Braun
von Richter reaction reacción de von
    Richter
von Richter synthesis síntesis de von
    Richter
vulcanite vulcanita
vulcanization vulcanización
vulcanize vulcanizar
vulpinite vulpinita

W

Wackenroder reaction reacción de
 Wackenroder
Wacker process proceso de Wacker
Wacker reaction reacción de Wacker
Wagner-Jauregg reaction reacción de
 Wagner-Jauregg
Wagner-Meerwein rearrangement
 reordenamiento de Wagner-Meerwein
Wagner reagent reactivo de Wagner
Wagner solution solución de Wagner
wagnerite wagnerita
Walden inversion inversión de Walden
Wallach rearrangement reordenamiento
 de Wallach
warfarin warfarina
warning odor olor de advertencia
wash bottle frasco lavador
washing lavado
washing bottle frasco lavador
washing soda sosa de lavar
waste desperdicios
waste control control de desperdicios
water agua
water analysis análisis de agua
water bath baño de agua, baño María
water constants constantes de agua
water detection detección de agua
water distillation destilación de agua
water gas gas de agua
water glass vidrio de agua
water of constitution agua de
 constitución
water of crystallization agua de
 cristalización
water of hydration agua de hidratación
water pollution contaminación de agua
water purification purificación de agua
water saturation saturación con agua
water softening ablandamiento de agua
water soluble soluble en agua,
 hidrosoluble
water-soluble gum goma soluble en agua
water-soluble oil aceite soluble en agua

water-soluble resin resina soluble en
 agua
water sterilization esterilización de agua
water still destilador de agua
waterproof a prueba de agua,
 impermeable
waterproofing agent agente
 impermeabilizante
waterproofing compound compuesto
 impermeabilizante
watertight hermético al agua
Watson equation ecuación de Watson
wave onda
wave function función de onda
wavellite wavellita
wax cera
wax tailings residuos de cera
weak débil
weak acid ácido débil
weak base base débil
weak electrolyte electrólito débil
weak salt sal débil
Weerman degradation degradación de
 Weerman
Weigert effect efecto de Weigert
weighing bottle recipiente de pesar
weight peso
weight buret bureta de pesar
Weissenberg effect efecto de
 Weissenberg
welding soldadura
Weldon process proceso de Weldon
Welter rule regla de Welter
Werner complex complejo de Werner
Werner theory teroría de Werner
Wessely-Moser rearrangement
 reordenamiento de Wessely-Moser
Weston cell pila de Weston
Westphalen-Lettre rearrangement
 reordenamiento de Westphalen-Lettre
wet analysis análisis húmedo
wet-bulb thermometer termómetro de
 ampolleta húmeda

wet deposition deposición húmeda
wetting agent agente humectante
Wharton reaction reacción de Wharton
whey suero de la leche
whiskers triquitas
whiskey whisky
white acid ácido blanco
white arsenic arsénico blanco
white copperas caparrosa blanca
white gasoline gasolina blanca
white gold oro blanco
white lead plomo blanco
white lotion loción blanca
white metal metal blanco
white oil aceite blanco
white phosphorus fósforo blanco
white precipitate precipitado blanco
white vitriol vitriolo blanco
whitener blanqueador
whitewash lechada de cal
whiting blanco de España
Whiting reaction reacción de Whiting
whitneyite whitneyita
Widman-Stoermer synthesis síntesis de
    Widman-Stoermer
Wien effect efecto de Wien
Wigner nuclides nucleidos de Wigner
Wijs solution solución de Wijs
Wilkinson catalyst catalizador de
    Wilkinson
willemite willemita
Willgerodt reaction reacción de
    Willgerodt
Williamson ether synthesis síntesis de
    éteres de Williamson
Williamson reaction reacción de
    Williamson
Williamson synthesis síntesis de
    Williamson
Wilzbach procedure procedimiento de
    Wilzbach
wine vino
wine ether éter de vino
wintergreen oil aceite de pirola
wire cloth tela de alambre
wire gauze gasa de alambre
Wiswesser line notation notación lineal
    de Wiswesser
Wiswesser notation notación de

Wiswesser
witherite witherita
Wittig ether rearrangement
    reordenamiento de éteres de Wittig
Wittig reaction reacción de Wittig
Wittig rearrangement reordenamiento
    de Wittig
Wohl degradation degradación de Wohl
Wohl reaction reacción de Wohl
Wohl-Ziegler reaction reacción de
    Wohl-Ziegler
Wöhler synthesis síntesis de Wöhler
wöhlerite wohlerita
Wohlwill process proceso de Wohlwill
Wolff-Kishner reaction reacción de
    Wolff-Kishner
Wolff-Kishner reduction reducción de
    Wolff-Kishner
Wolff rearrangement reordenamiento de
    Wolff
Wolffenstein-Boters reaction reacción
    de Wolffenstein-Boters
wolfram wolframio, volframio
wolfram white blanco de wolframio
wolframate wolframato
wolframic acid ácido wolfrámico
wolframite wolframita
wollastonite wollastonita
wood alcohol alcohol de madera
wood ash ceniza de madera
wood cellulose celulosa de madera
wood flour harina de madera
wood meal harina de madera
wood pulp pulpa de madera
wood rosin colofonia de madera
wood sugar azúcar de madera
wood tar alquitrán de madera
wood turpentine trementina de madera
Woodward-Hoffmann rules reglas de
    Woodward-Hoffmann
Woodward hydroxylation hidroxilación
    de Woodward
wool lana
wool fat grasa de lana
work function función trabajo
wormseed oil aceite de quenopondio
wormwood ajenjo, absintio
wulfenite wulfenita
Wulff process proceso de Wulff

Wullner law ley de Wullner
Wurster salt sal de Wurster
Wurtz-Fittig reaction reacción de Wurtz-Fittig

Wurtz flask matraz de Wurtz
Wurtz reaction reacción de Wurtz
Wurtz synthesis síntesis de Wurtz
wurtzite wurtzita

X

X-ray analysis análisis con rayos X
X-ray crystallography cristalografía por
  rayos X
X-ray diffraction difracción de rayos X
X-ray fluorescence fluorescencia de
  rayos X
X-ray fluorescence analysis análisis de
  fluorescencia de rayos X
X-ray image spectrography
  espectrografía de imagen de rayos X
X-ray microanalysis microanálisis con
  rayos X
X-ray spectrograph espectrógrafo de
  rayos X
X-ray spectrography espectrografía de
  rayos X
X-ray spectrometer espectrómetro de
  rayos X
X-ray spectrometry espectrometría de
  rayos X
X-ray spectrum espectro de rayos X
X-ray structure estructura por rayos X
X-ray tube tubo de rayos X
xanthan xantano
xanthan gum goma de xantano
xanthate xantato
xanthene xanteno
xanthene dye colorante de xanteno
xanthenol xantenol
xanthenyl xantenilo
xanthic acid ácido xántico
xanthic amide amida xántica
xanthine xantina
xanthine oxidase oxidasa de xantina
xantho xanto
xanthochromium xantocromo
xanthogenic acid ácido xantogénico
xanthone xantona
xanthophyll xantofila
xanthopterin xantopterina

xanthosiderite xantosiderita
xanthosine xantosina
xanthoxylene xantoxileno
xanthyl xantilo
xanthylium xantilio
xenate xenato
xenobiotic xenobiótico
xenol xenol
xenolite xenolita
xenon xenón
xenon effect efecto de xenón
xenon tetrafluoride tetrafluoruro de
  xenón
xenotime xenotima
xenyl xenilo
xenylamine xenilamina
xerogel xerogel
xerography xerografía
xi xi
xi particle partícula xi
xylan xilano
xylene xileno
xylenediol xilenodiol
xylenesulfonic acid ácido xilenosulfónico
xylenol xilenol
xylidene xilideno
xylidic acid ácido xilídico
xylidine xilidina
xylite xilita
xylitol xilitol
xylocaine xilocaína
xylol xilol
xylonic acid ácido xilónico
xylose xilosa
xyloyl xiloilo
xylyl xililo
xylyl bromide bromuro de xililo
xylyl chloride cloruro de xililo
xylyl dichloride dicloruro de xililo
xylylene xilileno

Y

yacca gum goma de yaca
yeast levadura
yellow acid ácido amarillo
yellow brass latón amarillo
yellow cake torta amarilla
yellow copperas caparrosa amarilla
yellow lake laca amarilla
yellow phosphorus fósforo amarillo
yellow precipitate precipitado amarillo
yellow prussiate of potash prusiato
    amarillo de potasa
yellow prussiate of soda prusiato
    amarillo de sosa
yellow rain lluvia amarilla
yellow resin resina amarilla
yellow salt sal amarilla
yield rendimiento
yl ilo
ylide iluro
ylidene ilideno
ylidyne ilidino
yne ino
yohimbine yohimbina
yperite iperita
ytterbia iterbia
ytterbium iterbio

ytterbium bromide bromuro de iterbio
ytterbium chloride cloruro de iterbio
ytterbium fluoride fluoruro de iterbio
ytterbium oxalate oxalato de iterbio
ytterbium oxide óxido de iterbio
ytterbium sulfate sulfato de iterbio
yttria itria
yttrialite itrialita
yttrium itrio
yttrium acetate acetato de itrio
yttrium antimonide antimoniuro de itrio
yttrium arsenide arseniuro de itrio
yttrium bromide bromuro de itrio
yttrium carbonate carbonato de itrio
yttrium chloride cloruro de itrio
yttrium hydroxide hidróxido de itrio
yttrium nitrate nitrato de itrio
yttrium oxalate oxalato de itrio
yttrium oxide óxido de itrio
yttrium phosphide fosfuro de itrio
yttrium sulfate sulfato de itrio
yttrium sulfide sulfuro de itrio
yttrium vanadate vanadato de itrio
yttrocerite itrocerita
yttrotantalite itrotantalita
Yukawa force fuerza de Yukawa

Z

zaratite zaratita
zeaxanthin zeaxantina
Zeeman displacement desplazamiento de Zeeman
Zeeman effect efecto de Zeeman
Zeeman energy energía de Zeeman
zein zeína
Zeise salt sal de Zeise
Zeisel determination determinación de Zeisel
Zeisel reaction reacción de Zeisel
zeolite zeolita
zeolite catalyst catalizador de zeolita
Zerewitinoff determination determinación de Zerewitinoff
Zerewitinoff reagent reactivo de Zerewitinoff
zero-order reaction reacción de orden cero
zero-point energy energía al cero absoluto
zeroth law of thermodynamics ley cero de termodinámica
zeta potential potencial zeta
Ziegler catalyst catalizador de Ziegler
Ziegler method método de Ziegler
Ziegler-Natta polymerization polimerización de Ziegler-Natta
Ziegler process proceso de Ziegler
Ziehl stain colorante de Ziehl
Ziervogel process proceso de Ziervogel
Ziesel reaction reacción de Ziesel
Zimmermann reaction reacción de Zimmermann
zinc cinc, zinc
zinc abietate abietato de cinc
zinc acetate acetato de cinc
zinc acetylacetonate acetilacetonato de cinc
zinc albuminate albuminato de cinc
zinc amalgam amalgama de cinc
zinc ammonium nitrite nitrito amónico de cinc

zinc antimonide antimoniuro de cinc
zinc arsenate arseniato de cinc
zinc arsenide arseniuro de cinc
zinc arsenite arsenito de cinc
zinc bacitracin bacitracina de cinc
zinc benzoate benzoato de cinc
zinc blende blenda de cinc
zinc borate borato de cinc
zinc bromate bromato de cinc
zinc bromide bromuro de cinc
zinc caprylate caprilato de cinc
zinc carbonate carbonato de cinc
zinc chlorate clorato de cinc
zinc chloride cloruro de cinc
zinc chloroiodide cloroyoduro de cinc
zinc chromate cromato de cinc
zinc citrate citrato de cinc
zinc cyanide cianuro de cinc
zinc dichromate dicromato de cinc
zinc dioxide dióxido de cinc
zinc dithionate ditionato de cinc
zinc dust polvo de cinc
zinc ethanoate etanoato de cinc
zinc ethylhexoate etilhexoato de cinc
zinc ethylsulfate etilsulfato de cinc
zinc ferrocyanide ferrocianuro de cinc
zinc fluoride fluoruro de cinc
zinc fluoroborate fluoroborato de cinc
zinc fluorosilicate fluorosilicato de cinc
zinc formate formiato de cinc
zinc gluconate gluconato de cinc
zinc hexacyanoferrate hexacianoferrato de cinc
zinc hexafluorosilicate hexafluorosilicato de cinc
zinc hydrosulfite hidrosulfito de cinc
zinc hydroxide hidróxido de cinc
zinc hypophosphite hipofosfito de cinc
zinc iodate yodato de cinc
zinc iodide yoduro de cinc
zinc lactate lactato de cinc
zinc laurate laurato de cinc
zinc linoleate linoleato de cinc

zinc malate   malato de cinc
zinc metaarsenite   metaarsenito de cinc
zinc molybdate   molibdato de cinc
zinc naphthenate   naftenato de cinc
zinc nitrate   nitrato de cinc
zinc nitride   nitruro de cinc
zinc octanoate   octanoato de cinc
zinc octoate   octoato de cinc
zinc oleate   oleato de cinc
zinc orthoarsenate   ortoarseniato de cinc
zinc orthophosphate   ortofosfato de cinc
zinc oxalate   oxalato de cinc
zinc oxide   óxido de cinc
zinc oxychloride   oxicloruro de cinc
zinc palmitate   palmitato de cinc
zinc perborate   perborato de cinc
zinc perchlorate   perclorato de cinc
zinc permanganate   permanganato de
   cinc
zinc peroxide   peróxido de cinc
zinc peroxoborate   peroxoborato de cinc
zinc phenate   fenato de cinc
zinc phenolsulfonate   fenolsulfonato de
   cinc
zinc phosphate   fosfato de cinc
zinc phosphide   fosfuro de cinc
zinc phosphonate   fosfonato de cinc
zinc potassium iodide   yoduro potásico
   de cinc
zinc propionate   propionato de cinc
zinc pyrithione   piritiona de cinc
zinc pyrophosphate   pirofosfato de cinc
zinc resinate   resinato de cinc
zinc ricinoleate   ricinoleato de cinc
zinc salicylate   salicilato de cinc
zinc selenide   seleniuro de cinc
zinc silicate   silicato de cinc
zinc stearate   estearato de cinc
zinc succinate   succinato de cinc
zinc sulfate   sulfato de cinc
zinc sulfate monohydrate   monohidrato
   de sulfato de cinc
zinc sulfide   sulfuro de cinc
zinc sulfite   sulfito de cinc
zinc sulfoxylate   sulfoxilato de cinc
zinc tartrate   tartrato de cinc
zinc telluride   telururo de cinc
zinc tetraborate   tetraborato de cinc
zinc thiocyanate   tiocianato de cinc

zinc thiophenate   tiofenato de cinc
zinc undecylenate   undecilenato de cinc
zinc vitriol   vitriolo de cinc
zinc white   blanco de cinc
zinc yellow   amarillo de cinc
zincate   cincato
zincite   cincita
Zincke nitration   nitración de Zincke
Zincke-Suhl reaction   reacción de
   Zincke-Suhl
zincous   cincoso
zingiberene   zingibereno
zinnwaldite   zinnwaldita
zippeite   zippeíta
ziram   ziram
zircon   zircón
zirconate   zirconato
zirconia   zirconia
zirconic   zircónico
zirconic acid   ácido zircónico
zirconic anhydride   anhídrido zircónico
zirconium   zirconio
zirconium acetate   acetato de zirconio
zirconium boride   boruro de zirconio
zirconium bromide   bromuro de zirconio
zirconium carbide   carburo de zirconio
zirconium carbonate   carbonato de
   zirconio
zirconium chloride   cloruro de zirconio
zirconium diboride   diboruro de zirconio
zirconium dioxide   dióxido de zirconio
zirconium disilicide   disiliciuro de
   zirconio
zirconium disulfide   disulfuro de zirconio
zirconium fluoride   fluoruro de zirconio
zirconium glycolate   glicolato de zirconio
zirconium hydride   hidruro de zirconio
zirconium hydroxide   hidróxido de
   zirconio
zirconium iodide   yoduro de zirconio
zirconium lactate   lactato de zirconio
zirconium naphthenate   naftenato de
   zirconio
zirconium nitrate   nitrato de zirconio
zirconium nitride   nitruro de zirconio
zirconium orthophosphate   ortofosfato
   de zirconio
zirconium oxide   óxido de zirconio

zirconium oxybromide oxibromuro de zirconio

zirconium oxychloride oxicloruro de zirconio

zirconium phosphate fosfato de zirconio

zirconium potassium fluoride fluoruro potásico de zirconio

zirconium pyrophosphate pirofosfato de zirconio

zirconium silicate siicato de zirconio

zirconium silicide siliciuro de zirconio

zirconium sulfate sulfato de zirconio

zirconium tetraacetylacetonate tetraacetilacetonato de zirconio

zirconium tetrachloride tetracloruro de zirconio

zirconium tetrafluoride tetrafluoruro de zirconio

zirconocene dichloride dicloruro de zirconoceno

zirconyl zirconilo

zirconyl acetate acetato de zirconilo

zirconyl bromide bromuro de zirconilo

zirconyl carbonate carbonato de zirconilo

zirconyl chloride cloruro de zirconilo

zirconyl hydroxide hidróxido de zirconilo

zirconyl hyroxychloride hidroxicloruro de zirconilo

zirconyl nitrate nitrato de zirconilo

zirconyl phosphate fosfato de zirconilo

zirconyl sulfate sulfato de zirconilo

zirlite zirlita

zoalene zoaleno

zoisite zoisita

zone electrophoresis electroforesis por zonas

zone refining refinamiento por zonas

zoochemistry zooquímica

Zsigmondy filters filtros de Zsigmondy

zwitterion zwitterion

zymase zimasa

zymochemistry zimoquímica

zymohexase zimohexasa

zymolysis zimólisis

zymose zimosa

zymosis zimosis

zymurgy zimurgia

ESPAÑOL-INGLÉS
SPANISH-ENGLISH

A

a prueba de ácido acidproof
a prueba de agua waterproof
a prueba de fuego fireproof
a prueba de gas gasproof
a prueba de humedad humidity-proof,
　　moisture-proof
abacá abaca
aberración aberration
aberración cromática chromatic
　　aberration
abietato cálcico calcium abietate
abietato de bencilo benzyl abietate
abietato de cinc zinc abietate
abietato de cobre copper abietate
abietato de etilo ethyl abietate
abietato de glicerilo glyceryl abietate
abietato de metilo methyl abietate
abietato potásico potassium abietate
abietato sódico sodium abietate
ablación ablation
ablandador softener
ablandamiento de agua water softening
abono orgánico compost
abrasión abrasion
abrasivo abrasive
abrazadera clamp
abrazadera de Bunsen Bunsen clamp
abrazadera de bureta buret clamp
abrillantador brightener
abrillantador óptico optical brightener
abrillantamiento brightening
absintio absinthium, absinthe,
　　wormwood
absoluto absolute
absorbancia absorbance
absorbedor de microondas microwave
　　absorber
absorbedor de neutrones neutron
　　absorber
absorbedor de radiación radiation
　　absorber
absorbedor neutrónico neutron absorber
absorbedor ultravioleta ultraviolet

absorber
absorbente absorbent
absorbente de oxígeno oxygen absorbent
absorciómetro absorptiometer
absorción absorption
absorción atómica atomic absorption
absorción de calor heat absorption
absorción de fotones photon absorption
absorción de gas gas absorption
absorción de luz absorption of light
absorción fotoeléctrica photoelectric
　　absorption
absorción neutrónica neutron absorption
absorción nuclear nuclear absorption
absorción térmica heat absorption
absorción ultravioleta ultraviolet
　　absorption
absortividad absorptivity
abstracción abstraction
abundancia abundance
abundancia isotópica isotope abundance
abundancia natural natural abundance
acacetina acacetin
acacia acacia
acadesina acadesine
acantita acanthite
accesorio accessory
acción action
acción amortiguadora buffer action
acción catalítica catalytic action
acción cíclica cyclic action
acción electrolítica electrolytic action
acción galvánica galvanic action
acción química chemical action
acción reversible reversible action
aceite oil
aceite blanco white oil
aceite carbólico carbolic oil
aceite combustible fuel oil
aceite comestible edible oil
aceite de aguacate avocado oil
aceite de ajo garlic oil
aceite de alcanfor camphor oil

aceite de alcaravea caraway oil
aceite de almendra almond oil
aceite de alquitrán tar oil
aceite de alquitrán de hulla coal-tar oil
aceite de alquitrán de pino pine-tar oil
aceite de anacardo cashew oil
aceite de anís anise oil
aceite de antraceno anthracene oil
aceite de babasú babassu oil
aceite de bergamota bergamot oil
aceite de betula betula oil
aceite de cacao cocoa oil
aceite de cada juniper tar
aceite de cálamo calamus oil
aceite de cananga cananga oil
aceite de canela cinnamon oil
aceite de cañamón hempseed oil
aceite de cardamomo cardamom oil
aceite de cártamo safflower oil
aceite de cascarilla cascarilla oil
aceite de casia cassia oil
aceite de castor castor oil
aceite de citronela citronella oil
aceite de clavero clove oil
aceite de cloronaftaleno
    chloronaphthalene oil
aceite de colofonia rosin oil
aceite de colza rapeseed oil, canola
aceite de copaiba copaiba oil
aceite de corteza de naranja orange-peel
    oil
aceite de crotón croton oil
aceite de estragón tarragon oil
aceite de fenilmostasa phenyl mustard
    oil
aceite de fusel fusel oil
aceite de girasol sunflower oil
aceite de hígado de bacalao cod-liver oil
aceite de hinojo fennel oil
aceite de hojas de laurel laurel leaves oil
aceite de horno furnace oil
aceite de huesos bone oil
aceite de huevo egg oil
aceite de hulla coal oil
aceite de jazmín jasmine oil
aceite de jengibre ginger oil
aceite de jojoba jojoba oil
aceite de lavanda lavender oil
aceite de lima lime oil

aceite de limón lemon oil
aceite de linaza linseed oil, flaxseed oil
aceite de luisa lemongrass oil
aceite de maíz corn oil
aceite de mandarina mandarin oil
aceite de maní peanut oil
aceite de manteca lard oil
aceite de manzana apple oil
aceite de manzanilla chamomile oil
aceite de menhaden menhaden oil
aceite de menta peppermint oil
aceite de menta verde spearmint oil
aceite de mostaza mustard oil
aceite de niobe niobe oil
aceite de nuez de coco coconut oil
aceite de oliva olive oil
aceite de pachulí patchouli oil
aceite de palma palm oil
aceite de palmarosa palmarosa oil
aceite de parafina paraffin oil
aceite de pata de vaca neatsfoot oil
aceite de perilla perilla oil
aceite de pescado fish oil
aceite de pino pine oil
aceite de pirola wintergreen oil
aceite de pizarra shale oil
aceite de plátano banana oil
aceite de poleo pennyroyal oil
aceite de quenopodio chenopodium oil,
    wormseed oil
aceite de ricino ricinus oil
aceite de salvado bran oil
aceite de salvia sage oil
aceite de sándalo sandalwood oil
aceite de sasafrás sassafras oil
aceite de sebo tallow oil
aceite de semilla de algodón cottonseed
    oil
aceite de semilla de perejil parsley oil
aceite de sésamo sesame oil
aceite de silicona silicone oil
aceite de soja soybean oil
aceite de tangerina tangerine oil
aceite de tiglio tiglium oil
aceite de transformador transformer oil
aceite de tung tung oil
aceite de tuya thuja oil
aceite de valeriana valerian oil
aceite de verbena verbena oil

aceite de vitriolo vitriol oil
aceite dulce sweet oil
aceite esencial essential oil
aceite estirenado styrenated oil
aceite fijo fixed oil
aceite hidrogenado hydrogenated oil
aceite ligero light oil
aceite lubricante lubricating oil
aceite mineral mineral oil
aceite neutro neutral oil
aceite pesado heavy oil
aceite residual residual oil
aceite rojo red oil
aceite secante drying oil
aceite sintético synthetic oil
aceite soluble soluble oil
aceite soluble en agua water-soluble oil
aceite soplado blown oil
aceite vegetal vegetable oil
aceite yodado iodized oil
aceleración acceleration
acelerador accelerator
acelerador de caucho rubber accelerator
acelerador de electrones electron
　accelerator
acelerador de partículas particle
　accelerator
acelerador de protones proton
　accelerator
acelerador electrónico electron
　accelerator
acelerador lineal electrónico electron
　linear accelerator
acelerador protónico proton accelerator
acenafteno acenaphthene
acenaftenoquinona acenaphthenequinone
acenaftileno acenaphthylene
aceptor acceptor
aceptor de ácido acid acceptor
aceptor de hidrógeno hydrogen acceptor
aceptor electrónico electron acceptor
aceptor protónico proton acceptor
acero steel
acero al boro boron steel
acero al carbono carbon steel
acero al cromo chromium steel
acero al níquel nickel steel
acero al tungsteno tungsten steel
acero aleado alloy steel

acero de horno eléctrico electric steel
acero inoxidable stainless steel
acetal acetal
acetaldehído acetaldehyde
acetaldol acetaldol
acetamida acetamide
acetamidina acetamidine
acetaminofén acetaminophen
acetaminofeno acetaminophen
acetaminosalol acetaminosalol
acetanilida acetanilide
acetato acetate
acetato alumínico aluminum acetate
acetato amónico ammonium acetate
acetato bárico barium acetate
acetato cálcico calcium acetate
acetato ceroso cerous acetate
acetato cobaltoso cobaltous acetate
acetato crómico chromic acetate
acetato cromoso chromous acetate
acetato cúprico cupric acetate
acetato de alilo allyl acetate
acetato de aluminio aluminum acetate
acetato de amilo amyl acetate
acetato de amonio ammonium acetate
acetato de anilina aniline acetate
acetato de anisilo anisyl acetate
acetato de bario barium acetate
acetato de bencilo benzyl acetate
acetato de benzaldoxima benzaldoxime
　acetate
acetato de benzoilo benzoyl acetate
acetato de berilio beryllium acetate
acetato de bismuto bismuth acetate
acetato de bornilo bornyl acetate
acetato de butanol butanol acetate
acetato de butilo butyl acetate
acetato de cadmio cadmium acetate
acetato de calcio calcium acetate
acetato de cedrilo cedryl acetate
acetato de celulosa cellulose acetate
acetato de cesio cesium acetate
acetato de cianometilo cyanomethyl
　acetate
acetato de ciclohexanol cyclohexanol
　acetate
acetato de cinamilo cinnamyl acetate
acetato de cinc zinc acetate
acetato de cloromadinona

chloromadinone acetate

acetato de cloruro de polivinilo
polyvinyl chloride acetate

acetato de cobalto cobalt acetate

acetato de cobre copper acetate

acetato de cresilo cresyl acetate

acetato de cromo chromium acetate

acetato de decilo decyl acetate

acetato de dicloroetilo dichloroethyl
acetate

acetato de erbio erbium acetate

acetato de estroncio strontium acetate

acetato de etilbutilo ethylbutyl acetate

acetato de etilhexilo ethylhexyl acetate

acetato de etilo ethyl acetate

acetato de fenetilo phenethyl acetate

acetato de feniletilo phenylethyl acetate

acetato de fenilmetilo phenylmethyl
acetate

acetato de fenilo phenyl acetate

acetato de fenilpropilo phenylpropyl
acetate

acetato de furfurilo furfuryl acetate

acetato de galio gallium acetate

acetato de geranilo geranyl acetate

acetato de glicol glycol acetate

acetato de guayacol guaiacol acetate

acetato de heptilo heptyl acetate

acetato de hexilo hexyl acetate

acetato de hidrocinamilo hydrocinnamyl
acetate

acetato de hierro iron acetate

acetato de isoamilo isoamyl acetate

acetato de isobornilo isobornyl acetate

acetato de isobutilo isobutyl acetate

acetato de isopropenilo isopropenyl
acetate

acetato de isopropilo isopropyl acetate

acetato de itrio yttrium acetate

acetato de lantano lanthanum acetate

acetato de linalilo linalyl acetate

acetato de litio lithium acetate

acetato de magnesio magnesium acetate

acetato de manganeso manganese acetate

acetato de mentilo menthyl acetate

acetato de mercurio mercury acetate

acetato de metilalilo methylallyl acetate

acetato de metilamilo methylamyl
acetate

acetato de metilo methyl acetate

acetato de neodimio neodymium acetate

acetato de níquel nickel acetate

acetato de nonilo nonyl acetate

acetato de octilo octyl acetate

acetato de pentilo pentyl acetate

acetato de plata silver acetate

acetato de plomo lead acetate

acetato de polivinilo polyvinyl acetate

acetato de potasio potassium acetate

acetato de praseodimio praseodymium
acetate

acetato de propilo propyl acetate

acetato de resorcina resorcinol acetate

acetato de rodinilo rhodinyl acetate

acetato de rubidio rubidium acetate

acetato de santalilo santalyl acetate

acetato de sodio sodium acetate

acetato de talio thallium acetate

acetato de terpinilo terpinyl acetate

acetato de undecilenilo undecylenyl
acetate

acetato de uranilo uranyl acetate

acetato de uranio uranium acetate

acetato de urea urea acetate

acetato de vinilo vinyl acetate

acetato de yodo iodine acetate

acetato de zirconilo zirconyl acetate

acetato de zirconio zirconium acetate

acetato estannoso stannous acetate

acetato estiralílico styralyl acetate

acetato fenilmercúrico phenylmercuric
acetate

acetato férrico ferric acetate

acetato ferroso ferrous acetate

acetato magnésico magnesium acetate

acetato manganoso manganous acetate

acetato mercúrico mercuric acetate

acetato mercurioso mercurous acetate

acetato niqueloso nickelous acetate

acetato potásico potassium acetate

acetato rojo red acetate

acetato sódico sodium acetate

acetatobutirato de celulosa cellulose
acetatobutyrate

acetatoisobutirato de sucrosa sucrose
acetate isobutyrate

acetazolamida acetazolamide

acetiamina acetiamine

acetilacetato de neodimio neodymium
acetylacetate

acetilacetona acetylacetone

acetilacetonato acetylacetonate

acetilacetonato de berilio beryllium
acetylacetonate

acetilacetonato de cinc zinc
acetylacetonate

acetilacetonato de cromo chromium
acetylacetonate

acetilacetonato de indio indium
acetylacetonate

acetilacetonato de molibdeno
molybdenum acetylacetonate

acetilacetonato de níquel nickel
acetylacetonate

acetilacetonato de titanio titanium
acetylacetonate

acetilacetonato de vanadio vanadium
acetylacetonate

acetilacetonato férrico ferric
acetylacetonate

acetilacetonato magnésico magnesium
acetylacetonate

acetilacetonato mangánico manganic
acetylacetonate

acetilación acetylation

acetilamino acetylamino

acetilbenceno acetylbenzene

acetilcarbromal acetylcarbromal

acetilcetena acetyl ketene

acetilcoenzima acetyl coenzyme

acetilcolina acetylcholine

acetilcolinesterasa acetylcholinesterase

acetilenilo acetylenyl

acetileno acetylene

acetileno de butilo butyl acetylene

acetilenógeno acetylenogen

acetilfenol acetylphenol

acetilferroceno acetylferrocene

acetilformiato sódico sodium
acetylformate

acetilmetilcarbinol acetylmethylcarbinol

acetilo acetyl

acetilricinoleato de metilo methyl
acetylricinoleate

acetilsalicilato cálcico calcium
acetylsalicylate

acetilsalicilato potásico potassium

acetylsalicylate

acetilsalicilato sódico sodium
acetylsalicylate

acetilurea acetylurea

acetiluro acetylide

acetiluro cálcico calcium acetylide

acetiluro cuproso cuprous acetylide

acetiluro de plata silver acetylide

acetiluro mercurioso mercurous
acetylide

acetina acetin

acetoacetanilida acetoacetanilide

acetoacetato acetoacetate

acetoacetato de butilo butyl acetoacetate

acetoacetato de etilo ethyl acetoacetate

acetoacetato de metilo methyl
acetoacetate

acetoarsenito de cobre copper
acetoarsenite

acetoestearina acetostearin

acetofenetinina acetophenetidine

acetofenona acetophenone

acetoína acetoin

acetol acetol

acetólisis acetolysis

acetona acetone

acetona clorada chlorinated acetone

acetona de acetofenona acetophenone
acetone

acetonilacetona acetonylacetone

acetonilideno acetonylidene

acetonitrilo acetonitrile

acetotartrato alumínico aluminum
acetotartrate

acetotoluida acetotoluide

acetotoluidida acetotoluidide

acetoxiestearato de butilo butyl
acetoxystearate

acetoxilación acetoxylation

acetoxima acetoxime

acetozona acetozone

acíclico acyclic

acicular acicular

acidez acidity

acídico acidic

acidificación acidification

acidificar acidify

acidimetría acidimetry

acidímetro acidimeter

ácido acid
ácido abienínico abieninic acid
ácido abietadienoico abietadienoic acid
ácido abiético abietic acid
ácido abscísico abscisic acid
ácido acético acetic acid
ácido acético glacial glacial acetic acid
ácido acetilacético acetylacetic acid
ácido acetilfórmico acetylformic acid
ácido acetilpropiónico acetylpropionic acid
ácido acetilsalicílico acetylsalicylic acid
ácido acetoacético acetoacetic acid
ácido acetonadicarboxílico acetone dicarboxylic acid
ácido acetonamonocarboxílico acetone monocarboxylic acid
ácido acónico aconic acid
ácido aconítico aconitic acid
ácido acridínico acridinic acid
ácido acrílico acrylic acid
ácido acroleico acroleic acid
ácido adenílico adenylic acid
ácido adípico adipic acid
ácido agárico agaric acid
ácido alánico allanic acid
ácido alántico alantic acid
ácido alantoico allantoic acid
ácido alantoxánico allantoxanic acid
ácido alantúrico allanturic acid
ácido alcanosulfónico alkanesulfonic acid
ácido aldónico aldonic acid
ácido aleurítico aleuritic acid
ácido algínico alginic acid
ácido alifático aliphatic acid
ácido alizárico alizaric acid
ácido alocinámico allocinnamic acid
ácido aloerésico aloeresic acid
ácido alofánico allophanic acid
ácido alomaleico allomaleic acid
ácido aloxánico alloxanic acid
ácido alumínico aluminic acid
ácido amalínico amalinic acid
ácido amarillo yellow acid
ácido amigdálico amygdalic acid
ácido amínico aminic acid
ácido aminoacético aminoacetic acid
ácido aminobencenosulfónico aminobenzenesulfonic acid
ácido aminobenzoico aminobenzoic acid
ácido aminobutanoico aminobutanoic acid
ácido aminobutírico aminobutyric acid
ácido aminocaproico aminocaproic acid
ácido aminoditiofórmico aminodithioformic acid
ácido aminoetilsulfúrico aminoethylsulfuric acid
ácido aminofenilacético aminophenylacetic acid
ácido aminofosfórico aminophosphoric acid
ácido aminoglutárico aminoglutaric acid
ácido aminohexanoico aminohexanoic acid
ácido aminohipúrico aminohippuric acid
ácido aminoisobutírico aminoisobutyric acid
ácido aminonaftolsulfónico aminonaphtholsulfonic acid
ácido aminonicotínico aminonicotinic acid
ácido aminopropanoico aminopropanoic acid
ácido aminosalicílico aminosalicylic acid
ácido amsónico amsonic acid
ácido anacárdico anacardic acid
ácido ancoico anchoic acid
ácido anemónico anemonic acid
ácido angélico angelic acid
ácido anilinosulfónico anilinesulfonic acid
ácido anísico anisic acid
ácido antimónico antimonic acid
ácido antranílico anthranilic acid
ácido antroico anthroic acid
ácido apiólico apiolic acid
ácido apocólico apocholic acid
ácido arabónico arabonic acid
ácido araquídico arachidic acid
ácido araquidónico arachidonic acid
ácido aromático aromatic acid
ácido arrénico arrhenic acid
ácido arsanílico arsanilic acid
ácido arsenénico arsenenic acid
ácido arsénico arsenic acid
ácido arsenioso arsenous acid

ácido arsínico arsinic acid
ácido arsónico arsonic acid
ácido asarónico asaronic acid
ácido ascarídico ascaridic acid
ácido ascaridólico ascaridolic acid
ácido ascórbico ascorbic acid
ácido asiático asiatic acid
ácido asparágico asparagic acid
ácido asparagínico asparaginic acid
ácido aspartámico aspartamic acid
ácido aspártico aspartic acid
ácido aspergílico aspergillic acid
ácido atroláctico atrolactic acid
ácido atrónico atronic acid
ácido atrópico atropic acid
ácido auribromhídrico auribromhydric
acid
ácido áurico auric acid
ácido azelaico azelaic acid
ácido azobenzoico azobenzoic acid
ácido barbitúrico barbituric acid
ácido behénico behenic acid
ácido behenólico behenolic acid
ácido bencenocarboxílico
benzenecarboxylic acid
ácido bencenodicarboxílico
benzenedicarboxylic acid
ácido bencenodisulfónico
benzenedisulfonic acid
ácido bencenofosfínico
benzenephosphinic acid
ácido bencenofosfónico
benzenephosphonic acid
ácido bencenomonosulfónico
benzenemonosulfonic acid
ácido bencenosulfínico benzenesulfinic
acid
ácido bencenosulfónico benzenesulfonic
acid
ácido bencenotetracarboxílico
benzenetetracarboxylic acid
ácido bencenotrisulfónico
benzenetrisulfonic acid
ácido bencidinodicarboxílico
benzidinedicarboxylic acid
ácido bencílico benzilic acid
ácido bencilsulfónico benzylsulfonic acid
ácido benzamidoacético benzamidoacetic
acid

ácido benzaminoacético benzaminoacetic
acid
ácido benzocarbólico benzocarbolic acid
ácido benzoglicólico benzoglycolic acid
ácido benzohidroxímico
benzohydroximic acid
ácido benzoico benzoic acid
ácido benzoilacético benzoylacetic acid
ácido benzoilacrílico benzoylacrylic acid
ácido benzoilaminoacético
benzoylaminoacetic acid
ácido benzoilbenzoico benzoylbenzoic
acid
ácido benzoilpropiónico
benzoylpropionic acid
ácido berberónico berberonic acid
ácido beta beta acid
ácido betulínico betulinic acid
ácido biliar bile acid
ácido bilirrubínico bilirubinic acid
ácido binario binary acid
ácido blanco white acid
ácido blando soft acid
ácido borácico boracic acid
ácido bórico boric acid
ácido borofosfórico borophosphoric acid
ácido borotúngstico borotungstic acid
ácido brasídico brassidic acid
ácido brasílico brassylic acid
ácido bromhídrico hydrobromic acid
ácido brómico bromic acid
ácido bromoacético bromoacetic acid
ácido bromoantrálico bromoanthrallic
acid
ácido bromoaúrico bromoauric acid
ácido bromobencenosulfónico
bromobenzenesulfonic acid
ácido bromobenzoico bromobenzoic acid
ácido bromobutírico bromobutyric acid
ácido bromopropiónico bromopropionic
acid
ácido bromosuccínico bromosuccinic
acid
ácido butanoico butanoic acid
ácido butenoico butenoic acid
ácido butilestánnico butylstannic acid
ácido butírico butyric acid
ácido cacodílico cacodylic acid
ácido cafeico caffeic acid

ácido cafetánico caffetannic acid
ácido cambopínico cambopinic acid
ácido canfocarboxílico
　　camphocarboxylic acid
ácido canfólico campholic acid
ácido canfórico camphoric acid
ácido cáprico capric acid
ácido caprílico caprylic acid
ácido caproico caproic acid
ácido carbámico carbamic acid
ácido carbanílico carbanilic acid
ácido carbázico carbazic acid
ácido carbazótico carbazotic acid
ácido carbílico carbylic acid
ácido carbólico carbolic acid
ácido carbónico carbonic acid
ácido carboxílico carboxylic acid
ácido carcinómico carcinomic acid
ácido cariofílico caryophyllic acid
ácido cárlico carlic acid
ácido carmínico carminic acid
ácido carnáubico carnaubic acid
ácido caróbico carobic acid
ácido carólico carolic acid
ácido cartámico carthamic acid
ácido cerebrónico cerebronic acid
ácido ceritámico ceritamic acid
ácido ceroplástico ceroplastic acid
ácido cerótico cerotic acid
ácido cerotínico cerotinic acid
ácido cetílico cetylic acid
ácido cetoglutárico ketoglutaric acid
ácido cetólico cetolic acid
ácido cetomalónico ketomalonic acid
ácido cetopropiónico ketopropionic acid
ácido cetovalérico ketovaleric acid
ácido cianhídrico hydrocyanic acid
ácido ciánico cyanic acid
ácido cianoacético cyanoacetic acid
ácido cianocarbónico cyanocarbonic acid
ácido cianoetanoico cyanoethanoic acid
ácido cianúrico cyanuric acid
ácido ciclámico cyclamic acid
ácido ciclohexanocarboxílico
　　cyclohexanecarboxylic acid
ácido ciclopentilpropiónico
　　cyclopentylpropionic acid
ácido cinámico cinnamic acid
ácido cincomerónico cinchomeronic acid

ácido cinconínico cinchoninic acid
ácido citacrónico citraconic acid
ácido citidílico cytidylic acid
ácido citramálico citramalic acid
ácido cítrico citric acid
ácido citrídico citridic acid
ácido citronélico citronellic acid
ácido clorhídrico hydrochloric acid
ácido clórico chloric acid
ácido cloroacético chloroacetic acid
ácido cloroáurico chloroauric acid
ácido cloroazótico chloroazotic acid
ácido clorobenzoico chlorobenzoic acid
ácido clorodifluoroacético
　　chlorodifluoroacetic acid
ácido cloroetanoico chloroethanoic acid
ácido cloroetilfosfónico
　　chloroethylphosphonic acid
ácido clorofórmico chloroformic acid
ácido clorogénico chlorogenic acid
ácido cloromalónico chloromalonic acid
ácido clorometilfosfónico
　　chloromethylphosphonic acid
ácido cloronítrico chloronitric acid
ácido cloronitrobenzoico
　　chloronitrobenzoic acid
ácido cloronitroso chloronitrous acid
ácido cloroperbenzoico chloroperbenzoic
　　acid
ácido cloroplatínico chloroplatinic acid
ácido cloropropiónico chloropropionic
　　acid
ácido clorosalicílico chlorosalicylic acid
ácido cloroso chlorous acid
ácido clorosulfónico chlorosulfonic acid
ácido clorosulfonilbenzoico
　　chlorosulfonylbenzoic acid
ácido clorosulfúrico chlorosulfuric acid
ácido clorovalérico chlorovaleric acid
ácido cójico kojic acid
ácido colaico cholaic acid
ácido coleico choleic acid
ácido cólico cholic acid
ácido colofólico colopholic acid
ácido colofónico colophonic acid
ácido coménico comenic acid
ácido complejo complex acid
ácido conjugado conjugate acid
ácido corísmico chorismic acid

ácido cosanoico cosanoic acid
ácido cresílico cresylic acid
ácido cresótico cresotic acid
ácido crisalícico chrysalicic acid
ácido crisanísico chrysanisic acid
ácido crisantemomonocarboxílico
    chrysanthemummonocarboxylic acid
ácido crisofánico chrysophanic acid
ácido crocónico croconic acid
ácido crómico chromic acid
ácido cromoso chromous acid
ácido cromotrópico chromotropic acid
ácido crotónico crotonic acid
ácido cumálico coumalic acid
ácido cumárico coumaric acid
ácido cúmico cumic acid
ácido cumínico cuminic acid
ácido chaulmúgrico chaulmoogric acid
ácido chi chi acid
ácido datúrico daturic acid
ácido de aceite de castor castor oil acid
ácido de alquitrán tar acid
ácido de Armstrong Armstrong acid
ácido de batería battery acid
ácido de Bayer Bayer acid
ácido de Broenner Broenner acid
ácido de Brönner Brönner acid
ácido de Brönsted Brönsted acid
ácido de cámara chamber acid
ácido de Caro Caro acid
ácido de Cassella Cassella acid
ácido de Cleve Cleve acid
ácido de contacto contact acid
ácido de Chicago Chicago acid
ácido de Dahl Dahl acid
ácido de Freund Freund acid
ácido de Koch Koch acid
ácido de Laurent Laurent acid
ácido de Lewis Lewis acid
ácido de manzana apple acid
ácido de Neville-Winter Neville-Winter
    acid
ácido de nuez de coco coconut acid
ácido de Reinecke Reinecke acid
ácido de Schäffer Schäffer acid
ácido de Shoelkopf Shoelkopf acid
ácido de Tobias Tobias acid
ácido débil weak acid
ácido decanocarboxílico

decanecarboxylic acid
ácido decanodioico decanedioic acid
ácido decanoico decanoic acid
ácido decatoico decatoic acid
ácido decíclico decyclic acid
ácido decílico decyl acid
ácido decoico decoic acid
ácido delfínico delphinic acid
ácido delta delta acid
ácido dérrico derric acid
ácido descarboxirrísico decarboxyrissic
    acid
ácido deshidroabiético dehydroabietic
    acid
ácido deshidroacético dehydroacetic acid
ácido deshidroascórbico
    dehydroascorbic acid
ácido deshidrocólico dehydrocholic acid
ácido desoxicólico deoxycholic acid
ácido desoxirribonucleico
    deoxyribonucleic acid
ácido desoxirribonucleico recombinante
    recombinant deoxyribonucleic acid
ácido deutérico deuteric acid
ácido diacético diacetic acid
ácido dialúrico dialuric acid
ácido diaminocaproico diaminocaproic
    acid
ácido diaminodifénico diaminodiphenic
    acid
ácido diazobencenosulfónico
    diazobenzenesulfonic acid
ácido diazoico diazoic acid
ácido diazosulfónico diazosulfonic acid
ácido dibásico dibasic acid
ácido dibromomalónico dibromomalonic
    acid
ácido dicarboxílico dicarboxylic acid
ácido dicloroacético dichloroacetic acid
ácido diclorobenzoico dichlorobenzoic
    acid
ácido dicloroetanoico dichloroethanoic
    acid
ácido diclorofenoxiacético
    dichlorophenoxyacetic acid
ácido dicloropropiónico
    dichloropropionic acid
ácido dicrómico dichromic acid
ácido dietilacético diethylacetic acid

ácido dietilbarbitúrico diethylbarbituric acid
ácido difénico diphenic acid
ácido difosfoglicérico diphosphoglyceric acid
ácido difosfórico diphosphoric acid
ácido digálico digallic acid
ácido diglicólico diglycolic acid
ácido dihidroxibenzoico dihydroxybenzoic acid
ácido dihidroxibutanodioico dihydroxybutanedioic acid
ácido dihidroxicinámico dihydroxycinnamic acid
ácido dihidroxiesteárico dihydroxystearic acid
ácido dihidroximalónico dihydroxymalonic acid
ácido dihidroxisuccínico dihydroxysuccinic acid
ácido diluido dilute acid
ácido dimetilaminobenzoico dimethylaminobenzoic acid
ácido dimetilarsínico dimethylarsinic acid
ácido dinitrosalicílico dinitrosalicylic acid
ácido dioxonítrico dioxonitric acid
ácido disalicílico disalicylic acid
ácido disulfónico disulfonic acid
ácido disulfúrico disulfuric acid
ácido ditiobenzoico dithiobenzoic acid
ácido ditiocarbámico dithiocarbamic acid
ácido ditiónico dithionic acid
ácido diyodoacético diiodoacetic acid
ácido docosanoico docosanoic acid
ácido docosenoico docosenoic acid
ácido dodecanoico dodecanoic acid
ácido dodecenilsuccinico dodecenylsuccinic acid
ácido durílico durylic acid
ácido duro hard acid
ácido eicosanoico eicosanoic acid
ácido elágico ellagic acid
ácido elaídico elaidic acid
ácido elatérico elateric acid
ácido electrolítico electrolyte acid
ácido elemónico elemonic acid

ácido eleoesteárico eleostearic acid
ácido embélico embelic acid
ácido embónico embonic acid
ácido enántico enanthic acid
ácido epsilón epsilon acid
ácido eritórbico erythorbic acid
ácido erúcico erucic acid
ácido estánnico stannic acid
ácido esteárico stearic acid
ácido estífnico styphnic acid
ácido etacético ethacetic acid
ácido etanodioico ethanedioic acid
ácido etanoico ethanoic acid
ácido etanoico glacial glacial ethanoic acid
ácido etanosulfónico ethanesulfonic acid
ácido etanotiólico ethanethiolic acid
ácido etérico etheric acid
ácido etilacético ethylacetic acid
ácido etilbencenosulfónico ethylbenzenesulfonic acid
ácido etilbutírico ethylbutyric acid
ácido etilendiaminotetraacético ethylenediaminetetraacetic acid
ácido etilendiaminotetraacético tetrasódico tetrasodium ethylenediaminetetraacetic acid
ácido etilendiaminotetraacético trisódico trisodium ethylenediaminetetraacetic acid
ácido etilensulfónico ethylenesulfonic acid
ácido etilfosfórico ethylphosphoric acid
ácido etilhexoico ethylhexoic acid
ácido etinoico ethinoic acid
ácido etisulfúrico ethylsulfuric acid
ácido eugénico eugenic acid
ácido eupitónico eupittonic acid
ácido fencólico fencholic acid
ácido fénico phenic acid
ácido fenilacético phenylacetic acid
ácido fenilarsónico phenylarsonic acid
ácido fenilbutanoico phenylbutanoic acid
ácido fenilbutírico phenylbutyric acid
ácido fenilesteárico phenylstearic acid
ácido feniletanoico phenylethanoic acid
ácido feniletilacético phenylethylacetic acid
ácido fenilfórmico phenylformic acid

ácido **fenilfosfínico** phenylphosphinic acid

ácido **fenilfosfónico** phenylphosphonic acid

ácido **fenilglicólico** phenylglycolic acid

ácido **fenílico** phenylic acid

ácido **fenilpropenoico** phenylpropenoic acid

ácido **fenilpropiónico** phenylpropionic acid

ácido **fenilsalicílico** phenylsalicylic acid

ácido **fenilsulfanílico** phenylsulfanilic acid

ácido **fenilsulfónico** phenylsulfonic acid

ácido **fenoldisulfónico** phenoldisulfonic acid

ácido **fenólico** phenolic acid

ácido **fenolsulfónico** phenolsulfonic acid

ácido **fenoxiacético** phenoxyacetic acid

ácido **ferriciánico** ferricyanic acid

ácido **ferrociánico** ferrocyanic acid

ácido **ferúlico** ferulic acid

ácido **filícico** filicic acid

ácido **filíxico** filixic acid

ácido **fitánico** phytanic acid

ácido **fítico** phytic acid

ácido **flaviánico** flavianic acid

ácido **fluobórico** fluoboric acid

ácido **fluohídrico** fluohydric acid

ácido **fluorénico** fluorenic acid

ácido **fluorgermánico** fluorgermanic acid

ácido **fluorhídrico** hydrofluoric acid

ácido **fluoroacético** fluoroacetic acid

ácido **fluorobórico** fluoroboric acid

ácido **fluoroetanoico** fluoroethanoic acid

ácido **fluorofosfórico** fluorophosphoric acid

ácido **fluoroplúmbico** fluoroplumbic acid

ácido **fluorosilícico** fluorosilicic acid

ácido **fluorosulfónico** fluorosulfonic acid

ácido **fluorosulfúrico** fluorosulfuric acid

ácido **fluosilícico** fluosilicic acid

ácido **fluosulfónico** fluosulfonic acid

ácido **fólico** folic acid

ácido **folínico** folinic acid

ácido **fórmico** formic acid

ácido **formilacético** formylacetic acid

ácido **formílico** formylic acid

ácido **formonitrólico** formonitrolic acid

ácido **fosfámico** phosphamic acid

ácido **fosfínico** phosphinic acid

ácido **fosfinoso** phosphinous acid

ácido **fosfomolíbdico** phosphomolybdic acid

ácido **fosfónico** phosphonic acid

ácido **fosfonoso** phosphonous acid

ácido **fosfórico** phosphoric acid

ácido **fosfórico anhidro** anhydrous phosphoric acid

ácido **fosforoso** phosphorous acid

ácido **fosfotúngstico** phosphotungstic acid

ácido **frangúlico** frangulic acid

ácido **frangulínico** frangulinic acid

ácido **ftalámico** phthalamic acid

ácido **ftálico** phthalic acid

ácido **ftioico** phthioic acid

ácido **fuerte** strong acid

ácido **fulgénico** fulgenic acid

ácido **fulmínico** fulminic acid

ácido **fulminúrico** fulminuric acid

ácido **fumarámico** fumaramic acid

ácido **fumárico** fumaric acid

ácido **furacrílico** furacrylic acid

ácido **furancarboxílico** furancarboxylic acid

ácido **furoico** furoic acid

ácido **furónico** furonic acid

ácido **fusídico** fusidic acid

ácido **gadoleico** gadoleic acid

ácido **gaídico** gaidic acid

ácido **galactárico** galactaric acid

ácido **galacturónico** galacturonic acid

ácido **gálico** gallic acid

ácido **galotánico** gallotannic acid

ácido **gamma** gamma acid

ácido **gammaaminobutírico** gamma-aminobutyric acid

ácido **gentísico** gentisic acid

ácido **geránico** geranic acid

ácido **germanofórmico** germanoformic acid

ácido **giberélico** gibberellic acid

ácido **gladiólico** gladiolic acid

ácido **glicérico** glyceric acid

ácido **glicerofosfórico** glycerophosphoric acid

ácido **glicídico** glycidic acid

ácido glicocólico glycocholic acid
ácido glicogénico glycogenic acid
ácido glicólico glycolic acid
ácido glicónico glyconic acid
ácido glioxálico glyoxalic acid
ácido glioxílico glyoxylic acid
ácido glucónico gluconic acid
ácido glucurónico glucuronic acid
ácido glutacónico glutaconic acid
ácido glutámico glutamic acid
ácido glutárico glutaric acid
ácido glutínico glutinic acid
ácido grafítico graphitic acid
ácido graso fatty acid
ácido graso conjugado conjugated fatty
    acid
ácido guanílico guanylic acid
ácido hemelítico hemellitic acid
ácido heneicosanoico heneicosanoic acid
ácido heptacosanoico heptacosanoic acid
ácido heptadecanoico heptadecanoic acid
ácido heptafluorobutírico
    heptafluorobutyric acid
ácido heptanodicarboxílico
    heptanedicarboxylic acid
ácido heptanodioico heptanedioic acid
ácido heptanoico heptanoic acid
ácido heptílico heptylic acid
ácido heptoico heptoic acid
ácido hexacosanoico hexacosanoic acid
ácido hexadecanoico hexadecanoic acid
ácido hexadecenoico hexadecenoic acid
ácido hexadienoico hexadienoic acid
ácido hexafluorofosfórico
    hexafluorophosphoric acid
ácido hexafluorosilícico hexafluorosilicic
    acid
ácido hexahídrico hexahydric acid
ácido hexahidrobenzoico
    hexahydrobenzoic acid
ácido hexahidroftálico
    hexahydrophthalic acid
ácido hexanodioico hexanedioic acid
ácido hexanoico hexanoic acid
ácido hexenoico hexenoic acid
ácido hexilacético hexylacetic acid
ácido hexílico hexylic acid
ácido hexoico hexoic acid
ácido hialurónico hyaluronic acid

ácido hidantoico hydantoic acid
ácido hidnocárpico hydnocarpic acid
ácido hidracrílico hydracrylic acid
ácido hidratrópico hydratropic acid
ácido hidrazoico hydrazoic acid
ácido hidrocinámico hydrocinnamic acid
ácido hidrofluorosilícico
    hydrofluorosilicic acid
ácido hidrofluosilícico hydrofluosilicic
    acid
ácido hidronítrico hydronitric acid
ácido hidrosulfúrico hydrosulfuric acid
ácido hidrosulfuroso hydrosulfurous
    acid
ácido hidroxámico hydroxamic acid
ácido hidroxiacético hydroxyacetic acid
ácido hidroxiazobencenosulfónico
    hydroxyazobenzenesulfonic acid
ácido hidroxibenzoico hydroxybenzoic
    acid
ácido hidroxibutírico hydroxybutyric
    acid
ácido hidroxiesterárico hydroxystearic
    acid
ácido hidroxietanosulfónico
    hydroxyethanesulfonic acid
ácido hidroxímico hydroximic acid
ácido hidroxinaftoico hydroxynaphthoic
    acid
ácido hidroxioleico hydroxyoleic acid
ácido hidroxipropiónico
    hydroxypropionic acid
ácido hidroxisalicílico hydroxysalicylic
    acid
ácido hidroxisuccinico hydroxysuccinic
    acid
ácido hidroxitoluico hydroxytoluic acid
ácido hígrico hygric acid
ácido hipobromoso hypobromous acid
ácido hipocloroso hypochlorous acid
ácido hipofosfórico hypophosphoric acid
ácido hipofosforoso hypophosphorous
    acid
ácido hiponitroso hyponitrous acid
ácido hiposulfúrico hyposulfuric acid
ácido hiposulfuroso hyposulfurous acid
ácido hipúrico hippuric acid
ácido homoftálico homophthalic acid
ácido homogentísico homogentisic acid

ácido homosalicílico homosalicylic acid
ácido húmico humic acid
ácido icosanoico icosanoic acid
ácido imídico imidic acid
ácido iminodiacético iminodiacetic acid
ácido indolacético indoleacetic acid
ácido indolbutírico indolebutyric acid
ácido inorgánico inorganic acid
ácido inosínico inosinic acid
ácido isetiónico isethionic acid
ácido isoascórbico isoascorbic acid
ácido isobutírico isobutyric acid
ácido isociánico isocyanic acid
ácido isocianúrico isocyanuric acid
ácido isocianúrico clorado chlorinated
    isocyanuric acid
ácido isocincomerónico
    isocinchomeronic acid
ácido isocrotónico isocrotonic acid
ácido isodecanoico isodecanoic acid
ácido isoftálico isophthalic acid
ácido isonicotínico isonicotinic acid
ácido isopentanoico isopentanoic acid
ácido isotiociánico isothiocyanic acid
ácido isovalérico isovaleric acid
ácido itacónico itaconic acid
ácido junipérico juniperic acid
ácido lactámico lactamic acid
ácido láctico lactic acid
ácido lactónico lactonic acid
ácido lanocérico lanoceric acid
ácido láurico lauric acid
ácido lepargílico lepargylic acid
ácido leucónico leuconic acid
ácido levulínico levulinic acid
ácido licánico licanic acid
ácido lignocérico lignoceric acid
ácido linoleico linoleic acid
ácido linolénico linolenic acid
ácido linólico linolic acid
ácido lisérgico lysergic acid
ácido lítico lithic acid
ácido litocólico lithocholic acid
ácido maleámico maleamic acid
ácido maleico maleic acid
ácido málico malic acid
ácido malonaldehídico malonaldehydic
    acid
ácido malónico malonic acid

ácido mandélico mandelic acid
ácido mangánico manganic acid
ácido margárico margaric acid
ácido melísico melissic acid
ácido melítico mellitic acid
ácido mercaptoacético mercaptoacetic
    acid
ácido mercaptobenzoico
    mercaptobenzoic acid
ácido mercaptopropiónico
    mercaptopropionic acid
ácido mercaptosuccínico
    mercaptosuccinic acid
ácido mercaptúrico mercapturic acid
ácido mesacónico mesaconic acid
ácido mesoxálico mesoxalic acid
ácido metabórico metaboric acid
ácido metacrílico methacrylic acid
ácido metafosfórico metaphosphoric acid
ácido metalínico metanilic acid
ácido metanocarboxílico
    methanecarboxylic acid
ácido metanodicarboxílico
    methanedicarboxylic acid
ácido metanoico methanoic acid
ácido metanosulfónico methanesulfonic
    acid
ácido metasilícico metasilicic acid
ácido metatitánico metatitanic acid
ácido metilacético methylacetic acid
ácido metilaminoacético
    methylaminoacetic acid
ácido metilbenzoico methylbenzoic acid
ácido metilenodisalicílico
    methylenedisalicylic acid
ácido metilfosfónico methylphosphonic
    acid
ácido metilfosfórico methylphosphoric
    acid
ácido metilmaleico methylmaleic acid
ácido metilpentanoico methylpentanoic
    acid
ácido metilsulfónico methylsulfonic acid
ácido metilsulfúrico methylsulfuric acid
ácido metiónico methionic acid
ácido metoxiacético methoxyacetic acid
ácido metoxibenzoico methoxybenzoic
    acid
ácido mevalónico mevalonic acid

ácido mineral mineral acid
ácido mirístico myristic acid
ácido mixto mixed acid
ácido molíbdico molybdic acid
ácido molibdosilícico molybdosilicic acid
ácido monoprótico monoprotic acid
ácido múcico mucic acid
ácido muriático muriatic acid
ácido naftalenacético naphthaleneacetic
  acid
ácido naftalencarboxílico
  naphthalenecarboxylic acid
ácido naftalendicarboxílico
  naphthalenedicarboxylic acid
ácido naftalendisulfónico
  naphthalenedisulfonic acid
ácido naftalensulfínico
  naphthalenesulfinic acid
ácido naftalensulfónico
  naphthalenesulfonic acid
ácido naftálico naphthalic acid
ácido nafténico naphthenic acid
ácido naftilacético naphthylacetic acid
ácido naftilaminosulfónico
  naphthylaminesulfonic acid
ácido naftiónico naphthionic acid
ácido naftoico naphthoic acid
ácido naftolsulfónico naphtholsulfonic
  acid
ácido naftoxiacético naphthoxyacetic
  acid
ácido nalidíxico nalidixic acid
ácido neodecanoico neodecanoic acid
ácido neopentanoico neopentanoic acid
ácido neotridecanoico neotridecanoic
  acid
ácido nervónico nervonic acid
ácido nicotínico nicotinic acid
ácido nióbico niobic acid
ácido nisínico nisinic acid
ácido nitranílico nitranilic acid
ácido nitrante nitrating acid
ácido nítrico nitric acid
ácido nítrico fumante fuming nitric acid
ácido nitrilotriacético nitrilotriacetic
  acid
ácido nitrobarbitúrico nitrobarbituric
  acid
ácido nitrobencenosulfónico

nitrobenzenesulfonic acid
ácido nitrobenzoico nitrobenzoic acid
ácido nitrocinámico nitrocinnamic acid
ácido nitroclorhídrico nitrohydrochloric
  acid
ácido nitrofenilacético nitrophenylacetic
  acid
ácido nitrofenilpropiólico
  nitrophenylpropiolic acid
ácido nitroftálico nitrophthalic acid
ácido nitrónico nitronic acid
ácido nitrosalicílico nitrosalicylic acid
ácido nitrosilsulfúrico nitrosylsulfuric
  acid
ácido nitroso nitrous acid
ácido nitrosobenzoico nitrosobenzoic
  acid
ácido nitroxántico nitroxanthic acid
ácido nitroxílico nitroxylic acid
ácido nonadecanoico nonadecanoic acid
ácido nonanodioico nonanedioic acid
ácido nonanoico nonanoic acid
ácido nonilfenoxiacético
  nonylphenoxyacetic acid
ácido nonílico nonylic acid
ácido nonoico nonoic acid
ácido nucleico nucleic acid
ácido octacosanoico octacosanoic acid
ácido octadecanoico octadecanoic acid
ácido octadecatrienoico octadecatrienoic
  acid
ácido octadecenoico octadecenoic acid
ácido octanodicarboxílico
  octanedicarboxylic acid
ácido octanodioico octanedioic acid
ácido octanoico octanoic acid
ácido octilacético octylacetic acid
ácido octoico octoic acid
ácido oleico oleic acid
ácido orgánico organic acid
ácido ornitúrico ornithuric acid
ácido orótico orotic acid
ácido orto ortho acid
ácido ortoacético orthoacetic acid
ácido ortoarsénico orthoarsenic acid
ácido ortobórico orthoboric acid
ácido ortocarbónico orthocarbonic acid
ácido ortofosfórico orthophosphoric acid
ácido ortosilícico orthosilicic acid

ácido ortotúngstico orthotungstic acid
ácido ósmico osmic acid
ácido oxalacético oxalacetic acid
ácido oxaldehídico oxaldehydic acid
ácido oxálico oxalic acid
ácido oxámico oxamic acid
ácido oxibezoico oxybenzoic acid
ácido oxidante oxidizing acid
ácido oxinaftoico oxynaphthoic acid
ácido oxoetanoico oxoethanoic acid
ácido palmítico palmitic acid
ácido palmitínico palmitinic acid
ácido palmitoleico palmitoleic acid
ácido palmitólico palmitolic acid
ácido palústrico palustric acid
ácido pamoico pamoic acid
ácido pánico pannic acid
ácido pantoténico pantothenic acid
ácido parabánico parabanic acid
ácido paraciánico paracyanic acid
ácido paracónico paraconic acid
ácido paraláctido paralactic acid
ácido parapéctico parapectic acid
ácido pararosólico pararosolic acid
ácido parasórbico parasorbic acid
ácido pariético parietic acid
ácido parinárico parinaric acid
ácido péctico pectic acid
ácido pelargónico pelargonic acid
ácido penicilánico penicillanic acid
ácido penicílico penicillic acid
ácido pentacosámico pentacosamic acid
ácido pentacosanoico pentacosanoic acid
ácido pentadecanoico pentadecanoic acid
ácido pentadecílico pentadecylic acid
ácido pentadienodioico pentadienedioic
    acid
ácido pentametilbenzoico
    pentamethylbenzoic acid
ácido pentanocarboxílico
    pentanecarboxylic acid
ácido pentanodicarboxílico
    pentanedicarboxylic acid
ácido pentanodioico pentanedioic acid
ácido pentanoico pentanoic acid
ácido penténico pentenic acid
ácido pentenodioico pentenedioic acid
ácido pentenoico pentenoic acid
ácido pentilfórmico pentylformic acid

ácido pentínico pentinic acid
ácido pentoico pentoic acid
ácido pentónico pentonic acid
ácido peracético peracetic acid
ácido perbenzoico perbenzoic acid
ácido percarbónico percarbonic acid
ácido perclórico perchloric acid
ácido percrómico perchromic acid
ácido perfluorobutírico perfluorobutyric
    acid
ácido perfluorosulfónico
    perfluorosulfonic acid
ácido perfórmico performic acid
ácido perfosfórico perphosphoric acid
ácido peri peri acid
ácido permangánico permanganic acid
ácido permonosulfúrico permonosulfuric
    acid
ácido pernítrico pernitric acid
ácido peroxiacético peroxyacetic acid
ácido peroxibenzoico peroxybenzoic acid
ácido peroxietanoico peroxyethanoic
    acid
ácido peroxifórmico peroxyformic acid
ácido peroximonosulfúrico
    peroxymonosulfuric acid
ácido peroxisulfúrico peroxysulfuric
    acid
ácido peroxomonosulfúrico
    peroxomonosulfuric acid
ácido peroxonítrico peroxonitric acid
ácido perrénico perrhenic acid
ácido persulfúrico persulfuric acid
ácido peryódico periodic acid
ácido picolínico picolinic acid
ácido picrámico picramic acid
ácido picramínico picraminic acid
ácido pícrico picric acid
ácido picrolónico picrolonic acid
ácido picronítrico picronitric acid
ácido pimárico pimaric acid
ácido pimélico pimelic acid
ácido pínico pinic acid
ácido piperonílico piperonylic acid
ácido piridinocarboxílico
    pyridinecarboxylic acid
ácido piridinodicarboxílico
    pyridinedicarboxylic acid
ácido piridinopentacarboxílico

pyridinepentacarboxylic acid

ácido piridinosulfónico pyridinesulfonic acid

ácido piroacémico pyroacemic acid

ácido pirobórico pyroboric acid

ácido pirofosfórico pyrophosphoric acid

ácido pirolígneo pyroligneous acid

ácido piromelítico pyromellitic acid

ácido piromúcico pyromucic acid

ácido pirosulfúrico pyrosulfuric acid

ácido pirosulfuroso pyrosulfurous acid

ácido pirotartárico pyrotartaric acid

ácido pirovanádico pyrovanadic acid

ácido pirúvico pyruvic acid

ácido piválico pivalic acid

ácido plúmbico plumbic acid

ácido poliacrílico polyacrylic acid

ácido polibásico polybasic acid

ácido policarboxílico polycarboxylic acid

ácido polifosfórico polyphosphoric acid

ácido poliinsaturado polyunsaturated acid

ácido politiónico polythionic acid

ácido prefénico prephenic acid

ácido propanocarboxílico propanecarboxylic acid

ácido propanoico propanoic acid

ácido propenoico propenoic acid

ácido propenotricarboxílico propenetricarboxylic acid

ácido propilfórmico propylformic acid

ácido propilmalónico propylmalonic acid

ácido propiólico propiolic acid

ácido propiónico propionic acid

ácido protónico protonic acid

ácido prúsico prussic acid

ácido quebulínico chebulinic acid

ácido quelidónico chelidonic acid

ácido quercínico quercinic acid

ácido quináldico quinaldic acid

ácido quínico quinic acid

ácido quinínico quininic acid

ácido quinolínico quinolinic acid

ácido racémico racemic acid

ácido rénico rhenic acid

ácido resorcílico resorcylic acid

ácido rho rho acid

ácido ribónico ribonic acid

ácido ribonucleico ribonucleic acid

ácido ribonucleico de transferencia transfer ribonucleic acid

ácido ribosfosfórico ribosephosphoric acid

ácido ricínico ricinic acid

ácido ricinoleico ricinoleic acid

ácido rísico rissic acid

ácido rizónico rhizonic acid

ácido rodánico rhodanic acid

ácido rodizónico rhodizonic acid

ácido rosólico rosolic acid

ácido ruberítrico ruberythric acid

ácido rubitánico rubitannic acid

ácido sabínico sabinic acid

ácido sacárico saccharic acid

ácido sacarínico saccharinic acid

ácido sacarónico saccharonic acid

ácido salicílico salicylic acid

ácido sarcoláctico sarcolactic acid

ácido sebácico sebacic acid

ácido sebacílico sebacylic acid

ácido seco dry acid

ácido sedanólico sedanolic acid

ácido selénico selenic acid

ácido selenínico seleninic acid

ácido selenioso selenious acid

ácido selenónico selenonic acid

ácido siálico sialic acid

ácido silícico silicic acid

ácido silicobenzoico silicobenzoic acid

ácido silicotúngstico silicotungstic acid

ácido sílvico sylvic acid

ácido sórbico sorbic acid

ácido subérico suberic acid

ácido succinámico succinamic acid

ácido succínico succinic acid

ácido sulfámico sulfamic acid

ácido sulfamídico sulfamidic acid

ácido sulfamínico sulfaminic acid

ácido sulfanílico sulfanilic acid

ácido sulfoacético sulfoacetic acid

ácido sulfobenzoico sulfobenzoic acid

ácido sulfocarbólico sulfocarbolic acid

ácido sulfocarbónico sulfocarbonic acid

ácido sulfociánico sulfocyanic acid

ácido sulfoleico sulfoleic acid

ácido sulfónico sulfonic acid

ácido sulfonohidrazónico sulfonohydrazonic acid

ácido sulfosalicílico sulfosalicylic acid
ácido sulfoxílico sulfoxylic acid
ácido sulfúrico sulfuric acid
ácido sulfúrico fumante fuming sulfuric acid
ácido sulfuroso sulfurous acid
ácido superfosfórico superphosphoric acid
ácido tánico tannic acid
ácido tantálico tantalic acid
ácido tartárico tartaric acid
ácido tartrónico tartronic acid
ácido taurocólico taurocholic acid
ácido teicoico teichoic acid
ácido telúrico telluric acid
ácido telurínico tellurinic acid
ácido teluroso tellurous acid
ácido tenoico thenoic acid
ácido tereftálico terephthalic acid
ácido tereftaloilbenzoico terephthaloylbenzoic acid
ácido tetracarboxílico tetracarboxylic acid
ácido tetracloroftálico tetrachlorophthalic acid
ácido tetracosanoico tetracosanoic acid
ácido tetradecanoico tetradecanoic acid
ácido tetrafosfórico tetraphosphoric acid
ácido tetrahidroxibenzoico tetrahydroxybenzoic acid
ácido tetraoxofosfórico tetraoxophosphoric acid
ácido tetraoxosulfúrico tetraoxosulfuric acid
ácido tetratiónico tetrathionic acid
ácido tetrólico tetrolic acid
ácido tíglico tiglic acid
ácido tímico thymic acid
ácido timidílico thymidylic acid
ácido tioacético thioacetic acid
ácido tiobarbitúrico thiobarbituric acid
ácido tiobenzoico thiobenzoic acid
ácido tiocarbónico thiocarbonic acid
ácido tiocarboxílico thiocarboxylic acid
ácido tiociánico thiocyanic acid
ácido tiodiglicólico thiodiglycolic acid
ácido tiodipropiónico thiodipropionic acid
ácido tiofosfórico thiophosphoric acid

ácido tioglicólico thioglycolic acid
ácido tioico thioic acid
ácido tioláctico thiolactic acid
ácido tiomálico thiomalic acid
ácido tiónico thionic acid
ácido tiosalicílico thiosalicylic acid
ácido tiosulfónico thiosulfonic acid
ácido tiosulfúrico thiosulfuric acid
ácido titánico titanic acid
ácido tolilacético tolylacetic acid
ácido toluensulfónico toluenesulfonic acid
ácido toluico toluic acid
ácido toluílico toluylic acid
ácido tranexámico tranexamic acid
ácido traumático traumatic acid
ácido triacontanoico triacontanoic acid
ácido tribásico tribasic acid
ácido tribromoacético tribromoacetic acid
ácido triciánico tricyanic acid
ácido tricloroacético trichloroacetic acid
ácido tricloroetanoico trichloroethanoic acid
ácido tricloroisocianúrico trichloroisocyanuric acid
ácido triclorometilfosfónico trichloromethylphosphonic acid
ácido tricosanoico tricosanoic acid
ácido tridecanoico tridecanoic acid
ácido tridecílico tridecylic acid
ácido trifenilacético triphenylacetic acid
ácido trifluoroacético trifluoroacetic acid
ácido trifluoroetanoico trifluoroethanoic acid
ácido trifosfórico triphosphoric acid
ácido trihidroxibenzoico trihydroxybenzoic acid
ácido trimelítico trimellitic acid
ácido trimetilacético trimethylacetic acid
ácido trimetiladípico trimethyladipic acid
ácido trimetilbenzoico trimethylbenzoic acid
ácido trinitrobenzoico trinitrobenzoic acid
ácido trioxobórico trioxoboric acid
ácido trioxosulfúrico trioxosulfuric acid

ácido tritiónico trithionic acid
ácido trópico tropic acid
ácido túngstico tungstic acid
ácido umbélico umbellic acid
ácido undecanoico undecanoic acid
ácido undecenoico undecenoic acid
ácido undecilénico undecylenic acid
ácido undecílico undecylic acid
ácido uránico uranic acid
ácido úrico uric acid
ácido uridílico uridylic acid
ácido urobenzoico urobenzoic acid
ácido urónico uronic acid
ácido úsnico usnic acid
ácido vaccénico vaccenic acid
ácido vainíllico vanillic acid
ácido valeriánico valerianic acid
ácido valérico valeric acid
ácido vanádico vanadic acid
ácido vanadioso vanadous acid
ácido verátrico veratric acid
ácido vínico vinic acid
ácido violúrico violuric acid
ácido wolfrámico wolframic acid
ácido xántico xanthic acid
ácido xantogénico xanthogenic acid
ácido xilenosulfónico xylenesulfonic acid
ácido xilídico xylidic acid
ácido xilónico xylonic acid
ácido yodhídrico hydriodic acid
ácido yódico iodic acid
ácido yodoacético iodoacetic acid
ácido yodogorgoico iodogorgoic acid
ácido zircónico zirconic acid
acidólisis acidolysis
acidulante acidulant
acilación acylation
acilo acyl
aciloína acyloin
acondicionador de suelo soil conditioner
aconitina aconitine
acónito aconite
acoplamiento coupling
acoplamiento de Glaser Glaser coupling
acoplamiento espín-espín spin-spin
    coupling
acoplamiento espín-orbital spin-orbit
    coupling
acoplamiento geminal geminal coupling

acoplamiento oxidativo oxidative
    coupling
acre pungent, acrid
acridina acridine
acriflavina acriflavine
acrilaldehído acrylaldehyde
acrilamida acrylamide
acrilato acrylate
acrilato cálcico calcium acrylate
acrilato de alilo allyl acrylate
acrilato de butilo butyl acrylate
acrilato de cianoetilo cyanoethyl acryate
acrilato de etilhexilo ethylhexyl acrylate
acrilato de etilo ethyl acrylate
acrilato de hidroxipropilo
    hydroxypropyl acrylate
acrilato de isobutilo isobutyl acrylate
acrilato de metilo methyl acrylate
acrílico acrylic
acrilonitrilo acrylonitrile
acrimonioso acrimonious
acroleína acrolein
acromático achromatic
acrómetro acrometer
actínico actinic
actínido actinide
actinio actinium
actinismo actinism
actinoide actinoid
actinometría actinometry
actinómetro actinometer
actinomicina actinomycin
actinón actinon
actinoquímica actinochemistry
actinquinol actinoquinol
activación activation
activación fotoquímica photochemical
    activation
activación térmica thermal activation
activado activated
activador activator
activador de flotación flotation activator
actividad activity
actividad beta beta activity
actividad específica specific activity
actividad iónica ion activity
actividad magnética magnetic activity
actividad óptica optical activity
actividad química chemical activity

actividad superficial surface activity
activo active
actuador actuator
acuación aquation
acumulador accumulator
acumulador de plomo lead accumulator
acuoso aqueous
adamantano adamantane
adamsita adamsite
adaptador adapter
adátomo adatom
adenina adenine
adenosina adenosine
adherencia adherence
adhesión adhesion
adhesión molecular molecular adhesion
adhesivo adhesive
adiabático adiabatic
adición addition
adición anormal de Kharasch Kharasch abnormal addition
adición electrofílica electrophilic addition
adición nucleofílica nucleophilic addition
adición oxidativa oxidative addition
adipato adipate
adipato de dialilo diallyl adipate
adipato de didecilo didecyl adipate
adipato de diisooctilo diisooctyl adipate
adipato de isooctilo isooctyl adipate
adipincetona adipinketone
adiponitrilo adiponitrile
aditamento de mechero burner attachment
aditivo additive
aditivo antigolpeteo antiknock additive
aditivo de alimento food additive
adrenalina adrenaline
adsorbato adsorbate
adsorbente adsorbent
adsorción adsorption
adsorción activada activated adsorption
adsorción apolar apolar adsorption
adsorción cromatográfica chromatographic adsorption
adsorción de gas gas adsorption
adsorción de van der Waals van der Waals adsorption
adsorción diferencial differential adsorption
adsorción homopolar homopolar adsorption
adsorción negativa negative adsorption
aducto adduct
aducto de urea urea adduct
adulteración adulteration
adulterado adulterated
adulterante adulterant
advección advection
advertencia de peligro hazard warning
adyuvante adjuvant
aeróbico aerobic
aerobio aerobic
aerogel aerogel
aerometría aerometry
aerómetro aerometer
aerosol aerosol
afecto de Auger Auger effect
afinidad affinity
afinidad electrónica electron affinity
afinidad química chemical affinity
aflatoxina aflatoxin
afrodina aphrodine
agar agar
agaricina agaricin
ágata agate
agente agent
agente acetilante acetylating agent
agente anibloqueo antiblock agent
agente antiespumante antifoam agent
agente antiestático antistatic agent
agente antigolpeteo antiknock agent
agente citotóxico cytotoxic agent
agente complejante complexing agent
agente de abrillantamiento brightening agent
agente de adhesión adhesion agent
agente de adición addition agent
agente de curado curing agent
agente de espesamiento thickening agent
agente de flotación flotation agent
agente de suspensión suspending agent
agente decolorante decolorizing agent
agente desespumante defoaming agent
agente disolvente dissolving agent
agente dispersante dispersing agent
agente emulsionante emulsifying agent
agente fijador fixing agent

agente floculante flocculating agent
agente halogenante halogenating agent
agente humectante wetting agent
agente impermeabilizante waterproofing agent
agente lipotrópico lipotropic agent
agente mutagénico mutagenic agent
agente oxidante oxidizing agent
agente quelante chelating agent
agente químico chemical agent
agente reductor reducing agent
agente retardante de teñido dye-retarding agent
agente secante drying agent
agente secuestrante sequestering agent
agente tensoactivo surface-active agent
agitador agitator, stirrer
agitador de vidrio glass stirrer
agitador magnético magnetic stirrer
aglicona aglycone
aglomeración agglomeration
aglucona aglucone
aglutinación agglutination, sinter, sintering
aglutinar agglutinate, sinterize
agmatina agmatine
agregación aggregation
agregación limitada por difusión diffusion limited aggregation
agregado aggregate
agrio sour
agrupamiento electrónico electron bunching
agua water
agua aireada aerated water
agua amoniacal ammonia water
agua blanda soft water
agua de bromo bromine water
agua de cal lime water
agua de cloro chlorine water
agua de constitución water of constitution
agua de cristalización water of crystallization
agua de hidratación water of hydration
agua destilada distilled water
agua dulce sweet water
agua dura hard water
agua ligera light water

agua pesada heavy water
agua purificada purified water
agua regia aqua regia
agua salina saline water
agua salobre brackish water
aire air
aire ambiente ambient air
aire líquido liquid air
airear aerate
aislado isolated
aislador insulator
aislamiento isolation
aislamiento térmico thermal insulation
aislar isolate
ajenjo wormwood, absinthium
alabandina alabandite
alabastro alabaster
alacena protectora hood
alacena protectora contra humo fume cupboard, fume hood
alacreatina alacreatine
alambique still
alanilo alanyl
alanina alanine
alanita allanite
alantoína allantoin
alarma de fuego fire alarm
alarma de llama flame alarm
alazopeptina alazopeptin
albaspidina albaspidin
albata albata
albedo albedo
albertita albertite
albita albite
albomicina albomycin
albumen albumen
albúmina albumin
albuminato albumitate
albuminato de cinc zinc albuminate
albuminoide albuminoid
albutoína albutoin
alcadieno alkadiene
alcalamida alkalamide
alcalescencia alkalescence
alcalescente alkalescent
álcali alkali
álcali mineral mineral alkali
alcalida alkalide
alcalimetría alkalimetry

alcalímetro alkalimeter
alcalinidad alkalinity
alcalino alkaline
alcalizar alkalize
alcaloidal alkaloidal
alcaloide alkaloid
alcalometría alkalometry
alcamina alkamine
alcanal alkanal
alcanfor camphor
alcanfor brominado brominated
   camphor
alcanfor de alquitrán tar camphor
alcanfor de menta peppermint camphor
alcanfor de trementina turpentine
   camphor
alcanina alkannin
alcanización alkanization
alcano alkane
alcanol alkanol
alcanolamina alkanolamine
alcasal alkasal
alcilo alcyl
alclofenac alclophenac
alcogel alcogel
alcohol alcohol
alcohol absoluto absolute alcohol
alcohol ácido acid alcohol
alcohol alílico allyl alcohol
alcohol amilcinámico amylcinnamic
   alcohol
alcohol amílico amyl alcohol
alcohol aminoetílico aminoethyl alcohol
alcohol aminopropílico aminopropyl
   alcohol
alcohol anhidro anhydrous alcohol
alcohol anísico anisic alcohol
alcohol anisílico anisyl alcohol
alcohol aromático aromatic alcohol
alcohol batílico batyl alcohol
alcohol behenílico behenyl alcohol
alcohol bencílico benzyl alcohol
alcohol benzoico benzoic alcohol
alcohol bornílico bornyl alcohol
alcohol bromoetílico bromoethyl alcohol
alcohol butílico butyl alcohol
alcohol butírico butyric alcohol
alcohol cáustico caustic alcohol
alcohol cerílico ceryl alcohol

alcohol cetílico cetyl alcohol
alcohol cinámico cinnamic alcohol
alcohol cinamílico cinnamyl alcohol
alcohol cloroetílico chloroethyl alcohol
alcohol cloroisopropílico
   chloroisopropyl alcohol
alcohol coniferilo coniferyl alcohol
alcohol crotílico crotyl alcohol
alcohol cúmico cumic alcohol
alcohol cumínico cuminic alcohol
alcohol de grano grain alcohol
alcohol de madera wood alcohol
alcohol decatílico decatyl alcohol
alcohol decílico decyl alcohol
alcohol desnaturalizado denatured
   alcohol
alcohol dicloroisopropílico
   dichloroisopropyl alcohol
alcohol dicloropropílico dichloropropyl
   alcohol
alcohol dihídrico dihydric alcohol
alcohol dodecílico dodecyl alcohol
alcohol erucílico erucyl alcohol
alcohol estearílico stearyl alcohol
alcohol estiralílico styralyl alcohol
alcohol etanílico ethanyl alcohol
alcohol etilbencílico ethylbenzyl alcohol
alcohol etilbutílico ethylbutyl alcohol
alcohol etilhexílico ethylhexyl alcohol
alcohol etílico ethyl alcohol
alcohol fenacílico phenacyl alcohol
alcohol fenetílico phenethyl alcohol
alcohol feniletílico phenylethyl alcohol
alcohol fenilpropílico phenylpropyl
   alcohol
alcohol fenquílico fenchyl alcohol
alcohol furfurílico furfuryl alcohol
alcohol glicílico glycyl alcohol
alcohol graso fatty alcohol
alcohol heptílico heptyl alcohol
alcohol hexacosílico hexacosyl alcohol
alcohol hexadecílico hexadecyl alcohol
alcohol hexahídrico hexahydric alcohol
alcohol hexílico hexyl alcohol
alcohol hidroabietílico hydroabietyl
   alcohol
alcohol hidrocinámico hydrocinnamic
   alcohol
alcohol hidroxibencílico hydroxybenzyl

alcohol
alcohol hidroxiestearílico hydroxystearyl alcohol
alcohol industrial industrial alcohol
alcohol isoamílico isoamyl alcohol
alcohol isobutílico isobutyl alcohol
alcohol isononílico isononyl alcohol
alcohol isooctílico isooctyl alcohol
alcohol isopentílico isopentyl alcohol
alcohol isopropílico isopropyl alcohol
alcohol laurílico lauryl alcohol
alcohol linolenílico linolenyl alcohol
alcohol melisílico melissyl alcohol
alcohol metalílico methallyl alcohol
alcohol metilalílico methylallyl alcohol
alcohol metilamílico methylamyl alcohol
alcohol metilbencílico methylbenzyl alcohol
alcohol metílico methyl alcohol
alcohol metoxibencílico methoxybenzyl alcohol
alcohol miristílico myristyl alcohol
alcohol monohídrico monohydric alcohol
alcohol naftílico naphthyl alcohol
alcohol nitrobencílico nitrobenzyl alcohol
alcohol nonílico nonyl alcohol
alcohol octadecílico octadecyl alcohol
alcohol octílico octyl alcohol
alcohol oleílico oleyl alcohol
alcohol palmitílico palmityl alcohol
alcohol pelargónico pelargonic alcohol
alcohol pentadecílico pentadecyl alcohol
alcohol pentílico pentyl alcohol
alcohol pirúvico pyruvic alcohol
alcohol polifurfurílico polyfurfuryl alcohol
alcohol polihídrico polyhydric alcohol
alcohol polivinílico polyvinyl alcohol
alcohol primario primary alcohol
alcohol propargílico propargyl alcohol
alcohol propenílico propenyl alcohol
alcohol propílico propyl alcohol
alcohol ricinoleílico ricinoleyl alcohol
alcohol salicílico salicyl alcohol
alcohol secundario secondary alcohol
alcohol seudohexílico pseudohexyl alcohol
alcohol tanacetílico tanacetyl alcohol

alcohol tenílico thenyl alcohol
alcohol terciario tertiary alcohol
alcohol tetrahidrofurfurílico tetrahydrofurfuryl alcohol
alcohol tioetílico thioethyl alcohol
alcohol tridecílico tridecyl alcohol
alcohol trihídrico trihydric alcohol
alcohol undecilénico undecylenic alcohol
alcohol undecílico undecyl alcohol
alcohol vinílico vinyl alcohol
alcoholato alcoholate
alcoholato alcalino alkali alcoholate
alcoholato de tántalo tantalum alcoholate
alcohólico alcoholic
alcohólisis alcoholysis
alcosol alcosol
alcoxi alkoxy
alcóxido alkoxide
alcóxido alumínico aluminum alkoxide
alcoxilato alkoxylate
aldehidina aldehydine
aldehído aldehyde
aldehído acético acetic aldehyde
aldehído acrílico acrylic aldehyde
aldehído alílico allyl aldehyde
aldehído amilcinámico amylcinnamic aldehyde
aldehído amílico amyl aldehyde
aldehído amónico ammonium aldehyde
aldehído anísico anisic aldehyde
aldehído aromático aromatic aldehyde
aldehído benzoico benzoic aldehyde
aldehído butílico butyl aldehyde
aldehído butírico butyric aldehyde
aldehído cáprico capric aldehyde
aldehído caproico caproic aldehyde
aldehído cinámico cinnamic aldehyde
aldehído cinamílico cinnamyl aldehyde
aldehído crotónico crotonic aldehyde
aldehído cúmico cumic aldehyde
aldehído cumínico cuminic aldehyde
aldehído decílico decyl aldehyde
aldehído fenilpropílico phenylpropyl aldehyde
aldehído fenólico phenol aldehyde
aldehído fórmico formic aldehyde
aldehído glicérico glyceric aldehyde
aldehído glicólico glycolic aldehyde
aldehído hidrocinámico hydrocinnamic

aldehyde

aldehído isobutílico isobutyl aldehyde

aldehído láurico lauric aldehyde

aldehído laurílico lauryl aldehyde

aldehído levulínico levulinic aldehyde

aldehído naftílico naphthyl aldehyde

aldehído naftoico naphthoic aldehyde

aldehído nonílico nonyl aldehyde

aldehído octadecenílico octadecenyl
aldehyde

aldehído octílico octyl aldehyde

aldehído oleílico oleyl aldehyde

aldehído pelargónico pelargonic
aldehyde

aldehído pentílico pentyl aldehyde

aldehído pirúvico pyruvic aldehyde

aldehído propílico propyl aldehyde

aldehído propiónico propionic aldehyde

aldehído salicílico salicylic aldehyde

aldehído toluico toluic aldehyde

aldehído vainíllico vanillic aldehyde

aldehído valérico valeric aldehyde

aldicarb aldicarb

aldocetena aldoketene

aldohexosa aldohexose

aldol aldol

aldolasa aldolase

aldopentosa aldopentose

aldosa aldose

aldosterona aldosterone

aldoxima aldoxime

aldrina aldrin

aleación alloy

aleación alumínica aluminum alloy

aleación binaria binary alloy

aleación cerámica ceramic alloy

aleación de Arndt Arndt alloy

aleación de boro boron alloy

aleación de cobalto cobalt alloy

aleación de cobre copper alloy

aleación de Devarda Devarda alloy

aleación de hierro-níquel iron-nickel
alloy

aleación de mercurio mercury alloy

aleación de níquel-hierro nickel-iron
alloy

aleación de oro gold alloy

aleación de plata silver alloy

aleación de platino-cobalto platinum-

cobalt alloy

aleación de platino-iridio platinum-
iridium alloy

aleación de platino-rodio platinum-
rhodium alloy

aleación de silicio silicon alloy

aleación de silicio-cobre silicon-copper
alloy

aleación de silicio-oro silicon-gold alloy

aleación de sodio-plomo sodium-lead
alloy

aleación de sodio-potasio sodium-
potassium alloy

aleación de titanio titanium alloy

aleación de tungsteno tungsten alloy

aleación eutéctica eutectic alloy

aleación fusible fusible alloy

aleación magnésica magnesium alloy

aleación pesada heavy alloy

alelomorfo allelomorph

aleoquímico allelochemical

alelotropismo allelotropism

alelótropo allelotrope

alemontita allemontite

aleno allene

aletrina allethrin

alexandrita alexandrite

alfa alpha

alfahidroxiácido alphahydroxy acid

alfatópico alphatopic

alfilo alphyl

alfin alfin

algarosa algarose

algas algae

algicida algicide

algina algin

alginato alginate

alginato amónico ammonium alginate

alginato cálcico calcium alginate

alginato potásico potassium alginate

alginato sódico sodium alginate

algodón cotton

algodón polvora guncotton

algulosa algulose

alheña henna

alicíclico alicyclic

alicina allicin

alícuota aliquot

alifático aliphatic

aligación alligation
alilacetona allylacetone
alilacetonitrilo allylacetonitrile
alilamina allylamine
alilanilina allylaniline
alileno allylene
alílico allylic
alilina allylin
alilo allyl
aliltiourea allylthiourea
aliltriclorosilano allyltrichlorosilane
alimentador feeder
alineación alignment
alipita alipite
alita alite
alizaramida alizaramide
alizarato alizarate
alizarina alizarin
almacenamiento storage
almacenamiento seguro safe storage
almandita almandite
almidón starch
almidón de maíz cornstarch
almidón soluble soluble starch
almizcle musk
alneón alneon
álnico alnico
alo allo
aloclasita alloclasite
alocromático allochromatic
áloe aloe
alófana allophane
alofanamida allophanamide
alofanato allophanate
alofánico allophanic
aloína aloin
aloisomerismo alloisomerism
alomérico allomeric
alomerismo allomerism
alomerización allomerization
alomorfismo allomorphism
alonógeno alunogen
alosa allose
alotoxina allotoxin
alotriomorfismo allotriomorphism
alotropía allotropy
alotrópico allotropic
alotropismo allotropism
alótropo allotrope

aloxana alloxan
aloxantina alloxanthin
aloxiprina aloxiprin
alpax alpax
alqueno alkene
alquilación alkylation
alquilación de sulfitos de Strecker
    Strecker sulfite alkylation
alquilamina alkylamine
alquilato alkylate
alquilbenceno alkylbenzene
alquileno alkylene
alquilfenol alkylphenol
alquilideno alkylidene
alquilo alkyl
alquilolamida alkylolamide
alquilolamina alkylolamine
alquiluro alkylide
alquino alkyne
alquitrán tar
alquitrán de hulla coal tar
alquitrán de madera wood tar
alquitrán de pino pine tar
alquitrán de turba peat tar
alquitrán mineral mineral tar
alstonidina alstonidine
alstonina alstonine
alta definición high definition
alta densidad high density
alta energía high energy
alta presión high pressure
alta temperatura high temperature
alteína altheine
alternación alternation
alto horno blast furnace
alto polímero high polymer
alto vacío high vacuum
altrosa altrose
alucinógeno hallucinogen
aludel aludel
alulosa allulose
alumbre alum
alumbre amónico ammonium alum
alumbre amónico férrico ferric
    ammonium alum
alumbre de cesio cesium alum
alumbre de cromo chrome alum
alumbre de filtro filter alum
alumbre de rubidio rubidium alum

alumbre de sosa soda alum
alumbre de talio thallium alum
alumbre negro black alum
alumbre potásico potassium alum, potash alum
alumbre sódico sodium alum
alúmina alumina
alúmina activada activated alumina
alúmina hidratada hydrated alumina
aluminato aluminate
aluminato bárico barium aluminate
aluminato cálcico calcium aluminate
aluminato cobaltoso cobaltous aluminate
aluminato de cobalto cobalt aluminate
aluminato de litio lithium aluminate
aluminato magnésico magnesium aluminate
aluminato potásico potassium aluminate
aluminato sódico sodium aluminate
aluminífero aluminiferous
aluminio aluminum
aluminita aluminite
aluminizado aluminized
aluminoférrico aluminoferric
aluminosilicato aluminosilicate
aluminosilicato potásico potassium aluminosilicate
aluminosilicato sódico sodium aluminosilicate
alumita alumite
alunita alunite
alunogenita alunogenite
alvita alvite
amalgama amalgam
amalgama de cadmio cadmium amalgam
amalgama de cinc zinc amalgam
amalgama de cobre copper amalgam
amalgama de mercurio mercury amalgam
amalgama de plata silver amalgam
amalgama de talio thallium amalgam
amalgama dental dental amalgam
amalgama sódica sodium amalgam
amalgamación amalgamation
amalgamador amalgamator
amanitina amanitine
amaranto amaranth
amargo bitter
amarillo brillante brilliant yellow

amarillo congo congo yellow
amarillo de alizarina alizarin yellow
amarillo de anilina aniline yellow
amarillo de antimonio antimony yellow
amarillo de bismuto bismuth yellow
amarillo de cadmio cadmium yellow
amarillo de cinc zinc yellow
amarillo de cobalto cobalt yellow
amarillo de cromo chrome yellow
amarillo de mantequilla butter yellow
amarillo de metilo methyl yellow
amarillo de naftol naphthol yellow
amarillo de óxido de hierro iron oxide yellow
amarillo de titanio titanellow
amarillo de uranina uranine yellow
amarina amarine
amaroide amaroid
amatista amethyst
amatol amatol
amazonita amazonite
ámbar amber
ambidente ambident
ambiente de laboratorio laboratory environment
ambiente peligroso hazardous environment
ambligonita amblygonite
ambrita ambrite
ameboide ameboid
amelina ammeline
americio americium
amesita amesite
ametocaína amethocaine
amicrón amicron
amíctico amyctic
amida amide
amida cíclica cyclic amide
amida de ácido acid amide
amida de ácido acético acetic acid amide
amida de ácido nicotínico nicotinic acid amide
amida de cloral chloral amide
amida de litio lithium amide
amida magnésica magnesium amide
amida málica malic amide
amida malónica malonic amide
amida salicílica salicylic amide
amida sódica sodium amide

amida xántica xanthic amide
amidación amidation
amidasa amidase
amidina amidine
amidino amidino
amido amido
amidol amidol
amidona amidone
amidrazona amidrazone
amigdalina amygdalin
amilamina amylamine
amilasa amylase
amilato amylate
amilbenceno amylbenzene
amilcarbinol amylcarbinol
amileno amylene
amilfenol amylphenol
amilo amyl
amiloide amyloid
amilólisis amylolysis
amilolítico amylolytic
amilopectina amylopectin
amilosa amylose
amiltriclorosilano amyltrichlorosilane
amina amine
amina aromática aromatic amine
amina de ácido acético acetic acid amine
amina de paladio palladium amine
amina de platino platinum amine
amina grasa fatty amine
amina primaria primary amine
amina secundaria secondary amine
amina terciaria tertiary amine
aminación amination
aminación de Voight Voight amination
aminilo aminyl
aminimida aminimide
amino amino
aminoacetanilida aminoacetanilide
aminoacetofenona aminoacetophenone
aminoacetona aminoacetone
aminoácido amino acid
aminoácido oxidasa amino acid oxidase
aminoalcohol amino alcohol
aminoantipirina aminoantipyrine
aminoantraquinona aminoanthraquinone
aminoazo aminoazo
aminoazobenceno aminoazobenzene
aminoazonaftaleno aminoazonaphthalene

aminoazotolueno aminoazotoluene
aminobenceno aminobenzene
aminobenzoato potásico potassium aminobenzoate
aminobenzotiazol aminobenzothiazole
aminobifenilo aminobiphenyl
aminobutano aminobutane
aminocetona aminoketone
aminociclohexano aminocyclohexane
aminodietilanilina aminodiethylaniline
aminodifenilamina aminodiphenylamine
aminodifenilo aminodiphenyl
aminodimetilanilina aminodimethylaniline
aminodimetilbenceno aminodimethylbenzene
aminoetano aminoethane
aminoetanol aminoethanol
aminofenol aminophenol
aminofilina aminophylline
aminoformo aminoform
aminohexano aminohexane
aminohipurato sódico sodium aminohippurate
aminoide aminoid
aminometano aminomethane
aminometilación aminomethylation
aminonaftaleno aminonaphthalene
aminonaftol aminonaphthol
aminooxidasa amino oxidase
aminopicolina aminopicoline
aminopiridina aminopyridine
aminopirina aminopyrine
aminopropano aminopropane
aminopropanol aminopropanol
aminopropeno aminopropene
aminopropilmorfolina aminopropylmorpholine
aminopterina aminopterin
aminopurina aminopurine
aminoquinolina aminoquinoline
aminorresina amino resin
aminosalicilato sódico sodium aminosalicylate
aminosulfonato amónico ammonium aminosulfonate
aminotiazol aminothiazole
aminotiopeno aminothiopene
aminotiourea aminothiourea

aminotolueno aminotoluene
aminotriazol aminotriazole
aminourea aminourea
aminoxileno aminoxylene
amirina amyrin
amital sódico sodium amytal
amitiozona amithiozone
amitrol amitrole
amobarbital amobarbital
amonación ammonation
amoniacal ammoniacal
amoníaco ammonia
amoníaco anhidro anhydrous ammonia
amoníaco saturado saturated ammonia
amonificación ammonification
amonio ammonium
amonio férrico ferric ammonium
amonio ferroso ferrous ammonium
amonólisis ammonolysis
amorfismo amorphism
amorfo amorphous
amortiguador buffer
amortiguador de fosfato phosphate
  buffer
amortiguamiento buffering
ampelita ampelite
amperometría amperometry
ampicilina ampicillin
amplitud molecular molecular amplitude
ampolla ampule
anabasina anabasine
anabólico anabolic
anabolismo anabolism
anaerobio anaerobic
anafilaxis anaphylaxis
anaforesis anaphoresis
análisis analysis
análisis absorciométrico absorptiometric
  analysis
análisis areamétrico areametric analysis
análisis colorimétrico colorimetric
  analysis
análisis complejométrico
  complexometric analysis
análisis con rayos X X-ray analysis
análisis con sonda electrónica electron
  probe analysis
análisis conductométrico conductometric
  analysis

análisis conformacional conformational
  analysis
análisis cualitativo qualitative analysis
análisis cuantitativo quantitative analysis
análisis de activación neutrónica
  neutron activation analysis
análisis de adsorción adsorption analysis
análisis de agua water analysis
análisis de fluorescencia de rayos X X-
  ray fluorescence analysis
análisis de gases gas analysis
análisis de gases despedidos evolved gas
  analysis
análisis de Podbielniak Podbielniak
  analysis
análisis de rayos positivos positive-ray
  analysis
análisis de trazas trace analysis
análisis electrográfico electrographic
  analysis
análisis electrolítico electrolytic analysis
análisis elemental elementary analysis
análisis espectral spectral analysis,
  spectrum analysis
análisis espectrofotométrico
  spectrophotometric analysis
análisis espectrométrico spectrometric
  analysis
análisis espectroscópico spectroscopic
  analysis
análisis físico physical analysis
análisis fluorométrico fluorometric
  analysis
análisis forense forensic analysis
análisis fotométrico photometric analysis
análisis gravimétrico gravimetric
  analysis
análisis húmedo wet analysis
análisis mecánico mechanical analysis
análisis orgánico organic analysis
análisis polarimétrico polarimetric
  analysis
análisis polarográfico polarographic
  analysis
análisis por activación activation
  analysis
análisis por difusión diffusion analysis
análisis por electrodeposición
  electrodeposition analysis

análisis por fluorescencia fluorescence analysis

análisis químico chemical analysis

análisis radiométrico radiometric analysis

análisis térmico thermal analysis

análisis térmico diferencial differential thermal analysis

análisis termocinético thermokinetic analysis

análisis termogravimétrico thermogravimetric analysis

análisis termométrico thermometric analysis

análisis turbidimétrico turbidimetric analysis

análisis volumétrico volumetric analysis

analítico analytical

analizador analyzer

analizador de Orsat Orsat analyzer

analizador espectral spectrum analyzer

análsis espectroquímico spectrochemical analysis

anamorfismo anamorphism

anaranjado de acridina acridine orange

anaranjado de antimonio antimony orange

anaranjado de bronce bronze orange

anaranjado de cromo chrome orange

anaranjado de metilo methyl orange

anaranjado de molibdeno molybdenum orange

anaranjado de naftol naphthol orange

anarcotina anarcotine

anarmonicidad anharmonicity

anarmónico anharmonic

anatasa anatase

andalucita andalusite

andesina andesine

andrógeno androgen

androstano androstane

androsterona androsterone

anemómetro anemometer

anemosita anemosite

anestesia anesthesia

anestésico anesthetic

anetol anethole

anfetamina amphetamine

anfibol amphibole

anfibolita amphibolite

anficroico amphichroic

anficromático amphichromatic

anfifático amphiphatic

anfifílico amphiphilic

anfífilo amphiphile

anfiprótico amphiprotic

anfolito ampholyte

anfolitoide ampholytoid

anfoterismo amphoterism

anfótero amphoteric

anglesita anglesite

angosturina angosturine

angstrom angstrom

angular angular

ángulo angle

ángulo de Bragg Bragg angle

ángulo de Brewster Brewster angle

ángulo de enlace bond angle

ángulo de valencia valence angle

ángulo dihedral dihedral angle

ángulo tetraédrico tetrahedral angle

anhidrasa carbónica carbonic anhydrase

anhídrido anhydride

anhídrido abiético abietic anhydride

anhídrido acético acetic anhydride

anhídrido antimónico antimonic anhydride

anhídrido arsénico arsenic anhydride

anhídrido básico basic anhydride

anhídrido benzoico benzoic anhydride

anhídrido bórico boric anhydride

anhídrido butírico butyric anhydride

anhídrido caprílico caprylic anhydride

anhídrido cíclico cyclic anhydride

anhídrido cinámico cinnamic anhydride

anhídrido citacrónico citraconic anhydride

anhídrido cloréndico chlorendic anhydride

anhídrido cloroacético chloroacetic anhydride

anhídrido clorocrómico chlorochromic anhydride

anhídrido cloroetanoico chloroethanoic anhydride

anhídrido cloromaleico chloromaleic anhydride

anhídrido crómico chromic anhydride

anhídrido de ácido acid anhydride

anhídrido de ácido benzoico benzoic acid anhydride

anhídrido de ácido perclórico perchloric acid anhydride

anhídrido de ácido vanádico vanadic acid anhydride

anhídrido de ácido yódico iodic acid anhydride

anhídrido de antimonio antimony anhydride

anhídrido de eritritol erythritol anhydride

anhídrido de estaño tin anhydride

anhídrido de sorbitol sorbitol anhydride

anhídrido de torio thorium anhydride

anhídrido de urea urea anhydride

anhídrido decanoico decanoic anhydride

anhídrido estánnico stannic anhydride

anhídrido estíbico stibic anhydride

anhídrido etanoico ethanoic anhydride

anhídrido fosfórico phosphoric anhydride

anhídrido ftálico phthalic anhydride

anhídrido glutárico glutaric anhydride

anhídrido isatoico isatoic anhydride

anhídrido isobutírico isobutyric anhydride

anhídrido láctico lactic anhydride

anhídrido maleico maleic anhydride

anhídrido memtetrahidroftálico memtetrahydrophthalic anhydride

anhídrido molíbdico molybdic anhydride

anhídrido ósmico osmic anhydride

anhídrido propiónico propionic anhydride

anhídrido succínico succinic anhydride

anhídrido sulfobenzoico sulfobenzoic anhydride

anhídrido sulfoftálico sulfophthalic anhydride

anhídrido sulfónico sulfonic anhydride

anhídrido sulfúrico sulfuric anhydride

anhídrido tetracloroftálico tetrachlorophthalic anhydride

anhídrido tetrahidroftálico tetrahydrophthalic anhydride

anhídrido trimelítico trimellitic anhydride

anhídrido túngstico tungstic anhydride

anhídrido valérico valeric anhydride

anhídrido vanádico vanadic anhydride

anhídrido zircónico zirconic anhydride

anhidrita anhydrite

anhidro anhydrous

anhidrona anhydrone

anhidrosíntesis anhydrosynthesis

anibina anibine

anilida anilide

anilida hidroxinaftoica hydroxynaphthoic anilide

anilina aniline

anilina de formaldehído formaldehyde aniline

anilinato anilinate

anilinio anilinium

anilino anilino

anillo ring

anillo bencénico benzene ring

anillo de Kekulé Kekulé ring

anillo de septanosa septanose ring

anillo fundido fused ring

anillo heteroatómico heteroatomic ring

anillo homoatómico homoatomic ring

anillos de Raschig Raschig rings

anión anion

anión ambidente ambident anion

aniónico anionic

anionotropía anionotropy

anisal anisal

anisaldehído anisaldehyde

anísico anisic

anisidina anisidine

anisilacetona anisylacetone

anisilo anisyl

anisoilo anisoyl

anisol anisole

anisomérico anisomeric

anisonitrilo anisonitrile

anisotónico anisotonic

anisotrópico anisotropic

anisótropo anisotropic

anitácido antacid

anitol anytol

anódico anodic

anodización anodizing

anodizar anodize

ánodo anode

ánodo acelerador accelerating anode
ánodo galvánico galvanic anode
ánodo reactivo reactive anode
anolito anolyte
anomalía anomaly
anómalo anomalous
anomérico anomeric
anómero anomer
anorgánico anorganic
anórtico anorthic
anortita anorthite
anoxinaftonato sódico sodium anoxynaphthonate
antagonismo antagonism
antagonista antagonist
antagonista estructural structural antagonist
antalcalino antalkaline
antemol anthemol
anteojos de seguridad safety goggles
antiadherente abherent
antiaromático antiaromatic
antibacterial antibacterial
antibacteriano antibacterial
antibiótico antibiotic
anticatalizador anticatalyzer
anticátodo anticathode
anticloro antichlor
anticoagulante anticoagulant
anticolinérgico anticholineric
anticomplemento anticomplement
anticongelante antifreeze
anticorrosivo anticorrosive
anticuerpo antibody
antideflagrante flameproof
antidetonante antiknock
antídoto antidote
antienlazado antibonding
antienzima antienzyme
antiestático antistatic
antiferromagnético antiferromagnetic
antiferromagnetismo antiferromagnetism
antiformina antiformin
antígeno antigen
antiglobulina antiglobulin
antigolpeteo antiknock
antigorita antigorite
antihistamina antihistamine

antiisómero antiisomer
antiisomorfismo antiisomorphism
antimateria antimatter
antímero antimer
antimetabolito antimetabolite
antimicina antimycin
antimoniato antimonate
antimoniato amónico ammonium antimoniate
antimoniato de plomo lead antimonate
antimoniato potásico potassium antimonate
antimoniato sódico sodium antimonate
antimónico antimonic
antimonilo antimonyl
antimonio antimony
antimonio cáustico caustic antimony
antimonioso antimonous
antimonita antimonite
antimoniuro alumínico aluminum antimonide
antimoniuro de bismuto bismuth antimonide
antimoniuro de cadmio cadmium antimonide
antimoniuro de cesio cesium antimonide
antimoniuro de cinc zinc antimonide
antimoniuro de galio gallium antimonide
antimoniuro de indio indium antimonide
antimoniuro de itrio yttrium antimonide
antimoniuro de lantano lanthanum antimonide
antimoniuro de manganeso crómico chromium manganese antimonide
antimoniuro de plata silver antimonide
antineutrón antineutron
antión anthion
antioxidante antioxidant
antioxígeno antioxygen
antiozonizante antiozonant
antiparasítico antiparasitic
antipartícula antiparticle
antiperspirante antiperspirant
antipireno antipyrene
antipirético antipyretic
antipirilo antipyryl
antipirina antipyrine
antiprotón antiproton
antipsicótico antipsychotic

antiquinol antiquinol
antirradiación antiradiation
antiséptico antiseptic
antitóxico antitoxic
antitoxina antitoxin
antitusivo antitussive
antiviral antiviral
antizimótico antizymotic
antocianidina anthocyanidin
antocianina anthocyanin
antofilita anthophyllite
antozono antozone
antraceno anthracene
antracita anthracite
antradiol anthradiol
antragalol anthragallol
antrahidroquinona anthrahydroquinone
antraldehído anthraldehyde
antralina anthralin
antralinato de butilo butyl anthranilate
antranil anthranil
antranilato potásico potassium
  anthranilate
antranol anthranol
antranona anthranone
antraparaceno anthraparazene
antrapurpurina anthrapurpurin
antraquinona anthraquinone
antrarrufina anthrarufin
antratetrol anthratetrol
antratriol anthratriol
antrileno anthrylene
antrilo anthryl
antrol anthrol
antrona anthrone
anular annular
anuleno annulene
añil indigo
apantallamiento screening
aparato apparatus
aparato de absorción absorption
  apparatus
aparato de análisis de gases gas-analysis
  apparatus
aparato de conductividad conductivity
  apparatus
aparato de destilación distillation
  apparatus
aparato de extracción extraction
  apparatus
aparato de Folin Folin apparatus
aparato de Kipp Kipp apparatus
aparato de laboratorio laboratory
  apparatus
aparato de Orsat Orsat apparatus
aparato de Pensky-Martens Pensky-
  Martens apparatus
aparato de punto de congelación
  freezing-point apparatus
aparato de punto de ebullición boiling-
  point apparatus
aparato de Thiele Thiele apparatus
apatita apatite
apionol apionol
apoatropina apoatropine
apocarotenal apocarotenal
apocodeína apocodeine
apofilita apophyllite
apomorfina apomorphine
aprótico aprotic
aproximación de Born-Oppenheimer
  Born-Oppenheimer approximation
aproximación de Hückel Hückel
  approximation
aproximado approximate
aquiral achiral
arabina arabin
arabinosa arabinose
arabinosida arabinoside
arabita arabite
arabitol arabitol
aragonita aragonite
aralquilo aralkyl
aramida aramid
araquidato de metilo methyl arachidate
arborescente arborescent
arbutina arbutin
arcilla clay
arcilla refractaria fireclay
área peligrosa hazardous area
área superficial surface area
arecaidina arecaidine
arecaína arecaine
arecolina arecoline
arena sand
arena de filtración filter sand
arena de sílice silica sand
areno arene

areómetro areometer
aresniato de cobalto cobalt arsenate
argenato argenate
argentamina argentamine
argentán argentan
argéntico argentic
argentita argentite
argentocianuro argentocyanide
argentocianuro potásico potassium argentocyanide
argentometría argentometry
argentopirita argentopyrite
arginasa arginase
arginina arginine
argirodita argyrodite
argol argol
argón argon
aribina aribine
arilalquilo arylalkyl
arilamina arylamine
arilarsonato arylarsonate
arileno arylene
arilo aryl
ariluro arylide
arino aryne
arnicina arnicin
aroilo aroyl
aromaticidad aromaticity
aromático aromatic
aromatización aromatization
arrurruz arrowroot
arsacetina arsacetin
arsanilato arsanilate
arsanilato sódico sodium arsanilate
arseniato arsenate
arseniato amónico ammonium arsenate
arseniato bárico barium arsenate
arseniato cálcico calcium arsenate
arseniato cobaltoso cobaltous arsenate
arseniato cúprico cupric arsenate
arseniato de cinc zinc arsenate
arseniato de cobre copper arsenate
arseniato de hierro iron arsenate
arseniato de litio lithium arsenate
arseniato de manganeso manganese arsenate
arseniato de mercurio mercury arsenate
arseniato de níquel nickel arsenate
arseniato de plata silver arsenate

arseniato de plomo lead arsenate
arseniato férrico ferric arsenate
arseniato ferroso ferrous arsenate
arseniato magnésico magnesium arsenate
arseniato magnésico amónico ammonium magnesium arsenate
arseniato manganoso manganous arsenate
arseniato mercúrico mercuric arsenate
arseniato niqueloso nickelous arsenate
arseniato potásico potassium arsenate
arseniato sódico sodium arsenate
arsenical arsenical
arsénico arsenic
arsénico blanco white arsenic
arsénico rojo red arsenic
arsenidina arsenidine
arsenilo arsenyl
arsenioso arsenous
arsenito arsenite
arsenito amónico ammonium arsenite
arsenito cálcico calcium arsenite
arsenito de cinc zinc arsenite
arsenito de cobre copper arsenite
arsenito de estroncio strontium arsenite
arsenito de plata silver arsenite
arsenito de plomo lead arsenite
arsenito férrico ferric arsenite
arsenito potásico potassium arsenite
arsenito sódico sodium arsenite
arseniuro arsenide
arseniuro de antimonio antimony arsenide
arseniuro de cesio cesium arsenide
arseniuro de cinc zinc arsenide
arseniuro de cromo chromium arsenide
arseniuro de galio gallium arsenide
arseniuro de indio indium arsenide
arseniuro de itrio yttrium arsenide
arseniuro de lantano lanthanum arsenide
arseniuro de plata silver arsenide
arseniuro sódico sodium arsenide
arseno arseno
arsenobenceno arsenobenzene
arsenolita arsenolite
arsenopirita arsenopyrite
arsina arsine
arsinoso arsinoso
arsonato arsonate

arsonio arsonium
arsono arsono
arsonoilo arsonoyl
arsonoso arsonoso
artesa boat
artesa de combustión combustion boat
artificial artificial
artinita artinite
artranitín arthranitin
asarilo asaryl
asarona asarone
asbestina asbestine
asbesto asbestos
asbolana asbolane
asbolita asbolite
ascaridol ascaridole
ascorbato cálcico calcium ascorbate
ascorbato de niacinamida niacinamide
    ascorbate
ascorbato sódico sodium ascorbate
asentamiento settling
aseptol aseptol
aseptolina aseptoline
asfalteno asphaltene
asfaltita asphaltite
asfalto asphalt
asfalto de petróleo petroleum asphalt
asimetría asymmetry
asimétrico asymmetrical, unsymmetrical
asimilación assimilation
asimilación de carbono carbon
    assimilation
asiminina asiminine
asistente de teñido dyeing assistant
askarel askarel
asociación association
asociación molecular molecular
    association
asparagina asparagine
asparaginasa asparaginase
asparamida asparamide
aspariginilo aspariginyl
aspartama aspartame
aspartamida aspartamide
aspartasa aspartase
aspartato aspartate
aspartato sódico sodium aspartate
aspartilo aspartyl
aspartoilo aspartoyl

aspidinol aspidinol
aspidosamina aspidosamine
aspidospermina aspidospermine
aspirador aspirator
aspirina aspirin
astaceno astacene
astátco astatic
astatino astatine
ástato astatine
asterisco asterisk
astorismo astorism
astrofilita astrophyllite
astroquímica astrochemistry
atacamita atacamite
atáctico atactic
atenolol atenolol
atenuación attenuation
atenuador de haz beam attenuator
atisina atisine
atmólisis atmolysis
atmómetro atmometer
atmósfera atmosphere
atmósfera controlada controlled
    atmosphere
atmósfera de hidrógeno hydrogen
    atmosphere
atmósfera iónica ionic atmosphere
atmósfera oxidante oxidizing
    atmosphere
atmósfera reactiva reactive atmosphere
atmósfera reductora reducing
    atmosphere
atmosférico atmospheric
ato ate
atómicamente atomically
atomicidad atomicity
atómico atomic
atomización atomization
atomizador atomizer
átomo atom
átomo asimétrico asymmetric atom
átomo clave key atom
átomo de Bohr Bohr atom
átomo de carbono carbon atom
átomo de carbono asimétrico
    asymmetric carbon atom
átomo de carbono cuaternario
    quaternary carbon atom
átomo de carbono secundario secondary

carbon atom
átomo de carbono simétrico
    symmetrical carbon atom
átomo de carbono terciario tertiary
    carbon atom
átomo de hidrógeno hydrogen atom
átomo de Lewis Lewis atom
átomo donador donor atom
átomo espiro spiro atom
átomo excitado excited atom
átomo exótico exotic atom
átomo-gramo gram-atom
átomo heterocíclico heterocyclic atom
átomo libre free atom
átomo marcado tagged atom
átomo neutro neutral atom
átomo nuclear nuclear atom
átomo nuclear de Rutherford
    Rutherford nuclear atom
átomo pesado heavy atom
átomo radiactivo radioactive atom
átomo substituyente substituent
átomo tetraédrico tetrahedral atom
átomo trazador tracer atom
atomología atomology
atomsita atomsite
atopita atopite
atracción attraction
atracción atómica atomic attraction
atracción de van der Waals van der
    Waals attraction
atracción interiónica interionic attraction
atracción molecular molecular attraction
atracción nuclear nuclear attraction
atrazina atrazine
atroneno atronene
atropina atropine
atropoilo atropoyl
attapulgita attapulgite
aubepina aubepine
augita augite
auramina auramine
aurato aurate
aurato potásico potassium aurate
aurelita aurelite
aureolina aureolin
auribromuro auribromide
auribromuro sódico sodium auribromide
auricalcita aurichalcite

auricloruro aurichloride
auricloruro sódico sodium aurichloride
áurico auric
aurífero auriferous
aurina aurin
auro aurous
auroáurico auroauric
aurobromuro aurobromide
aurocianuro aurous cyanide
aurocloruro aurochloride
auromina auromine
aurona aurone
aurosulfuro aurous sulfide
aurotiomalato sódico sodium
    aurothiomalate
auroyoduro aurous iodide
austenita austenite
autoabsorción self-absorption
autocatálisis autocatalysis
autocatalítico autocatalytic
autoclave autoclave
autodescomposición autodecomposition
autoexcitación autoexcitation
autoionización autoionization
autólisis autolysis
autoluminiscencia autoluminescence
automático automatic
automatización automation
autoorganización self-organization
autooxidación self-oxidation
autoprotólisis autoprotolysis
autorreducción self-reduction
autunita autunite
auxiliar auxiliary
auxina auxin
auxocromo auxochrome
avalancha electrónica electron avalanche
avalancha iónica ion avalanche
avalita avalite
avenina avenine
avidina avidin
axial axial
azaguanina azaguanine
azauridina azauridine
azelato azelate
azeotrópico azeotropic
azeótropo azeotrope
azetidina azetidine
azida azide

azida bárica barium azide
azida de bencilo benzyl azide
azida de benzaldehído benzaldehyde azide
azida de bromo bromine azide
azida de cianógeno cyanogen azide
azida de fenilo phenyl azide
azida de hidrógeno hydrogen azide
azida de plata silver azide
azida de plomo lead azide
azida sódica sodium azide
azido azido
azimida azimide
azimido azimido
azimidobenceno azimidobenzene
azimino azimino
aziminobenceno aziminobenzene
azina azine
azino azino
aziridina aziridine
azlón azlon
azo azo
azobenceno azobenzene
azobisisobutironitrilo azobisisobutyronitrile
azocianuro azocyanide
azocolorante azo dye
azodicabornamida azodicarbonamide
azodina azodine
azofenileno azophenylene
azofenilo azophenyl
azofenol azophenol
azogue quicksilver
azoico azoic
azoimida azoimide
azol azole
azometina azomethine
azonaftaleno azonaphthalene
azonio azonium
azorita azorite
azosulfamida azosulfamide
azotolueno azotoluene
azotómetro azotometer
azoxi azoxy
azoxibenceno azoxybenzene
azoxinaftaleno azoxynaphthalene
azoxitoluidina azoxytoluidine

azúcar sugar
azúcar de caña cane sugar
azúcar de leche milk sugar
azúcar de madera wood sugar
azúcar de pectina pectin sugar
azúcar de plomo sugar of lead
azúcar de remolacha beet sugar
azúcar de uva grape sugar
azúcar invertido invert sugar
azúcar no reductor nonreducing sugar
azúcar reductor reducing sugar
azufre sulfur, sulphur
azufre libre free sulfur
azufre rómbico rhombic sulfur
azul alcalino alkali blue
azul cerúleo cerulean blue
azul congo congo blue
azul de alizarina alizarin blue
azul de anilina aniline blue
azul de antraceno anthracene blue
azul de bromofenol bromophenol blue
azul de bromotimol bromothymol blue
azul de bronce bronze blue
azul de cobalto cobalt blue
azul de cobre copper blue
azul de cresilo cresyl blue
azul de China Chinese blue
azul de hierro iron blue
azul de índigo indigo blue
azul de metileno methylene blue
azul de metilo methyl blue
azul de molibdeno molybdenum blue
azul de naftol naphthol blue
azul de Prusia Prussian blue
azul de quinolina quinoline blue
azul de resorcina resorcinol blue
azul de Thenard Thenard blue
azul de timol thymol blue
azul de Turnbull Turnbull blue
azul fenileno phenylene blue
azul pavo real peacock blue
azul ultramarino ultramarine blue
azul Victoria Victoria blue
azulamiento blueing
azuleno azulene
azulina azulin
azurita azurite

B

bacarina baccarin
bacilo bacillus
bacitracina bacitracin
bacitracina de cinc zinc bacitracin
bacteria bacterium
bacterial bacterial
bacteriano bacterial
bactericida bactericide
bacterina bacterin
bacteriófago bacteriophage
bacteriólisis bacteriolysis
bacteriológico bacteriological
bacteriostático bacteriostatic
badeleyita baddeleyite
badequita baddeckite
bafle baffle
bagazo bagasse
bagrationita bagrationite
baicaleína baicalein
baicalina baicalin
baikalita baikalite
bainita bainite
baja energía low energy
balance balance
balanceado balanced
balanza scale, balance
balanza analítica analytical balance
balanza de astil beam balance
balanza de registro recording balance
balanza de torsión torsion balance
balata balata
baloeléctrico balloelectric
balómetro balometer
bálsamo balsam
bálsamo de Perú balsam of Peru
bálsamo de Tolú balsam of Tolu
banalsita banalsite
banda band
banda de absorción absorption band
banda de conducción conduction band
banda de energía energy band
banda de energía prohibida forbidden
    energy band

banda de valencia valence band
banda electrónica electron band
banda permitida allowed band
banda prohibida forbidden band
banisterina banisterine
baño bath
baño de agua water bath
baño de aire air bath
baño de arena sand bath
baño de vapor vapor bath, steam bath
baño galvánico galvanic bath
baño María water bath
baptisina baptisin
baptitoxina baptitoxine
baquerita bakerite
barbaloína barbaloin
barbana barban
barberita barberite
barbital barbital
barbitona barbitone
barbiturato barbiturate
barbiturato sódico sodium barbiturate
barcenita barcenite
baria barye
bario barium
barión baryon
barisilita barysilite
barita baryte
barita cáustica caustic baryta
barkevita barkevite
barn barn
barniz varnish, glaze
baroforesis barophoresis
baroluminiscencia baroluminescence
barómetro barometer
baroscopio baroscope
barosimina barosimin
barostato barostat
barra bar
barrandita barrandite
barrera barrier
barrera de Coulomb Coulomb barrier
barrera de Gamow Gamow barrier

306

barrido scan, scanning
barril barrel
bartrina barthrin
basanita basanite, bassanite
base base
base blanda soft base
base conjugada conjugate base
base de alquitrán tar base
base de Bandrowski Bandrowski base
base de colina choline base
base de Lewis Lewis base
base de Millon Millon base
base de Schiff Schiff base
base débil weak base
base diluida dilute base
base dura hard base
base fuerte strong base
base natural natural base
base nitrogenosa nitrogenous base
base orgánica organic base
basicidad basicity
básico basic
basisterol bassisterol
basorina bassorin
bastnasita bastnasite
batería battery
batería de acumuladores storage battery
batería de plomo lead battery
batocromático bathochromatic
batocrómico bathochromic
batocuproína bathocuproine
batofenantrolina bathophenanthroline
batracotoxinina batrachotoxinin
batu batu
bauxita bauxite
bauxita calcinada calcined bauxite
bayerita bayerite
becarita beccarite
becquerel becquerel
becquerelita becquerelite
behenato de metilo methyl behenate
behenolilo behenolyl
behenona behenone
beiriquita beyrichite
bel bel
beldongrita beldongrite
belita belite
belonesita belonesite
belladona belladonna

belladonina belladonnine
bemberg bemberg
benazolina benazolin
bencenida benzenide
bencenilo benzenyl
bencenio benzenium
benceno benzene
bencenoazimida benzene azimide
bencenoazoanilida benzeneazoanilide
bencenoazoanilina benzeneazoaniline
bencenoazobenceno benzeneazobenzene
bencenocarbaldehído
    benzenecarbaldehyde
bencenocarbinol benzenecarbinol
bencenocarbonitrilo benzenecarbonitrile
bencenocarboxilato benzenecarboxylate
bencenodicarbinol benzenedicarbinol
bencenodicarbonal benzenedicarbonal
bencenodicarbonitrilo
    benzenedicarbonitrile
bencenodisulfonato potásico potassium
    benzenedisulfonate
bencenodisulfóxido benzenedisulfoxide
bencenoditiol benzenedithiol
bencenoide benzenoid
bencenona benzenone
bencenosulfamida benzenesulfamide
bencenosulfonamida
    benzenesulfonamide
bencenosulfonato amónico ammonium
    benzenesulfonate
bencenosulfonato sódico sodium
    benzenesulfonate
bencenosulfonilo benzenesulfonyl
bencenosulfóxido benzenesulfoxide
bencenosulfuro benzenesulfide
bencenotriol benzenetriol
bencidina benzidine
bencidino benzidino
bencilacetamida benzylacetamide
bencilación benzylation
bencilamina benzylamine
bencilanilina benzylaniline
bencilbenceno benzylbenzene
bencilcarbinol benzyl carbinol
bencilcianamida benzylcyanamide
bencildianilo benzyldianil
bencildimetilamina benzyldimethylamine
bencileno benzylene

benciletanolamina benzylethanolamine
bencilfenantreno benzylphenanthrene
bencilfenilamina benzylphenylamine
bencilfenol benzylphenol
bencilformamida benzylformamide
bencilhidrazina benzylhydrazine
bencilhidroquinona benzylhydroquinone
bencilidenacetofenona
    benzylideneacetophenone
bencilidenacetona benzylideneacetone
bencilideno benzylidene
bencilidietanolamina
    benzylidiethanolamine
bencilidina benzylidine
bencilisoeugenol benzyl isoeugenol
bencilisopropilamina
    benzylisopropylamine
bencilisotiocianato benzyl isothiocyanate
bencilmercaptano benzyl mercaptan
bencilmetilamina benzylmethylamine
bencilnaftaleno benzylnaphthalene
bencilo benzyl, benzil
benciloxifenol benzyloxyphenol
bencilpiridina benzylpyridine
benciltiol benzylthiol
bencimidazol benzimidazole
bencimidazolilo benzimidazolyl
bencimidoilo benzimidoyl
bencina benzin, benzine
bencina de petróleo petroleum benzin
bencindol benzindole
bencino benzyne
bencisoxasol benzisoxasole
bendiocarb bendiocarb
beneficio beneficiation
benefina benefin
benomilo benomyl
bensulida bensulide
bentazona bentazone
bentonita bentonite
benzacetina benzacetin
benzaconina benzaconine
benzacridina benzacridine
benzal benzal
benzalacetona benzalacetone
benzalazina benzalazine
benzalcloruro benzal chloride
benzalcohol benzalcohol
benzalconio benzalkonium

benzaldehído benzaldehyde
benzaldoxima benzaldoxime
benzalftalida benzalphthalide
benzamida benzamide
benzamida de plata silver benzamide
benzamidina benzamidine
benzamido benzamido
benzanilida benzanilide
benzantracina benzanthracene
benzantrona benzanthrone
benzatina benzathine
benzazida benzazide
benzazimida benzazimide
benzazina benzazine
benzfurano benzfuran
benzhidrazoína benzhydrazoin
benzhidrilo benzhydryl
benzhidrol benzhydrol
benzoato benzoate
benzoato alumínico aluminum benzoate
benzoato amónico ammonium benzoate
benzoato bárico barium benzoate
benzoato cálcico calcium benzoate
benzoato ceroso cerous benzoate
benzoato cobaltoso cobaltous benzoate
benzoato cúprico cupric benzoate
benzoato de alilo allyl benzoate
benzoato de amilo amyl benzoate
benzoato de antipireno antipyrene
    benzoate
benzoato de bencilo benzyl benzoate
benzoato de benzoilo benzoyl benzoate
benzoato de butilo butyl benzoate
benzoato de cafeína caffeine benzoate
benzoato de cesio cesium benzoate
benzoato de cinc zinc benzoate
benzoato de cobre copper benzoate
benzoato de cocaína cocaine benzoate
benzoato de denatonio denatonium
    benzoate
benzoato de etileno ethylene benzoate
benzoato de etilo ethyl benzoate
benzoato de hidroxiquinolina
    hydroxyquinoline benzoate
benzoato de hierro iron benzoate
benzoato de isoamilo isoamyl benzoate
benzoato de isobutilo isobutyl benzoate
benzoato de litio lithium benzoate
benzoato de mercurio mercury benzoate

benzoato de metilo methyl benzoate
benzoato de níquel nickel benzoate
benzoato de plata silver benzoate
benzoato de plomo lead benzoate
benzoato de propilo propyl benzoate
benzoato de tetrahidrofurfurilo
     tetrahydrofurfuryl benzoate
benzoato de uranilo uranyl benzoate
benzoato fenilmercúrico phenylmercuric
     benzoate
benzoato férrico ferric benzoate
benzoato magnésico magnesium
     benzoate
benzoato mercúrico mercuric benzoate
benzoato potásico potassium benzoate
benzoato sódico sodium benzoate
benzocaína benzocaine
benzodiantrona benzodianthrone
benzodiazina benzodiazine
benzodiazol benzodiazole
benzodihidropirona benzodihydropyrone
benzodioxazina benzodioxazine
benzofenol benzophenol
benzofenona benzophenone
benzofluoreno benzofluorene
benzofuranilo benzofuranyl
benzofurano benzofuran
benzofuranona benzofuranone
benzoguanamina benzoguanamine
benzohidrilo benzohydryl
benzohidrol benzohydrol
benzoilacetaldehído benzoylacetaldehyde
benzoilacetato de etilo ethyl
     benzoylacetate
benzoilacetona benzoilacetone
benzoilación benzoylation
benzoilamida benzoylamide
benzoilamino benzoylamino
benzoilanilina benzoylaniline
benzoilazida benzoylazide
benzoilciclopropano
     benzoylcyclopropane
benzoilferroceno benzoylferrocene
benzoilglicina benzoylglycin
benzoilhidrazina benzoylhydrazine
benzoilo benzoyl
benzoiloxi benzoyloxy
benzoilpiridina benzoylpyridine
benzoína benzoin

benzol benzol
benzomorfolina benzomorpholine
benzonafteno benzonaphthene
benzonaftol benzonaphthol
benzonitrilo benzonitrile
benzopentazol benzopentazole
benzoperóxido benzoperoxide
benzopinacol benzopinacol
benzopiranilo benzopyranyl
benzopirano benzopyran
benzopireno benzopyrene
benzopiridina benzopyridine
benzopirona benzopyrone
benzopirrol benzopyrrole
benzopurpurina benzopurpurine
benzoquinol benzoquinol
benzoquinona benzoquinone
benzoselenofurano benzoselenofuran
benzosulfimida benzosulfimide
benzosulfimida sódica sodium
     benzosulfimide
benzotetrazina benzotetrazine
benzotetrazol benzotetrazole
benzotiacilo benzothiazyl
benzotiazina benzothiazine
benzotiazol benzothiazole
benzotiofurano benzothiofuran
benzotiopirano benzothiopyran
benzotiopirona benzothiopyrone
benzotoluida benzotoluide
benzotriazepina benzotriazepine
benzotriazina benzotriazine
benzotriazol benzotriazole
benzotricloruro benzotrichloride
benzotrifluoruro benzotrifluoride
benzotrifurano benzotrifuran
benzoxanteno benzoxanthene
benzoxazina benzoxazine
benzoxazol benzoxazole
benzoxazolilo benzoxazolyl
benzoxazolona benzoxozolone
benzoxtetrazina benzoxtetrazine
benzoxtriazina benzoxtriazine
benzozono benzozone
benzpirazol benzpyrazole
benztiofeno benzthiophene
bequelita beckelite
berberina berberine
beresovita beresovite

berilato beryllate
berilia beryllia
berilio beryllium
berilio-cobre beryllium-copper
beriliuro beryllide
berilo beryl
berilonita beryllonite
berkelio berkelium
bermellón vermillion
bertierina berthierine
bertolida berthollide
bertolita bertholite
bertrandita bertrandite
beta beta
betaína betaine
betametasona betamethasone
betatópico betatopic
betatrón betatron
betulina betulin
betulinol betulinol
beumlerita baeumlerite
bi bi
biaceno biacene
biacetileno biacetylene
bialilo biallyl
bianilina bianiline
bianisidina bianisidine
biantrilo bianthryl
biaxial biaxial
bibencilo bibenzyl
biborato biborate
biborato de trihexilenglicol trihexylene
    glycol biborate
bicarbonato bicarbonate
bicarbonato amónico ammonium
    bicarbonate
bicarbonato bárico barium bicarbonate
bicarbonato de cesio cesium bicarbonate
bicarbonato de colina choline
    bicarbonate
bicarbonato de estroncio strontium
    bicarbonate
bicarbonato de litio lithium bicarbonate
bicarbonato de sosa bicarbonate of soda,
    baking soda
bicarbonato potásico potassium
    bicarbonate, bicarbonate of potash
bicarbonato sódico sodium bicarbonate,
    bicarbonate of soda, baking soda

bicíclico bicyclic
biciclodecano bicyclodecane
biciclohexilo bicyclohexyl
bicloruro bichloride
bicloruro de azufre sulfur bichloride
bicloruro de bencilo benzyl bichloride
bicloruro de benzol benzol bichloride
bicloruro de mercurio mercury
    bichloride
bicloruro de paladio palladium
    bichloride
bicloruro de platino platinum bichloride
bicromato amónico ammonium
    bichromate
bicromato bárico barium bichromate
bicromato potásico potassium
    bichromate
bicromato sódico sodium bichromate
bieberita bieberite
biestabilidad bistability
bietileno biethylene
bifenilamina biphenylamine
bifenildiamina biphenyldiamine
bifenileno biphenylene
bifenililamina biphenylylamine
bifenililo biphenylyl
bifenilo biphenyl
bifenilo policlorado polychlorinated
    biphenyl
biflavonilo biflavonyl
bifluoruro amónico ammonium
    bifluoride
bifluoruro potásico potassium bifluoride
bifluoruro sódico sodium bifluoride
biformina biformin
bifosfato amónico ammonium
    biphosphate
bifosfato magnésico magnesium
    biphosphate
bifosfato sódico sodium biphosphate
biftalato potásico potassium biphthalate
bifurcación bifurcation
biguanida biguanide
bihexilo bihexyl
biimino biimino
bilateral bilateral
bilifuscina bilifuscin
bilirrubina bilirubin
bimetal bimetal

bimolecular bimolecular
binaftalenilo binaphthalenyl
binaftaleno binaphthalene
binaftilo binaphthyl
binapacrilo binapacryl
binario binary
binita binnite
binitro binitro
binoxalato potásico potassium binoxalate
binoxalato sódico sodium binoxalate
binóxido binoxide
binóxido bárico barium binoxide
binóxido de manganeso manganese binoxide
binóxido sódico sodium binoxide
binuclear binuclear
bioactivación bioactivation
bioacumulación bioaccumulation
bioanálisis bioanalysis
biocatalizador biocatalyst
biocida biocide
biocitina biocytin
biocoloide biocolloid
bioconversión bioconversion
biocromo biochrome
biodegradable biodegradable
bioelectroquímica bioelectrochemistry
bioelemento bioelement
bioensayo bioassay
bioética bioethics
biofila biophyl
bioflavonoide bioflavonoid
biofosfato cálcico calcium biophosphate
biogas biogas
biogénico biogenic
biogeoquímica biogeochemistry
bioingeniería bioengineering
biología biology
biología molecular molecular biology
biológico biological
bioluminiscencia bioluminescence
biomasa biomass
biomaterial biomaterial
biomineral biomineral
biomolécula biomolecule
biopolímero biopolymer
bioquímica biochemistry
bioquímica de radiación radiation biochemistry

bioquímico biochemical
biorización biorization
biorreactor bioreactor
biorresmetrina bioresmethrin
biortofosfato de plomo lead biorthophosphate
biosa biose
biosfera biosphere
biosíntesis biosynthesis
biostático biostat
biosterol biosterol
biotecnología biotechnology
biótico biotic
biotina biotin
biotita biotite
bioxalato amónico ammonium bioxalate
bioxalato bárico barium bioxalate
bioxilo bioxyl
biperiden biperiden
bipirámide trigonal trigonal bipyramid
bipiridilo bipyridyl
bipropenilo bipropenyl
biquinolilo biquinolyl
biquinolina biquinoline
birradical biradical
birrectificación birectification
birrefrigencia birefringence
birrefrigencia circular circular birefringence
birrefringente birefringent
birrotación birotation
bis bis
bisaboleno bisabolene
bisamida bisamide
bisdiazo bisdiazo
bisfenol bisphenol
bishidroximetilurea bishydroxymethylurea
bismanol bismanol
bismita bismite
bismutato sódico sodium bismuthate
bismutida bismuthide
bismutilo bismuthyl
bismutina bismuthine
bismutinita bismuthinite
bismutita bismuthite
bismuto bismuth
bismuto de fenol phenol bismuth
bismutoso bismuthous

bistetrazol bistetrazole
bistriazol bistriazole
bisulfato bisulfate
bisulfato amónico ammonium bisulfate
bisulfato bárico barium bisulfate
bisulfato de acetona sódico sodium
    acetone bisulfate
bisulfato potásico potassium bisulfate
bisulfato sódico sodium bisulfate
bisulfato sódico de acetona acetone
    sodium bisulfate
bisulfito bisulfite
bisulfito amónico ammonium bisulfite
bisulfito cálcico calcium bisulfite
bisulfito de formaldehído sódico sodium
    formaldehyde bisulfite
bisulfito potásico potassium bisulfite
bisulfito sódico sodium bisulfite
bisulfuro cálcico calcium bisulfide
bisulfuro de carbono carbon bisulfide
bisulfuro de estaño tin bisulfide
bisulfuro de yodo iodine bisulfide
bisulfuro sódico sodium bisulfide
bitartrato bitartrate
bitartrato amónico ammonium bitartrate
bitartrato de colina choline bitartrate
bitartrato potásico potassium bitartrate
bitartrato sódico sodium bitartrate
bitionol bithionol
bitionolato sódico sodium bithionolate
bitrex bitrex
bittern bittern
bitumen bitumen
bitumen asfáltico asphaltic bitumen
bituminoso bituminous
biurea biurea
biuret biuret
bivalencia bivalence
bivalente bivalent
bivinilo bivinyl
bizaso bisazo
blanco de cinc zinc white
blanco de China Chinese white
blanco de España Spanish white
blanco de titanio titanium white
blanco de wolframio wolfram white
blanco fijo blanc fixe
blanco perla pearl white
blanco permanente permanent white

blanqueador whitener
blanquear bleach
blanqueo bleaching
blenda blende
blenda de cinc zinc blende
bleomicina bleomycin
blindaje shielding
bloque block
bloqueo block
boehmita boehmite
boghead boghead coal
bohidruro sódico sodium borohydride
boleíta boleite
bolómetro bolometer
bolsa de filtración filter bag
bomba bomb, pump
bomba calorimétrica bomb calorimeter
bomba centrífuga centrifugal pump
bomba criogénica cryogenic pump
bomba de aire air pump
bomba de difusión diffusion pump
bomba de succión suction pump
bomba de vacío vacuum pump
bomba iónica ion pump
bomba para filtración filter pump
bombardeo bombardment
bombardeo con átomos rápidos fast-
    atom bombardment
bombardeo electrónico electron
    bombardment
bombardeo neutrónico neutron
    bombardment
bombardeo nuclear nuclear
    bombardment
bombardeo protónico proton
    bombardment
bondur bondur
boracita boracite
borano borane
boranodiilo boranediyl
borato borate
borato alumínico aluminum borate
borato amónico ammonium borate
borato bárico barium borate
borato cálcico calcium borate
borato de bismuto bismuth borate
borato de butilo butyl borate
borato de cinc zinc borate
borato de cobre copper borate

borato de cocaína cocaine borate
borato de litio lithium borate
borato de manganeso manganese borate
borato de metilo methyl borate
borato de plata silver borate
borato de plomo lead borate
borato de tributilo tributyl borate
borato de trietilo triethyl borate
borato de trimetilo trimethyl borate
borato fenilmercúrico phenylmercuric borate
borato magnésico magnesium borate
borato mercúrico mercuric borate
borato potásico potassium borate
borato sódico sodium borate
bórax borax
borazina borazine
borazol borazole
borde de absorción absorption edge
borina borine
bornano bornane
borneno bornene
borneol borneol
bornilo bornyl
bornita bornite
boro boron
boroetano boroethane
borofluoruro potásico potassium borofluoride
boroformiato alumínico aluminum boroformate
boroformiato sódico sodium boroformate
borohidruro borohydride
borohidruro alumínico aluminum borohydride
borohidruro de litio lithium borohydride
borohidruro potásico potassium borohydride
borol borol
borosalicilato borosalicylate
borosilicato borosilicate
borosilicato de plomo lead borosilicate
borotungstato bárico barium borotungstate
borotungstato de cadmio cadmium borotungstate
boroxina boroxine
bort bort

boruro boride
boruro alumínico aluminum boride
boruro bárico barium boride
boruro cobáltico cobaltic boride
boruro de cromo chromium boride
boruro de hafnio hafnium boride
boruro de manganeso manganese boride
boruro de molibdeno molybdenum boride
boruro de silicio silicon boride
boruro de titanio titanium boride
boruro de tungsteno tungsten boride
boruro de zirconio zirconium boride
botella bottle
botella para gotear dropping bottle
brasilina brasilin
braunita braunite
brea tar, pitch
brea de alquitrán de hulla coal-tar pitch
brea de alquitrán de pino pine-tar pitch
brea mineral mineral pitch
bretonita bretonite
breunerita breunnerite
brillante brilliant
brillo luster
britolita britholite
brocanita brochanite
bromacetato bromacetate
bromacil bromacil
bromal bromal
bromalina bromalin
bromato bromate
bromato alumínico aluminum bromate
bromato bárico barium bromate
bromato cálcico calcium bromate
bromato ceroso cerous bromate
bromato cobaltoso cobaltous bromate
bromato de alcanfor camphor bromate
bromato de cadmio cadmium bromate
bromato de cesio cesium bromate
bromato de cinc zinc bromate
bromato de estroncio strontium bromate
bromato de plata silver bromate
bromato de plomo lead bromate
bromato magnésico magnesium bromate
bromato potásico potassium bromate
bromato sódico sodium bromate
bromelaína bromelain
bromelina bromelin

bromeosina bromeosin
brometilo bromethyl
bromhidrato hydrobromide
bromhidrato de etilamina ethylamine
  hydrobromide
brominación bromination
brominado bromated
bromito sódico sodium bromite
bromo bromine
bromoacetamida bromoacetamide
bromoacetato bromoacetate
bromoacetato de etilo ethyl
  bromoacetate
bromoacetona bromoacetone
bromoalileno bromoallylene
bromoanilina bromoaniline
bromobenceno bromobenzene
bromobenzaldehído bromobenzaldehyde
bromobutano bromobutane
bromociclopentano bromocyclopentane
bromocloroetano bromochloroethane
bromoclorometano bromochloromethane
bromocloropropano
  bromochloropropane
bromododecano bromododecane
bromoestireno bromostyrene
bromoetano bromoethane
bromofenilacetonitrilo
  bromophenylacetonitrile
bromofenol bromophenol
bromofluorobenceno
  bromofluorobenzene
bromoformo bromoform
bromoformo de acetona acetone
  bromoform
bromofosgeno bromophosgene
bromohexano bromohexane
bromohidrina de etileno ethylene
  bromohydrin
bromohidrina de glicol glycol
  bromohydrin
bromol bromol
bromometano bromomethane
bromometría bromometry
bromonaftaleno bromonaphthalene
bromopentano bromopentane
bromopicrina bromopicrin
bromopiridina bromopyridine
bromopirina bromopyrine

bromopropeno bromopropene
bromosuccinimida bromosuccinimide
bromotiofenol bromothiophenol
bromotolueno bromotoluene
bromotriclorometano
  bromotrichloromethane
bromotrifluoroetileno
  bromotrifluoroethylene
bromotrifluorometano
  bromotrifluoromethane
bromuro alumínico aluminum bromide
bromuro amónico ammonium bromide
bromuro amónico de cadmio cadmium
  ammonium bromide
bromuro antimonioso antimonous
  bromide
bromuro arsenioso arsenous bromide
bromuro áurico auric bromide
bromuro bárico barium bromide
bromuro bárico mercúrico mercuric
  barium bromide
bromuro cálcico calcium bromide
bromuro ceroso cerous bromide
bromuro cobaltoso cobaltous bromide
bromuro crómico chromic bromide
bromuro cromoso chromous bromide
bromuro cúprico cupric bromide
bromuro cuproso cuprous bromide
bromuro de acetilo acetyl bromide
bromuro de alilo allyl bromide
bromuro de antimonio antimony
  bromide
bromuro de arsénico arsenic bromide
bromuro de azufre sulfur monobromide,
  sulfur bromide
bromuro de bencilo benzyl bromide
bromuro de benzhidrilo benzhydryl
  bromide
bromuro de benzoilo benzoyl bromide
bromuro de benzol benzol bromide
bromuro de berilio beryllium bromide
bromuro de bismutilo bismuthyl
  bromide
bromuro de bismuto bismuth bromide
bromuro de boro boron bromide
bromuro de butilo butyl bromide
bromuro de cadmio cadmium bromide
bromuro de carbonilo carbonyl bromide
bromuro de cesio cesium bromide

bromuro de cetilo   cetyl bromide
bromuro de cetilpiridinio
    cetylpyridinium bromide
bromuro de cetiltrimetilamonio
    cetyltrimethylammonium bromide
bromuro de cianógeno   cyanogen
    bromide
bromuro de ciclohexilo   cyclohexyl
    bromide
bromuro de cinc   zinc bromide
bromuro de cloro   chlorine bromide
bromuro de cobalto   cobalt bromide
bromuro de cobre   copper bromide
bromuro de cromo   chromium bromide
bromuro de difenilmetilo
    diphenylmethyl bromide
bromuro de estaño   tin bromide
bromuro de estroncio   strontium bromide
bromuro de etileno   ethylene bromide
bromuro de etilhexilo   ethylhexyl
    bromide
bromuro de etilmagnesio
    ethylmagnesium bromide
bromuro de etilo   ethyl bromide
bromuro de fenilmagnesio
    phenylmagnesium bromide
bromuro de fenilo   phenyl bromide
bromuro de formilo   formyl bromide
bromuro de fosfonio   phosphonium
    bromide
bromuro de fosforilo   phosphoryl
    bromide
bromuro de fósforo   phosphorus bromide
bromuro de gadolinio   gadolinium
    bromide
bromuro de galio   gallium bromide
bromuro de hexilo   hexyl bromide
bromuro de hidrógeno   hydrogen
    bromide
bromuro de hierro   iron bromide
bromuro de hierro amónico   ammonium
    iron bromide
bromuro de indio   indium bromide
bromuro de iridio   iridium bromide
bromuro de isopropilo   isopropyl
    bromide
bromuro de iterbio   ytterbium bromide
bromuro de itrio   yttrium bromide
bromuro de litio   lithium bromide

bromuro de manganeso   manganese
    bromide
bromuro de mercurio   mercury bromide
bromuro de metileno   methylene bromide
bromuro de metilmagnesio
    methylmagnesium bromide
bromuro de metilo   methyl bromide
bromuro de niobio   niobium bromide
bromuro de níquel   nickel bromide
bromuro de nitrobencilo   nitrobenzyl
    bromide
bromuro de nitrosilo   nitrosyl bromide
bromuro de nonilo   nonyl bromide
bromuro de octilo   octyl bromide
bromuro de oro   gold bromide
bromuro de pentametileno
    pentamethylene bromide
bromuro de pentilo   pentyl bromide
bromuro de plata   silver bromide
bromuro de plomo   lead bromide
bromuro de propargilo   propargyl
    bromide
bromuro de propilmagnesio
    propylmagnesium bromide
bromuro de propilo   propyl bromide
bromuro de radio   radium bromide
bromuro de rubidio   rubidium bromide
bromuro de rutenio   ruthenium bromide
bromuro de selenio   selenium bromide
bromuro de silicio   silicon bromide
bromuro de sulfinilo   sulfinyl bromide
bromuro de talio   thallium bromide
bromuro de tántalo   tantalum bromide
bromuro de telurio   tellurium bromide
bromuro de tionilo   thionyl bromide
bromuro de titanio   titanium bromide
bromuro de trimetileno   trimethylene
    bromide
bromuro de uranio   uranium bromide
bromuro de vanadilo   vanadyl bromide
bromuro de vanadio   vanadium bromide
bromuro de vinilo   vinyl bromide
bromuro de xililo   xylyl bromide
bromuro de yodo   iodine bromide
bromuro de zirconilo   zirconyl bromide
bromuro de zirconio   zirconium bromide
bromuro estánnico   stannic bromide
bromuro estannoso   stannous bromide
bromuro férrico   ferric bromide

bromuro ferroso ferrous bromide
bromuro fosfórico phosphoric bromide
bromuro gálico gallic bromide
bromuro galoso gallous bromide
bromuro magnésico magnesium bromide
bromuro manganoso manganous
    bromide
bromuro mercúrico mercuric bromide
bromuro mercúrico bárico barium
    mercury bromide
bromuro mercurioso mercurous bromide
bromuro niqueloso nickelous bromide
bromuro potásico potassium bromide
bromuro samárico samaric bromide
bromuro sódico sodium bromide
bromuro talioso thallous bromide
bromuro telúrico telluric bromide
bromuro teluroso tellurous bromide
bronce bronze
bronce de aluminio aluminum bronze
bronce de campana bell metal
bronce de cañón gun metal
bronce de estaño tin bronze
bronce de manganeso manganese bronze
bronce de oro gold bronze
bronce de plata silver bronze
bronce de silicio silicon bronze
bronce de tungsteno tungsten bronze
bronce duro hard bronze
bronce fosforado phosphor bronze
broranotriilo boranetriyl
brucina brucine
brucita brucite
bruto crude
bufanina buphanine
bufotanina bufotanine
bufotenina bufotenine
bunamiodilo bunamiodyl
bunsenita bunsenite
burbuja bubble
bureta buret
bureta automática automatic buret
bureta de pesar weight buret
bureta graduada graduated buret
butadieno butadiene
butaldehído butaldehyde
butanal butanal
butano butane
butanoato cálcico calcium butanoate

butanoato de butilo butyl butanoate
butanoato de etilo ethyl butanoate
butanoato de pentilo pentyl butanoate
butanoato de propilo propyl butanoate
butanoato potásico potassium butanoate
butanodial butanedial
butanodiamida butanediamide
butanodiamina butanediamine
butanodiol butanediol
butanodiolamina butanediolamine
butanodiona butanedione
butanol butanol
butanona butanone
butanonitrilo butanenitrile
butanotiol butanethiol
butanotriol butanetriol
butenal butenal
butenilideno butenylidene
butenilo butenyl
buteno butene
butesina butesin
butetal butethal
butil-litio butyllithium
butilacrilamida butylacrylamide
butilamina butylamine
butilaminoetanol butylaminoethanol
butilanilina butylaniline
butilantraquinona butylanthraquinone
butilbenceno butylbenzene
butilbencenosulfonamida
    butylbenzenesulfonamide
butilcarbinol butyl carbinol
butildiamilamina butyldiamylamine
butilenglycol butylene glycol
butileno butylene
butiletanolamina butylethanolamine
butiletilacetaldehído
    butylethylacetaldehyde
butilfenol butylphenol
butilformamida butylformamide
butilhidroquinona butylhydroquinone
butilideno butylidene
butilisocianato butylisocyanate
butilmercaptano butylmercaptan
butilmetacrilato butylmethacrylate
butilo butyl
butilparabeno butylparaben
butilperoxipivalato butylperoxypivalate
butiltiofenol butylthiophenol

butiltolueno butyltoluene
butiltriclorosilano butyltrichlorosilane
butilurea butylurea
butino butyne
butinodiol butynediol
butiraldehído butyraldehyde
butirato bárico barium butyrate
butirato cobaltoso cobaltous butyrate
butirato de alilo allyl butyrate
butirato de amilo amyl butyrate
butirato de bencilo benzyl butyrate
butirato de butilo butyl butyrate
butirato de etilo ethyl butyrate
butirato de geranilo geranyl butyrate
butirato de isoamilo isoamyl butyrate
butirato de isopropilo isopropyl butyrate
butirato de metilo methyl butyrate
butirato de plomo lead butyrate
butirato de propilo propyl butyrate
butirato de vinilo vinyl butyrate

butirato sódico sodium butyrate
butírico butyric
butirilo butyryl
butirina butyrin
butirolactama butyrolactam
butirolactona butyrolactone
butirona butyrone
butironitrilo butyronitrile
butonato butonate
butopironoxilo butopyronoxyl
butoxi butoxy
butóxido alumínico aluminum butoxide
butóxido de piperonilo piperonyl
　　butoxide
butóxido potásico potassium butoxide
butoxietanol butoxyethanol
butoxifenol butoxyphenol
butoxipropanol butoxypropanol
butoxitriglicol butoxytriglycol

C

cabrerita cabrerite
cacao cocoa
cacodilato cacodylate
cacodilato sódico sodium cacodylate
cacodilo cacodyl
cacotelina cacotheline
cactina cactine
cadaleno cadalene
cadaverina cadaverine
cadena chain
cadena abierta open chain
cadena cerrada closed chain
cadena de carbonos carbon chain
cadena de desintegración decay chain
cadena de transporte electrónico
    electron transport chain
cadena lateral side chain, lateral chain
cadena periódica periodic chain
cadena radiactiva radioactive chain
cadena ramificada branched chain
cadena recta straight chain
cadineno cadinene
cadión cadion
cadmio cadmium
cadmio anaranjado orange cadmium
cadmio de dimetilo dimethyl cadmium
cadmoso cadmous
cafeína caffeine
cainita kainite
caja con atmósfera inerte inert
    atmosphere box
caja con guantes glove box
caja de Petri petri dish
cal lime
cal apagada slaked lime
cal cáustica caustic lime
cal clorada chlorinated lime
cal hidratada hydrated lime
cal sodada soda lime
cal sulfurada sulfurated lime
cal viva quicklime, burnt lime
calabarina calabarine
calamina calamine

calandria calender
calaverita calaverite
calcantita chalcanthite
calcedonia chalcedony
calceína calcein
calciferol calciferol
calcímetro calcimeter
calcimina calcimine
calcina calcine
calcinación calcination, roasting
calcinado calcined
calcinita calcinite
calcinol calcinol
calcio calcium
calcita calcite
calcofanita chalcophanite
calcógeno chalcogen
calcona chalcone
calcopirita chalcopyrite
calcosiderita chalcosiderite
calcosina chalcocite
caldeo heating
caledonita caledonite
calentamiento heating
calentamiento adiabático adiabatic
    heating
calentamiento espontáneo spontaneous
    heating
calibración calibration
calibrado calibrated
calibrar calibrate
calicantina calycanthine
caliche caliche
californio californium
californita californite
calinita kalinite
calixareno calixarene
caliza limestone
caliza magnésica magnesium limestone
calomel calomel
calor heat
calor atómico atomic heat
calor de absorción heat of absorption

calor de activación   heat of activation
calor de adhesión   heat of adhesion
calor de adsorción   heat of adsorption
calor de agregación   heat of aggregation
calor de asociación   heat of association
calor de atomización   heat of atomization
calor de combinación   heat of combination
calor de combustible   heat of fuel
calor de combustión   heat of combustion
calor de compresión   heat of compression
calor de condensación   heat of condensation
calor de cristalización   heat of crystallization
calor de descomposición   heat of decomposition
calor de dilución   heat of dilution
calor de disociación   heat of dissociation
calor de disolución   heat of dissolution
calor de emisión   heat of emission
calor de enfriamiento   heat of cooling
calor de evaporación   heat of evaporation
calor de explosión   heat of explosion
calor de formación   heat of formation
calor de fusión   heat of fusion
calor de hidratación   heat of hydration
calor de ionización   heat of ionization
calor de isomerización   heat of isomerization
calor de neutralización   heat of neutralization
calor de oxidación   heat of oxidation
calor de racemización   heat of racemization
calor de reacción   heat of reaction
calor de solidificación   heat of solidification
calor de solución   heat of solution
calor de sublimación   heat of sublimation
calor de transición   heat of transition
calor de vaporización   heat of vaporization
calor específico   specific heat
calor específico molal   molal specific heat
calor específico molar   molar specific heat
calor específico molecular   molecular specific heat
calor latente   latent heat
calor latente molar   molar latent heat
calor molar   molar heat
calor nuclear   nuclear heat
calor total   total heat
caloría   calorie
calorimetría   calorimetry
calorimetría de exploración diferencial   differential scanning calorimetry
calorimétrico   calorimetric
calorímetro   calorimeter
calorímetro adiabático   adiabatic calorimeter
calorímetro de Atwater   Atwater calorimeter
calorización   calorizing
calotropina   calotropin
calutrón   calutron
cámara   chamber
cámara de fisión   fission chamber
cámara de ionización   ionization chamber
cambio adiabático   adiabatic change
cambio de energía   energy change
cambio de estado   change of state
cambio de éster   ester interchange
cambio de fase   change of phase
cambio químico   chemical change
campana   bell glass, bell jar, hood
campana de humos   fume cupboard, fume hood
campo   field
campo atómico   atomic field
campo molecular   molecular field
canadina   canadine
canalización   channeling
canavanina   canavanine
cancrinita   cancrinite
canfano   camphane
canfanol   camphanol
canfanona   camphanone
canfeno   camphene
canfeno clorado   chlorinated camphene
canforoilo   camphoroyl
canforoquinona   camphoroquinone
cannabidiol   cannabidiol
cannabinol   cannabinol
cannabis   cannabis
canonita   cannonite

cantaridina cantharidin
cantaxantina canthaxanthin
cantidad quantity
cantidad de factor factor quantity
cantidad de substancia amount of
   substance
cantidad física physical quantity
cantidad molal parcial partial molal
   quantity
cáñamo hemp
caolin kaolin
caolinita kaolinite
caos chaos
capa layer, shell
capa abierta open shell
capa anódica anodic layer
capa catódica cathodic layer
capa de barrera barrier layer
capa de Beilby Beilby layer
capa de difusión diffusion layer
capa de Gouy Gouy layer
capa de Helmholtz Helmholtz layer
capa de níquel nickel layer
capa de ozono ozone layer
capa de valencia valence shell
capa doble double layer
capa doble eléctrica electric double layer
capa electrónica electron shell
capa exterior outer shell
capa fina thin layer
capacidad capacity
capacidad amortiguadora buffer
   capacity
capacidad básica basic capacity
capacidad térmica heat capacity
capacidad térmica específica specific
   heat capacity
capacidad térmica molar molar heat
   capacity
capacidad térmica molecular molecular
   heat capacity
caparrosa copperas
caparrosa amarilla yellow copperas
caparrosa azul blue copperas
caparrosa blanca white copperas
capilaridad capillarity
capraldehído capraldehyde
caprato caprate
caprato de etilo ethyl caprate

caprato de metilo methyl caprate
capreomicina capreomycin
caprilamida caprylamide
caprilato caprylate
caprilato amónico ammonium caprylate
caprilato de cinc zinc caprylate
caprilato de etilo ethyl caprylate
caprilato de metilo methyl caprylate
caprilato sódico sodium caprylate
caprililo caprylyl
caprilo capryl
caproato de etilo ethyl caproate
caproato de metilo methyl caproate
caproilo caproyl
caprolactama caprolactam
caprolactona caprolactone
capronilo capronyl
cápsula capsule
cápsula de combustión combustion
   capsule
captafol captafol
captano captan
captura capture
captura electrónica electron capture
captura neutrónica neutron capture
captura neutrónica radiativa radiative
   neutron capture
captura nuclear nuclear capture
captura protónica proton capture
captura radiativa radiative capture
caracolita caracolite
característica de electrodo electrode
   characteristic
caramelano caramelan
caramelización caramelization
caramelo caramel
carano carane
caraborano carbaborane
carbacol carbachol
carbaldehído carbaldehyde
carbamato carbamate
carbamato amónico ammonium
   carbamate
carbamato de alilo allyl carbamate
carbamato de butilo butyl carbamate
carbamato de diglicol diglycol
   carbamate
carbamato de etilo ethyl carbamate
carbamato de fenilo phenyl carbamate

carbamato de hidroxietilo hydroxyethyl carbamate
carbamato de pentilo pentyl carbamate
carbamato de propilo propyl carbamate
carbámico carbamic
carbamida carbamide
carbamidina carbamidine
carbamido carbamido
carbamilo carbamyl
carbamilurea carbamylurea
carbamita carbamite
carbamoilo carbamoyl
carbanilida carbanilide
carbanilo carbanil
carbanión carbanion
carbarilo carbaryl
carbato carbate
carbazato de etilo ethyl carbazate
carbazida carbazide
carbazoilo carbazoyl
carbazol carbazole
carbazolilo carbazolyl
carbazona carbazone
carbazotato amónico ammonium carbazotate
carbenio carbenium
carbeno carbene
carbenoide carbenoid
carbetoxiciclohexanona carbethoxycyclohexanone
carbetoxiciclopentanona carbethoxycyclopentanone
carbetoxipiperazina carbethoxypiperazine
carbilamina carbylamine
carbiloxima carbyloxime
carbino carbyne
carbinol carbinol
carbitol carbitol
carbobenzoilo carbobenzoyl
carbocatión carbocation
carbocíclico carbocyclic
carbodiazona carbodiazone
carbodihidrazida carbodihydrazide
carbodiimida carbodiimide
carboditioico carbodithioic
carboestirilo carbostyril
carbofenotión carbophenothion
carbofurano carbofuran

carbohidrasa carbohydrase
carbohidrato carbohydrate
carbohidrazida carbohydrazide
carbohidruro carbohydride
carboide carboid
carbolato sódico sodium carbolate
carbolfucsina carbolfuchsin
carbómero carbomer
carbometeno carbomethene
carbometoxi carbomethoxy
carbomicina carbomycin
carbón carbon, charcoal, coal
carbón activado activated carbon, activated charcoal
carbón activo active carbon
carbón bituminoso bituminous coal
carbón de bujía cannel coal
carbón industrial industrial carbon
carbón mineral mineral carbon
carbonación carbonation
carbonado carbonado
carbonatación carbonatation
carbonato carbonate
carbonato ácido amónico ammonium acid carbonate
carbonato ácido potásico potassium acid carbonate
carbonato ácido sódico sodium acid carbonate
carbonato alumínico aluminum carbonate
carbonato amónico ammonium carbonate
carbonato bárico barium carbonate
carbonato cálcico calcium carbonate
carbonato ceroso cerous carbonate
carbonato cobaltoso cobaltous carbonate
carbonato cromoso chromous carbonate
carbonato cúprico cupric carbonate
carbonato cuproso cuprous carbonate
carbonato de berilio beryllium carbonate
carbonato de bismutilo bismuthyl carbonate
carbonato de bismuto bismuth carbonate
carbonato de cadmio cadmium carbonate
carbonato de cerio cerium carbonate
carbonato de cesio cesium carbonate
carbonato de cinc zinc carbonate

carbonato de cobalto cobalt carbonate
carbonato de cobre copper carbonate
carbonato de dietilo diethyl carbonate
carbonato de difenilo diphenyl carbonate
carbonato de estroncio strontium carbonate
carbonato de etilbutilo ethylbutyl carbonate
carbonato de etileno ethylene carbonate
carbonato de etilo ethyl carbonate
carbonato de fenilo phenyl carbonate
carbonato de glicerina glycerin carbonate
carbonato de glicol glycol carbonate
carbonato de guanidina guanidine carbonate
carbonato de hidrógeno hydrogen carbonate
carbonato de hidrógeno amónico ammonium hydrogen carbonate
carbonato de hidrógeno de estroncio strontium hydrogen carbonate
carbonato de hidrógeno de litio lithium hydrogen carbonate
carbonato de hidrógeno magnésico magnesium hydrogen carbonate
carbonato de hidrógeno potásico potassium hydrogen carbonate
carbonato de hidrógeno sódico sodium hydrogen carbonate
carbonato de hidroximetileno hydroxymethylene carbonate
carbonato de hierro iron carbonate
carbonato de itrio yttrium carbonate
carbonato de lantano lanthanum carbonate
carbonato de litio lithium carbonate
carbonato de manganeso manganese carbonate
carbonato de metilo methyl carbonate
carbonato de metilo sódico sodium methyl carbonate
carbonato de neodimio neodymium carbonate
carbonato de níquel nickel carbonate
carbonato de plata silver carbonate
carbonato de plomo lead carbonate
carbonato de propileno propylene carbonate

carbonato de radio radium carbonate
carbonato de rubidio rubidium carbonate
carbonato de talio thallium carbonate
carbonato de zirconilo zirconyl carbonate
carbonato de zirconilo amónico ammonium zirconyl carbonate
carbonato de zirconio zirconium carbonate
carbonato ferroso ferrous carbonate
carbonato magnésico magnesium carbonate
carbonato manganoso manganous carbonate
carbonato mercúrico mercuric carbonate
carbonato niqueloso nickelous carbonate
carbonato potásico potassium carbonate
carbonato sódico sodium carbonate
carbonato sódico potásico potassium sodium carbonate
carbonato talioso thallous carbonate
carbonatomineral carbonate mineral
carbónico carbonic
carbonilación carbonylation
carbonilato carbonylate
carbonildiimidazol carbonyldiimidazole
carbonilo carbonyl
carbonilo de cobalto cobalt carbonyl
carbonilo de cromo chromium carbonyl
carbonilo de hierro iron carbonyl
carbonilo de manganeso manganese carbonyl
carbonilo de níquel nickel carbonyl
carbonilo de rutenio ruthenium carbonyl
carbonilo de tungsteno tungsten carbonyl
carbonimidoilo carbonimidoyl
carbonio carbonium
carbonita carbonite
carbonitrilo carbonitrile
carbonización carbonization
carbonizado carbonized
carbonizar carbonize
carbonldioxi carbonyldioxy
carbono carbon
carbono asimétrico asymmetric carbon
carbono combinado combined carbon
carbono cuaternario quaternary carbon

carbono divalente divalent carbon
carbono fijo fixed carbon
carbono secundario secondary carbon
carbono terciario tertiary carbon
carbono tetraédrico tetrahedral carbon
carbonohidrazida carbonohydrazide
carbonohidrazido carbonohydrazido
carbonómetro carbometer
carborano carborane
carbotioico carbothioic
carbotiónico carbothionic
carboxamida carboxamide
carboxamidina carboxamidine
carboxi carboxy
carboxibenceno carboxybenzene
carboxilación carboxylation
carboxilación de Koch-Haaf Koch-Haaf carboxylation
carboxilasa carboxylase
carboxílico carboxylic
carboxilo carboxyl
carboximetilcelulosa carboxymethylcellulose
carboximetilcelulosa sódica sodium carboxymethylcellulose
carboxipeptidasa carboxypeptidase
carburita carburite
carburización carburization
carburo carbide
carburo alumínico aluminum carbide
carburo bárico barium carbide
carburo cálcico calcium carbide
carburo crómico chromic carbide
carburo de berilio beryllium carbide
carburo de boro boron carbide
carburo de cromo chromium carbide
carburo de ditungsteno ditungsten carbide
carburo de hafnio hafnium carbide
carburo de hierro iron carbide
carburo de litio lithium carbide
carburo de manganeso manganese carbide
carburo de niobio niobium carbide
carburo de silicio silicon carbide
carburo de tántalo tantalum carbide
carburo de titanio titanium carbide
carburo de torio thorium carbide
carburo de tungsteno tungsten carbide

carburo de uranio uranium carbide
carburo de vanadio vanadium carbide
carburo de zirconio zirconium carbide
carburo sódico sodium carbide
carcinógeno carcinogen
cardenillo verdigris
cardenillo azul blue verdigris
cardenolida cardenolide
careno carene
carga charge
carga atómica atomic charge
carga de electrón charge of electron
carga efectiva effective charge
carga electrónica electron charge
carga elemental elementary charge
carga equivalente equivalent charge
carga específica specific charge
carga específica de electrón electron specific charge
carga formal formal charge
carga iónica ionic charge
carga libre free charge
carga nuclear nuclear charge
carga química chemical load
cargado charged
cariofileno caryophyllene
carmín carmine
carminato sódico sodium carminate
carminita carminite
carnalita carnallite
carnegina carnegine
carnina carnine
carnitina carnitine
carnosina carnosine
carnotita carnotite
carobina carobine
carona carone
caroteno carotene
carotenoide carotenoid
carpaína carpain
carraginina carrageenan
carreno carreen
cartamina carthamin
carvacrol carvacrol
carveol carveol
carvestreno carvestrene
carvona carvone
cascada cascade
caseína casein

caseinasa caseinase
caseinato cálcico calcium caseinate
caseinato sódico sodium caseinate
casiterita cassiterite, tinstone
catabiótico catabiotic
catabólico catabolic
catabolismo catabolism
cataforesis cataphoresis
catalasa catalase
catálisis catalysis
catálisis ácido-base acid-base catalysis
catálisis de radiación radiation catalysis
catálisis fotoquímica photochemical catalysis
catálisis heterogénea heterogeneous catalysis
catálisis homogénea homogeneous catalysis
catálisis negativa negative catalysis
catálisis protolítica protolytic catalysis
catalítico catalytic
catalizador catalyst
catalizador alfin alfin catalyst
catalizador de Adams Adams catalyst
catalizador de Raney Raney catalyst
catalizador de Wilkinson Wilkinson catalyst
catalizador de zeolita zeolite catalyst
catalizador de Ziegler Ziegler catalyst
catalizador estereoespecífico stereospecific catalyst
catalizador redox redox catalyst
catalizador selectivo de tamaño shape-selective catalyst
catarómetro catharometer
catartina cathartin
catecol catechol
catecolborano catecholborane
catena catena
catenación catenation
catenano catenane
catenario catenary
catenilo catenyl
catequina catechin
catetómetro cathetometer
catina cathine
catión cation
catiónico cationic
cationotropía cationotropy

cationtrofía cationtrophy
catlinita catlinite
catódico cathodic
cátodo cathode
cátodo electrolítico electrolytic cathode
católito catholyte
caucho rubber
caucho artificial artificial rubber
caucho butilo butyl rubber
caucho clorado chlorinated rubber
caucho de clorohidrina chlorohydrin rubber
caucho de estireno-butadieno styrene-butadiene rubber
caucho de poliuretano polyurethane rubber
caucho de silicio silicon rubber
caucho de silicona silicone rubber
caucho frío cold rubber
caucho nitrilo nitrile rubber
caucho poliéster polyester rubber
caucho sintético synthetic rubber
causticidad causticity
cáustico caustic
cavitación cavitation
cebadilla sabadilla
cedazo sieve, screen
cedazo molecular molecular sieve
cedrina cedrin
cedrol cedrol
cefalina cephalin
cefalosporina cephalosporin
cefamicina cephamycin
celasa cellase
celda de plomo lead cell
celda electrolítica electrolytic cell
celestina celestine
celestita celestite
celobiosa cellobiose
celodextrina cellodextrin
celofán cellophane
celoidina celloidin
celosía lattice
célula cell
célula de absorción absorption cell
célula de Bénard Bénard cell
célula de redox redox cell
célula electroquímica electrochemical cell

célula fotoeléctrica photoelectric cell
célula fotovoltaica photovoltaic cell
célula osmótica osmotic cell
célula solar solar cell
celulasa cellulase
celulosa cellulose
celulosa alfa alpha cellulose
celulosa de madera wood cellulose
celulosato sódico sodium cellulosate
celulósico cellulosic
cementación cementation
cementita cementite
cemento cement
cemento portland portland cement
ceniza ash
ceniza caustificada causticized ash
ceniza de estaño tin ash
ceniza de huesos bone ash
ceniza de humero fly ash
ceniza de madera wood ash
ceniza de sosa soda ash
cenizas de perla pearl ash
centelleador scintillator
centelleante scintillating
centelleo scintillation
centi centi
centígrado centigrade
centinormal centinormal
centrado en el cubo cubic-centered
centrífuga centrifuge
centrifugación centrifugation
centrífugo centrifugal
centrípeto centripetal
centro activo active center
centro de simetría center of symmetry
control quiral chiral center
centrón centron
cera wax
cera de abejas beeswax
cera de Brasil Brazil wax
cera de candelilla candelilla wax
cera de caña de azúcar sugarcane wax
cera de carnauba carnauba wax
cera de ceresina ceresin wax
cera de China Chinese wax
cera de parafina paraffin wax
cera de petrolato petrolatum wax
cera de petróleo petroleum wax
cera de tierra earth wax

cera microcristalina microcrystalline wax
cera mineral mineral wax
cera montana montan wax
cerámica ceramic
ceramida ceramide
cerato cerate
cerebrosa cerebrose
cerebrósidos cerebrosides
ceresina ceresin
ceria ceria
cérico ceric
cerílico cerylic
cerilo ceryl
cerina cerin
cerio cerium
cerita cerite
cermet cermet
cero absoluto absolute zero
ceroato de metilo methyl ceroate
cerosina cerosin
cerosinilo cerosinyl
ceroso cerous
ceroteno cerotene
cerusita cerussite
cervantita cervantite
cerveza beer
cesio cesium
cespitina cespitine
cetal ketal
cetano cetane
cetena ketene
ceteno cetene
cetilato cetylate
cetilbetamina cetylbetamine
cetileno cetylene
cetilmercaptano cetyl mercaptan
cetilo cetyl
cetimida ketimide
cetimina ketimine
cetina ketine, cetin
ceto keto
cetoácido keto acid
cetoamina ketoamine
cetobenzotriazina ketobenzotriazine
cetohexosa ketohexose
cetol ketol, cetol
cetona ketone
cetona aromática aromatic ketone

cetona de aldehído aldehyde ketone
cetona etílica bencílica benzyl ethyl
   ketone
cetopentosa ketopentose
cetosa ketose
cetosis ketosis
cetoxima ketoxime
cetrimida cetrimide
cianamida cyanamide
cianamida cálcica calcium cyanamide
cianamida de dialilo diallyl cyanamide
cianamida de dimetilo dimethyl
   cyanamide
cianamida sódica sodium cyanamide
cianato cyanate
cianato bárico barium cyanate
cianato de clorobecilo chlorobenzyl
   cyanate
cianato de etilo ethyl cyanate
cianato de metilo methyl cyanate
cianato de plata silver cyanate
cianato de plomo lead cyanate
cianato mercúrico mercuric cyanate
cianato potásico potassium cyanate
cianato sódico sodium cyanate
cianaurita cyanaurite
cianazina cyanazine
cianhidrina cyanohydrin
cianhidrina de acetaldehído
   acetaldehyde cyanohydrin
cianhidrina de acetona acetone
   cyanohydrin
cianhidrina de benzaldehído
   benzaldehyde cyanohydrin
cianhidrina de bromoacetona
   bromoacetone cyanohydrin
cianhidrina de etileno ethylene
   cyanohydrin
cianidina cyanidine
cianina cyanine
cianina de alizarina alizarin cyanine
cianita cyanite
ciano cyano
cianoacetamida cyanoacetamide
cianoacetato de etilhexilo ethylhexyl
   cyanoacetate
cianoacetato de etilo ethyl cyanoacetate
cianoacetato de metilo methyl
   cyanoacetate

cianoacrilato cyanoacrylate
cianoacrilato de isobutilo isobutyl
   cyanoacrylate
cianoaldehído cyanoaldehyde
cianoanilina cyanoaniline
cianoaurita sódica sodium cyanoaurite
cianobenceno cyanobenzene
cianobencilo cyanobenzyl
cianobenzamida cyanobenzamide
cianoborohidruro sódico sodium
   cyanoborohydride
cianocarburo cyanocarbon
cianocobalamina cyanocobalamin
cianocuprato sódico sodium
   cyanocuprate
cianoetanoato de etilo ethyl
   cyanoethanoate
cianoetanosulfonato sódico sodium
   cyanoethanesulfonate
cianoetilación cyanoethylation
cianoferrato cyanoferrate
cianoformiato cyanoformate
cianoformiato de metilo methyl
   cyanoformate
cianógeno cyanogen
cianopiridina cyanopyridine
cianoplatinato cyanoplatinate
cianoplatinita bárica barium
   cyanoplatinite
cianurato de trialilo triallyl cyanurate
cianúrico cyanuric
cianuro cyanide
cianuro amónico ammonium cyanide
cianuro áurico auric cyanide
cianuro bárico barium cyanide
cianuro cálcico calcium cyanide
cianuro cobaltoso cobaltous cyanide
cianuro cúprico cupric cyanide
cianuro cuproso cuprous cyanide
cianuro de alilo allyl cyanide
cianuro de bencilo benzyl cyanide
cianuro de benzoilo benzoyl cyanide
cianuro de bromo bromine cyanide
cianuro de bromobencilo bromobenzyl
   cyanide
cianuro de butilo butyl cyanide
cianuro de cadmio cadmium cyanide
cianuro de carbonilo carbonyl cyanide
cianuro de cesio cesium cyanide

cianuro de cinc zinc cyanide
cianuro de cloro chlorine cyanide
cianuro de clorometilo chloromethyl
    cyanide
cianuro de cobalto cobalt cyanide
cianuro de cobre copper cyanide
cianuro de cobre sódico sodium copper
    cyanide
cianuro de etileno ethylene cyanide
cianuro de etilo ethyl cyanide
cianuro de fenilo phenyl cyanide
cianuro de flúor fluorine cyanide
cianuro de hidrógeno hydrogen cyanide
cianuro de isopropilo isopropyl cyanide
cianuro de mercurio mercury cyanide
cianuro de metilmercurio
    methylmercury cyanide
cianuro de metilo methyl cyanide
cianuro de níquel nickel cyanide
cianuro de nitrobencilo nitrobenzyl
    cyanide
cianuro de oro gold cyanide
cianuro de oro sódico sodium gold
    cyanide
cianuro de pentilo pentyl cyanide
cianuro de plata silver cyanide
cianuro de plata sódico sodium silver
    cyanide
cianuro de plomo lead cyanide
cianuro de propilo propyl cyanide
cianuro de vinilo vinyl cyanide
cianuro de yodo iodine cyanide
cianuro ferripotásico ferripotassium
    cyanide
cianuro mercúrico mercuric cyanide
cianuro negro black cyanide
cianuro palmítico palmitic cyanide
cianuro potásico potassium cyanide
cianuro potásico cuproso cuprous
    potassium cyanide
cianuro potásico de cadmio cadmium
    potassium cyanide
cianuro potásico de plata silver
    potassium cyanide
cianuro potásico mercúrico mercuric
    potassium cyanide
cianuro sódico sodium cyanide
cianuro sódico de plata silver sodium
    cyanide

cibotáctico cybotactic
ciclamato cyclamate
ciclamato cálcico calcium cyclamate
ciclamato potásico potassium cyclamate
ciclamato sódico sodium cyclamate
cicleno cyclene
cíclico cyclic
ciclitol cyclitol
ciclización cyclization
ciclización de Volhard-Erdmann
    Volhard-Erdmann cyclization
ciclo cycle
ciclo de ácido cítrico citric acid cycle
ciclo de ácido tricarboxílico
    tricarboxylic acid cycle
ciclo de azufre sulfur cycle
ciclo de Born-Haber Born-Haber cycle
ciclo de carbono carbon cycle
ciclo de carbono-nitrógeno carbon-
    nitrogen cycle
ciclo de Carnot Carnot cycle
ciclo de fósforo phosphorus cycle
ciclo de glioxilato glyoxylate cycle
ciclo de Krebs Krebs cycle
ciclo de nitrógeno nitrogen cycle
ciclo de oxidación-reducción oxidation-
    reduction cycle
ciclo de urea urea cycle
ciclo neutrónico neutron cycle
ciclo protónico proton cycle
ciclo térmico heat cycle
cicloadición cycloaddition
cicloalcano cycloalkane
cicloalifático cycloaliphatic
cicloalqueno cycloalkene
cicloalquilo cycloalkyl
ciclobarbital cyclobarbital
ciclobarbitona cyclobarbitone
ciclobutadieno cyclobutadiene
ciclobutano cyclobutane
ciclobutanol cyclobutanol
ciclobuteno cyclobutene
ciclobutilo cyclobutyl
ciclofano cyclophane
ciclofosfamida cyclophosphamide
cicloheptadeceno cycloheptadecene
cicloheptano cycloheptane
cicloheptanol cycloheptanol
cicloheptanona cycloheptanone

ciclohepteno cycloheptene
cicloheptilo cycloheptyl
ciclohexano cyclohexane
ciclohexanodiol cyclohexanediol
ciclohexanol cyclohexanol
ciclohexanona cyclohexanone
ciclohexeniltriclorosilano
    cyclohexenyltrichlorosilane
ciclohexeno cyclohexene
ciclohexenol cyclohexenol
ciclohexilamina cyclohexylamine
ciclohexilbenceno cyclohexylbenzene
ciclohexilfenol cyclohexylphenol
ciclohexilideno cyclohexylidene
ciclohexilpiperdina cyclohexylpiperdine
cicloheximida cycloheximide
ciclón cyclone
ciclonita cyclonite
ciclononano cyclononane
ciclooctadieno cyclooctadiene
ciclooctano cyclooctane
ciclooctatetraeno cyclooctatetraene
ciclootteno cyclooctene
cicloolefina cycloolefin
cicloparafina cycloparaffin
ciclopentadecanona cyclopentadecanone
ciclopentadienida sódica sodium
    cyclopentadienide
ciclopentadieno cyclopentadiene
ciclopentano cyclopentane
ciclopentanol cyclopentanol
ciclopentanona cyclopentanone
ciclopentenilacetona
    cyclopentenylacetone
ciclopentenilo cyclopentenyl
ciclopenteno cyclopentene
ciclopentilacetona cyclopentylacetone
ciclopentileno cyclopentylene
ciclopentilo cyclopentyl
ciclopropano cyclopropane
ciclopropilo cyclopropyl
ciclosilano cyclosilane
ciclotetraeno cyclotetraene
ciclotrón cyclotron
cicuta hemlock
cicutoxina cicutoxine
cierre de anillo ring closure
ciguatoxina ciguatoxin
cilindrita cylindrite

cilindro cylinder
cilindro de filtración filter cylinder
cilíndro de gas gas cylinder
cilindro graduado graduated cylinder
cimeno cymene
cimetidina cimetidine
cimilo cymyl
cinabrio cinnabar
cinamaldehído cinnamaldehyde
cinamaldehído de hexilo hexyl
    cinnamaldehyde
cinamamida cinnamamide
cinamato cinnamate
cinamato cálcico calcium cinnamate
cinamato de alilo allyl cinnamate
cinamato de bencilo benzyl cinnamate
cinamato de cadmio cadmium cinnamate
cinamato de cinamilo cinnamyl
    cinnamate
cinamato de etilo ethyl cinnamate
cinamato de isobutilo isobutyl cinnamate
cinamato de metilo methyl cinnamate
cinamato potásico potassium cinnamate
cinamato sódico sodium cinnamate
cinameína cinnamein
cinamenilo cinnamenyl
cinameno cinnamene
cinamida cinnamide
cinamilideno cinnamylidene
cinamoilo cinnamoyl
cinc zinc
cincato zincate
cincato cobalto cobalt zincate
cincita zincite
cincocaína cinchocain
cinconamina cinchonamine
cinconidina cinchonidin
cinconina cinchonine
cincoso zincous
cineno cinene
cineol cineol
cinerina cinerin
cinerolona cinerolone
cinética kinetics
cinética de enzima enzyme kinetics
cinética de reacciones reaction kinetics
cinética química chemical kinetics
cinético kinetic
cinolina cinnoline

ciprita cyprite
cis cis
cisco breeze
cisoide cisoid
cistamina cystamine
cisteína cysteine
cistina cystine
citidina cytidine
citiolona citiolone
citocromo cytochrome
citoplasma cytoplasm
citoquímica cytochemistry
citosina cytosine
citotóxico cytotoxic
citotoxina cytotoxin
citraconoilo citraconoyl
citral citral
citramida citramide
citrato citrate
citrato ácido magnésico acid magnesium
     citrate
citrato amónico ammonium citrate
citrato amónico férrico ferric
     ammonium citrate
citrato bárico barium citrate
citrato cálcico calcium citrate
citrato ceroso cerous citrate
citrato cobaltoso cobaltous citrate
citrato cúprico cupric citrate
citrato de bismuto bismuth citrate
citrato de butilo butyl citrate
citrato de cafeína caffeine citrate
citrato de cinc zinc citrate
citrato de codeína codeine citrate
citrato de etilo ethyl citrate
citrato de fentanilo fentanyl citrate
citrato de hierro iron citrate
citrato de hierro amónico ammonium
     iron citrate
citrato de litio lithium citrate
citrato de manganeso manganese citrate
citrato de plata silver citrate
citrato de plomo lead citrate
citrato de tributilo tributyl citrate
citrato de trietilo triethyl citrate
citrato disódico disodium citrate
citrato estannoso stannous citrate
citrato férrico ferric citrate
citrato magnésico magnesium citrate

citrato manganoso manganous citrate
citrato potásico potassium citrate
citrato sódico sodium citrate
citrato tricálcico tricalcium citrate
citrato trisódico trisodium citrate
citreno citrene
citresia citresia
citrinina citrinin
citronelal citronellal
citronelol citronellol
citronilo citronyl
citrulina citrulline
civetal civettal
civeto civet
clarificación clarification
clarificador clarifier
clasificación classification
clasificación de Ehrenfest Ehrenfest
     classification
clasificación periódica periodic
     classification
clatrato clathrate
clebrio clebrium
clinohedrita clinohedrite
clinoptilolita clinoptilolite
clioquinol clioquinol
clomercuriferroceno
     chloromercuriferrocene
clomifeno clomiphene
clonazepam clonazepam
clopidol clopidol
clorado chlorinated
cloral chloral
clorálcali chloralkali
cloralosa chloralose
cloramina chloramine
cloranfenicol chloramphenicol
cloranil chloranil
cloranilato chloranilate
clorapatita chlorapatite
clorato chlorate
clorato alumínico aluminum chlorate
clorato amónico ammonium chlorate
clorato bárico barium chlorate
clorato cálcico calcium chlorate
clorato cobaltoso cobaltous chlorate
clorato de cadmio cadmium chlorate
clorato de cinc zinc chlorate
clorato de estroncio strontium chlorate

clorato de estroncio-potasio strontium-potassium chlorate
clorato de litio lithium chlorate
clorato de plata silver chlorate
clorato de plomo lead chlorate
clorato magnésico magnesium chlorate
clorato mercurioso mercurous chlorate
clorato potásico potassium chlorate
clorato potásico bárico barium potassium chlorate
clorato sódico sodium chlorate
cloraurita chloraurite
clorazina chlorazine
clorbensida chlorbenside
clorbutanol chlorbutanol
clordano chlordane
clordimeformo chlordimeform
clorendato de dialilo diallyl chlorendate
clorfenvinfos chlorfenvinphos
clorhexidina chlorhexidine
clorhidrato hydrochloride
clorhidrato de acetamida acetamide hydrochloride
clorhidrato de acetamidina acetamidine hydrochloride
clorhidrato de aminoazobenceno aminoazobenzene hydrochloride
clorhidrato de anilina aniline hydrochloride
clorhidrato de betaína betaine hydrochloride
clorhidrato de buclizina buclizine hydrochloride
clorhidrato de butetamina butethamine hydrochloride
clorhidrato de cafeína caffeine hydrochloride
clorhidrato de caucho rubber hydrochloride
clorhidrato de ciclizina cyclizine hydrochloride
clorhidrato de clordiazepóxido chlordiazepoxide hydrochloride
clorhidrato de cocaína cocaine hydrochloride
clorhidrato de codeína codeine hydrochloride
clorhidrato de diaminofenol diaminophenol hydrochloride

clorhidrato de dipenteno dipentene hydrochloride
clorhidrato de efedrina ephedrine hydrochloride
clorhidrato de etilmorfina ethylmorphine hydrochloride
clorhidrato de fenacaína phenacaine hydrochloride
clorhidrato de fenilciclideno phenylcyclidene hydrochloride
clorhidrato de galactosamina galactosamine hydrochloride
clorhidrato de guanidina guanidine hydrochloride
clorhidrato de hidralazina hydralazine hydrochloride
clorhidrato de hidroxilamina hydroxylamine hydrochloride
clorhidrato de meclizina meclizine hydrochloride
clorhidrato de meperidina meperidine hydrochloride
clorhidrato de metadona methadone hydrochloride
clorhidrato de metopón metopon hydrochloride
clorhidrato de naftilamina naphthylamine hydrochloride
clorhidrato de nicotina nicotine hydrochloride
clorhidrato de nilidrina nylidrin hydrochloride
clorhidrato de papaverina papaverine hydrochloride
clorhidrato de pineno pinene hydrochloride
clorhidrato de piridoxal pyridoxal hydrochloride
clorhidrato de procaína procaine hydrochloride
clorhidrato de promazina promazine hydrochloride
clorhidrato de semicarbazida semicarbazide hydrochloride
clorhidrato de tetracaína tetracaine hydrochloride
clorhidrato de tiamina thiamine hydrochloride
clorhidrato de vancomicina vancomycin

hydrochloride
clorhídrico hydrochloric
cloridión chloridion
clorilo chloryl
clorimida chlorimide
clorinación chlorination
clorinidad chlorinity
clorinólisis chlorinolysis
cloritión chlorition
clorito chlorite
clorito cálcico calcium chlorite
clorito de plata silver chlorite
clorito sódico sodium chlorite
clormetazanona chlormethazanone
cloro chlorine
cloroacetal de etilo ethyl chloroacetal
cloroacetaldehído chloroacetaldehyde
cloroacetamida chloroacetamide
cloroacetato chloroacetate
cloroacetato de cadmio cadmium
　　chloroacetate
cloroacetato de etilo ethyl chloroacetate
cloroacetato de metilo methyl
　　chloroacetate
cloroacetato sódico sodium chloroacetate
cloroacetilo chloroacetyl
cloroacetiluretano chloroacetylurethane
cloroacetoacetanilida
　　chloroacetoacetanilide
cloroacetoacetato de etilo ethyl
　　chloroacetoacetate
cloroacetofenona chloroacetophenone
cloroacetona chloroacetone
cloroacetonitrilo chloroacetonitrile
cloroacrilato chloroacrylate
cloroacrilonitrilo chloroacrylonitrile
cloroacroleína chloroacrolein
cloroaluminato sódico sodium
　　chloroaluminate
cloroanilato de lantano lanthanum
　　chloroanilate
cloroanilina chloroaniline
cloroantraquinona chloroanthraquinone
cloroargentato chloroargentate
cloroaurato sódico sodium chloroaurate
clorobecilpiridina chlorobenzylpyridine
clorobenceno chlorobenzene
clorobencenosulfonamida
　　chlorobenzenesulfonamide

clorobencenotiol chlorobenzenethiol
clorobenzal chlorobenzal
clorobenzaldehído chlorobenzaldehyde
clorobenzantrona chlorobenzanthrone
clorobenzhidrol chlorobenzhydrol
clorobenzoato chlorobenzoate
clorobenzofenona chlorobenzophenone
clorobenzohidrol chlorobenzohydrol
clorobenzoilo chlorobenzoyl
clorobenzotricloruro
　　chlorobenzotrichloride
clorobenzotrifluoruro
　　chlorobenzotrifluoride
clorobromo chlorobromo
clorobromuro de etileno ethylene
　　chlorobromide
clorobutadieno chlorobutadiene
clorobutano chlorobutane
clorobutanol chlorobutanol
clorobutirilo chlorobutyryl
clorocarbonato chlorocarbonate
clorocarbonato de bencilo benzyl
　　chlorocarbonate
clorocarbonato de etilo ethyl
　　chlorocarbonate
clorocarbonato de pentilo pentyl
　　chlorocarbonate
clorocarbónico chlorocarbonic
clorocarburo chlorocarbon
clorocianógeno chlorocyanogen
clorocianuro chlorocyanide
clorocosano chlorocosane
clorocromato chlorochromate
clorocromato potásico potassium
　　chlorochromate
clorocumarina chlorocoumarin
clorodecona chlorodecone
clorodifenilo chlorodiphenyl
clorodifluoroetano chlorodifluoroethane
clorodifluorometano
　　chlorodifluoromethane
clorodinitrobenceno
　　chlorodinitrobenzene
cloroestireno chlorostyrene
cloroetanamida chloroethanamide
cloroetano chloroethane
cloroetanol chloroethanol
cloroeteno chloroethene
cloroetilcloroformiato

chloroethylchloroformate

cloroetileno chloroethylene

clorofenetol chlorofenethol

clorofenicita chlorophoenicite

clorofeniltriclorosilano
chlorophenyltrichlorosilane

clorofenol chlorophenol

clorofila chlorophyll

clorofilina chlorophyllin

clorofluorocarburo chlorofluorocarbon

clorofluoruro chlorofluoride

cloroformiato chloroformate

cloroformiato de bencilo benzyl
chloroformate

cloroformiato de etilo ethyl
chloroformate

cloroformiato de isobutilo isobutyl
chloroformate

cloroformiato de metilo methyl
chloroformate

cloroformiato de triclorometilo
trichloromethyl chloroformate

cloroformo chloroform

cloroformo de acetona acetone
chloroform

cloroformoxima chloroformoxime

clorofosfato de dibutilo dibutyl
chlorophosphate

clorofosfato de dioctilo dioctyl
chlorophosphate

cloroftalato sódico sodium
chlorophthalate

clorohidrato alumínico aluminum
chlorohydrate

clorohidrina chlorohydrin

clorohidrina de etileno ethylene
chlorohydrin

clorohidrina de propileno propylene
chlorohydrin

clorohidrina de trimetileno trimethylene
chlorohydrin

clorohidroquinona chlorohydroquinone

clorohidroxibenceno
chlorohydroxybenzene

clorohidroxitolueno
chlorohydroxytoluene

clorometano chloromethane

clorometilación chloromethylation

clorometilanilina chloromethylaniline

clorometilbenceno chloromethylbenzene

clorometilcloroformiato
chloromethylchloroformate

clorometilclorosulfonato
chloromethylchlorosulfonate

clorometilnaftaleno
chloromethylnaphthalene

clorometilo chloromethyl

cloronaftaleno chloronaphthalene

cloronitroanilina chloronitroaniline

cloronitrobenceno chloronitrobenzene

cloronitrotolueno chloronitrotoluene

cloropentafluoroacetona
chloropentafluoroacetone

cloropentano chloropentane

cloropicrina chloropicrin

cloropiridina chloropyridine

cloroplatinato chloroplatinate

cloroplatinato bárico barium
chloroplatinate

cloroplatinato potásico potassium
chloroplatinate

cloroplatinato sódico sodium
chloroplatinate

cloroplatinita sódica sodium
chloroplatinite

cloropreno chloroprene

cloropropano chloropropane

cloropropeno chloropropene

cloropropileno chloropropylene

cloropropionato de etilo ethyl
chloropropionate

cloropropionitrilo chloropropionitrile

cloroquina chloroquine

cloroquinaldina chloroquinaldine

clorosalicilanilida chlorosalicylanilide

clorosfosfato de dietilo diethyl
chlorophosphate

clorosilano de metilo methyl
chlorosilane

clorosilo chlorosyl

cloroso chlorous

clorosuccinimida chlorosuccinimide

clorosulfonato de etilo ethyl
chlorosulfonate

clorosulfonato de metilo methyl
chlorosulfonate

clorosulfonato de propilo propyl
chlorosulfonate

clorosulfonato sódico sodium chlorosulfonate
clorotiazida chlorothiazide
clorotimol chlorothymol
clorotiofenol chlorothiophenol
clorotolueno chlorotoluene
clorotrifluoroetileno chlorotrifluoroethylene
clorotrifluorometano chlorotrifluoromethane
clorovinildicloroarsina chlorovinyldichloroarsine
cloroxileno chloroxylene
cloroxina chloroxine
cloroyoduro chloroiodide
cloroyoduro de cinc zinc chloroiodide
clorpromazina chlorpromazine
clorquinaldol chlorquinaldol
clortetraciclina chlortetracycline
cloruro chloride
cloruro alumínico aluminum chloride
cloruro amónico ammonium chloride
cloruro amónico de níquel nickel ammonium chloride
cloruro amónico de platino platinum ammonium chloride
cloruro amónico platínico platinic ammonium chloride
cloruro antimónico antimonic chloride
cloruro antimonioso antimonous chloride
cloruro arsenioso arsenous chloride
cloruro áurico auric chloride
cloruro bárico barium chloride
cloruro cálcico calcium chloride
cloruro ceroso cerous chloride
cloruro cianofórmico cyanoformic chloride
cloruro cianúrico cyanuric chloride
cloruro cobáltico cobaltic chloride
cloruro cobaltoso cobaltous chloride
cloruro crómico chromic chloride
cloruro cromoso chromous chloride
cloruro cúprico cupric chloride
cloruro cuproso cuprous chloride
cloruro de acetamida acetamide chloride
cloruro de acetilo acetyl chloride
cloruro de ácido acid chloride
cloruro de acriloilo acryloyl chloride

cloruro de alilo allyl chloride
cloruro de amida amide chloride
cloruro de amilo amyl chloride
cloruro de anilina aniline chloride
cloruro de anisoilo anisoyl chloride
cloruro de antimonio antimony chloride
cloruro de arsénico arsenic chloride
cloruro de azufre sulfur chloride
cloruro de bencenocarbonilo benzenecarbonyl chloride
cloruro de bencenodiazonio benzenediazonium chloride
cloruro de bencetonio benzethonium chloride
cloruro de bencilo benzyl chloride
cloruro de benciltrimetilamonio benzyltrimethylammonium chloride
cloruro de benzalconio benzalkonium chloride
cloruro de benzhidrilo benzhydryl chloride
cloruro de benzoilo benzoyl chloride
cloruro de benzol benzol chloride
cloruro de berilio beryllium chloride
cloruro de bismutilo bismuthyl chloride
cloruro de bismuto bismuth chloride
cloruro de boro boron chloride
cloruro de bromo bromine chloride
cloruro de butanoilo butanoyl chloride
cloruro de butilmagnesio butylmagnesium chloride
cloruro de butilo butyl chloride
cloruro de butirilo butyryl chloride
cloruro de butiroilo butyroyl chloride
cloruro de cacodilo cacodyl chloride
cloruro de cadmio cadmium chloride
cloruro de carbamida carbamide chloride
cloruro de carbonilo carbonyl chloride
cloruro de cerio cerium chloride
cloruro de cesio cesium chloride
cloruro de cetilpiridinio cetylpyridinium chloride
cloruro de cetiltrimetilamonio cetyltrimethylammonium chloride
cloruro de cianógeno cyanogen chloride
cloruro de ciclohexilo cyclohexyl chloride
cloruro de cinamoilo cinnamoyl chloride

cloruro de cinc zinc chloride

cloruro de cloroacetilo chloroacetyl chloride

cloruro de clorobecilo chlorobenzyl chloride

cloruro de clorobenzoilo chlorobenzoyl chloride

cloruro de cloroformilo chloroformyl chloride

cloruro de cobalto cobalt chloride

cloruro de cobre copper chloride

cloruro de cobre sódico sodium copper chloride

cloruro de cocaína cocaine chloride

cloruro de colina choline chloride

cloruro de cromilo chromyl chloride

cloruro de cromo chromium chloride

cloruro de decanoilo decanoyl chloride

cloruro de diazobenceno diazobenzene chloride

cloruro de dicloroacetilo dichloroacetyl chloride

cloruro de diclorobencilo dichlorobenzyl chloride

cloruro de diclorobenzoilo dichlorobenzoyl chloride

cloruro de didimio didymium chloride

cloruro de dimetilbencilo dimethylbenzyl chloride

cloruro de dinitrobenzoilo dinitrobenzoyl chloride

cloruro de escandio scandium chloride

cloruro de estaño tin chloride

cloruro de estearoilo stearoyl chloride

cloruro de estroncio strontium chloride

cloruro de etanoilo ethanoyl chloride

cloruro de etilbencilo ethylbenzyl chloride

cloruro de etileno ethylene chloride

cloruro de etilhexilo ethylhexyl chloride

cloruro de etilideno ethylidene chloride

cloruro de etilmagnesio ethylmagnesium chloride

cloruro de etilo ethyl chloride

cloruro de europio europium chloride

cloruro de fenacilo phenacyl chloride

cloruro de fenarsazina phenarsazine chloride

cloruro de fenilacetilo phenylacetyl chloride

chloride

cloruro de fenilcarbilamina phenylcarbylamine chloride

cloruro de fenilmagnesio phenylmagnesium chloride

cloruro de fenilo phenyl chloride

cloruro de fenilpropilo phenylpropyl chloride

cloruro de ferrocenoilo ferrocenoyl chloride

cloruro de formilo formyl chloride

cloruro de fosfonio phosphonium chloride

cloruro de fosforilo phosphoryl chloride

cloruro de fósforo phosphorus chloride

cloruro de ftaloilo phthaloyl chloride

cloruro de fumarilo fumaryl chloride

cloruro de fumaroilo fumaroyl chloride

cloruro de furoilo furoyl chloride

cloruro de gadolinio gadolinium chloride

cloruro de glicerilo glyceryl chloride

cloruro de heptanoilo heptanoyl chloride

cloruro de hexadecanoilo hexadecanoyl chloride

cloruro de hexametonio hexamethonium chloride

cloruro de hexanoilo hexanoyl chloride

cloruro de hexilo hexyl chloride

cloruro de hidrazina hydrazine chloride

cloruro de hidrógeno hydrogen chloride

cloruro de hidrógeno anhidro anhydrous hydrogen chloride

cloruro de hidroxilamonio hydroxylammonium chloride

cloruro de hierro iron chloride

cloruro de hierro amónico ammonium iron chloride

cloruro de holmio holmium chloride

cloruro de indio indium chloride

cloruro de iridio iridium chloride

cloruro de isoamilo isoamyl chloride

cloruro de isobornilo isobornyl chloride

cloruro de isobutiroilo isobutyroyl chloride

cloruro de isodecilo isodecyl chloride

cloruro de isoftaloilo isophthaloyl chloride

cloruro de isopropenilo isopropenyl chloride

cloruro de isopropilo isopropyl chloride

cloruro de iterbio ytterbium chloride

cloruro de itrio yttrium chloride

cloruro de lantano lanthanum chloride

cloruro de laurilpiridinio
laurylpyridinium chloride

cloruro de lauroilo lauroyl chloride

cloruro de litio lithium chloride

cloruro de lutecio lutetium chloride

cloruro de manganeso manganese
chloride

cloruro de mercurio mercury chloride

cloruro de mesilo mesyl chloride

cloruro de mesitilensulfonilo
mesitylenesulfonyl chloride

cloruro de metacriloilo methacryloyl
chloride

cloruro de metanosulfonilo
methanesulfonyl chloride

cloruro de metilalilo methylallyl
chloride

cloruro de metileno methylene chloride

cloruro de metilmagnesio
methylmagnesium chloride

cloruro de metilo methyl chloride

cloruro de miristilo myristyl chloride

cloruro de molibdeno molybdenum
chloride

cloruro de neodimio neodymium
chloride

cloruro de niobio niobium chloride

cloruro de níquel nickel chloride

cloruro de níquel amónico ammonium
nickel chloride

cloruro de nitrilo nitryl chloride

cloruro de nitrobenzoilo nitrobenzoyl
chloride

cloruro de nitrógeno nitrogen chloride

cloruro de nitrosilo nitrosyl chloride

cloruro de nonanoilo nonanoyl chloride

cloruro de nonilo nonyl chloride

cloruro de octanoilo octanoyl chloride

cloruro de octilmagnesio
octylmagnesium chloride

cloruro de octilo octyl chloride

cloruro de oleoilo oleoyl chloride

cloruro de oro gold chloride

cloruro de oro sódico sodium gold
chloride

cloruro de osmio osmium chloride

cloruro de oxalilo oxalyl chloride

cloruro de paladio palladium chloride

cloruro de paladio sódico sodium
palladium chloride

cloruro de palmitoilo palmitoyl chloride

cloruro de pelargonilo pelargonyl
chloride

cloruro de pentilo pentyl chloride

cloruro de picrilo picryl chloride

cloruro de pirosulfurilo pyrosulfuryl
chloride

cloruro de plata silver chloride

cloruro de plata sódico sodium silver
chloride

cloruro de platino platinum chloride

cloruro de plomo lead chloride

cloruro de polietilenglicol polyethylene
glycol chloride

cloruro de polivinilideno polyvinylidene
chloride

cloruro de polivinilo polyvinyl chloride

cloruro de praseodimio praseodymium
chloride

cloruro de propanoilo propanoyl
chloride

cloruro de propargilo propargyl chloride

cloruro de propileno propylene chloride

cloruro de propilo propyl chloride

cloruro de propionilo propionyl chloride

cloruro de radio radium chloride

cloruro de renio rhenium chloride

cloruro de rodio rhodium chloride

cloruro de rodiocarbonilo rhodium
carbonyl chloride

cloruro de rubidio rubidium chloride

cloruro de rutenio ruthenium chloride

cloruro de samario samarium chloride

cloruro de santalilo santalyl chloride

cloruro de sebacoilo sebacoyl chloride

cloruro de selenio selenium chloride

cloruro de silicio silicon chloride

cloruro de succinilo succinyl chloride

cloruro de sulfinilo sulfinyl chloride

cloruro de sulfonilo sulfonyl chloride

cloruro de sulfurilo sulfuryl chloride

cloruro de suxametonio suxamethonium
chloride

cloruro de talio thallium chloride

cloruro de tántalo tantalum chloride
cloruro de telurio tellurium chloride
cloruro de terbio terbium chloride
cloruro de tereftaloilo terephthaloyl
  chloride
cloruro de tetrabutilamonio
  tetrabutylammonium chloride
cloruro de tetradecilo tetradecyl chloride
cloruro de tetrametilamonio
  tetramethylammonium chloride
cloruro de tionilo thionyl chloride
cloruro de titanio titanium chloride
cloruro de tolilo tolyl chloride
cloruro de toluensulfonilo
  toluenesulfonyl chloride
cloruro de torio thorium chloride
cloruro de tributilestaño tributyltin
  chloride
cloruro de triclorometilsulfenilo
  trichloromethylsulfenyl chloride
cloruro de trifenilestaño triphenyltin
  chloride
cloruro de tulio thulium chloride
cloruro de tungsteno tungsten chloride
cloruro de uranilo uranyl chloride
cloruro de uranio uranium chloride
cloruro de vanadilo vanadyl chloride
cloruro de vanadio vanadium chloride
cloruro de vinilideno vinylidene chloride
cloruro de vinilmagnesio
  vinylmagnesium chloride
cloruro de vinilo vinyl chloride
cloruro de xililo xylyl chloride
cloruro de yodo iodine chloride
cloruro de zirconilo zirconyl chloride
cloruro de zirconio zirconium chloride
cloruro de zirconio potásico potassium
  zirconium chloride
cloruro dietilalumínico diethylaluminum
  chloride
cloruro estánnico stannic chloride
cloruro estannoso stannous chloride
cloruro fenilmercúrico phenylmercuric
  chloride
cloruro férrico ferric chloride
cloruro ferroso ferrous chloride
cloruro fosfórico phosphoric chloride
cloruro gálico gallic chloride
cloruro germánico germanic chloride

cloruro hidroxifenilmercúrico
  hydroxyphenylmercuric chloride
cloruro irídico iridic chloride
cloruro magnésico magnesium chloride
cloruro magnésico amónico ammonium
  magnesium chloride
cloruro manganoso manganous chloride
cloruro mercúrico mercuric chloride
cloruro mercúrico amoniacal
  ammoniated mercury chloride
cloruro mercúrico amónico ammonium
  mercuric chloride
cloruro mercurioso mercurous chloride
cloruro niqueloso nickelous chloride
cloruro osmoso osmous chloride
cloruro paladioso palladous chloride
cloruro platínico platinic chloride
cloruro platinoso platinous chloride
cloruro potásico potassium chloride
cloruro potásico alumínico aluminum
  potassium chloride
cloruro potásico de iridio iridium
  potassium chloride
cloruro potásico de osmio osmium
  potassium chloride
cloruro potásico de paladio palladium
  potassium chloride
cloruro potásico de platino platinum
  potassium chloride
cloruro potásico paladioso palladous
  potassium chloride
cloruro ruténico ruthenic chloride
cloruro samárico samaric chloride
cloruro samaroso samarous chloride
cloruro sódico sodium chloride
cloruro sódico alumínico aluminum
  sodium chloride
cloruro sódico de paladio palladium
  sodium chloride
cloruro sódico de plata silver sodium
  chloride
cloruro sódico de platino platinum
  sodium chloride
cloruro sódico paladioso palladous
  sodium chloride
cloruro sódico platínico platinic sodium
  chloride
cloruro sulfúrico sulfuric chloride
cloruro talioso thallous chloride

cloruro tantálico tantalic chloride
cloruro tantaloso tantalous chloride
cloruro teluroso tellurous chloride
cloruro titánico titanic chloride
cloruro titanoso titanous chloride
cloruro uránico uranic chloride
cloruro uranoso uranous chloride
cloruro vanadílico vanadylic chloride
cloruro vanadioso vanadous chloride
closo closo
coacervación coacervation
coagulación coagulation
coagulante coagulant
coalescencia coalescence
coalescente coalescent
cobalamina cobalamin
cobaltamina cobaltammine
cobáltico cobaltic
cobaltita cobaltite
cobalto cobalt
cobaltoceno cobaltocene
cobre copper
cobre azul blue copper
cobustible de cohete rocket fuel
coca coca
cocaína cocaine
cocarboxilasa cocarboxylase
cocer roast
cocido cooking, roasting
coco coccus
cocromo cochrome
coculina cocculin
cochinilina cochinilin
cochinilla cochineal
codeína codeine
coeficiente coefficient
coeficiente crítico critical coefficient
coeficiente de absorción absorption
    coefficient
coeficiente de absorción atómica atomic
    absorption coefficient
coeficiente de acidez acidity coefficient
coeficiente de actividad activity
    coefficient
coeficiente de adsorción adsorption
    coefficient
coeficiente de atenuación attenuation
    coefficient
coeficiente de conversión conversion

    coefficient
coeficiente de distribución distribution
    coefficient
coeficiente de Einstein Einstein
    coefficient
coeficiente de expansión térmica
    thermal expansion coefficient
coeficiente de extinción extinction
    coefficient
coeficiente de extracción extraction
    coefficient
coeficiente de fenol phenol coefficient
coeficiente de ionización ionization
    coefficient
coeficiente de Joule-Thomas Joule-
    Thomas coefficient
coeficiente de partición partition
    coeffcient
coeficiente de sedimentación
    sedimentation coefficient
coeficiente de selectividad selectivity
    coefficient
coeficiente de solubilidad solubility
    coefficient
coeficiente de temperatura temperature
    coefficient
coeficiente de transferencia transfer
    coefficient
coeficiente de transporte transport
    coefficient
coeficiente estequiométrico
    stoichiometric coefficient
coeficiente osmótico osmotic coefficient
coenzima coenzyme
coercitividad coercivity
cofactor cofactor
coherencia coherence
coherente coherent
cohesión cohesion
cohesión dieléctrica dielectric cohesion
cohobación cohobation
cola glue, cola
colagenasa collagenase
colágeno collagen
colamina colamine
colano cholane
colantreno cholanthrene
colapez isinglass
colargol collargol

colas tailings
colecalciferol cholecalciferol
colector collector
colemanita colemanite
colestérico cholesteric
colesterol cholesterol
colestipol colestipol
colestiramina cholestyramine
colidina collidine
coligar colligate
coligativo colligative
colina choline
colinesterasa cholinesterase
colisión collision
colisión de intercambio exchange collision
colisión desactivante deactivating collision
colisión elástica elastic collision
colisión electrónica electron collision
colisión inelástica inelastic collision
colisión iónica molecular molecular ion collision
colisión nuclear nuclear collision
colisión térmica thermal collision
colistina colistin
colmena beehive
colodión collodion
colofonia colophony, rosin
colofonia de madera wood rosin
colofonita colophonite
coloidal colloidal
coloide colloid
coloide hidrofílico hydrophilic colloid
coloide hidrofóbico hydrophobic colloid
coloide irreversible irreversible colloid
coloide molecular molecular colloid
colometría visual visual colorimetry
color certificado certified color
color de alimento food color
coloración coloration
colorante colorant, dye, stain
colorante ácido acid dye
colorante al azufre sulfur dye
colorante al sulfuro sulfide dye
colorante azoico azo dye
colorante azoico primario primary azo dye
colorante de alizarina alizarin dye

colorante de alquitrán de hulla coal-tar dye
colorante de anilina aniline dye
colorante de antraquinona anthraquinone dye
colorante de azina azine dye
colorante de bencidina benzidine dye
colorante de cianina cyanine dye
colorante de cromo chrome dye
colorante de estilbeno stilbene dye
colorante de quinolina quinoline dye
colorante de tiazol thiazole dye
colorante de tina vat dye
colorante de triarilmetano triarylmethane dye
colorante de trifenilmetano triphenylmethane dye
colorante de xanteno xanthene dye
colorante de Ziehl Ziehl stain
colorante desarrollado developed dye
colorante diazoico diazo dye
colorante directo direct dye
colorante fluorescente fluorescent dye
colorante incoloro colorless dye
colorante indirecto indirect dye
colorante nitro nitro dye
colorante nitroso nitroso dye
colorante oxiazoico oxyazo dye
colorante reactivo reactive dye
colorante substantivo substantive dye
colorante vegetal vegetable dye
colorimetría colorimetry
colorimétrico colorimetric
colorímetro colorimeter
colorímetro fotoeléctrico photoelectric colorimeter
colorización colorization
colquicina colchicine
columbato columbate
columbio columbium
columbita columbite
columna column
columna de Clusius Clusius column
columna de destilación distillation column
columna de gradiente de densidad density-gradient column
columna de intercambio iónico ion-exchange column

columna de placas plate column
columna empaquetada packed column
columna fraccionante fractionating column
columna rectificante rectifying column
columna térmica thermal column
colza colza
combinación combination
combinación molecular molecular combination
combinación química chemical combination
combinado combined
combustibilidad combustibility
combustible fuel
combustible de boro boron fuel
combustible fósil fossil fuel
combustible nuclear nuclear fuel
combustible sin humo smokeless fuel
combustible sintético synthetic fuel
combustión combustion
combustión espontánea spontaneous combustion
combustión fraccionaria fractional combustion
combustión homogénea homogeneous combustion
combustión lenta slow combustion
combustión orgánica organic combustion
combustión superficial surface combustion
comicelización comicellization
comirina comirin
comparador comparator
comparador de colores color comparator
comparador de densidad density comparator
compatibilidad compatibility
compatible compatible
compensación interna internal compensation
compensación intramolecular intramolecular compensation
complejación complexing
complejo complex
complejo activado activated complex
complejo coordinado coordinated complex

complejo de acetileno acetylene complex
complejo de adición addition complex
complejo de inclusión inclusion complex
complejo de Meisenheimer Meisenheimer complex
complejo de Werner Werner complex
complejo enzima-substrato enzyme-substrate complex
complejo excitado exciplex
complejo inerte inert complex
complejo lábil labile complex
complejo pi pi complex
complejo vitamínico vitamin complex
complexona complexone
componente component
componente activo active component
composición composition
composición constante constant composition
composición isotópica isotopic composition
compresión adiabática adiabatic compression
compresor de aire air compressor
comprimido compressed
comprobación de calidad quality assurance
compuesto compound
compuesto acíclico acyclic compound
compuesto alicíclico alicyclic compound
compuesto alifático aliphatic compound
compuesto antimónico antimonic compound
compuesto antimonioso antimonous compound
compuesto argéntico argentic compound
compuesto aromático aromatic compound
compuesto asimétrico asymmetric compound
compuesto áurico auric compound
compuesto bicíclico bicyclic compound
compuesto binario binary compound
compuesto cíclico cyclic compound
compuesto complejo complex compound
compuesto de acabado finishing compound
compuesto de carbono carbon compound
compuesto de coordinación coordination

compound

compuesto de encerramiento enclosure compound

compuesto de Herz Herz compound

compuesto de inclusión inclusion compound

compuesto de intercalación intercalation compound

compuesto de Vaska Vaska compound

compuesto electrovalente electrovalent compound

compuesto en anillo ring compound

compuesto en jaula cage compound

compuesto estequiométrico stoichiometric compound

compuesto exotérmico exothermic compound

compuesto fotocrómico photochromic compound

compuesto heterocíclico heterocyclic compound

compuesto impermeabilizante waterproofing compound

compuesto insaturado unsaturated compound

compuesto intermetálico intermetallic compound

compuesto marcado labeled compound

compuesto monístico monistic compound

compuesto monofuncional monofunctional compound

compuesto nativo native compound

compuesto neutro neutral compound

compuesto no polar nonpolar compound

compuesto orgánico organic compound

compuesto orgánico volátil volatile organic compound

compuesto organoalumínico organoaluminum compound

compuesto organocíncico organozinc compound

compuesto organofosforado organophosphorus compound

compuesto organolítico organolithium compound

compuesto organomagnesiano organomagnesium compound

compuesto organomercúrico organomercury compound

compuesto organometálico organometallic compound

compuesto polar polar compound

compuesto químico chemical compound

compuesto saturado saturated compound

compuesto simétrico symmetrical compound

compuesto superficial surface compound

compuesto ternario ternary compound

compuesto tetraédrico tetrahedral compound

compuesto trazador tracer compound

compuesto volátil volatile compound

cóncavo concave

concentración concentration

concentración de ion hidrógeno hydrogen-ion concentration

concentración de substancia substance concentration

concentración de volmen de pigmento pigment volume concentration

concentración electrónica electron concentration

concentración molal molal concentration

concentración molecular molecular concentration

concentrado concentrated

concentrador concentrator

concentrar concentrate

concoidal conchoidal

concreto concrete

condensación condensation

condensación aciloínica acyloin condensation

condensación aldólica aldol condensation

condensación capilar capillary condensation

condensación de aldehído aldehyde condensation

condensación de benzoína benzoin condensation

condensación de Claisen Claisen condensation

condensación de Claisen-Schmidt Claisen-Schmidt condensation

condensación de Darzens Darzens condensation

condensación de Friedel-Crafts Friedel-Crafts condensation

condensación de Guareschi-Thorpe
Guareschi-Thorpe condensation
condensación de Michael Michael
condensation
condensación de pinacol pinacol
condensation
condensación de Stobbe Stobbe
condensation
condensación fraccionaria fractional
condensation
condensación intramolecular
intramolecular condensation
condensación retrógrada retrograde
condensation
condensado condensed
condensador condenser
condensador de Friedrichs Friedrichs
condenser
condensador de Liebig Liebig condenser
condensador de reflujo reflux condenser
condensador en espiral coil condenser
condensador parcial partial condenser
condestilación condistillation
condiciones ambientales environmental
conditions
condiciones críticas critical conditions
condiciones de laboratorio laboratory
conditions
condiciones normales normal conditions,
standard conditions
condiciones óptimas optimum conditions
conducción conduction
conducción electrolítica electrolytic
conduction
conducción electrónica electron
conduction
conducción metálica metallic conduction
conductancia conductance
conductancia de electrodo electrode
conductance
conductancia electrolítica electrolytic
conductance
conductancia equivalente equivalent
conductance
conductancia específica specific
conductance
conductancia iónica ionic conductance
conductancia térmica thermal
conductance

conductividad conductivity
conductividad de solución solution
conductivity
conductividad eléctrica electric
conductivity
conductividad electrolítica electrolytic
conductivity
conductividad equivalente equivalent
conductivity
conductividad específica specific
conductivity
conductividad molal molal conductivity
conductividad molar molar conductivity
conductividad molecular molecular
conductivity
conductividad térmica thermal
conductivity, heat conductivity
conductométrico conductometric
conductor conductor
conductor electrónico electron conductor
conesina conessine
configuración configuration
configuración absoluta absolute
configuration
configuración electrónica electron
configuration
conformación conformation
conformación eclipsada eclipsed
conformation
conformación en bote boat conformation
conformación en cadena chain
conformation
conformación en silla chair
conformation
congelación freezing, congelation
congelar freeze
conglomerado conglomerate
coniceína conicein
coniferilo coniferyl
coniferina coniferin
coniferol coniferol
conina conine
conjugación conjugation
conjugado conjugated
cono cone
cono de filtración filter cone
cono de Seger Seger cone
cono pirométrico pyrometric cone
conproporción conproportionation

conquiolina conchiolin
conservación de energía conservation of energy
conservación de materia conservation of matter
consistencia consistency
consistómetro consistometer
consoluto consolute
constantán constantan
constante constant
constante atómica atomic constant
constante crioscópica cryoscopic constant
constante crítica critical constant
constante de absorción absorption constant
constante de acidez acidity constant
constante de acoplamiento coupling constant
constante de apantallamiento screening constant
constante de Avogadro Avogadro constant
constante de basicidad basicity constant
constante de Boltzmann Boltzmann constant
constante de célula cell constant
constante de desintegración decay constant
constante de desintegración radiactiva radioactive decay constant
constante de difusión diffusion constant
constante de disociación dissociation constant
constante de dissociación básica base dissociation constant
constante de equilibrio equilibrium constant
constante de estabilidad stability constant
constante de Faraday Faraday constant
constante de Fermi Fermi constant
constante de formación formation constant
constante de fuerza force constant
constante de gas gas constant
constante de ionización ionization constant
constante de Madelung Madelung constant

constante de masa atómica unificada unified atomic mass constant
constante de Michaelis Michaelis constant
constante de Planck Planck constant
constante de producto de solubilidad solubility-product constant
constante de reacción reaction constant
constante de Rydberg Rydberg constant
constante de sedimentación sedimentation constant
constante de van der Waals van der Waals constant
constante de velocidad velocity constant
constante dieléctrica dielectric constant
constante ebulloscópica ebullioscopic constant
constante elástica elastic constant
constante física physical constant
constante fundamental fundamental constant
constante química chemical constant
constante radiactiva radioactive constant
constante rotacional rotational constant
constante térmica thermal constant
constantes de agua water constants
constitución química chemical constitution
constitucional constitutional
constituyente constituent
contacto contact
contador con halógeno halogen counter
contador de burbujas bubble counter
contador de centelleos scintillation counter
contador de Geiger Geiger counter
contador de Geiger-Müller Geiger-Müller counter
contador de radiación radiation counter
contaminación contamination, pollution
contaminación ambiental environmental contamination
contaminación atmosférica atmospheric pollution
contaminación de agua water pollution
contaminación de aire air pollution
contaminación química chemical pollution

contaminación radiactiva radioactive contamination
contaminación térmica thermal pollution
contaminante contaminant
contenedor container
contenido ácido acid content
contenido de uranio uranium content
contenido térmico heat content
conteo de partículas particle counting
continuidad de estado continuity of state
continuo continuous
continuo de Lyman Lyman continuum
contracción contraction
contracción de actinoides actinoid contraction
contracción lantánida lanthanide contraction
contracolorante counterstain
contracorriente countercurrent
contraion counterion
contraste contrast
control control
control automático automatic control
control de calidad quality control
control de desperdicios waste control
control de flujo de gas gas-flow control
control de llama flame control
control de reacción reaction control
control de temperatura temperature control
controlado por cuarzo quartz-controlled
convección convection
conversión conversion
conversión de bencidina benzidine conversion
conversión de energía energy conversion
conversión geométrica geometric conversion
conversión interna internal conversion
convertidor converter
convertidor catalítico catalytic converter
convertidor de energía energy converter
convertidor de neutrones neutron converter
convexo convex
coordinación coordination
coordinación bipiramidal trigonal trigonal bipyramidal coordination
coordinación tetraédrica tetrahedral coordination
coordinado coordinate
copal copal
copelación cupellation
copigmento copigment
copirina copyrine
copolimerización copolymerization
copolímero copolymer
copolímero aleatorio random copolymer
copolímero con injerto graft copolymer
copolímero de acrilamida acrylamide copolymer
copolímero de acrilonitrilo acrylonitrile copolymer
copolímero de bloques block copolymer
copolímero olefínico olefin copolymer
coprastanol coprastanol
coprecipitación coprecipitation
coprosterol coprosterol
coque coke
coque de petróleo petroleum coke
coquización coking
coral coral
corcho cork
cordita cordite
coriandrol coriandrol
corindón corundum
coritoide choritoid
cornezuelo ergot
corona corona
coroneno coronene
corrección de Rydberg Rydberg correction
corrección radiativa radiative correction
correlación correlation
corriente constante constant current
corriente de difusión diffusion current
corriente de electrodo electrode current
corriente de gas gas current
corriente de ionización ionization current
corriente de migración migration current
corriente de polarización polarization current
corriente eléctrica electric current
corriente electrónica electron current, electron stream
corriente galvánica galvanic current

corriente neutrónica neutron current
corrosión corrosion
corrosión electrolítica electrolytic corrosion
corrosión electroquímica electrochemical corrosion
corrosión galvánica galvanic corrosion
corrosión química chemical corrosion
corrosivo corrosive
corsita corsite
corticoide corticoid
corticosterona corticosterone
corticotropina corticotropin
cortisol cortisol
cortisona cortisone
cosanol cosanol
cosilo cosyl
cosmético cosmetic
costra scale
coulombio coulomb
covalencia covalence
covalencia coordinada coordinate covalence
covalente covalent
covelina coveline
covelita covellite
covolumen de van der Waals van der Waals covolume
craquear crack
craqueo cracking
craqueo catalítico catalytic cracking
craqueo térmico thermal cracking
creatina creatine
creatinasa creatinase
creatinina creatinine
creatininasa creatininase
crecimiento de cristal crystal growth
crémor tártaro cream of tartar
creosol creosol
creosota creosote
cresilato cresylate
cresilo cresyl
cresol cresol
cresolftaleína cresolphthalein
cresorcina cresorcin
creta chalk
criba screen
cribado screening
criogenia cryogenics

criogénica cryogenics
criogénico cryogenic
criogenina cryogenin
criohidrato cryohydrate
criolita cryolite
criómetro cryometer
crioquímica cryochemistry
crioscopia cryoscopy
crioscopio de Hortvet Hortvet cryoscope
crióstato cryostat
criptando cryptand
criptato cryptate
criptidina cryptidine
criptocianina cryptocyanine
criptohalita cryptohalite
criptón krypton
criptoxantina cryptoxanthin
crisanilina chrysaniline
criseno chrysene
crisol crucible
crisol de filtración filter crucible
crisol de Gooch Gooch crucible
crisólito chrysolite
crisoquinona chrysoquinone
crisotilo chrysotile
cristal crystal
cristal covalente covalent crystal
cristal cúbico cubic crystal
cristal cultivado grown crystal
cristal de cuarzo quartz crystal
cristal de roca rock crystal
cristal homopolar homopolar crystal
cristal ideal ideal crystal
cristal iónico ionic crystal
cristal líquido liquid crystal
cristal macromolecular macromolecular crystal
cristal metálico metallic crystal
cristal mixto mixed crystal
cristal nemático nematic crystal
cristal no polar nonpolar crystal
cristal óptico optical crystal
cristal oscilante oscillating crystal
cristal positivo positive crystal
cristalino crystalline
cristalito crystallite
cristalización crystallization
cristalización fraccionaria fractional crystallization

cristalizador al vacío vacuum crystallizer

cristalizar crystallize

cristalografía crystallography

cristalografía electrónica electron crystallography

cristalografía por rayos X X-ray crystallography

cristalograma crystallogram

cristalograma de Bragg Bragg crystallogram

cristaloide crystalloid

cristaloluminiscencia crystalloluminescence

cristobalita cristobalite

crítico critical

croceína brillante brilliant crocein

crocetina crocetin

crocidolita crocidolite

croma chroma

cromano chroman

cromaticidad chromaticity

cromático chromatic

cromatina chromatin

cromato chromate

cromato ácido sódico sodium acid chromate

cromato amónico ammonium chromate

cromato bárico barium chromate

cromato cálcico calcium chromate

cromato cobaltoso cobaltous chromate

cromato cúprico cupric chromate

cromato de bismuto bismuth chromate

cromato de butilo butyl chromate

cromato de cesio cesium chromate

cromato de cinc zinc chromate

cromato de cobalto cobalt chromate

cromato de cobre copper chromate

cromato de estaño tin chromate

cromato de estroncio strontium chromate

cromato de hierro iron chromate

cromato de hierro amónico ammonium iron chromate

cromato de litio lithium chromate

cromato de manganeso manganese chromate

cromato de mercurio mercury chromate

cromato de plata silver chromate

cromato de plomo lead chromate

cromato de rubidio rubidium chromate

cromato de talio thallium chromate

cromato estánnico stannic chromate

cromato estannoso stannous chromate

cromato férrico ferric chromate

cromato magnésico magnesium chromate

cromato manganoso manganous chromate

cromato mercúrico mercuric chromate

cromato mercurioso mercurous chromate

cromato potásico potassium chromate

cromato potásico bárico barium potassium chromate

cromato potásico rojo red potassium chromate

cromato sódico sodium chromate

cromatografía chromatography

cromatografía bidimensional two-dimensional chromatography

cromatografía circular circular chromatography

cromatografía de capa fina thin-layer chromatography

cromatografía de exclusión iónica ion-exclusion chromatography

cromatografía de gases gas chromatography

cromatografía de gel gel chromatography

cromatografía de intercambio iónico ion-exchange chromatography

cromatografía de partición partition chromatography

cromatografía de película fina thin-film chromatography

cromatografía de permeación en gel gel-permeation chromatography

cromatografía descendente descending chromatography

cromatografía en fase vapor vapor-phase chromatography

cromatografía gas-líquido gas-liquid chromatography

cromatografía horizontal horizontal chromatography

cromatografía líquida liquid

chromatography
cromatografía líquida de alto rendimiento high-performance liquid chromatography
cromatografía por adsorción adsorption chromatography
cromatografía por afinidad affinity chromatography
cromatografía por desplazamiento displacement chromatography
cromatografía radial radial chromatography
cromatografía sobre papel paper chromatography
cromatograma chromatogram
crómico chromic
cromilo chromyl
crominancia chrominance
cromita chromite
cromita sódica sodium chromite
cromitita chromitite
cromo chromium, chrome
cromo-cobre chromium-copper
cromofílico chromophilic
cromofórico chromophoric
cromóforo chromophore
cromogénico chromogenic
cromógeno chromogen
cromoisómero chromoisomer
cromona chromone
cromosana chromosan
cromoso chromous
cromosoma chromosome
cromotropía chromotropy
cronopotenciometría chronopotentiometry
crotonal crotonal
crotonaldehído crotonaldehyde
crotonato de butilo butyl crotonate
crotonato de etilo ethyl crotonate
crotonileno crotonylene
crotonilo crotonyl
crotonoide crotonoid
crotonoilo crotonoyl
crotoxina crotoxin
crudo crude
cuadri quadri
cuadrimolecular quadrimolecular
cuadripolo quadrupole

cuadrivalencia quadrivalence
cuadrivalente quadrivalent
cuajo rennet
cualitativo qualitative
cuantificación quantization
cuantificado quantized
cuantificar quantize
cuantitativo quantitative
cuantivalencia quantivalence
cuanto quantum
cuanto de energía quantum of energy
cuanto de luz quantum of light
cuanto de radiación quantum of radiation
cuanto virtual virtual quantum
cuarcita quartzite, quartz rock
cuarteto quartet
cuarzo quartz
cuarzo fundido fused quartz
cuasia quassia
cuasicristal quasicrystal
cúbico cubical
cúbico centrado en el cuerpo body-centered cubic
cúbico centrado en las caras face-centered cubic
cubierto coated
cubo cube
cuello de vidrio esmerilado ground-glass neck
cuenta count
cuero leather
cuerpo body
cuerpo cetónico ketone body
cumal cumal
cumaldehído cumaldehyde
cumarina coumarin
cumarona coumarone
cumenilo cumenyl
cumeno cumene
cumidina cumidine
cumilo cumyl
cuminol cuminol
cumol cumol
cumuleno cumulene
cupferrón cupferron
cuprato cuprate
cuprato amónico ammonium cuprate
cupreína cupreine

cupreno cuprene
cúprico cupric
cuprita cuprite
cuprocianuro cuprocyanide
cuprocianuro potásico potassium cuprocyanide
cuprocúprico cuprocupric
cuproína cuproine
cuproníquel cupronickel
cuproso cuprous
cuprotungsteno cuprotungsten
cúpula dome
curado curing
curare curare
curativo curative
curcumina curcumin

curie curie
curio curium
curtido tanning
curva de actividad activity curve
curva de afinidad affinity curve
curva de Bragg Bragg curve
curva de energía electrónica electronic energy curve
curva de excitación excitation curve
curva de ionización de Bragg Bragg ionization curve
curva de Sargent Sargent curve
curva de solubilidad solubility curve
curva espinodal spinodal curve
cusparina cusparine

CH

chabasita chabazite
chapado cladding
chavicol chavicol
chi chi

chicle chicle
chispa spark
chisporroteo sparking, sputter,
    sputtering

D

dacita dacite
dactilina dactylin
dactinomicina dactinomycin
dafnetina daphnetin
dafnita daphnite
dahllita dahllite
dalapón dalapon
dalina dahlin
dalton dalton
damascinina damascinine
dambonita dambonite
damenita dahmenite
damourita damourite
danaíta danaite
danalita danalite
dansilo dansyl
dapsona dapsone
darapskita darapskite
datación dating
datación por carbono carbon dating
datación por radiocarbono radiocarbon
    dating
datación química chemical dating
datación radiactiva radioactive dating
datación radiométrica radiometric
    dating
datolita datolite
datos químicos chemical data
daubreelita daubreelite
daunomicina daunomycin
daviesita daviesite
dawsonita dawsonite
deanol deanol
débil weak
debye debye
decaborano decaborane
decahidrato decahydrate
decahidrato de carbonato sódico sodium
    carbonate decahydrate
decahidrato de pentaborato sódico
    sodium pentaborate decahydrate
decahidrato de sulfato sódico sodium
    sulfate decahydrate

decahidronaftaleno
    decahydronaphthalene
decaimiento decay
decalina decalin
decametiltetrasiloxano
    decamethyltetrasiloxane
decametrina decamethrin
decanal decanal
decanamida decanamide
decano decane
decanoato decanoate
decanoato de etilo ethyl decanoate
decanodiol decanediol
decanoilo decanoyl
decanol decanol
decanonitrilo decanenitrile
decantación decantation
decantar decant
decapado pickling
decatilo decatyl
deceno decene
decicaína decicaine
decilamida decylamide
decilamina decylamine
decileno decylene
deciletileno decylethylene
decilmercaptano decyl mercaptan
decilo decyl
deciltriclorosilano decyltrichlorosilane
decimolar decimolar
decino decyne
decinormal decinormal
decocción decoction
decoloración decoloration
decolorante decolorant
decolorización decolorization
decrepitación decrepitation
decrepitar decrepitate
defecto defect
defecto de Frenkel Frenkel defect
defecto de masa mass defect
defecto de Schottky Schottky defect
defervescencia defervescence

deficiencia electrónica electron deficiency
deflagración deflagration
deflegmación dephlegmation
deflexión magnética magnetic deflection
deformabilidad deformability
deformación deformation, strain
degeneración degeneration
degenerado degenerate
degradación degradation
degradación de Barbier-Wieland Barbier-Wieland degradation
degradación de Bergmann Bergmann degradation
degradación de Edman Edman degradation
degradación de Emde Emde degradation
degradación de Grignard Grignard degradation
degradación de Hofmann Hofmann degradation
degradación de Krafft Krafft degradation
degradación de Miescher Miescher degradation
degradación de Ruff-Fenton Ruff-Fenton degradation
degradación de Strecker Strecker degradation
degradación de Weerman Weerman degradation
degradación de Wohl Wohl degradation
degradación térmica thermal degradation
degradar degrade
delanio delanium
delcosina delcosine
delessita delessite
delfinina delphinin
delicuescencia deliquescence
delicuescente deliquescent
delta delta
delvauxita delvauxite
demal demal
demanda química de oxígeno chemical oxygen demand
demeclociclina demeclocycline
demetileno demethylene
demetona demeton

demolización demolization
demulsibilidad demulsibility
dendrita dendrite
dendrobina dendrobine
densidad density
densidad absoluta absolute density
densidad aparente apparent density
densidad crítica critical density
densidad de carga charge density
densidad de carga electrónica electron-charge density
densidad de colisiones collision density
densidad de corriente current density
densidad de energía energy density
densidad de gas gas density
densidad de neutrones neutron density
densidad de vapor vapor density
densidad electrónica electron density
densidad limitadora limiting density
densidad molecular molecular density
densidad nuclear nuclear density
densidad óptica optical density
densidad relativa relative density
densitometría ultravioleta ultraviolet densitometry
densitómetro densitometer
deoxi deoxy
deposición deposition
deposición al vacío vacuum deposition
deposición electrolítica electrolytic deposition
deposición electrónica sputter, sputtering
deposición electroquímica electrochemical deposition
deposición húmeda wet deposition
deposición metálica metal deposition
deposición química chemical deposition
deposición rítmica rhythmic deposition
deposición seca dry deposition
depositado plated
depósito deposit, plating
depósito radiactivo radioactive deposit
depresión depression
depresión de punto de congelación freezing-point depression
depresión de punto de fusión melting-point depression
depresión molecular molecular depression

depresor depressor
depsida depside
depurador depurator, scavenger, scrubber
depurador de gas getter
derivación derivation
derivado derivative
derretir melt
derritol derritol
des des
desacetilación deacetylation
desacoplar decouple
desactivación deactivation
desactivador deactivator
desaglomeración disagglomeration
desalinación desalination
desamidación deamidation
desaminación deamination
desarrollador developer
desarrollo development
descalcificación decalcification
descarbonizar decarbonize
descarboxilación decarboxylation
descarboxilar decarboxylate
descarboxilasa decarboxylase
descarburización decarburization
descarburizar decarburize
descarga discharge
descarga gaseosa gaseous discharge
desclizita desclizite
desclorinación dechlorination
descomponer decompose
descomposición decomposition
descomposición doble double decomposition
descomposición electrolítica electrolytic decomposition
descomposición subatómica subatomic decomposition
descomposición térmica thermal decomposition
descontaminación decontamination
desecado desiccated
desecador desiccator
desecador de Fresenius Fresenius desiccator
desecante desiccant
deseleniuro diselenide
desemulsificación demulsification

desespumante defoamer
desfloculación deflocculation
desfloculante defloculant
desgaseamiento degassing, outgassing
desgasificación degasification
deshidratación dehydration
deshidratado dehydrated
deshidratador dehydrator
deshidratante dehydrant
deshidro dehydro
deshidroacetato cálcico calcium dehydroacetate
deshidroacetato sódico sodium dehydroacetate
deshidrociclización dehydrocyclization
deshidrocolesterol dehydrocholesterol
deshidrogenación dehydrogenation
deshidrogenasa dehydrogenase
deshidroisoandrosterona dehydroisoandrosterone
deshidrólisis dehydrolysis
deshumidificación dehumidification
desilo desyl
desinfección disinfection
desinfectante disinfectant
desintegración decay, disintegration
desintegración alfa alpha decay
desintegración atómica atomic disintegration
desintegración beta beta decay
desintegración en cadena chain decay
desintegración espontánea spontaneous decay
desintegración exponencial exponential decay
desintegración gamma gamma decay
desintegración múltiple multiple decay
desintegración nuclear nuclear decay
desintegración radiactiva radioactive decay
desionización deionization
desionizado deionized
deslanosida deslanoside
deslizamiento shift, shifting
deslizamiento batocromático bathochromatic shift
deslizamiento de Stokes Stokes shift
deslizamiento espectral spectral shift
deslizamiento isomérico isomeric shift

deslizamiento isotópico isotope shift
deslizamiento químico chemical shift
deslocalización delocalization
desmagnetización demagnetization
desmetilación demethylation
desmina desmin
desmineralización demineralization
desmotropismo desmotropism
desmótropo desmotrope
desmuriación demuriation
desnaturalización denaturation
desnaturalizado denatured
desnaturalizante denaturant
desnaturalizar denature
desnitración denitration
desnitrificación denitrification
desnitrogenación denitrogenation
desodorante deodorant
desorción desorption
desoxi deoxy, desoxy
desoxianisoína deoxyanisoin
desoxibenzoína deoxybenzoin
desoxidación deoxidation
desoxidante deoxidizer
desoxidar deoxidize
desoxigenación deoxygenation
desoxirribonucleasa deoxyribonuclease
desoxirribosa deoxyribose
desperdicios waste
desperdicios agrícolas agricultural waste
desperdicios industriales industrial
   waste
desperdicios nucleares nuclear waste
desperdicios peligrosos hazardous waste
desperdicios químicos chemical waste
desperdicios radiactivos radioactive
   waste
desplazamiento displacement, shift,
   shifting
desplazamiento de Zeeman Zeeman
   displacement
desplazamiento doble double
   displacement
despolarización depolarization
despolarizador depolarizer
despolmerización depolymerization
desproporción disproportionation
destello flash
destello de luz light flash

destilación distillation
destilación a alto vacío high-vacuum
   distillation
destilación al vacío vacuum distillation
destilación al vapor steam distillation
destilación analítica analytical
   distillation
destilación azeotrópica azeotropic
   distillation
destilación continua continuous
   distillation
destilación de agua water distillation
destilación destructiva destructive
   distillation
destilación extractiva extractive
   distillation
destilación fraccionaria fractional
   distillation
destilación instantánea flash distillation
destilación molecular molecular
   distillation
destilación seca dry distillation
destilación simple simple distillation
destilado distilled
destilado de alquitrán de hulla coal-tar
   distillate
destilador distiller, still
destilador de agua water still
destilador molecular molecular still
destoxificación detoxification
desulfuración desulfuration
desvitrificación devitrification
detección de agua water detection
detector de fuego fire detector
detector de gases gas detector, gas
   sniffer
detector de humedad humidity detector
detector de infrarrojo infrared detector
detector de ionización por llama flame-
   ionization detector
detector de llama flame detector
detector de radiación radiation detector
detector de umbral threshold detector
detector iónico ion detector
detector químico chemical detector
detector térmico heat detector
detergente detergent
detergente iónico ionic detergent
detergente sintético synthetic detergent

deterioro electroquímico
electrochemical deterioration
determinación determination
determinación de ion hidrógeno
hydrogen-ion determination
determinación de Van Slyke Van Slyke
determination
determinación de Zeisel Zeisel
determination
determinación de Zerewitinoff
Zerewitinoff determination
detonación detonation
detonador detonator
deuteración deuteration
deuterado deuterated
deuterar deuterate
deuterio deuterium
deutero deutero
deuterón deuteron
deuteroxilo deuteroxyl
deuteruro deuteride
deuteruro alumínico de litio lithium
aluminum deuteride
deuteruro de litio lithium deuteride
deweylita deweylite
dexanfetamina dexamphetamine
dextranasa dextranase
dextrano dextran
dextrina dextrin
dextrinosa dextrinose
dextro dextro
dextrógiro dextrorotatory, dextrogyric
dextrosa dextrose
di di
diabasa diabase
diacetamida diacetamide
diacetanilida diacetanilide
diacetato diacetate
diacetato alumínico aluminum diacetate
diacetato cálcico calcium diacetate
diacetato de celulosa cellulose diacetate
diacetato de dibutilestaño dibutyltin
diacetate
diacetato de etilenglicol ethylene glycol
diacetate
diacetato de etileno ethylene diacetate
diacetato de etilideno ethylidene
diacetate
diacetato de glicerilo glyceryl diacetate

diacetato de glicerina glycerol diacetate
diacetato de glicerol glycerol diacetate
diacetato de metalidina methallydine
diacetate
diacetato de paladio palladium diacetate
diacetato sódico sodium diacetate
diacetilamida diacetylamide
diacetilanilida diacetylanilide
diacetileno diacetylene
diacetiletano diacetylethane
diacetilmorfina diacetylmorphine
diacetilo diacetyl
diacetilperóxido diacetylperoxide
diacetilurea diacetylurea
diacetina diacetin
diacetona diacetone
diacetonaacrilamida diacetone
acrylamide
diacetonaalcohol diacetone alcohol
diacetonamina diacetonamine
diácido diacid
diacolación diacolation
diadoquita diadochite
diaestereoisómero diastereoisomer
diaestereotópico diastereotopic
diaforita diaphorite
diafragma diaphragm
diagénesis diagenesis
diagrama centroide centroid diagram
diagrama de correlación correlation
diagram
diagrama de Cox Cox chart
diagrama de Ellingham Ellingham
diagram
diagrama de equilibrio equilibrium
diagram
diagrama de fases phase diagram
diagrama de flujo flow diagram
diagrama de Grotrian Grotrian diagram
diagrama de Jablonski Jablonski
diagram
diagrama de Moseley Moseley diagram
diagrama de niveles de energía energy-
level diagram
diagrama de Orgel Orgel diagram
diagrama de puntos de congelación
freezing-point diagram
diagrama de puntos de ebullición
boiling-point diagram

diagrama de Tanabe-Sugano Tanabe-Sugano diagram

diagrama molecular molecular diagram

diagrama triangular triangular diagram

dialdehído dialdehyde

dialdehído de almidón starch dialdehyde

dialilamina diallylamine

dialilmelanina diallylmelamine

dialilo diallyl

dialilurea diallylurea

diálisis dialysis

diálisis de equilibrio equilibrium dialysis

dializado dialyzed

dializador dialyzer

dialozita dialozite

dialqueno dialkene

dialquileno dialkylene

dialquilo dialkyl

diamagnetismo diamagnetism

diamagnetismo atómico atomic diamagnetism

diamagnetismo molecular molecular diamagnetism

diamante diamond

diamante industrial industrial diamond

diámetro atómico atomic diameter

diámetro molecular molecular diameter

diámetro molecular efectivo effective molecular diameter

diamida diamide

diamidina diamidine

diamido diamido

diamilamina diamylamine

diamilanilina diamylaniline

diamileno diamylene

diamilfenol diamyl phenol

diamilo diamyl

diamina diamine

diamino diamino

diaminoacridina diaminoacridine

diaminoazobenceno diaminoazobenzene

diaminoazoxitolueno diaminoazoxytoluene

diaminobenceno diaminobenzene

diaminobencidina diaminobenzidine

diaminobenzofenona diaminobenzophenone

diaminobutano diaminobutane

diaminodifenilamina

diaminodiphenylamine

diaminodifenilmetano diaminodiphenylmethane

diaminodifenilo diaminodiphenyl

diaminoetano diaminoethane

diaminofenazina diaminophenazine

diaminofenol diaminophenol

diaminohexano diaminohexane

diaminopentano diaminopentane

diaminopropano diaminopropane

diaminotolueno diaminotoluene

diamonio diammonium

diamorfina diamorphine

dianhídrido piromelítico pyromellitic dianhydride

dianilina dianiline

dianilinoetano dianilinoethane

dianisidina dianisidine

diarabinosa diarabinose

diarsano diarsane

diarsenito diarsenite

diarsilo diarsyl

diasólisis diasolysis

diásporo diaspore

diastasa diastase

diatermia diathermy

diatómico diatomic

diatomita diatomite

diatrizoato sódico sodium diatrizoate

diazaciclo diazacyclo

diazepam diazepam

diazina diazine

diazinón diazinon

diazo diazo

diazoacetato diazoacetate

diazoacetato de etilo ethyl diazoacetate

diazoamina diazoamine

diazoamino diazoamino

diazoaminobenceno diazoaminobenzene

diazoaminonaftaleno diazoaminonaphthalene

diazoato diazoate

diazobenceno diazobenzene

diazoetano diazoethane

diazoetanoato de etilo ethyl diazoethanoate

diazofenol diazophenol

diazohidrato diazohydrate

diazohidróxido diazohydroxide

diazoimida diazoimide
diazoimido diazoimido
diazol diazole
diazometano diazomethane
diazonio diazonium
diazonitrofenol diazonitrophenol
diazosulfonato diazosulfonate
diazotización diazotization
diazoxi diazoxy
diazóxido diazoxide
dibásico dibasic
dibenalacetona dibenalacetone
dibencenilo dibenzenyl
dibenceno dibenzene
dibencenocromo dibenzenechromium
dibencilamina dibenzylamine
dibencilideno dibenzylidene
dibencilidino dibenzylidyne
dibencilmetilamina dibenzylmethylamine
dibencilo dibenzyl
dibenzamida dibenzamide
dibenzantraceno dibenzanthracene
dibenzantrona dibenzanthrone
dibenzantronilo dibenzanthronyl
dibenzo dibenzo
dibenzoantraceno dibenzoanthracene
dibenzociclohepatadienona
    dibenzocyclohepatadienone
dibenzoiletano dibenzoylethane
dibenzoiletileno dibenzoylethylene
dibenzoilo dibenzoyl
dibenzopirona dibenzopyrone
dibenzopirrol dibenzopyrrole
dibenzotiacina dibenzothiazine
diborano diborane
diboruro de titanio titanium diboride
diboruro de zirconio zirconium diboride
dibromhidrato dihydrobromide
dibromo dibromo
dibromoacetato de etilo ethyl
    dibromoacetate
dibromoacetileno dibromoacetylene
dibromoantraceno dibromoanthracene
dibromobenceno dibromobenzene
dibromoclorometano
    dibromochloromethane
dibromocloropropano
    dibromochloropropane
dibromodifluorometano

dibromodifluoromethane
dibromoetano dibromoethane
dibromofluorobenceno
    dibromofluorobenzene
dibromometano dibromomethane
dibromopentano dibromopentane
dibromopropano dibromopropane
dibromopropanol dibromopropanol
dibromuro dibromide
dibromuro de acetileno acetylene
    dibromide
dibromuro de benceno benzene
    dibromide
dibromuro de bencilideno benzylidene
    dibromide
dibromuro de etileno ethylene dibromide
dibromuro de etilideno ethylidene
    dibromide
dibromuro de galio gallium dibromide
dibromuro de germanio germanium
    dibromide
dibromuro de paladio palladium
    dibromide
dibromuro de pentametileno
    pentamethylene dibromide
dibromuro de platino platinum
    dibromide
dibromuro de selenio selenium
    dibromide
dibromuro de telurio tellurium
    dibromide
dibucaína dibucaine
dibultilamina dibutylamine
dibutilditiocarbamato sódico sodium
    dibutyldithiocarbamate
dibutilestaño dibutyltin
dibutilo dibutyl
dibutiltiourea dibutylthiourea
dibutirato de etilenglicol ethylene glycol
    dibutyrate
dibutoxianilina dibutoxyaniline
dicaprilo dicapryl
dicarbilamina dicarbylamine
dicarbocianina dicarbocyanine
dicarbonato dicarbonate
dicarboxilo dicarboxyl
dicarburo dicarbide
dicarburo cálcico calcium dicarbide
dicarburo de uranio uranium dicarbide

dicetena diketene
dicetilo dicetyl
dicetona diketone
dicetopiperazina diketopiperazine
dicetopirrolidina diketopyrrolidine
diciandiamida dicyandiamide
diciano dicyano
dicianoargenato potásico potassium dicyanoargenate
dicianoaurato potásico potassium dicyanoaurate
dicianoaurato sódico sodium dicyanoaurate
dicianógeno dicyanogen
dicianuro dicyanide
dicianuro de etileno ethylene dicyanide
dicianuro de paladio palladium dicyanide
dicíclico dicyclic
diciclohexilamina dicyclohexylamine
diciclohexilcarbodiimida dicyclohexylcarbodiimide
diciclohexilo dicyclohexyl
diciclopentadienilo dicyclopentadienyl
diciclopentadieno dicyclopentadiene
diclona dichlone
dicloramina dichloramine
diclorhidrato dihydrochloride
diclorhidrato de hidrazina hydrazine dihydrochloride
diclorhidrato de piperazina piperazine dihydrochloride
dicloro dichlorine
dicloroacetal dichloroacetal
dicloroacetato de etilo ethyl dichloroacetate
dicloroacetato de metilo methyl dichloroacetate
dicloroacetileno dichloroacetylene
dicloroanilina dichloroaniline
dicloroantraceno dichloroanthracene
diclorobenceno dichlorobenzene
diclorobencidina dichlorobenzidine
diclorobenzaldehído dichlorobenzaldehyde
diclorobutano dichlorobutane
diclorobuteno dichlorobutene
diclorocarbeno dichlorocarbene
diclorodifluorurometano dichlorodifluoromethane

dicloroestearato de metilo methyl dichlorostearate
dicloroetano dichloroethane
dicloroeteno dichloroethene
dicloroetileno dichloroethylene
diclorofeno dichlorophene
diclorofenol dichlorophenol
diclorohidrina dichlorohydrin
diclorohidrina de glicerol glycerol dichlorohydrin
dicloroisocianurato potásico potassium dichloroisocyanurate
dicloroisocianurato sódico sodium dichloroisocyanurate
diclorometano dichloromethane
diclorometilsilano dichloromethylsilane
dicloronaftaleno dichloronaphthalene
dicloronitrobenceno dichloronitrobenzene
dicloropentano dichloropentane
dicloropropano dichloropropane
dicloropropeno dichloropropene
dicloropropionato sódico sodium dichloropropionate
diclororiofeno dichlorothiophene
diclorotolueno dichlorotoluene
diclorotrifluoruro de antimonio antimony dichlorotrifluoride
dicloruro dichloride
dicloruro benceno benzene dichloride
dicloruro clorometilfosfónico chloromethylphosphonic dichloride
dicloruro de acetileno acetylene dichloride
dicloruro de amileno amylene dichloride
dicloruro de azufre sulfur dichloride
dicloruro de bencilideno benzylidene dichloride
dicloruro de bencilo benzyl dichloride
dicloruro de bismuto bismuth dichloride
dicloruro de carbono carbon dichloride
dicloruro de dibutilestaño dibutyltin dichloride
dicloruro de dimetilestaño dimethyltin dichloride
dicloruro de estaño tin dichloride
dicloruro de etilaluminio ethylaluminum dichloride

dicloruro de etileno ethylene dichloride
dicloruro de ferrocenoilo ferrocenoyl dichloride
dicloruro de galio gallium dichloride
dicloruro de germanio germanium dichloride
dicloruro de glicol glycol dichloride
dicloruro de hierro iron dichloride
dicloruro de indio indium dichloride
dicloruro de metileno methylene dichloride
dicloruro de molibdeno molybdenum dichloride
dicloruro de oro gold dichloride
dicloruro de osmio osmium dichloride
dicloruro de oxalilo oxalyl dichloride
dicloruro de platino platinum dichloride
dicloruro de polivinilo polyvinyl dichloride
dicloruro de propileno propylene dichloride
dicloruro de rutenio ruthenium dichloride
dicloruro de selenio selenium dichloride
dicloruro de telurio tellurium dichloride
dicloruro de tetraglicol tetraglycol dichloride
dicloruro de titanio titanium dichloride
dicloruro de titanoceno titanocene dichloride
dicloruro de trietilenglicol triethylene glycol dichloride
dicloruro de triglicol triglycol dichloride
dicloruro de tungsteno tungsten dichloride
dicloruro de vanadio vanadium dichloride
dicloruro de xililo xylyl dichloride
dicloruro de zirconoceno zirconocene dichloride
dicrilo dicryl
dicroico dichroic
dicroismo dichroism
dicroismo circular circular dichroism
dicromático dichromatic
dicromato dichromate
dicromato alumínico aluminum dichromate
dicromato amónico ammonium dichromate
dicromato bárico barium dichromate
dicromato cálcico calcium dichromate
dicromato de cesio cesium dichromate
dicromato de cinc zinc dichromate
dicromato de litio lithium dichromate
dicromato de mercurio mercury dichromate
dicromato de plata silver dichromate
dicromato de plomo lead dichromate
dicromato de rubidio rubidium dichromate
dicromato férrico ferric dichromate
dicromato magnésico magnesium dichromate
dicromato mercúrico mercuric dichromate
dicromato potásico potassium dichromate
dicromato sódico sodium dichromate
dicrómico dichromic
didecanoato de trietilenglicol triethylene glycol didecanoate
didimio didymium
didimolita didymolite
diectilo diectyl
dieldrina dieldrin
dieléctrico dielectric
dienestrol dienestrol
dieno diene
dienófilo dienophile
diestearato alumínico aluminum distearate
diestearato de poliglicol polyglycol distearate
diestearato de propilenglicol propylene glycol distearate
diéster diester
diesterato de glicerol glycerol distearate
diestireno distyrene
dietanolamina diethanolamine
dietanolamina de butilo butyl diethanolamine
diéter diether
dietilacetal diethylacetal
dietilamida de ácido lisérgico lysergic acid diethylamide
dietilamina diethylamine
dietilamino diethylamino

dietilaminobenzaldehído
diethylaminobenzaldehyde
dietilaminoetanol diethylaminoethanol
dietilaminopropilamina
diethylaminopropylamine
dietilanilina diethylaniline
dietilbenceno diethylbenzene
dietilcadmio diethylcadmium
dietilcarbamazina diethylcarbamazine
dietilcarbinol diethylcarbinol
dietilcetona diethyl ketone
dietilciclohexilamina
diethylcyclohexylamine
dietilcinc diethylzinc
dietilditiocarbamato de cadmio
cadmium diethyldithiocarbamate
dietilditiocarbamato de selenio selenium
diethyldithiocarbamate
dietilendiamina diethylenediamine
dietilenglicol diethylene glycol
dietileno diethylene
dietilenotriamina diethylenetriamine
dietilestilbestrol diethylstilbestrol
dietiletileno diethylethylene
dietilideno diethylidene
dietilmalonato de dietilo diethyl
diethylmalonate
dietilmetilmetano diethylmethylmethane
dietilo diethyl
dietiloxamida diethyloxamide
dietilurea diethylurea
dietoxianilina diethoxyaniline
dietoxietano diethoxyethane
dietzeita dietzeite
difenilamina diphenylamine
difenilamino diphenylamino
difenilaminosulfonato bárico barium
diphenylamine sulfonate
difenilbenceno diphenylbenzene
difenilcarbazida diphenylcarbazide
difenilcetona diphenyl ketone
difenilcloroarsina diphenylchloroarsine
difenildiclorosilano
diphenyldichlorosilane
difenildiimida diphenyldiimide
difenileno diphenylene
difenilestaño diphenyltin
difeniletano diphenylethane
difeniletileno diphenylethylene

difenilglioxal diphenylglyoxal
difenilguanidina diphenylguanidine
difenilhidantoína sódica sodium
diphenylhydantoin
difenilmetano diphenylmethane
difenilmetilo diphenylmethyl
difenilo diphenyl
difenilo clorado chlorinated diphenyl
difenilsulfona diphenylsulfone
difenilurea diphenylurea
difenol diphenol
difenolato diphenolate
difluoro difluoro
difluoroamino difluoroamino
difluorobenceno difluorobenzene
difluorodiazina difluorodiazine
difluoroetano difluoroethane
difluorometano difluoromethane
difluoruro difluoride
difluoruro de cobalto cobalt difluoride
difluoruro de estaño tin difluoride
difluoruro de hidrógeno amónico
ammonium hydrogen difluoride
difluoruro de hidrógeno sódico sodium
hydrogen difluoride
difluoruro de vanadio vanadium
difluoride
diformiato alumínico aluminum
diformate
diformiato de etilenglicol ethylene
glycol diformate
diformiato de glicol glycol diformate
difosfano diphosphane
difosfato diphosphate
difosfato de adenosina adenosine
diphosphate
difosfato de fructosa fructose
diphosphate
difosfato disódico disodium diphosphate
difosfato potásico potassium diphosphate
difosfato tetrasódico tetrasodium
diphosphate
difosfina diphosphine
difosgeno diphosgene
difracción diffraction
difracción de rayos X X-ray diffraction
difracción electrónica electron
diffraction
difracción electrónica de baja energía

low-energy electron diffraction
difracción neutrónica neutron diffraction
difracción por transmisión transmission diffraction
difusión diffusion
difusión electroquímica electrochemical diffusion
difusión gaseosa gaseous diffusion
difusión térmica thermal diffusion
difuso diffuse
digestión digestion
digitalina digitalin
digitalis digitalis
digitalosa digitalose
digitonina digitonin
digitoxigenina digitoxigenin
digitoxina digitoxin
diglicérido diglyceride
diglicerol diglycerol
diglicol diglycol
diglicólido diglycolide
diglime diglyme
diglime de butilo butyl diglyme
digol digol
digoxina digoxin
dihedral dihedral
dihedrón dihedron
dihexilamina dihexylamine
dihexilo dihexyl
dihidoabietato de trietilenglicol
    triethylene glycol dihydroabietate
dihidrato dihydrate
dihidrazona dihydrazone
dihidro dihydro
dihidroazirina dihydroazirine
dihidrobenceno dihydrobenzene
dihidrocalcona dihydrochalcone
dihidrocolesterol dihydrocholesterol
dihidrogenofosfato sódico sodium dihydrogenphosphate
dihidrogenoortoperyodato sódico
    sodium dihydrogenorthoperiodate
dihidrogenosulfato magnésico
    magnesium dihydrogensulfate
dihidronaftaleno dihydronaphthalene
dihidropirano dihydropyran
dihidropropeno dihydropropene
dihidroxi dihydroxy
dihidroxiacetona dihydroxyacetone

dihidroxiamina dihydroxyamine
dihidroxiantraquinona
    dihydroxyanthraquinone
dihidroxibenceno dihydroxybenzene
dihidroxidifenilsulfona
    dihydroxydiphenylsulfone
dihidroxietano dihydroxyethane
dihidroxietilglicina sódica sodium dihydroxyethylglycine
dihidroxinaftaleno dihydroxynaphthalene
dihidroxipropano dihydroxypropane
dihidruro dihydride
diimida diimide
diimina diimine
diimino diimino
diisoamilo diisoamyl
diisobutilamina diisobutylamine
diisobutilcetona diisobutyl ketone
diisobutileno diisobutylene
diisobutilo diisobutyl
diisobutilofenol diisobutylphenol
diisocianato diisocyanate
diisocianato de hexametileno
    hexamethylene diisocyanate
diisopentilo diisopentyl
diisopropilbenceno diisopropylbenzene
diisopropilo diisopropyl
diisopropóxido de cloroaluminio
    chloroaluminum diisopropoxide
dilactona dilactone
dilatación dilatation
dilatómetro dilatometer
dilaurato de dibutilestaño dibutyltin dilaurate
dilaurilamina dilaurylamine
dilaurilfosfito dilauryl phosphite
dilaurilo dilauryl
dilaurilperóxido dilauryl peroxide
dilaurilsulfuro dilauryl sulfide
dilución dilution
dilución infinita infinite dilution
diluido dilute
diluyente diluent
dimedona dimedone
dimercaptoacetato de glicol glycol dimercaptoacetate
dimérico dimeric
dimerización dimerization
dímero dimer

dímero de acrilonitrilo acrylonitrile dimer

dímero de acroleína acrolein dimer

dimetacrilato de tetraetilenglicol tetraethylene glycol dimethacrylate

dimetano dimethano

dimetilacetal dimethylacetal

dimetilacetal de fenilacetaldehído phenylacetaldehyde dimethylacetal

dimetilamina dimethylamine

dimetilamino dimethylamino

dimetilaminoetanol dimethylaminoethanol

dimetilaminopropanol dimethylaminopropanol

dimetilán dimetilan

dimetilanilina dimethylaniline

dimetilantraceno dimethylanthracene

dimetilarseniato sódico sodium dimethylarsenate

dimetilarsina dimethylarsine

dimetilbenceno dimethylbenzene

dimetilberilio dimethylberyllium

dimetilbutadieno dimethylbutadiene

dimetilbutano dimethylbutane

dimetilbuteno dimethylbutene

dimetilcarbinol dimethylcarbinol

dimetilciclohexano dimethylcyclohexane

dimetilciclopentano dimethylcyclopentane

dimetilcloroacetal dimethylchloroacetal

dimetildiclorosilano dimethyldichlorosilane

dimetilditiocarbamato sódico sodium dimethyldithiocarbamate

dimetileno dimethylene

dimetiletileno dimethylethylene

dimetilfenol dimethylphenol

dimetilformamida dimethylformamide

dimetilfosfina dimethylphosphine

dimetilglioxima dimethylglyoxime

dimetilglioxima de níquel nickel dimethylglyoxime

dimetilhepteno dimethylheptene

dimetilhexadieno dimethylhexadiene

dimetilhexanodiol dimethylhexanediol

dimetilhexinodiol dimethylhexynediol

dimetilhidantoína dimethylhydantoin

dimetilhidrazina dimethylhydrazine

dimetilhidroxibenceno dimethylhydroxybenzene

dimetilisopropanolamina dimethylisopropanolamine

dimetilmercurio dimethylmercury

dimetilmetano dimethylmethane

dimetilnitrosamina dimethylnitrosamine

dimetilo dimethyl

dimetiloctadieno dimethyloctadiene

dimetilpentaldehído dimethylpentaldehyde

dimetilpentano dimethylpentane

dimetilpropano dimethylpropane

dimetilurea dimethylurea

dimetilxantina dimethylxanthine

dimetoato dimethoate

dimetoxi dimethoxy

dimetoxianilina dimethoxyaniline

dimetoxibenceno dimethoxybenzene

dimetoxietano dimethoxyethane

dimetoximetano dimethoxymethane

dimetoxitetraglicol dimethoxytetraglycol

dimetracrilato de butileno butylene dimethacrylate

dimétrico dimetric

dimórfico dimorphic

dimorfismo dimorphism

dinada dynad

dinaftileno dinaphthylene

dinaftilo dinaphthyl

dinafto dinaphtho

dinaftol dinaphthol

dinamita dynamite

dinenciléter dibenzylether

dinérico dineric

dineutrón dineutron

dinitrato dinitrate

dinitrato de dietilenglicol diethylene glycol dinitrate

dinitrato de etilenglicol ethylene glycol dinitrate

dinitrato de etileno ethylene dinitrate

dinitrato de glicol glycol dinitrate

dinitrato de paladio palladium dinitrate

dinitrato de propilenglicol propylene glycol dinitrate

dinitrilo malónico malonic dinitrile

dinitrito dinitrite

dinitro dinitro

dinitroanilina dinitroaniline
dinitroantraquinona
  dinitroanthraquinone
dinitrobenceno dinitrobenzene
dinitrocresol dinitrocresol
dinitrocresolato sódico sodium
  dinitrocresolate
dinitrofenilhidrazina
  dinitrophenylhydrazine
dinitrofenol dinitrophenol
dinitrofluorobenceno
  dinitrofluorobenzene
dinitrógeno dinitrogen
dinitronaftaleno dinitronaphthalene
dinitronaftol dinitronaphthol
dinitroso dinitroso
dinitrotolueno dinitrotoluene
dinucleótido dinucleotide
dioctilamina dioctylamine
dioctilmetano dioctylmethane
dioctilo dioctyl
dioctoato de trietilenglicol triethylene
  glycol dioctoate
diol diol
diolefina diolefin
diona dione
diópsido diopside
dioptasa dioptase
diorita diorite
dioxadieno dioxadiene
dioxalano dioxalane
dioxana dioxan
dioxano dioxane
dioxazol dioxazole
dioxdiazol dioxdiazole
dioxi dioxy
dióxido dioxide
dióxido bárico barium dioxide
dióxido cálcico calcium dioxide
dióxido de azufre sulfur dioxide
dióxido de carbono carbon dioxide
dióxido de cerio cerium dioxide
dióxido de cesio cesium dioxide
dióxido de cinc zinc dioxide
dióxido de cloro chlorine dioxide
dióxido de cromo chromium dioxide
dióxido de dietileno diethylene dioxide
dióxido de dipenteno dipentene dioxide
dióxido de estaño tin dioxide

dióxido de estroncio strontium dioxide
dióxido de germanio germanium dioxide
dióxido de hidrógeno hydrogen dioxide
dióxido de manganeso manganese
  dioxide
dióxido de molibdeno molybdenum
  dioxide
dióxido de neptunio neptunium dioxide
dióxido de nitrógeno nitrogen dioxide
dióxido de osmio osmium dioxide
dióxido de paladio palladium dioxide
dióxido de platino platinum dioxide
dióxido de plomo lead dioxide
dióxido de selenio selenium dioxide
dióxido de silicio silicon dioxide
dióxido de telurio tellurium dioxide
dióxido de titanio titanium dioxide
dióxido de torio thorium dioxide
dióxido de tricarbono tricarbon dioxide
dióxido de uranio uranium dioxide
dióxido de vanadio vanadium dioxide
dióxido de vinilciclohexeno
  vinylcyclohexene dioxide
dióxido de yodo iodine dioxide
dióxido de zirconio zirconium dioxide
dióxido líquido liquid dioxide
dióxido magnésico magnesium dioxide
dióxido sódico sodium dioxide
dioxigenilo dioxygenyl
dioxima dioxime
dioxina dioxin
dioxo dioxo
dioxolano dioxolane
dipalmitato dipalmitate
dipalmitilamina dipalmitylamine
dipelargonato de trietilenglicol
  triethylene glycol dipelargonate
dipentaeritritol dipentaerythritol
dipenteno dipentene
dipentenoglicol dipentene glycol
dipentilo dipentyl
dipéptido dipeptide
dipiridilo dipyridyl
dipiridina dipyridine
diplomo dilead
dipnona dypnone
dipolar dipolar
dipolo dipole
dipolo molecular molecular dipole

dipropilberilio dipropylberyllium
dipropilenglicol dipropylene glycol
dipropilo dipropyl
dipropionato de trietilenglicol
   triethylene glycol dipropionate
diprotón diproton
dirradical diradical
dirresorcina diresorcinol
disacárido disaccharide
disacrilo discaryl
disco de estallido bursting disk
discrasita dyscrasite
discriminación de filtro filter
   discrimination
diselano diselane
diseleniuro de niobio niobium diselenide
diseleniuro de tungsteno tungsten
   diselenide
disilano disilane
disilicato disilicate
disiliciuro disilicide
disiliciuro de molibdeno molybdenum
   disilicide
disiliciuro de titanio titanium disilicide
disiliciuro de zirconio zirconium
   disilicide
disililo disilyl
disiloxano disiloxane
disimetría dissymmetry
disimétrico dissymmetrical
disinalilo disilanyl
disipación dissipation
disipación de electrodo electrode
   dissipation
disipación térmica heat dissipation
dislfuro de dietilo diethyl disulfide
dislocación dislocation
dismutación dismutation
dismutasa superóxida superoxide
   dismutase
disociación dissociation
disociación de ácido acid dissociation
disociación electrolítica electrolytic
   dissociation
disociación fotoquímica photochemical
   dissociation
disociación hidrolítica hydrolytic
   dissociation
disociación molecular molecular

dissociation
disociado dissociated
disociar dissociate
disolución solution, dissolution
disolvente dissolvent, solvent
disolver dissolve
dispersador dispersator
dispersante dispersing
dispersar disperse
dispersión dispersion, scattering
dispersión anómala anomalous
   dispersion
dispersión coloidal colloidal dispersion
dispersión de Delbrück Delbrück
   scattering
dispersión de luz light scattering
dispersión de partículas particle
   scattering
dispersión de Rutherford Rutherford
   scattering
dispersión molecular molecular
   dispersion
dispersión nuclear nuclear scattering
dispersión rotatoria rotatory dispersion
dispersión rotatoria óptica optical
   rotatory dispersion
dispersión sódica sodium dispersion
disperso dispersed, scattered
dispersoide dispersoid
disponibilidad availability
disponibilidad biológica biological
   availability
disponible available
dispositivo de laboratorio laboratory
   device
dispositivo de seguridad safety device
disprosio dysprosium
distancia distance
distancia atómica atomic distance
distancia de enlace bond distance
distancia de enlazado bonding distance
distancia interatómica interatomic
   distance
distribución de Maxwell-Boltzmann
   Maxwell-Boltzmann distribution
distribución de partículas particle
   distribution
distribución electrónica electron
   distribution

disuelto dissolved
disulfato disulfate
disulfato de hierro iron disulfate
disulfato potásico potassium disulfate
disulfato sódico sodium disulfate
disulfiram disulfiram
disulfito disulfite
disulfito potásico potassium disulfite
disulfito sódico sodium disulfite
disulfo disulfo
disulfonato disulfonate
disulfuro disulfide
disulfuro de arsénico arsenic disulfide
disulfuro de bencilo benzyl disulfide
disulfuro de benzoilo benzoyl disulfide
disulfuro de benzotiacilo benzothiazyl
   disulfide
disulfuro de carbono carbon disulfide
disulfuro de dibencilo dibenzyl disulfide
disulfuro de dimetilo dimethyl disulfide
disulfuro de germanio germanium
   disulfide
disulfuro de hafnio hafnium disulfide
disulfuro de hidrógeno hydrogen
   disulfide
disulfuro de hierro iron disulfide
disulfuro de iridio iridium disulfide
disulfuro de molibdeno molybdenum
   disulfide
disulfuro de osmio osmium disulfide
disulfuro de paladio palladium disulfide
disulfuro de silicio silicon disulfide
disulfuro de tántalo tantalum disulfide
disulfuro de telurio tellurium disulfide
disulfuro de titanio titanium disulfide
disulfuro de torio thorium disulfide
disulfuro de tungsteno tungsten disulfide
disulfuro de vanadio vanadium disulfide
disulfuro de yodo iodine disulfide
disulfuro de zirconio zirconium disulfide
ditanato de bismuto bismuth ditannate
diterpeno diterpene
diterpenoide diterpenoid
ditiano dithiane
ditiazina dithiazine
ditienilo dithienyl
ditieno dithiene
ditio dithio
ditiocarbamato dithiocarbamate

ditiocarbamato amónico ammonium
   dithiocarbamate
ditiocarbonato dithiocarbonate
ditiocarbonato de mercurio mercury
   dithiocarbonate
ditiocarboxi dithiocarboxy
ditiol dithiol
ditioleno dithiolene
ditionato dithionate
ditionato amónico ammonium dithionate
ditionato bárico barium dithionate
ditionato de cinc zinc dithionate
ditionato de estroncio strontium
   dithionate
ditionato de manganeso manganese
   dithionate
ditionato de plata silver dithionate
ditionato de plomo lead dithionate
ditionato potásico potassium dithionate
ditionato sódico sodium dithionate
ditionita dithionite
ditionito sódico sodium dithionite
ditiooxamida dithiooxamide
ditiosalicilato sódico sodium
   dithiosalicylate
ditizona dithizone
ditolilo ditolyl
diuranato sódico sodium diuranate
diuriniato bárico barium diurinate
divalencia divalence
divalente divalent
divanadato sódico sodium divanadate
divinilacetileno divinylacetylene
divinilbenceno divinylbenzene
divinilo divinyl
división splitting, cleavage
división de grasas fat splitting
dixililetano dixylylethane
dixililo dixylyl
diyodo diiodo
diyodoacetileno diiodoacetylene
diyodometano diiodomethane
diyoduro diiodide
diyoduro de arsénico arsenic diiodide
diyoduro de etileno ethylene diiodide
diyoduro de etilideno ethylidene diiodide
diyoduro de galio gallium diiodide
diyoduro de germanio germanium
   diiodide

diyoduro de metileno methylene diiodide
diyoduro de paladio palladium diiodide
diyoduro de platino platinum diiodide
doblete doublet
docosano docosane
docosanol docosanol
dodecaedral dodecahedral
dodecaedro dodecahedron
dodecahedro rómbico rhombic dodecahedron
dodecahidrato dodecahydrate
dodecanal dodecanal
dodecano dodecane
dodecanoato de butilo butyl dodecanoate
dodecanoilo dodecanoyl
dodecanol dodecanol
dodecenal dodecenal
dodeceno dodecene
dodecilamina dodecylamine
dodecilbenceno dodecylbenzene
dodecilbencenosulfonato sódico sodium dodecylbenzenesulfonate
dodecileno dodecylene
dodecilfenol dodecylphenol
dodecilo dodecyl
dodeciltriclorosilano dodecyltrichlorosilane
dolerita dolerite
dolicol dolichol
dolomita dolomite
dolomol dolomol
domesticina domesticine
dominguita domingite
donador donor

donador electrónico electron donor
donador protónico proton donor
donarita donarite
dopado doping
dopado con oro gold doping
dopante dope
dorado gold-plating
dosis de radiación radiation dose
dosis letal lethal dose
dotriacontano dotriacontane
draconilo draconyl
drimina drimin
droga drug
droga psicotrópica psychotropic drug
droga sulfa sulfa drug
dúctil ductile
ductilidad ductility
dufrenita dufrenite
dufrenoisita dufrenoysite
dulcitol dulcitol
dumortierita dumortierite
dunita dunnite
duplete duplet
duraluminio duralumin
dureno durene
dureza hardness
dureza de agua hardness of water
dureza de Brinell Brinell hardness
dureza de Knoop Knoop hardness
dureza permanente permanent hardness
durileno durylene
durilo duryl
duro hard
durol durol

E

easina easin
ebonita ebonite
ebullición ebullition, boiling
ebullidor ebullator
ebulliómetro ebulliometer
ebullometría ebulliometry
ebullómetro diferencial differential
   ebulliometer
ebulloscopia ebullioscopy
ebulloscopio ebullioscope
ecdisona ecdysone
ecgonina ecgonine
eclipsado eclipsed
eclogita eclogite
ecología ecology
economizador economizer
ecosistema ecosystem
ecuación equation
ecuación de adsorción de Gibbs Gibbs
   adsorption equation
ecuación de Arrhenius Arrhenius
   equation
ecuación de Beattie-Bridgman Beattie-
   Bridgman equation
ecuación de Berthelot Berthelot equation
ecuación de Boltzmann Boltzmann
   equation
ecuación de Bragg Bragg equation
ecuación de Butler-Volmer Butler-
   Volmer equation
ecuación de Clapeyron Clapeyron
   equation
ecuación de Clapeyron-Clausius
   Clapeyron-Clausius equation
ecuación de Clausius-Mosotti Clausius-
   Mosotti equation
ecuación de de Broglie de Broglie
   equation
ecuación de Dieterici Dieterici equation
ecuación de Duhem Duhem equation
ecuación de estado equation of state
ecuación de Eyring Eyring equation
ecuación de Flood Flood equation

ecuación de Fokker-Planck Fokker-
   Planck equation
ecuación de gases gas equation
ecuación de Gibbs-Duhem Gibbs-
   Duhem equation
ecuación de Gibbs-Helmholtz Gibbs-
   Helmholtz equation
ecuación de Gibbs-Poynting Gibbs-
   Poynting equation
ecuación de Haggenmacher
   Haggenmacher equation
ecuación de Hammett Hammett equation
ecuación de Helmholtz Helmholtz
   equation
ecuación de Henderson-Hasselbach
   Henderson-Hasselbach equation
ecuación de Ilkovic Ilkovic equation
ecuación de Karplus Karplus equation
ecuación de Kohlrausch Kohlrausch
   equation
ecuación de Mark-Houwink Mark-
   Houwink equation
ecuación de Morse Morse equation
ecuación de Nernst Nernst equation
ecuación de Nernst-Einstein Nernst-
   Einstein equation
ecuación de Onsager Onsager equation
ecuación de Perrin Perrin equation
ecuación de Pitzer Pitzer equation
ecuación de Ramsay-Shields Ramsay-
   Shields equation
ecuación de Ramsay-Young Ramsay-
   Young equation
ecuación de reacción reaction equation
ecuación de Sackur-Tetrode Sackur-
   Tetrode equation
ecuación de Schomaker-Stevenson
   Schomaker-Stevenson equation
ecuación de Schrödinger Schrödinger
   equation
ecuación de van der Waals van der
   Waals equation
ecuación de van't Hoff van't Hoff

equation
ecuación de Watson Watson equation
ecuación maestra master equation
ecuación molecular molecular equation
ecuación nuclear nuclear equation
ecuación química chemical equation
ecuatorial equatorial
edeato edeate
edenita edenite
edestina edestin
edetato edetate
edetato disódico edetate disodium
edetato disódico cálcico calcium
    disodium edetate
edetato sódico sodium edetate
edingtonita edingtonite
efectividad de filtro filter effectiveness
efecto effect
efecto alostérico allosteric effect
efecto anódico anodic effect
efecto celostérico cellosteric effect
efecto cinético kinetic effect
efecto de apantallamiento screening
    effect
efecto de Back-Goudsmit Back-
    Goudsmit effect
efecto de Baker-Nathan Baker-Nathan
    effect
efecto de Baker-Venkataraman Baker-
    Venkataraman effect
efecto de Becquerel Becquerel effect
efecto de campo eléctrico electric field
    effect
efecto de Cerenkov Cerenkov effect
efecto de Compton Compton effect
efecto de Cotton Cotton effect
efecto de Chadwick-Goldhaber
    Chadwick-Goldhaber effect
efecto de Draper Draper effect
efecto de empaquetamiento packing
    effect
efecto de Faraday Faraday effect
efecto de Hall Hall effect
efecto de invernadero greenhouse effect
efecto de ion común common-ion effect
efecto de ionización ionization effect
efecto de Jahn-Teller Jahn-Teller effect
efecto de Joule Joule effect
efecto de Joule-Kelvin Joule-Kelvin

effect
efecto de Joule-Thomson Joule-
    Thomson effect
efecto de Kelvin Kelvin effect
efecto de Kundt Kundt effect
efecto de Mössbauer Mössbauer effect
efecto de orientación orientation effect
efecto de Overhauser Overhauser effect
efecto de Paschen-Back Paschen-Back
    effect
efecto de Pasteur Pasteur effect
efecto de Pfeiffer Pfeiffer effect
efecto de presión pressure effect
efecto de quelato chelate effect
efecto de radiación radiation effect
efecto de Raman Raman effect
efecto de Ramsauer Ramsauer effect
efecto de relajación relaxation effect
efecto de saturación saturation effect
efecto de Soret Soret effect
efecto de Stark Stark effect
efecto de Stark-Luneland Stark-
    Luneland effect
efecto de Szilard-Chalmers Szilard-
    Chalmers effect
efecto de transición transition effect
efecto de Tyndall Tyndall effect
efecto de Weigert Weigert effect
efecto de Weissenberg Weissenberg
    effect
efecto de Wien Wien effect
efecto de xenón xenon effect
efecto de Zeeman Zeeman effect
efecto electroforético electrophoretic
    effect
efecto electroquímico electrochemical
    effect
efecto estérico steric effect
efecto fotoquímico photochemical effect
efecto inductivo inductive effect
efecto isotópico isotope effect
efecto piezoeléctrico piezoelectric effect
efecto químico chemical effect
efecto térmico heat effect, thermal effect
efecto termiónico thermionic effect
efecto termomagnético thermomagnetic
    effect
efecto trans trans effect
efecto túnel tunnel effect

efedrina ephedrine
efervescencia effervescence
efervescente effervescent
eficiencia cuántica quantum efficiency
eficiencia de fluorescencia fluorescence efficiency
eficiencia de placas plate efficiency
eficiencia química chemical efficiency
eflorescencia efflorescence
efluente effluent
efusiómetro effusiometer
efusión effusion
eicosano eicosane
eicosanoide eicosanoid
eicosanol eicosanol
einstein einstein
einstenio einsteinium
eje axis
eje de simetría axis of symmetry
eje óptico optical axis
eje principal principal axis
eka eka
ekahafnio ekahafnium
elaboración elaboration
elaidato de metilo methyl elaidate
elaidinización elaidinization
elastasa elastase
elasticidad elasticity
elástico elastic
elastina elastin
elastómero elastomer
elaterita elaterite
electodializador electrodialyzer
electreto electret
electricidad atómica atomic electricity
electricidad galvánica galvanic electricity
eléctrico electric
electrificación electrification
electroanálisis electroanalysis
electrocinética electrokinetics
electrocrático electrocratic
electrocromatografía electrochromatography
electrocromatografía sobre papel paper electrochromatography
electrodeposición electrodeposition, plating
electrodepositado plated

electrodepósito electrodeposit
electrodiálisis electrodialysis
electrodisolución electrodissolution
electrodispersión electrodispersion
electrodo electrode
electrodo acelerador accelerating electrode
electrodo con gas gas electrode
electrodo de cadmio cadmium electrode
electrodo de calomel calomel electrode
electrodo de carbono carbon electrode
electrodo de Clark Clark electrode
electrodo de hidrógeno hydrogen electrode
electrodo de oxígeno oxygen electrode
electrodo de quinhidrona quinhydrone electrode
electrodo de referencia reference electrode
electrodo de vidrio glass electrode
electrodo metálico metal electrode
electrodo polarizado polarized electrode
electroenfoque electrofocusing
electroestricción electrostriction
electrofílico electrophilic
electrófilo electrophile
electroflotación electroflotation
electroforesis electrophoresis
electroforesis de gel gel electrophoresis
electroforesis por zonas zone electrophoresis
electroforesis sobre papel paper electrophoresis
electroforético electrophoretic
electroforetograma electrophoretogram
electroformación electroforming
electroforograma electrophorogram
electrografito electrographite
electrogravimetría electrogravimetry
electrólisis electrolysis
electrolítico electrolytic
electrólito electrolyte
electrólito anfótero amphoteric electrolyte
electrólito coloidal colloidal electrolyte
electrólito de batería battery electrolyte
electrólito débil weak electrolyte
electrólito fuerte strong electrolyte
electrólito fundido fused electrolyte

electrolizador electrolyzer
electrolizar electrolyze
electroluminiscencia
  electroluminescence
electromagnético electromagnetic
electromerismo electromerism
electrometalurgia electrometallurgy
electrómetro electrometer
electromigración electromigration
electrón electron
electrón de Auger Auger electron
electrón de conducción conduction
  electron
electrón de conversión conversion
  electron
electrón de enlazado bonding electron
electrón disponible available electron
electrón exterior outer electron
electrón giratorio spinning electron
electrón libre free electron
electrón ligado bound electron
electrón metaestable metastable electron
electrón metastásico metastasic electron
electrón orbital orbital electron
electrón pesado heavy electron
electrón pi pi electron
electrón positivo positive electron
electrón primario primary electron
electrón sigma sigma electron
electrón-volt electron-volt
electronegatividad electronegativity
electronegativo electronegative
electrones de valencia valence electrons
electrones dispersos scattered electrons
electrones emparejados paired electrons
electrones equivalentes equivalent
  electrons
electrones no emparejados unpaired
  electrons
electrones ópticos optical electrons
electrones planetarios planetary
  electrons
electroósmosis electroosmosis
electropositivo electropositive
electropotencial electropotential
electropulido electropolishing
electroquímica electrochemistry
electroquímico electrochemical
electrorrecubrimiento electroplating

electrorrevestimiento electrocoating
electroscopia electroscopy
electroscopio electroscope
electrosíntesis electrosynthesis
electrosol electrosol
electrovalencia electrovalence
electrovalente electrovalent
electroviscocidad electroviscosity
electrum electrum
electruro electride
elemento element
elemento activo active element
elemento artificial artificial element
elemento combustible fuel element
elemento de simetría element of
  symmetry
elemento de tierras raras rare-earth
  element
elemento de transición transition
  element
elemento electronegativo electronegative
  element
elemento electropositivo electropositive
  element
elemento enmascarado masked element
elemento inestable unstable element
elemento magnético magnetic element
elemento nativo native element
elemento negativo negative element
elemento neutro neutral element
elemento pesado heavy element
elemento positivo positive element
elemento químico chemical element
elemento radiactivo radioactive element
elemento radiactivo artificial artificial
  radioactive element
elemento siderófilo siderophile element
elemento superpesado superheavy
  element
elemento transactínido transactinide
  element
elemento transuránico transuranic
  element
elemento trazador tracer element
eleonorita eleonorite
elevación de punto de congelación
  freezing-point elevation
elevación de punto de ebullición
  boiling-point elevation

elevación molecular molecular elevation
elíxir elixir
elpidita elpidite
elución elution
eluido eluate
elutriación elutriation
eluyente eluant
emanación emanation
emanación radiactiva radioactive emanation
embolita embolite
embudo funnel
embudo de Buchner Buchner funnel
embudo de Bunsen Bunsen funnel
embudo de filtración filter funnel
embudo de separación separatory funnel
emetamina emetamine
emetina emetine
emisión emission
emisión alfa alpha emission
emisión de fotones photon emission
emisión de luz light emission
emisión de partículas particle emission
emisión de positrón positron emission
emisión electrónica electron emission
emisión espontánea spontaneous emission
emisión gamma gamma emission
emisión inducida induced emission
emisión iónica de campo field-ion emission
emisión primaria primary emission
emisión radiactiva radioactive emission
emisión secundaria secondary emission
emisor alfa alpha emitter
emisor beta beta emitter
emisor gamma gamma emitter
emodina emodin
empaque packaging
empaquetado packaging
empaquetado ampollado blister packaging
empaquetamiento packing
empírico empirical
emplectita emplectite
emulagador emulgator
emulsión emulsion
emulsionador emulsifier
emulsionamiento emulsification

emulsionar emulsify
en polvo powdered
enamina enamine
enantilo enanthyl
enantiomérico enantiomeric
enantiómero enantiomer
enantiomórfico enantiomorphic
enantiomorfismo enantiomorphism
enantiomorfo enantiomorph
enantiotópico enantiotopic
enantiotropía enantiotropy
enantiotrópico enantiotropic
enantiótropo enantiotropic
enantol enanthol
enargita enargite
encadenamiento cruzado cross-linking
encapsulación encapsulation
encefalina enkephalin
endo endo
endógeno endogenous
endomicina endomycin
endopeptidasa endopeptidase
endorfina endorphin
endósmosis endosmosis
endotérmico endothermic
endrina endrin
endulcorante sweetener
endulcorante no nutritivo nonnutritive sweetner
endulzador sweetener
endulzar sweeten
endurecedor hardener
endurecimiento hardening
endurecimiento de grasa fat hardening
endurecimiento electroquímico electrochemical hardening
endurecimiento por llama flame hardening
energía energy
energía al cero absoluto zero-point energy
energía atómica atomic energy
energía cinética kinetic energy
energía de activación activation energy
energía de deformación deformation energy
energía de disociación dissociation energy
energía de enlace bond energy

energía de estabilización stabilization
energy

energía de fisión fission energy

energía de fusión fusion energy

energía de ionización ionization energy

energía de partícula particle energy

energía de reacción reaction energy

energía de resonancia resonance energy

energía de separación separation energy

energía de separación de neutrón
neutron binding energy

energía de unión nuclear nuclear
binding energy

energía de Zeeman Zeeman energy

energía electrónica electron energy

energía equipotencial equipotential
energy

energía específica specific energy

energía interna internal energy

energía latente latent energy

energía libre free energy

energía libre de Gibbs Gibbs free
energy

energía libre de Helmholtz Helmholtz
free energy

energía neutrónica neutron energy

energía nuclear nuclear energy

energía potencial nuclear nuclear
potential energy

energía química chemical energy

energía reticular lattice energy

energía rotacional rotational energy

energía solar solar energy

energía superficial surface energy

energía superficial interfacial interfacial
surface energy

energía térmica thermal energy, heat
energy

energía termonuclear thermonuclear
energy

energía umbral threshold energy

energía vibracional vibrational energy

enflurano enflurane

enfriamiento cooling

enfriamiento adiabático adiabatic
cooling

enilo enyl

enlace bond, linkage

enlace coordinado coordinate bond

enlace covalente covalent bond

enlace covalente dativo dative covalent
bond

enlace covalente polar polar covalent
bond

enlace dativo dative bond

enlace de alta energía high-energy bond

enlace de carbono carbon bond

enlace de hidrógeno hydrogen bond

enlace de par de electrones electron-pair
bond

enlace de péptido peptide bond, peptide
linkage

enlace de valencia valence bond

enlace delta delta bond

enlace doble double bond

enlace doble aislado isolated double
bond

enlace electrostático electrostatic bond

enlace electrovalente electrovalent bond

enlace glicosídico glycosidic bond

enlace heteropolar heteropolar bond

enlace homopolar homopolar bond

enlace intermedio intermediate bond

enlace iónico ionic bond

enlace metal-metal metal-metal bond

enlace metálico metallic bond

enlace múltiple multiple bond

enlace pi pi bond

enlace químico chemical bond

enlace reactivo reactive bond

enlace sencillo single bond

enlace sigma sigma bond

enlace triple triple bond

enlaces dobles acumulativos cumulalive
double bonds

enlaces dobles conjugados conjugated
double bonds

enlazado bonding

enlazado delta delta bonding

enlazado hidrofóbico hydrophobic
bonding

enlazado iónico ionic bonding

enlazado metálico metallic bonding

enlazado múltiple multiple bonding

enlazado pi pi bonding

enmascaramiento masking

eno ene

enodiol enediol

enol enol
enolasa enolase
enriquecimiento enrichment
enriquecimiento de uranio uranium enrichment
ensayo assay, test
enstatita enstatite
entalpía enthalpy
entalpía de evaporación enthalpy of evaporation
entalpía de fusión enthalpy of fusion
entalpía de reacción enthalpy of reaction
entidad química chemical entity
entrada de gas gas inlet
entrada de hidrógeno hydrogen inlet
entrada de oxígeno oxygen inlet
entropía entropy
entropía de activación entropy of activation
entropía de mezcla entropy of mixing
entropía de transición entropy of transition
envejecimiento aging
enzima enzyme
enzima alostérica allosteric enzyme
enzima proteolítica proteolytic enzyme
eosina eosin
epi epi
epialcanfor epicamphor
epicianhidrina epicyanohydrin
epiclorita epichlorite
epiclorohidrina epichlorohydrin
epididimita epididymite
epidioxi epidioxy
epidióxido epidioxide
epiditio epidithio
epiestilbita epistilbite
epigenita epigenite
epihidrina epihydrin
epimerasa epimerase
epimérico epimeric
epimerida epimeride
epimerismo epimerism
epimerización epimerization
epímero epimer
epinefrina epinephrine
epitaxia epitaxy
epitio epithio
epitioximino epithioximino

epoxi epoxy
epoxibutano epoxybutane
epoxidación epoxidation
epóxido epoxide
epóxido de hexafluoropropileno hexafluoropropylene epoxide
epóxido de tetrafluoroetileno tetrafluoroethylene epoxide
epoxiestearato de butilo butyl epoxystearate
epoxietano epoxyethane
epoxiimino epoxyimino
epoxinitrilo epoxynitrilo
epoxipropano epoxypropane
epoxipropanol epoxypropanol
epoxitio epoxythio
epsilón epsilon
epsomita epsomite
equilenina equilenin
equilibrio equilibrium
equilibrio ácido-base acid-base equilibrium
equilibrio de Donnan Donnan equilibrium
equilibrio de fases phase equilibrium
equilibrio de Gibbs-Donnan Gibbs-Donnan equilibrium
equilibrio de oxígeno oxygen balance
equilibrio de redox redox equilibrium
equilibrio de sedimentación sedimentation equilibrium
equilibrio dinámico dynamic equilibrium
equilibrio energético energy balance
equilibrio estable stable equilibrium
equilibrio incompleto incomplete equilibrium
equilibrio iónico ionic equilibrium
equilibrio líquido-sólido liquid-solid equilibrium
equilibrio líquido-vapor liquid-vapor equilibrium
equilibrio químico chemical equilibrium
equilibrio radiactivo radioactive equilibrium
equilibrio sólido-líquido solid-liquid equilibrium
equilibrio térmico thermal equilibrium
equilibrio vapor-líquido vapor-liquid equilibrium

equinomicina echinomycin
equinopsina echinopsine
equipartición equipartition
equipartición de energía equipartition of energy
equipo de laboratorio laboratory equipment
equipo de manejo remoto remote-handling equipment
equivalencia equivalence
equivalencia cuántica quantum equivalence
equivalente equivalent
equivalente coloidal colloid equivalent
equivalente de hidrógeno hydrogen equivalent
equivalente de neutralización neutralization equivalent
equivalente de nitrógeno nitrogen equivalent
equivalente de saponificación saponification equivalent
equivalente electroquímico electrochemical equivalent
equivalente fotoquímico photochemical equivalent
equivalente-gramo gram-equivalent
equivalente osmótico osmotic equivalent
equivalente químico chemical equivalent
equivalente térmico heat equivalent
erbia erbia
erbio erbium
erbón erbon
ergina ergine
ergocalciferol ergocalciferol
ergódico ergodic
ergometrina ergometrine
ergonovina ergonovine
ergosinina ergosinine
ergosterol ergosterol
ergotamina ergotamine
ergotinina ergotinine
ergotoxina ergotoxine
erilita erilite
erinita erinite
eriodictiol eriodictyol
eritorbato sódico sodium erythorbate
eritrina erythrin
eritrita erythrite

eritritilo erythrityl
eritritol erythritol
eritro erythro
eritroidina erythroidine
eritrol erythrol
eritromicina erythromycin
eritrosa erythrose
eritrosiderita erythrosiderite
eritrosina erythrosin
erosión erosion
erosión química chemical erosion
error absoluto absolute error
error de calibración calibration error
error de medida measurement error
error de paralaje parallax error
error determinado determinate error
error relativo relative error
erubescita erubescite
escala scale
escala atómica atomic scale
escala Celsius Celsius scale
escala centígrada centigrade scale
escala cromática chromatic scale
escala de Knoop Knoop scale
escala de Mohs Mohs scale
escala de pH pH scale
escala de temperatura temperature scale
escala de temperatura absoluta absolute temperature scale
escala Fahrenheit Fahrenheit scale
escala Kelvin Kelvin scale
escala Rankine Rankine scale
escala termométrica thermometer scale
escandio scandium
escape de gas gas escape
escapolita scapolite
escatol skatole
escisión scission
escleroscopio scleroscope
escolecita scolecite
escopolamina scopolamine
escopolina scopoline
escoria slag
escoria de fosfato phosphate slag
escorodita scorodite
escualano squalane
escualeno squalene
esencia de perla pearl essence
esencial essential

eserina eserine
esfalerita sphalerite
esfingomielina sphingomyelin
esfingosina sphingosine
esmalte enamel
esmalte de porcelana porcelain enamel
esmalte de vidrio glass enamel
esmaltín smalt
esméctico smectic
esmeralda emerald
esmeril emery
esmerilado ground
espacio space
espacio de configuración configuration
    space
espacio de Hilbert Hilbert space
espacio elemental elementary space
espacio interatómico interatomic space
espacio intraatómico intraatomic space
espalación spallation
espandex spandex
espato spar
espato de Islandia Iceland spar
espato flúor fluorspar
espato pesado heavy spar
espato satinado satin spar
especialidad química chemical specialty
especie species
especie atómica atomic species
especie nuclear nuclear species
específico specific
espectral spectral
espectro spectrum
espectro atómico atomic spectrum
espectro continuo continuous spectrum
espectro de absorción absorption
    spectrum
espectro de absorción de infrarrojo
    infrared absorption spectrum
espectro de acción action spectrum
espectro de arco arc spectrum
espectro de Aston Aston spectrum
espectro de bandas band spectrum
espectro de difracción diffraction
    spectrum
espectro de emisión emission spectrum
espectro de excitación excitation
    spectrum
espectro de explosión explosion

    spectrum
espectro de fisión fission spectrum
espectro de Fraunhofer Fraunhofer
    spectrum
espectro de hidrógeno hydrogen
    spectrum
espectro de infrarrojo infrared spectrum
espectro de líneas line spectrum
espectro de líneas brillantes bright-line
    spectrum
espectro de luz light spectrum
espectro de llama flame spectrum
espectro de masa mass spectrum
espectro de microondas microwave
    spectrum
espectro de primer orden first-order
    spectrum
espectro de Raman Raman spectrum
espectro de rayos gamma gamma-ray
    spectrum
espectro de rayos X X-ray spectrum
espectro de resonancia resonance
    spectrum
espectro de Rydberg Rydberg spectrum
espectro discreto discrete spectrum
espectro electromagnético
    electromagnetic spectrum
espectro electrónico electron spectrum
espectro electroquímico electrochemical
    spectrum
espectro magnético magnetic spectrum
espectro molecular molecular spectrum
espectro neutrónico neutron spectrum
espectro nuclear nuclear spectrum
espectro óptico optical spectrum
espectro paramagnético paramagnetic
    spectrum
espectro químico chemical spectrum
espectro rotacional rotational spectrum
espectro ultravioleta ultraviolet
    spectrum
espectro vibracional vibrational
    spectrum
espectro visible visible spectrum
espectroanálisis spectroanalysis
espectroanalítco spectroanalytical
espectrofluorometría spectrofluorometry
espectrofluorométrico
    spectrofluorometric

espectrofluorómetro spectrofluorometer
espectrofotometría spectrophotometry
espectrofotometría de absorción ultravioleta ultraviolet-absorption spectrophotometry
espectrofotometría de absorción visible visible absorption spectrophotometry
espectrofotometría de llama flame spectrophotometry
espectrofotometría de matriz matrix spectrophotometry
espectrofotometría de Raman Raman spectrophotometry
espectrofotometría diferencial differential spectrophotometry
espectrofotometría ultravioleta ultraviolet spectrophotometry
espectrofotometría visible visible spectrophotometry
espectrofotométrico spectrophotometric
espectrofotómetro spectrophotometer
espectrofotómetro de doble haz double-beam spectrophotometer
espectrofotómetro de filtro filter spectrophotometer
espectrofotómetro de llama flame spectrophotometer
espectrografía spectrography
espectrografía de imagen de rayos X X-ray image spectrography
espectrografía de rayos X X-ray spectrography
espectrográficamente spectrographically
espectrográfico spectrographic
espectrógrafo spectrograph
espectrógrafo de masa mass spectrograph
espectrógrafo de rayos X X-ray spectrograph
espectrógrafo óptico optical spectrograph
espectrograma spectrogram
espectrometría spectrometry
espectrometría de llama flame spectrometry
espectrometría de masa mass spectrometry
espectrometría de masa de iones secundarios secondary-ion mass

spectrometry
espectrometría de neutrones lentos slow-neutron spectrometry
espectrometría de neutrones rápidos fast-neutron spectrometry
espectrometría de rayos X X-ray spectrometry
espectrometría neutrónica neutron spectrometry
espectrométricamente spectrometrically
espectrométrico spectrometric
espectrómetro spectrometer
espectrómetro de Bragg Bragg spectrometer
espectrómetro de centelleos scintillation spectrometer
espectrómetro de helio helium spectrometer
espectrómetro de infrarrojo infrared spectrometer
espectrómetro de llama flame spectrometer
espectrómetro de masa mass spectrometer
espectrómetro de masa de helio helium mass spectrometer
espectrómetro de microondas microwave spectrometer
espectrómetro de radiofrecuencia radiofrequency spectrometer
espectrómetro de rayos beta beta-ray spectrometer
espectrómetro de rayos gamma gamma-ray spectrometer
espectrómetro de rayos X X-ray spectrometer
espectrómetro de rejilla grid spectrometer
espectrómetro neutrónico neutron spectrometer
espectrómetro óptico optical spectrometer
espectrómetro ultravioleta ultraviolet spectrometer
espectromicroscopia de fotoelectrones photoelectron spectromicroscopy
espectromicroscópico spectromicroscopic
espectromicroscopio spectromicroscope

espectroquímica spectrochemistry
espectroquímico spectrochemical
espectrorradiómetro spectroradiometer
espectroscopia spectroscopy
espectroscopia atómica atomic spectroscopy
espectroscopia de absorción absorption spectroscopy
espectroscopia de absorción atómica atomic absorption spectroscopy
espectroscopia de correlación correlation spectroscopy
espectroscopia de destello flash spectroscopy
espectroscopia de emisión emission spectroscopy
espectroscopia de emisión atómica atomic emission spectroscopy
espectroscopia de emisión de llama flame-emission spectroscopy
espectroscopia de fluorescencia fluorescence spectroscopy
espectroscopia de fotoelectrones photoelectron spectroscopy
espectroscopia de infrarrojo infrared spectroscopy
espectroscopia de ionización en resonancia resonance ionization spectroscopy
espectroscopia de láser laser spectroscopy
espectroscopia de masa mass spectroscopy
espectroscopia de microondas microwave spectroscopy
espectroscopia de modulación modulation spectroscopy
espectroscopia de Mössbauer Mössbauer spectroscopy
espectroscopia de radiofrecuencia radiofrequency spectroscopy
espectroscopia de Raman Raman spectroscopy
espectroscopia de rayos gamma gamma-ray spectroscopy
espectroscopia de saturación saturation spectroscopy
espectroscopia electrónica electron spectroscopy

espectroscopia fotoacústica photoacoustic spectroscopy
espectroscopia gamma gamma spectroscopy
espectroscopia molecular molecular spectroscopy
espectroscopia neutrónica neutron spectroscopy
espectroscopia nuclear nuclear spectroscopy
espectroscopia óptica optical spectroscopy
espectroscopia quiroóptica chirooptical spectroscopy
espectroscopia ultravioleta ultraviolet spectroscopy
espectroscopia ultravioleta-visible ultraviolet-visible spectroscopy
espectroscópicamente spectroscopically
espectroscópico spectroscopic
espectroscopio spectroscope
espectroscopio de masa mass spectroscope
espectroscopio de microondas microwave spectroscope
espectroscopio óptico optical spectroscope
especular specular
esperrilita sperrylite
espesador thickener
espesamiento thickening
espetropolarimétrico spectropolarimetric
espetropolarímetro spectropolarimeter
espín spin
espín antiparalelo antiparallel spin
espín de electrón electron spin
espín isobárico isobaric spin
espín isotópico isotopic spin
espín nuclear nuclear spin
espinela spinel
espines paralelos parallel spins
espirano spiran
espíritu spirit
espironolactona spironolactone
espiropentano spiropentane
espiroqueta spirochete
espita unidireccional one-way tap
espodumena spodumene
esponja sponge

esponja de caucho rubber sponge
esponja de hierro iron sponge
esponja de platino platinum sponge
esponja de titanio titanium sponge
espontáneo spontaneous
espuma foam, froth
espuma de plástico plastic foam
espuma poliéter polyether foam
estabilidad stability
estabilidad ambiental environmental
    stability
estabilidad limitada limited stability
estabilidad nuclear nuclear stability
estabilidad térmica thermal stability
estabilidad termodinámica
    thermodynamic stability
estabilización stabilization
estabilizador stabilizer
estabilizador de emulsión emulsion
    stabilizer
estabilizador ultravioleta ultraviolet
    stabilizer
estable stable
estadística cuántica quantum statistics
estadística de Bose-Einstein Bose-
    Einstein statistics
estadística de Fermi-Dirac Fermi-Dirac
    statistics
estado state
estado activado activated state
estado amorfo amorphous state
estado coloidal colloidal state
estado correspondiente corresponding
    state
estado cristalino crystalline state
estado crítico critical state
estado cuántico quantum state
estado de oxidación oxidation state
estado de transición transition state
estado disociado dissociated state
estado estacionario steady state,
    stationary state
estado excitado excited state
estado fundamental ground state
estado fundamental nuclear nuclear
    ground state
estado gaseoso gaseous state
estado ligado bound state
estado líquido liquid state

estado mesomórfico mesomorphic state
estado metaestable metastable state
estado normal normal state
estado orbital orbital state
estado reducido reduced state
estado sólido solid state
estado virtual virtual state
estalagmometría stalagmometry
estalagmómetro stalagmometer
estándar standard
estandarización standardization
estandarizado standardized
estandarizar standardize
estannano stannane
estannato stannate
estannato bárico barium stannate
estannato cálcico calcium stannate
estannato cobaltoso cobaltous stannate
estannato de bismuto bismuth stannate
estannato de níquel nickel stannate
estannato de plomo lead stannate
estannato magnésico magnesium
    stannate
estannato potásico potassium stannate
estannato sódico sodium stannate
estánnico stannic
estannilo stannyl
estannita stannite
estannita sódica sodium stannite
estanniuro magnésico magnesium
    stannide
estannoso stannous
estañado tin-plating, tinning
estañar tin-plate
estaño tin, stannum
estático static
estearaldehído stearaldehyde
estearamida stearamide
estearamida de butilo butyl stearamide
estearato stearate
estearato alumínico aluminum stearate
estearato amónico ammonium stearate
estearato bárico barium stearate
estearato cálcico calcium stearate
estearato de atropina atropine stearate
estearato de butilo butyl stearate
estearato de butoxietilo butoxyethyl
    stearate
estearato de cadmio cadmium stearate

estearato de cerio cerium stearate
estearato de ciclohexilo cyclohexyl stearate
estearato de cinc zinc stearate
estearato de cobre copper stearate
estearato de dietilenglicol diethylene glycol stearate
estearato de diglicol diglycol stearate
estearato de glicol glycol stearate
estearato de isobutilo isobutyl stearate
estearato de isocetilo isocetyl stearate
estearato de litio lithium stearate
estearato de metilo methyl stearate
estearato de plata silver stearate
estearato de plomo lead stearate
estearato de trihidroxietilamina trihydroxyethylamine stearate
estearato de vinilo vinyl stearate
estearato férrico ferric steareate
estearato magnésico magnesium stearate
estearato mercúrico mercuric stearate
estearato potásico potassium stearate
estearato sódico sodium stearate
estearilmercaptano stearyl mercaptan
estearilo stearyl
estearina stearin
estearoilaminofenol stearoylaminophenol
estearoilo stearoyl
estearoilolactilato sódico sodium stearoyl lactylate
estearona stearone
estearonitrilo stearonitrile
esteatita steatite, soapstone
estequiometría stoichiometry
estequiométrico stoichiometric
éster ester
éster acético acetic ester
éster acetoacético acetoacetic ester
éster acrílico acrylic ester
éster amilacético amylacetic ester
éster carbónico carbonic ester
éster cianoacético cyanoacetic ester
éster cianúrico cyanuric ester
éster de ácido bórico boric acid ester
éster de brosilato brosylate ester
éster de poliglicerol polyglycerol ester
éster de titanio titanium ester
éster diazoacético diazoacetic ester
éster ftálico phthalic ester

éster glicólico glycol ester
éster graso fatty ester
éster malónico malonic ester
éster malónico sódico sodium malonic ester
éster metílico methyl ester
éster ortofórmico orthoformic ester
esterasa esterase
estereoespecífico stereospecific
estereoisomerismo stereoisomerism
estereoisómero stereoisomer
estereoquímica stereochemistry
estereoquímica absoluta absolute stereochemistry
estereoquímica relativa relative stereochemistry
estereorregular stereoregular
estereoscopio stereoscope
estereoselectividad stereoselectivity
estérico steric
esterificación esterification
esteril sterile
esterilización sterilization
esterilización de agua water sterilization
esteriode steroid
esterol sterol
estersil estersil
estibina stibine
estibino stibino
estibinoso stibinoso
estibnita stibnite
estibo stibo
estibonio stibonium
estibono stibono
estibonoso stibonoso
estiboso stiboso
estigmasterol stigmasterol
estilbeno stilbene
estilbestrol stilbestrol
estilbita stilbite
estimulación stimulation
estiracina styracin
estirenglicol styrene glycol
estireno styrene
estirenosulfonato sódico sodium styrenesulfonate
estirilo styryl
estocástico stochastic
estradiol estradiol

estragol estragole
estrano estrane
estreptolina streptolin
estreptomicina streptomycin
estrés stress
estricnidina strychnidine
estricnina strychnine
estriol estriol
estrógeno estrogen
estrona estrone
estroncia strontia
estroncianita strontianite
estroncio strontium
estructura structure
estructura antifluorita antifluorite
    structure
estructura asimétrica asymmetric
    structure
estructura atómica atomic structure
estructura centrada en el cuerpo body-
    centered structure
estructura centrada en las caras face-
    centered structure
estructura con defectos defect structure
estructura cristalina crystal structure
estructura cúbica centrada en el cuerpo
    body-centered cubic structure
estructura cúbica centrada en las caras
    face-centered cubic structure
estructura de anillo ring structure
estructura de benceno benzene structure
estructura de capas shell structure
estructura de Dewar Dewar structure
estructura de Kekulé Kekulé structure
estructura de Lewis Lewis structure
estructura de van der Waals van der
    Waals structure
estructura fina fine structure
estructura hiperfina hyperfine structure
estructura molecular molecular
    structure
estructura nuclear nuclear structure
estructura por rayos X X-ray structure
estructural structural
eta eta
etal ethal
etaldehído ethaldehyde
etamina ethamine
etanal ethanal

etanaldehído ethanaldehyde
etanamida ethanamide
etanamidina ethanamidine
etanilo ethanyl
etanita ethanite
etano ethane
etanoato ethanoate
etanoato alumínico aluminum ethanoate
etanoato cálcico calcium ethanoate
etanoato de amilo amyl ethanoate
etanoato de cinc zinc ethanoate
etanoato de etilo ethyl ethanoate
etanoato de hierro iron ethanoate
etanoato de plomo lead ethanoate
etanoato fenilmercúrico phenylmercuric
    ethanoate
etanoato potásico potassium ethanoate
etanoato sódico sodium ethanoate
etanodial ethanedial
etanodiamida ethanediamide
etanodiamina ethanediamine
etanodinitrilo ethanedinitrile
etanoditiol ethanedithiol
etanoilación ethanoylation
etanoilo ethanoyl
etanol ethanol
etanolamina ethanolamine
etanolato ethanolate
etanolato de talio thallium ethanolate
etanolato sódico sodium ethanolate
etanolurea ethanolurea
etanonitrilo ethanenitrile
etanotial ethanethial
etanotiol ethanethiol
etenilo ethenyl
etenio ethenium
eteno ethene
etenol ethenol
etenona ethenone
éter ether
éter acético acetic ether
éter alílico allyl ether
éter amílico amyl ether
éter bencílico benzyl ether
éter bencílico de hidroquinona
    hydroquinone benzyl ether
éter bencílico de isoamilo isoamyl
    benzyl ether
éter butilfenílico butylphenyl ether

éter butílico   butyl ether
éter carbónico   carbonic ether
éter celulósico   cellulose ether
éter cetílico   cetyl ether
éter clorhídrico   hydrochloric ether
éter de petróleo   petroleum ether
éter de vino   wine ether
éter diacético   diacetic ether
éter dicloroetílico   dichloroethyl ether
éter didecílico   didecyl ether
éter diectílico   diectyl ether
éter dietílico   diethyl ether
éter dietílico de hidroquinona
    hydroquinone diethyl ether
éter difenílico   diphenyl ether
éter diisopropílico   diisopropyl ether
éter dimetílico   dimethyl ether
éter dimetílico de glicol   glycol dimethyl
    ether
éter dimetílico de hidroquinona
    hydroquinone dimethyl ether
éter dimetílico de resorcina   resorcinol
    dimethyl ether
éter dimetílico de trietilenglicol
    triethylene glycol dimethyl ether
éter dioctílico   dioctyl ether
éter dipropílico   dipropyl ether
éter divinílico   divinyl ether
éter en corona   crown ether
éter etílico   ethyl ether
éter etílico bencílico   benzyl ethyl ether
éter etílico de dietilenglicol   diethylene
    glycol ethyl ether
éter etílico polivinílico   polyvinyl ethyl
    ether
éter etilmetílico   ethyl methyl ether
éter fenílico   phenyl ether
éter fenílico de propilenglicol   propylene
    glycol phenyl ether
éter fenólico   phenol ether
éter fórmico   formic ether
éter glicidílico   glycidyl ether
éter glicólico   glycol ether
éter hexaclorometílico   hexachloromethyl
    ether
éter hexílico   hexyl ether
éter isoamílico   isoamyl ether
éter isobutílico   isobutyl ether
éter isobutílico polivinílico   polyvinyl

isobutyl ether
éter isopropílico   isopropyl ether
éter metilbencílico   methylbenzyl ether
éter metilfenílico   methylphenyl ether
éter metílico   methyl ether
éter metílico polivinílico   polyvinyl
    methyl ether
éter monobutílico de etilenglicol
    ethylene glycol monobutyl ether
éter monoetílico de etilenglicol   ethylene
    glycol monoethyl ether
éter monoetílico de hidroquinona
    hydroquinone monoethyl ether
éter monometílico de etilenglicol
    ethylene glycol monomethyl ether
éter nítrico   nitric ether
éter pentilacético   pentylacetic ether
éter pentílico   pentyl ether
éter polivinílico   polyvinyl ether
éter propiónico   propionic ether
etéreo   ethereal
eterificación   etherification
etida   ethide
etidina   ethidine
etil-litio   ethyllithium
etilacetamida   ethylacetamide
etilacetanilida   ethylacetanilide
etilacetileno   ethylacetylene
etilacetoacetato de cobre   copper
    ethylacetoacetate
etilacetona   ethylacetone
etilación   ethylation
etilamilcetona   ethyl amyl ketone
etilamina   ethylamine
etilamino   ethylamino
etilaminoacetato   ethylaminoacetate
etilanilina   ethylaniline
etilantraquinona   ethylanthraquinone
etilato   ethylate
etilato alumínico   aluminum ethylate
etilato de vanadio   vanadium ethylate
etilato sódico   sodium ethylate
etilaziridina   ethylaziridine
etilbenceno   ethylbenzene
etilbencilanilina   ethylbenzylaniline
etilbutanol   ethylbutanol
etilbuteno   ethylbutene
etilbutilamina   ethylbutylamine
etilbutilcetona   ethylbutyl ketone

etilbutilo ethylbutyl
etilbutiraldehído ethylbutyraldehyde
etilcarbazol ethylcarbazole
etilcarbinol ethylcarbinol
etilcelulosa ethylcellulose
etilciclohexano ethylcyclohexane
etilciclohexilamina ethylcyclohexylamine
etilciclopentano ethylcyclopentane
etilciclopentanona ethylcyclopentanone
etildiclorosilano ethyldichlorosilane
etildietanolamina ethyldiethanolamine
etilenamina ethyleneamine
etilendiamina ethylenediamine
etilendiaminotetraacetonitrilo
    ethylenediaminetetraacetonitrile
etilenditiol ethylenedithiol
etilenglicol ethylene glycol
etilenimina ethyleneimine
etileno ethylene
etileno-propileno ethylene-propylene
etilenonaftaleno ethylenenaphthalene
etiletanolamina ethylethanolamine
etiletilenimina ethylethyleneimine
etiletileno ethylethylene
etilfenilcetona ethyl phenyl ketone
etilfenol ethylphenol
etilfluorosulfonato ethylfluorosulfonate
etilfurano ethylfuran
etilheptano ethylheptane
etilhexaldehído ethylhexaldehyde
etilhexanal ethylhexanal
etilhexanodiol ethylhexanediol
etilhexanol ethylhexanol
etilhexenal ethylhexenal
etilhexilamina ethylhexylamine
etilhexilanilina ethylhexylaniline
etilhexilo ethylhexyl
etilhexoato cálcico calcium ethylhexoate
etilhexoato de cinc zinc ethylhexoate
etilhexoato de cobalto cobalt
    ethylhexoate
etilhexoato estannoso stannous
    ethylhexoate
etilhidrazina ethylhydrazine
etilideno ethylidene
etilidino ethylidyne
etilina ethylin
etilisobutilmetano ethylisobutylmethane
etilisohexanol ethylisohexanol

etilmalonato de dietilo diethyl
    ethylmalonate
etilmercaptano ethyl mercaptan
etilmetilacetileno ethylmethylacetylene
etilmetilcetona ethyl methyl ketone
etilmorfolina ethylmorpholine
etilnaftilamina ethylnaphthylamine
etilo ethyl
etiloico ethyloic
etilol ethylol
etilparabeno ethylparaben
etilpentano ethylpentane
etilsalicilato de bencilo benzyl
    ethylsalicylate
etilsulfato bárico barium ethylsulfate
etilsulfato de cinc zinc ethylsulfate
etiltriclorosilano ethyltrichlorosilane
etiluretano ethylurethane
etilviniléter ethyl vinyl ether
etina ethine
etinilación ethynylation
etinilestradiol ethynylestradiol
etinilo ethynyl
etino ethyne
etión ethion
etionamida ethionamide
etiqueta label
etisterona ethisterone
etohexadiol ethohexadiol
etolida etholide
etoxalilo ethoxalyl
etoxi ethoxy
etoxiacetanilida ethoxyacetanilide
etoxiacetona ethoxyacetone
etoxianilina ethoxyaniline
etoxicarbonilo ethoxycarbonyl
etóxido ethoxide
etóxido alumínico aluminum ethoxide
etoxido sódico sodium ethoxide
etoxietanol ethoxyethanol
etoxifenol ethoxyphenol
etoxilo ethoxyl
etoximetilenomalononitrilo
    ethoxymethylenemalononitrile
etoxitriglicol ethoxytriglycol
ettringita ettringite
eucairita eucairite
eucaliptol eucalyptol
eucarvona eucarvone

eucriptita eucryptite
eucroíta euchroite
eudaleno eudalene
eudialita eudialite
eudiómetro eudiometer
eudiómetro de Bunsen Bunsen
    eudiometer
eugenol eugenol
eulitina eulytin
eupirina eupyrine
europio europium
eutéctico eutectic
eutectoide eutectoid
eutroficación eutrophication
eutropía eutropy
euxenita euxenite
evansita evansite
evaporación evaporation
evaporación catódica cathodic
    evaporation, cathodic sputtering,
    sputter, sputtering
evaporación retrógrada retrograde
    evaporation
evaporador evaporator
evaporar evaporate
evaporímetro evaporimeter
evaporita evaporite
evolución evolution
exactitud accuracy
examen microscópico microscopic
    examination
excéntrico eccentric
exceso de neutrones neutron excess
excímero excimer
excitación excitation
excitación atómica atomic excitation
excitación de Coulomb Coulomb
    excitation
excitación fotoquímica photochemical
    excitation
excitación nuclear nuclear excitation
excitación por colisión collision
    excitation
excitación térmica thermal excitation
excitado excited
excitador exciter
excitar excite
excitón exciton
exclusión exclusion

exclusión iónica ion exclusion
exilxantato sódico sodium ethylxanthate
exo exo
exocarburo exocarbon
exocondensación exocondensation
exógeno exogenous
exopeptidasa exopeptidase
exósmosis exosmosis
exotérmico exothermic
expandirse expand
expansión expansion
expansión adiabática adiabatic
    expansion
expansión térmica thermal expansion
expansor expander
experimentación experimentation
experimental experimental
experimento experiment
experimento de laboratorio laboratory
    experiment
exploración scan, scanning
explosión explosion
explosión de Coulomb Coulomb
    explosion
explosivo explosive
explosivo aceptable acceptable explosive
explosivo de Nobel Nobel explosive
explosivo iniciador initiating explosive
explosivo permisible permissible
    explosive
explotar explode
exposición exposure
expresión expression
expresión fraccionaria fractional
    expression
extensor extender
extinción extinction
extinguir extinguish, quench
extintor de fuego fire extinguisher
extracción extraction
extracción analítica analytical extraction
extracción electrolítica electrowinning
extracción líquido-líquido liquid-liquid
    extraction
extracción por solvente solvent
    extraction
extractante extractant
extracto extract
extracto de Göulard Göulard extract

extracto de malta malt extract
extractor extractor
extractor de Soxhlet Soxhlet extractor
extranuclear extranuclear
extrapolación extrapolation

extrapolación de Birge-Sponer Birge-Sponer extrapolation
extrusión extrusion
exudar exude

F

fabianol fabianol
fabricación de cerveza brewing
fabulita fabulite
fac fac
factor factor
factor de atenuación attenuation factor
factor de Cabannes Cabannes factor
factor de conversión conversion factor
factor de dispersión atómica atomic
   scattering factor
factor de van't Hoff van't Hoff factor
factor limitador limiting factor
factor volumétrico volumetric factor
factores ambientales environmental
   factors
fagarina fagarine
familia de desintegración decay family
familia radiactiva radioactive family
fantasma de Rowland Rowland ghost
farad farad
faraday faraday
faradiol faradiol
farinosa farinose
farmacéutico pharmaceutical
farmacocinética pharmacokinetics
farmacolita pharmacolite
farmacología pharmacology
farneseno farnesene
farnesol farnesol
fase phase
fase continua continuous phase
fase discontinua discontinuous phase
fase dispersa dispersed phase
fase esméctica smectic phase
fase estacionaria stationary phase
fase interna internal phase
fase metaestable metastable phase
fase nemática nematic phase
fasotropía phasotropy
faujasita faujasite
fayalita fayalite
felandreno phellandrene
feldespato feldspar

feldespato potásico potassium feldspar
feldespatoide feldspathoid
felita felite
felvación felvation
fémico femic
femtoquímica femtochemistry
fen phen, fen
fenacetina phenacetin
fenacetol phenacetol
fenacilideno phenacylidene
fenacilo phenacyl
fenantraquinona phenanthraquinone
fenantrenilo phenanthrenyl
fenantreno phenanthrene
fenantrenoquinona phenanthrenequinone
fenantridina phenanthridine
fenantrileno phenanthrylene
fenantrilo phenanthryl
fenantrofenacina phenanthrophenazine
fenantrol phenanthrol
fenantrolina phenanthroline
fenantrona phenanthrone
fenato phenate
fenato de cinc zinc phenate
fenato sódico sodium phenate
fenazina phenazine
fenazona phenazone
fenazonio phenazonium
fencano fenchane
fencanol fenchanol
fencanona fenchanone
fencol fenchol
fencona fenchone
fencoxima fenchoxime
fenenilo phenenyl
fenetidina phenetidine
fenetidino phenetidino
fenetilamina phenethylamine
fenetileno phenethylene
fenetilo phenethyl
fenetol phenetole
fenidona phenidone
fenil-litio phenyllithium

fenilacetaldehído phenylacetaldehyde
fenilacetamida phenylacetamide
fenilacetato de bencilo benzyl phenylacetate
fenilacetato de etilo ethyl phenylacetate
fenilacetato de isobutilo isobutyl phenylacetate
fenilacetato sódico sodium phenylacetate
fenilacetilo phenylacetyl
fenilacetonitrilo phenylacetonitrile
fenilación phenylation
fenilalanina phenylalanine
fenilamina phenylamine
fenilanilina phenylaniline
fenilato cálcico calcium phenylate
fenilazo phenylazo
fenilbarbital phenylbarbital
fenilbenceno phenylbenzene
fenilbencilo phenylbenzyl
fenilbenzamida phenylbenzamide
fenilbenzoato phenylbenzoate
fenilbenzoilo phenylbenzoyl
fenilbutano phenylbutane
fenilbutazona phenylbutazone
fenilbuteno phenylbutene
fenilbutinol phenylbutynol
fenilcarbamoilo phenylcarbamoyl
fenilcarbetoxipirazolona phenylcarbethoxypyrazolone
fenilcarbimida phenylcarbimide
fenilcarbinol phenylcarbinol
fenilcetona phenyl ketone
fenilciclohexano phenylcyclohexane
fenilciclohexanol phenylcyclohexanol
fenilcloroformo phenylchloroform
fenildicloroarsina phenyldichloroarsine
fenildietanolamina phenyldiethanolamine
fenildimetilurea phenyldimethylurea
fenilendiamina phenylenediamine
fenileno phenylene
feniletano phenylethane
feniletilamina phenylethylamine
feniletilenglicol phenylethylene glycol
feniletileno phenylethylene
feniletiletanolamina phenylethylethanolamine
feniletiléter phenylethylether
feniletilmercaptano phenylethyl mercaptan

feniletilo phenylethyl
fenilfenato sódico sodium phenylphenate
fenilfenol phenylphenol
fenilfenolato sódico sodium phenylphenolate
fenilformamida phenylformamide
fenilfosfina phenylphosphine
fenilfosfinato sódico sodium phenylphosphinate
fenilglicina phenylglycine
fenilglicol phenylglycol
fenilhidrazina phenylhydrazine
fenilhidrazina de pentilo pentyl phenylhydrazine
fenilhidrazona phenylhydrazone
fenilhidrazona de benzaldehído benzaldehyde phenylhydrazone
fenilhidroxilamina phenylhydroxylamine
fenílico phenylic
fenilideno phenylidene
fenilmagnesio phenylmagnesium
fenilmercaptano phenyl mercaptan
fenilmercúrico phenylmercuric
fenilmetano phenylmethane
fenilmetanol phenylmethanol
fenilmetilcarbinol phenylmethylcarbinol
fenilmetilcetona phenyl methyl ketone
fenilmetileno phenylmethylene
fenilmetiletanolamina phenylmethylethanolamine
fenilmetilo phenylmethyl
fenilmorfolina phenylmorpholine
fenilnaftaleno phenylnaphthalene
fenilnaftilamina phenylnaphthylamine
fenilnitroamina phenylnitroamine
fenilnonano phenylnonane
fenilo phenyl
fenilpentano phenylpentane
fenilpiperazina phenylpiperazine
fenilpirazolida phenylpyrazolide
fenilpirazolidona phenylpyrazolidone
fenilpiridina phenylpyridine
fenilpropano phenylpropane
fenilpropanol phenylpropanol
fenilpropanona phenylpropanone
fenilpropenal phenylpropenal
fenilpropenol phenylpropenol
fenilpropilcetona phenylpropyl ketone
fenilpropilo phenylpropyl

fenilpropilpiridina phenylpropylpyridine
fenilsulfonilo phenylsulfonyl
feniltiourea phenylthiourea
feniltolueno phenyltoluene
feniltriclorosilano phenyltrichlorosilane
feniltridecano phenyltridecane
feniluretano phenylurethane
fenilvinilcetona phenyl vinyl ketone
fenindiona phenindione
fenitoína phenytoin
fenitrotión fenitrothion
fenixina phenixin
fenmetilo phenmethyl
fenoacetato sódico sodium phenoacetate
fenobarbital phenobarbital
fenobarbital sódico sodium
    phenobarbital
fenobarbitona phenobarbitone
fenodiazina phenodiazine
fenol phenol
fenol polihídrico polyhydric phenol
fenolato phenolate
fenolato alumínico aluminum phenolate
fenolato cálcico calcium phenolate
fenolato de hierro iron phenolate
fenolato de resorcina resorcinol
    phenolate
fenolato férrico ferric phenolate
fenolato sódico sodium phenolate
fenolftaleína phenolphthalein
fenolftalida phenolphthalide
fenólico phenolic
fenolsafranina phenolsafranin
fenolsulfato sódico mercúrico mercuric
    sodium phenolsulfate
fenolsulfonaftaleína
    phenolsulfonephthalein
fenolsulfonato phenolsulfonate
fenolsulfonato amónico ammonium
    phenolsulfonate
fenolsulfonato de cinc zinc
    phenolsulfonate
fenolsulfonato de cobre copper
    phenolsulfonate
fenolsulfonato sódico sodium
    phenolsulfonate
fenómeno físico physical phenomenon
fenómeno lambda lambda phenomenon
fenómeno sigma sigma phenomenon

fenona phenone
fenotiazina phenothiazine
fenotol phenotole
fenotrina phenothrin
fenoxazina phenoxazine
fenoxi phenoxy
fenoxibenceno phenoxybenzene
fenoxidihidroxipropano
    phenoxydihydroxypropane
fenóxido phenoxide
fenoxietanol phenoxyethanol
fenoxietilpenicilina
    phenoxyethylpenicillin
fenoxipropanodiol phenoxypropanediol
fenqueno fenchene
fenquilo fenchyl
fensulfotiona fensulfothion
fentión fenthion
ferberita ferberite
fergusonita fergusonite
fermentable fermentable
fermentación fermentation
fermentación alcohólica alcoholic
    fermentation
fermentar ferment
fermio fermium
feromona pheromone
ferrato ferrate
ferrato bárico barium ferrate
ferrato potásico potassium ferrate
ferredoxina ferredoxin
ferri ferri
ferricianuro ferricyanide
ferricianuro de plata silver ferricyanide
ferricianuro férrico ferric ferricyanide
ferricianuro ferroso ferrous ferricyanide
ferricianuro potásico potassium
    ferricyanide
ferricianuro sódico sodium ferricyanide
férrico ferric
ferricromo ferrichrome
ferriferroso ferriferrous
ferrimagnetismo ferrimagnetism
ferrimanganeso ferrimanganese
ferrimangánico ferrimanganic
ferripirina ferripyrine
ferripotasio ferripotassium
ferrisodio ferrisodium
ferrita ferrite

ferrita bárica barium ferrite
ferrita cobaltosa cobaltous ferrite
ferrita ferrosa ferrous ferrite
ferrita potásica potassium ferrite
ferrítico ferritic
ferro ferro
ferroaleación ferroalloy
ferroaluminio ferroaluminum
ferroamonio ferroammonium
ferroboro ferroboron
ferrocenilborano ferrocenylborane
ferroceno ferrocene
ferroceno de clorocarbonilo
   chlorocarbonyl ferrocene
ferroceno de titanio titanium ferrocene
ferrocenoilo ferrocenoyl
ferrocerio ferrocerium
ferrocianato de etilo ethyl ferrocyanate
ferrocianuro ferrocyanide
ferrocianuro amónico ammonium
   ferrocyanide
ferrocianuro bárico barium ferrocyanide
ferrocianuro cálcico calcium
   ferrocyanide
ferrocianuro de cinc zinc ferrocyanide
ferrocianuro de cobre copper
   ferrocyanide
ferrocianuro de hierro iron ferrocyanide
ferrocianuro férrico ferric ferrocyanide
ferrocianuro ferroso ferrous
   ferrocyanide
ferrocianuro potásico potassium
   ferrocyanide
ferrocianuro sódico sodium ferrocyanide
ferrocromo ferrochromium
ferrodinámico ferrodynamic
ferrodolomita ferrodolomite
ferroelectricidad ferroelectricity
ferroeléctrico ferroelectric
ferroferricianuro ferroferricyanide
ferroférrico ferroferric
ferroferrocianuro ferroferrocyanide
ferrofósforo ferrophosphorus
ferroína ferroin
ferromagnesio ferromagnesium
ferromagnesita ferromagnesite
ferromagnético ferromagnetic
ferromagnetismo ferromagnetism
ferromanganeso ferromanganese

ferromangánico ferromanganic
ferromanganoso ferromanganous
ferromolibdeno ferromolybdenum
ferrón ferron
ferroniobio ferroniobium
ferroníquel ferronickel
ferropirina ferropyrine
ferropotasio ferropotassium
ferrosilicio ferrosilicon
ferrosilita ferrosilite
ferroso ferrous
ferrosoférrico ferrosoferric
ferrotitanio ferrotitanium
ferrotungsteno ferrotungsten
ferrovanadio ferrovanadium
ferroverdina ferroverdin
ferroxilo ferroxyl
ferrozirconio ferrozirconium
ferruginoso ferruginous
ferrum ferrum
fertilizante fertilizer
ferulaldehído ferulaldehyde
feruleno ferulene
fervanita fervanite
fibra fiber
fibra de carbono carbon fiber
fibra de celulosa cellulose fiber
fibra de fluorocarburo fluorocarbon
   fiber
fibra de grafito graphite fiber
fibra de vidrio glass fiber, fiberglass
fibra olefínica olefin fiber
fibra óptica optical fiber
fibra poliacrílica polyacrylic fiber
fibra poliéster polyester fiber
fibra sintética synthetic fiber
fibrido fibrid
fibrilla fibril
fibrina fibrin
fibrinógeno fibrinogen
fibroína fibroin
fibrolita fibrolite
fibroóptico fiber-optic
ficina ficin
ficocoloide phycocolloid
fictelita fichtelite
fíctil fictile
fijación fixation
fijación de nitrógeno nitrogen fixation

fijar fix, bind
fijativo fixative
filamento filament
filipita phillipite
filtración filtration
filtración al vacío vacuum filtration
filtración bajo presión filtration under pressure
filtración de gel gel filtration
filtración fraccionaria fractional filtration
filtración por bordes edge filtration, streamline filtration
filtrado filtrate
filtro filter
filtro acelerador accelofilter
filtro de aire air filter
filtro de Bechhold Bechhold filter
filtro de Berkefeld Berkefeld filter
filtro de gas gas filter
filtro de gelatina gelatin filter
filtro de Pasteur Pasteur filter
filtro de presión pressure filter
filtros de Zsigmondy Zsigmondy filters
fisetina fisetin
física electrónica electron physics
física nuclear nuclear physics
fisicoquímica physical chemistry
fisicoquímico physicochemical
físil fissile
fisión fission
fisión atómica atomic fission
fisión de uranio uranium fission
fisión espontánea spontaneous fission
fisión homolítica homolytic fission
fisión inducida induced fission
fisión nuclear nuclear fission
fisión por neutrones rápidos fast-neutron fission
fisión radiactiva radioactive fission
fisión térmica thermal fission
fisionable fissionable
fisioquímica fissiochemistry
fisostigmina physostigmine
fitano phytane
fitato cálcico calcium phytate
fitato hexacálcico hexacalcium phytate
fitato sódico sodium phytate
fitilo phytyl

fitina phytin
fitol phytol
fitoquímica phytochemistry
fitotóxico phytotoxic
flagstafita flagstaffite
flavanilina flavaniline
flavanol flavanol
flavanona flavanone
flavazina flavazine
flavilio flavylium
flavina flavin
flavofenina flavophenine
flavoglaucina flavoglaucin
flavol flavol
flavona flavone
flavonoide flavonoid
flavonol flavonol
flavopanina flavopannin
flavoproteína flavoprotein
flavopurpurina flavopurpurin
flavoxantina flavoxanthin
floculación flocculation
floculado flocculate
floculante flocculant
floculento flocculent
flor flower
floroglucinol phloroglucinol
floruro de alilo allyl fluoride
flotación flotation
flox flox
flucloxacilina flucloxacillin
fluctuación fluctuation
fluellita fluellite
fluidez fluidity
fluidización fluidization
fluido fluid
fluido de Bouin Bouin fluid
fluido de Carnoy Carnoy fluid
fluido de Condy Condy fluid
fluido hidráulico hydraulic fluid
fluido newtoniano Newtonian fluid
fluido supercrítico supercritical fluid
fluido volátil volatile fluid
flujo flow, flux
flujo de gas gas flow
flujo de Knudsen Knudsen flow
flujo electrónico electron flow
flujo en frío cold flow
flujo molecular molecular flow

flujo neutrónico neutron flux
flujo plástico plastic flow
flujo térmico heat flow
fluo fluo
fluocerita fluocerite
fluoflavina fluoflavine
flúor fluorine
fluoracetamida fluoracetamide
fluoración fluoration
fluorán fluoran
fluorandiol fluorandiol
fluoranteno fluoranthene
fluorantraquinona fluoranthraquinone
fluorapatita fluorapatite
fluorenilamina fluorenylamine
fluorenilideno fluorenylidene
fluorenilo fluorenyl
fluoreno fluorene
fluorenol fluorenol
fluorenona fluorenone
fluoresceína fluorescein
fluoresceína sódica sodium fluorescein
fluorescencia fluorescence
fluorescencia de rayos X X-ray
    fluorescence
fluorescencia de resonancia resonance
    fluorescence
fluorescente fluorescent
fluorescina fluorescin
fluorgermanato fluorgermanate
fluorilidina fluorylidine
fluorilo fluoryl
fluorimetría fluorimetry
fluorímetro fluorimeter
fluorinión fluorinion
fluorión fluorion
fluorita fluorite
fluorización fluoridation
fluoro fluoro
fluoroacetamida fluoroacetamide
fluoroacetanilida fluoroacetanilide
fluoroacetato fluoroacetate
fluoroacetato sódico sodium
    fluoroacetate
fluoroacetofenona fluoroacetophenone
fluoroalcano fluoroalkane
fluoroanilina fluoroaniline
fluoroapatita fluoroapatite
fluorobenceno fluorobenzene

fluoroborato fluoroborate
fluoroborato amónico ammonium
    fluoroborate
fluoroborato de cinc zinc fluoroborate
fluoroborato de plata silver fluoroborate
fluoroborato de plomo lead fluoroborate
fluoroborato férrico ferric fluoroborate
fluoroborato potásico potassium
    fluoroborate
fluoroborato sódico sodium fluoroborate
fluorocarburo fluorocarbon
fluorocromato fluorochromate
fluorodiclorometano
    fluorodichloromethane
fluoroelastómero fluoroelastomer
fluorofenol fluorophenol
fluoroformo fluoroform
fluoróforo fluorophore
fluorofosfato cálcico calcium
    fluorophosphate
fluorofosfato de litio lithium
    fluorophosphate
fluorofosfato sódico sodium
    fluorophosphate
fluorofosfonato fluorophosphonate
fluorofotómetro fluorophotometer
fluorógeno fluorogen
fluorohidrocarburo fluorohydrocarbon
fluorometano fluoromethane
fluorometolona fluorometholone
fluorometría fluorometry
fluorómetro fluorometer
fluorona fluorone
fluoronio fluoronium
fluoropolímero fluoropolymer
fluoroquímico fluorochemical
fluoroscopio fluoroscope
fluorosilicato fluorosilicate
fluorosilicato cálcico calcium
    fluorosilicate
fluorosilicato de cinc zinc fluorosilicate
fluorosilicato de cobalto cobalt
    fluorosilicate
fluorosilicato de plomo lead
    fluorosilicate
fluorosilicato férrico ferric fluorosilicate
fluorosilicato ferroso ferrous
    fluorosilicate
fluorosilicato sódico sodium

fluorosilicate

fluorosulfonato de metilo methyl fluorosulfonate

fluoroteno fluorothene

fluorotriclorometano fluorotrichloromethane

fluoroyodato fluoroiodate

fluoruro fluoride

fluoruro ácido amónico ammonium acid fluoride

fluoruro ácido potásico potassium acid fluoride

fluoruro ácido sódico sodium acid fluoride

fluoruro alumínico aluminum fluoride

fluoruro alumínico potásico potassium aluminum fluoride

fluoruro amónico ammonium fluoride

fluoruro antimónico antimonic fluoride

fluoruro antimonioso antimonous fluoride

fluoruro arsenioso arsenous fluoride

fluoruro bárico barium fluoride

fluoruro cálcico calcium fluoride

fluoruro cérico ceric fluoride

fluoruro ceroso cerous fluoride

fluoruro cobáltico cobaltic fluoride

fluoruro cobaltoso cobaltous fluoride

fluoruro crómico chromic fluoride

fluoruro cromoso chromous fluoride

fluoruro cúprico cupric fluoride

fluoruro de acetilo acetyl fluoride

fluoruro de anilina aniline fluoride

fluoruro de antimonio antimony fluoride

fluoruro de arsénico arsenic fluoride

fluoruro de azufre sulfur fluoride

fluoruro de bencilo benzyl fluoride

fluoruro de benzoilo benzoyl fluoride

fluoruro de berilio beryllium fluoride

fluoruro de berilio sódico sodium beryllium fluoride

fluoruro de bismutilo bismuthyl fluoride

fluoruro de boro boron fluoride

fluoruro de bromo bromine fluoride

fluoruro de cadmio cadmium fluoride

fluoruro de carbonilo carbonyl fluoride

fluoruro de carbono carbon fluoride

fluoruro de cesio cesium fluoride

fluoruro de cianógeno cyanogen fluoride

fluoruro de cinc zinc fluoride

fluoruro de cobre copper fluoride

fluoruro de cromilo chromyl fluoride

fluoruro de cromo chromium fluoride

fluoruro de escandio scandium fluoride

fluoruro de estaño tin fluoride

fluoruro de estroncio strontium fluoride

fluoruro de europio europium fluoride

fluoruro de fenilo phenyl fluoride

fluoruro de formilo formyl fluoride

fluoruro de fosforilo phosphoryl fluoride

fluoruro de gadolinio gadolinium fluoride

fluoruro de hidrógeno hydrogen fluoride

fluoruro de hidrógeno amónico ammonium hydrogen fluoride

fluoruro de hidrógeno potásico potassium hydrogen fluoride

fluoruro de hidrógeno sódico sodium hydrogen fluoride

fluoruro de hierro iron fluoride

fluoruro de holmio holmium fluoride

fluoruro de iterbio ytterbium fluoride

fluoruro de lantano lanthanum fluoride

fluoruro de litio lithium fluoride

fluoruro de lutecio lutetium fluoride

fluoruro de manganeso manganese fluoride

fluoruro de metilo methyl fluoride

fluoruro de neodimio neodymium fluoride

fluoruro de niobio niobium fluoride

fluoruro de níquel nickel fluoride

fluoruro de nitrilo nitryl fluoride

fluoruro de nitrógeno nitrogen fluoride

fluoruro de nitrosilo nitrosyl fluoride

fluoruro de oxígeno oxygen fluoride

fluoruro de perclorilo perchloryl fluoride

fluoruro de plata silver fluoride

fluoruro de plomo lead fluoride

fluoruro de polivinilideno polyvinylidene fluoride

fluoruro de polivinilo polyvinyl fluoride

fluoruro de renio rhenium fluoride

fluoruro de rubidio rubidium fluoride

fluoruro de rutenio ruthenium fluoride

fluoruro de silicio silicon fluoride

fluoruro de sulfinilo sulfinyl fluoride

fluoruro de sulfonilo sulfonyl fluoride
fluoruro de sulfurilo sulfuryl fluoride
fluoruro de talio thallium fluoride
fluoruro de tántalo tantalum fluoride
fluoruro de telurio tellurium fluoride
fluoruro de titanio titanium fluoride
fluoruro de titanio potásico potassium titanium fluoride
fluoruro de torio thorium fluoride
fluoruro de tungsteno tungsten fluoride
fluoruro de uranilo uranyl fluoride
fluoruro de uranio uranium fluoride
fluoruro de vanadio vanadium fluoride
fluoruro de vinilideno vinylidene fluoride
fluoruro de vinilo vinyl fluoride
fluoruro de yodo iodine fluoride
fluoruro de zirconio zirconium fluoride
fluoruro estánnico stannic fluoride
fluoruro estannoso stannous fluoride
fluoruro férrico ferric fluoride
fluoruro ferroso ferrous fluoride
fluoruro magnésico magnesium fluoride
fluoruro mangánico manganic fluoride
fluoruro manganoso manganous fluoride
fluoruro mercúrico mercuric fluoride
fluoruro monocálcico monocalcium fluoride
fluoruro potásico potassium fluoride
fluoruro potásico berílico beryllium potassium fluoride
fluoruro potásico de germanio germanium potassium fluoride
fluoruro potásico de tántalo tantalum potassium fluoride
fluoruro potásico de titanio titanium potassium fluoride
fluoruro potásico de zirconio zirconium potassium fluoride
fluoruro sódico sodium fluoride
fluoruro sódico alumínico aluminum sodium fluoride
fluoruro sódico berílico beryllium sodium fluoride
fluoryodo fluoriodine
fluosilicato fluosilicate
fluosilicato amónico ammonium fluosilicate
fluosilicato bárico barium fluosilicate

fluosilicato de plata silver fluosilicate
fluosilicato magnésico magnesium fluosilicate
fluosilicato potásico potassium fluosilicate
fluoximesterona fluoxymesterone
fluozirconato potásico potassium fluozirconate
folacina folacin
folato sódico sodium folate
fomómetro fomometer
fómula de Briet-Wigner Briet-Wigner formula
fongisterol fongisterol
fonoquímica phonochemistry
forato phorate
forbol phorbol
forilo phoryl
forma canónica canonical form
forma cristalina crystal form
formacilo formazyl
formación formation
formación al vacío vacuum forming
formación de anillo ring formation
formal formal
formaldehído formaldehyde
formaldoxima formaldoxime
formalidad formality
formalina formalin
formamida formamide
formamida de etanol ethanol formamide
formamida de etilo ethyl formamide
formamidina formamidine
formamido formamido
formamilo formamyl
formamina formamine
formanilida formanilide
formanilina formaniline
formazán formazan
formiato formate, formiate
formiato alumínico aluminum formate
formiato amónico ammonium formate
formiato bárico barium formate
formiato cálcico calcium formate
formiato cobaltoso cobaltous formate
formiato cromoso chromous formate
formiato de amilo amyl formate
formiato de bencilo benzyl formate
formiato de bornilo bornyl formate

formiato de butilo butyl formate
formiato de cadmio cadmium formate
formiato de cinc zinc formate
formiato de estroncio strontium formate
formiato de etilo ethyl formate
formiato de geranilo geranyl formate
formiato de heptilo heptyl formate
formiato de hexilo hexyl formate
formiato de hidrazina hydrazine formate
formiato de hierro iron formate
formiato de isoamilo isoamyl formate
formiato de linalilo linalyl formate
formiato de metilo methyl formate
formiato de níquel nickel formate
formiato de octilo octyl formate
formiato de pentilo pentyl formate
formiato de plomo lead formate
formiato de propilo propyl formate
formiato de talio thallium formate
formiato de uranilo uranyl formate
formiato férrico ferric formate
formiato magnésico magnesium formate
formiato manganoso manganous formate
formiato mercurioso mercurous formate
formiato niqueloso nickelous formate
formiato potásico potassium formate
formiato sódico sodium formate
formilación formylation
formilamida formylamide
formilamina formylamine
formilamino formylamino
formileno formylene
formilhidrazina formylhydrazine
formilo formyl
formiloxi formyloxy
formilpiperidina formylpiperidine
formimidoilo formimidoyl
formina formine
formohidrazida formohydrazide
formoilo formoyl
formol formol
formolita formolite
formonitrilo formonitrile
formotión formothion
formoxi formoxy
formoxilo formoxyl
formoxima formoxime
fórmula formula
fórmula constitucional constitutional

formula
fórmula de Balmer Balmer formula
fórmula de Bamberger Bamberger formula
fórmula de Cook Cook formula
fórmula de London London formula
fórmula de Meyer Meyer formula
fórmula de reacción reaction formula
fórmula de Rydberg Rydberg formula
fórmula de van't Hoff van't Hoff formula
fórmula electrónica electron formula
fórmula empírica empirical formula
fórmula estructural structural formula
fórmula general general formula
fórmula genérica generic formula
fórmula gráfica graphic formula
fórmula molecular molecular formula
fórmula química chemical formula
fórmula racional rational formula
formulación formulation
forona phorone
forsterita forsterite
fosalona phosalone
fosfa phospha
fosfaceno phosphazene
fosfamida phosphamide
fosfamidón phosphamidon
fosfano phosphane
fosfatasa phosphatase
fosfatasa ácida acid phosphatase
fosfátido phosphatide
fosfato phosphate
fosfato ácido acid phosphate
fosfato ácido amónico ammonium acid phosphate
fosfato ácido de amilo amyl acid phosphate
fosfato ácido potásico potassium acid phosphate
fosfato ácido sódico sodium acid phosphate
fosfato alumínico aluminum phosphate
fosfato alumínico sódico sodium aluminum phosphate
fosfato amónico ammonium phosphate
fosfato amónico cobaltoso cobaltous ammonium phosphate
fosfato amónico magnésico magnesium

ammonium phosphate

fosfato amónico sódico sodium ammonium phosphate

fosfato bárico barium phosphate

fosfato bicálcico bicalcium phosphate

fosfato cálcico calcium phosphate

fosfato cálcico monobásico monobasic calcium phosphate

fosfato cálcico tribásico tribasic calcium phosphate

fosfato ceroso cerous phosphate

fosfato cobaltoso cobaltous phosphate

fosfato crómico chromic phosphate

fosfato cúprico cupric phosphate

fosfato de adenosina adenosine phosphate

fosfato de almidón starch phosphate

fosfato de betaína betaine phosphate

fosfato de bismuto bismuth phosphate

fosfato de boro boron phosphate

fosfato de cadmio cadmium phosphate

fosfato de cinc zinc phosphate

fosfato de cobalto cobalt phosphate

fosfato de cobre copper phosphate

fosfato de codeína codeine phosphate

fosfato de cresilo cresyl phosphate

fosfato de cromo chromium phosphate

fosfato de dibutilo dibutyl phosphate

fosfato de dimetilo dimethyl phosphate

fosfato de estroncio strontium phosphate

fosfato de etilo ethyl phosphate

fosfato de galio gallium phosphate

fosfato de hidrógeno potásico potassium hydrogen phosphate

fosfato de hidrógeno sódico sodium hydrogen phosphate

fosfato de hierro iron phosphate

fosfato de huesos bone phosphate

fosfato de litio lithium phosphate

fosfato de manganeso manganese phosphate

fosfato de manganeso amónico ammonium manganese phosphate

fosfato de mercurio mercury phosphate

fosfato de neodimio neodymium phosphate

fosfato de níquel nickel phosphate

fosfato de octilo octyl phosphate

fosfato de piridoxal pyridoxal phosphate

fosfato de plata silver phosphate

fosfato de plomo lead phosphate

fosfato de riboflavina riboflavin phosphate

fosfato de tetraetilo tetraethyl phosphate

fosfato de trialilo triallyl phosphate

fosfato de tributilo tributyl phosphate

fosfato de tricresilo tricresyl phosphate

fosfato de trietilo triethyl phosphate

fosfato de trifenilo triphenyl phosphate

fosfato de trimagnesio trimagnesium phosphate

fosfato de trimetilo trimethyl phosphate

fosfato de trioctilo trioctyl phosphate

fosfato de trisetilhexilo trisethylhexyl phosphate

fosfato de tritolilo tritolyl phosphate

fosfato de urea-amonio urea-ammonium phosphate

fosfato de uridina uridine phosphate

fosfato de zirconilo zirconyl phosphate

fosfato de zirconio zirconium phosphate

fosfato dicálcico dicalcium phosphate

fosfato disódico disodium phosphate

fosfato férrico ferric phosphate

fosfato ferroso ferrous phosphate

fosfato magnésico magnesium phosphate

fosfato magnésico amónico ammonium magnesium phosphate

fosfato manganoso manganous phosphate

fosfato mercúrico mercuric phosphate

fosfato mercurioso mercurous phosphate

fosfato monocálcico monocalcium phosphate

fosfato niqueloso nickelous phosphate

fosfato potásico potassium phosphate

fosfato sódico sodium phosphate

fosfato sódico amónico ammonium sodium phosphate

fosfato tricálcico tricalcium phosphate

fosfato tripotásico tripotassium phosphate

fosfato trisódico trisodium phosphate

fosfato trisódico clorado chlorinated trisodium phosphate

fosfenilo phosphenyl

fosfina phosphine

fosfina de dietilo diethyl phosphine

fosfina de etilo ethyl phosphine

fosfina de tributilo tributyl phosphine
fosfinato phosphinate
fosfinato alumínico aluminum phosphinate
fosfinato bárico barium phosphinate
fosfinato cálcico calcium phosphinate
fosfinato de manganeso manganese phosphinate
fosfinato de plomo lead phosphinate
fosfinato magnésico magnesium phosphinate
fosfinato potásico potassium phosphinate
fosfinato sódico sodium phosphinate
fosfinico phosphinico
fosfinita phosphinite
fosfino phosphino
fosfinoilo phosphinoyl
fosfinoso phosphinoso
fosfito phosphite
fosfito amónico ammonium phosphite
fosfito cálcico calcium phosphite
fosfito de dialilo diallyl phosphite
fosfito de dibutilo dibutyl phosphite
fosfito de dietilo diethyl phosphite
fosfito de difenilo diphenyl phosphite
fosfito de dioctilo dioctyl phosphite
fosfito de fenildidecilo phenyldidecyl phosphite
fosfito de fenilneopentilo phenylneopentyl phosphite
fosfito de plomo lead phosphite
fosfito de tributilo tributyl phosphite
fosfito de tricresilo tricresyl phosphite
fosfito de trietilo triethyl phosphite
fosfito de trifenilo triphenyl phosphite
fosfito de triisooctilo triisooctyl phosphite
fosfito de triisopropilo triisopropyl phosphite
fosfito de trimetilo trimethyl phosphite
fosfito de trioctadecilo trioctadecyl phosphite
fosfito de triscloroetilo trischloroethyl phosphite
fosfito de trisetilhexilo trisethylhexyl phosphite
fosfito potásico potassium phosphite
fosfito sódico sodium phosphite
fosfo phospho

fosfoaluminato sódico sodium phosphoaluminate
fosfobenceno phosphobenzene
fosfocreatina phosphocreatine
fosfodiéster phosphodiester
fosfodiesterasa phosphodiesterase
fosfoglicérido phosphoglyceride
fosfolipasa phospholipase
fosfolípido phospholipid
fosfomicina fosfomycin
fosfomolibdato phosphomolybdate
fosfomolibdato amónico ammonium phosphomolybdate
fosfomolibdato sódico sodium phosphomolybdate
fosfonato phosphonate
fosfonato de cinc zinc phosphonate
fosfonato de dibutilbutilo dibutylbutyl phosphonate
fosfonato magnésico magnesium phosphonate
fosfonato potásico potassium phosphonate
fosfonato sódico sodium phosphonate
fosfonio phosphonium
fosfonita phosphonite
fosfonitrilo phosphonitrile
fosfono phosphono
fosfonoilo phosphonoyl
fosfonoso phosphonoso
fosfoproteína phosphoprotein
fosforado phosphorated
fosforano phosphorane
fosforato phosphorate
fosforescencia phosphorescence
fosforescente phosphorescent
fosfórico phosphoric
fosforilación phosphorylation
fosforilasa phosphorylase
fosforilo phosphoryl
fosforimetría phosphorimetry
fosforita phosphorite
fósforo phosphorus, phosphor
fósforo amarillo yellow phosphorus
fósforo blanco white phosphorus
fósforo de Baldwin Baldwin phosphorus
fósforo rojo red phosphorus
fosforoso phosphoroso
fosforotioato de trietilo triethyl

phosphorothioate

fosforotiolato de tributilo tributyl phosphorothiolate

fosfosilicato bárico barium phosphosilicate

fosfotungstato phosphotungstate

fosfotungstato amónico ammonium phosphotungstate

fosfotungstato sódico sodium phosphotungstate

fosfovanadato sódico sodium phosphovanadate

fosfuro phosphide

fosfuro alumínico aluminum phosphide

fosfuro arsenioso arsenous phosphide

fosfuro bárico barium phosphide

fosfuro cálcico calcium phosphide

fosfuro cuproso cuprous phosphide

fosfuro de boro boron phosphide

fosfuro de cesio cesium phosphide

fosfuro de cinc zinc phosphide

fosfuro de cobalto cobalt phosphide

fosfuro de cobre copper phosphide

fosfuro de cromo chromium phosphide

fosfuro de galio gallium phosphide

fosfuro de hidrógeno hydrogen phosphide

fosfuro de hierro iron phosphide

fosfuro de indio indium phosphide

fosfuro de itrio yttrium phosphide

fosfuro de lantano lanthanum phosphide

fosfuro de plata silver phosphide

fosfuro estánnico stannic phosphide

fosfuro ferroso ferrous phosphide

fosfuro magnésico magnesium phosphide

fosfuro potásico potassium phosphide

fosfuro sódico sodium phosphide

fosgeno phosgene

foshagita foshagite

fósil fossil

fosmet phosmet

fostato diamónico diammonium phosphate

fostión fostion

fotocatálisis photocatalysis

fotocatalizador photocatalyst

fotocátodo photocathode

fotoconducción photoconduction

fotocrómico photochromic

fotocromismo photochromism

fotodegradación photodegradation

fotodeposición photodeposition

fotodescomposición photodecomposition

fotodesintegración photodisintegration

fotodimerización photodimerization

fotodisociación photodissociation

fotoeléctrico photoelectric

fotoelectrón photoelectron

fotoemisión photoemission

fotoemisivo photoemissive

fotoemisor photoemitter

fotoestabilización photostabilization

fotográfico photographic

fotoionización photoionization

fotólisis photolysis

fotólisis instantánea flash photolysis

fotolítico photolytic

fotólito photolyte

fotoluminiscencia photoluminescence

fotometría con filtro filter photometry

fotometría de llama flame photometry

fotométrico photometric

fotómetro photometer

fotómetro de llama flame photometer

fotón photon

fotoneutrón photoneutron

fotonuclear photonuclear

fotooxidación photooxidation

fotopolimerización photopolymerization

fotopolímero photopolymer

fotoprotón photoproton

fotoquímica photochemistry

fotoquímico photochemical

fotorreacción photoreaction

fotorreducción photoreduction

fotosensibilidad photosensitivity

fotosensibilización photosensitization

fotosensibilizador photosensitizer

fotosensible photosensitive

fotosíntesis photosynthesis

fototropia phototropy

fototropismo phototropism

fotovoltaico photovoltaic

fowlerita fowlerite

foyaíta foyaite

fracción fraction

fracción de empaquetamiento packing fraction

fracción de masa mass fraction
fracción molar mole fraction
fraccionación fractionation
fraccionamiento cracking
fraccionante fractionating
fraccionar crack
fraccionario fractional
fractura fracture
fracturación hidráulica hydraulic
    fracturing
fragancia fragrance
fragilidad fragility
fragilización cáustica caustic cracking
fragmentación térmica thermal
    fragmentation
fragmento atómico atomic fragment
fragmento nuclear nuclear fragment
francio francium
francolita francolite
frangulina frangulin
franklinita franklinite
franqueíta franckeite
frasco lavador wash bottle
fraunhofer fraunhofer
fraxetina fraxetin
fraxina fraxin
fraxinol fraxinol
frecuencia frequency
frecuencia atómica atomic frequency
frecuencia molecular molecular
    frequency
freibergita freibergite
freierslebenita freierslebenite
freyalita freyalite
fricción friction
frigorífico frigorific
frita frit
fritado fritting
fructofuranosidasa fructofuranosidase
fructosa fructose
fructosida fructoside
ftalamida phthalamide
ftalato phthalate
ftalato de bistridecilo bistridecyl
    phthalate
ftalato de butilbencilo butylbenzyl
    phthalate
ftalato de butilo butyl phthalate
ftalato de dialilo diallyl phthalate

ftalato de dibutilo dibutyl phthalate
ftalato de dietilo diethyl phthalate
ftalato de difenilo diphenyl phthalate
ftalato de dimetilo dimethyl phthalate
ftalato de dinonilo dinonyl phthalate
ftalato de dioctilo dioctyl phthalate
ftalato de etilo ethyl phthalate
ftalato de fenilo phenyl phthalate
ftalato de glicerilo glyceryl phthalate
ftalato de hidrógeno potásico potassium
    hydrogen phthalate
ftalato de isoamilo isoamyl phthalate
ftalato de pentilo pentyl phthalate
ftalato de plomo lead phthalate
ftalato sódico sodium phthalate
ftalazina phthalazine
ftaldiamida phthaldiamide
ftaleína phthalein
ftalida phthalide
ftalidilideno phthalidylidene
ftalidilo phthalidyl
ftalilo phthalyl
ftalimida phthalimide
ftalimido phthalimido
ftalocianina phthalocyanine
ftalodinitrilo phthalodinitrile
ftaloilo phthaloyl
ftalonitrilo phthalonitrile
fuberidazol fuberidazole
fucitol fucitol
fucosa fucose
fucosita fucosite
fucsina fuchsin
fucsina ácida acid fuchsin
fucusina fucusine
fucusol fucusol
fuchsita fuchsite
fuente source
fuente de electrones electron source
fuente de energía energy source
fuente de infrarrojo infrared source
fuente de neutrones neutron source
fuente de radiación radiation source
fuente de rayos gamma gamma-ray
    source
fuente iónica ion source
fuente radiactiva radioactive source
fuerte strong
fuerza force, strength

fuerza catalítica catalytic force
fuerza centrífuga centrifugal force
fuerza de campo field strength
fuerza de Majorana Majorana force
fuerza de Yukawa Yukawa force
fuerza dieléctrica dielectric strength
fuerza electromotriz electromotive force
fuerza intermolecular intermolecular
    force
fuerza iónica ionic strength
fuerza nuclear nuclear force
fuerzas de dispersión dispersion forces
fuerzas de van der Waals van der Waals
    forces
fuerzas ligantes metálicas metallic
    binding forces
fuga leakage, leak
fuga de gas gas leak
fugacidad fugacity
fulereno fullerene, buckyballs
fulerita fullerite
fulguración fulguration
fulgurador fulgurator
fulgurita fulgurite
fulminato fulminate
fulminato de mercurio mercury
    fulminate
fulminato de plata silver fulminate
fulvaleno fulvalene
fulveno fulvene
fumagilina fumagillin
fumante fuming
fumaramida fumaramide
fumarato fumarate
fumarato cálcico calcium fumarate
fumarato de bencilo benzyl fumarate
fumarato de cadmio cadmium fumarate
fumarato de dibutilo dibutyl fumarate
fumarato de plomo lead fumarate
fumarato ferroso ferrous fumarate
fumarhidrazida fumarhydrazide
fumarilo fumaryl
fumaroilo fumaroyl
fumigante fumigant
función de correlación correlation
    function
función de excitación excitation function
función de Gibbs Gibbs function
función de onda wave function

función de partición partition function
función de Patterson Patterson function
función sigma sigma function
función trabajo work function
fundamental fundamental, ground
fundente magnésico magnesium flux
fundido fused, molten
fundir fuse, melt, smelt
fungicida fungicide
fural fural
furaldehído furaldehyde
furamida furamide
furancarbinol furancarbinol
furandiona furandione
furanilmetilo furanylmethyl
furanmetanotiol furanmethanethiol
furanmetilamina furanmethylamine
furano furan
furanosa furanose
furanosida furanoside
furazán furazan
furazanilo furazanyl
furazolidona furazolidone
furfural furfural
furfuraldehído furfuraldehyde
furfuramida furfuramide
furfurano furfuran
furfurilamida furfurylamide
furfurilamina furfurylamine
furfurilideno furfurylidene
furfurilo furfuryl
furfurol furfurol
furil furil
furilcarbinol furylcarbinol
furildioxima furildioxime
furilidina furylidine
furilmetilo furylmethyl
furilo furyl
furoamida furoamide
furoato furoate
furoato de butilo butyl furoate
furoato de etilo ethyl furoate
furoato de isoamilo isoamyl furoate
furoato de isobutilo isobutyl furoate
furoato de pentilo pentyl furoate
furoato de propilo propyl furoate
furoilo furoyl
furoína furoin
furol furol

fusibilidad fusibility
fusible fusible
fusidato sódico sodium fusidate
fusión fusion
fusión atómica atomic fusion

fusión nuclear nuclear fusion
fusión termonuclear thermonuclear
  fusion
fusionable fusionable

G

gadinina gadinine
gadolinio gadolinium
gadolinita gadolinite
gafas de protección goggles
gafas protectoras safety goggles
gahnita gahnite
gailusita gaylussite
galacetofenona gallacetophenone
galactano galactan
galactosa galactose
galactosidasa galactosidase
galamina gallamine
galato gallate
galato amónico ammonium gallate
galato de bismuto bismuth gallate
galato de metilo methyl gallate
galato de octilo octyl gallate
galato de propilo propyl gallate
galeína gallein
galena galena
galenita galenite
gálico gallic
galina gallin
galio gallium
galocianina gallocyanine
galógeno gallogen
galoilo galloyl
galoso gallous
galotanino gallotannin
galvánico galvanic
galvanismo galvanism
galvanización galvanizing
galvanizado galvanized
galvanizar galvanize
galvanomagnético galvanomagnetic
galvanometría galvanometry
galvanométrico galvanometric
galvanómetro galvanometer
galvanoplasteado plated
galvanoplastia galvanoplasty,
    electroplating, plating
gamexano gammexane
gamilo gammil

gamma gamma
gammaglobulina gamma-globulin
ganga gangue
ganofilita ganophyllite
ganomalita ganomalite
ganomatita ganomatite
garnierita garnierite
gas gas
gas asfixiante asphyxiant gas
gas atmosférico atmospheric gas
gas comprimido compressed gas
gas de agua water gas
gas de coque coke gas
gas de hidrocarburos hydrocarbon gas
gas de hulla coal gas
gas de pantano marsh gas
gas de petróleo petroleum gas, oil gas
gas de petróleo licuado liquefied
    petroleum gas
gas de refinería refinery gas
gas de síntesis synthesis gas
gas detonante detonating gas
gas electrolítico electrolytic gas
gas electrónico electron gas
gas hilarante laughing gas
gas ideal ideal gas
gas inerte inert gas
gas ionizado ionized gas
gas licuado liquefied gas
gas líquido liquid gas
gas molecular molecular gas
gas monoatómico monoatomic gas
gas mostaza mustard gas
gas natural natural gas
gas natural licuado liquefied natural gas
gas natural sintético synthetic natural
    gas
gas neurotóxico nerve gas
gas noble noble gas
gas nocivo noxious gas
gas perfecto perfect gas
gas permanente permanent gas
gas pobre producer gas

gas poliatómico polyatomic gas
gas portador carrier gas
gas raro rare gas
gas real real gas
gas rico rich gas
gas sintético synthetic gas
gas venenoso poison gas
gasa de alambre wire gauze
gaseamiento gassing
gaseoso gaseous
gasificación gasification
gasohol gasohol
gasol gasol
gasóleo gas oil
gasolina gasoline
gasolina blanca white gasoline
gasométrico gasometric
gastrina gastrin
gaulterina gaultherin
gehlenita gehlenite
geisina geissine
gel gel
gel de alúmina alumina gel
gel de sílice silica gel
gel incendiario incendiary gel
gel iónico ionic gel
gelatina gelatin
gelatina explosiva blasting gelatin
gelatinización gelatinization
gelatinizar gelatinize
gelatinoso gelatinous
gelificación gelling
gelignita gelignite
gema gem, precious stone
geminal geminal
gen gene
genalcaloide genalkaloid
generación generation
generador de gas gas generator
generar generate
genérico generic
genético genetic
genistina genistin
gentamicina gentamicin
gentiobiosa gentiobiose
geocronita geocronite
geometría estérica steric geometry
geoquímica geochemistry
geranial geranial

geranialdehído geranialdehyde
geranilo geranyl
geraniol geraniol
germanato germanate
germanato de litio lithium germanate
germanato sódico sodium germanate
germánico germanic
germanio germanium
germanita germanite
germaniuro germanide
germano germane
germanoso germanous
germicida germicidal
germinación germination
gibbsita gibbsite
giberelina gibberellin
gibrel gibrel
gigatolita gigantolite
gilpinita gilpinite
gilsonita gilsonite
ginocardina gynocardine
girolita gyrolite
gismondita gismondite
gitonina gitonin
gitoxina gitoxin
glacial glacial
glaserita glaserite
glauberita glauberite
glaucocroíta glaucochroite
glaucodot glaucodot
glaucofana glaucophane
glauconita glauconite
glaucopicrina glaucopicrine
gliadina gliadin
glical glycal
gliceral glyceral
gliceraldehído glyceraldehyde
glicerato glycerate
glicerato de etilo ethyl glycerate
glicérico glyceric
glicérido glyceride
glicerilo glyceryl
glicerina glycerin, glycerol
glicerinato glycerinate
glicero glycero
glicerofosfato glycerophosphate
glicerofosfato cálcico calcium
　　glycerophosphate
glicerofosfato de manganeso manganese

glycerophosphate
glicerofosfato férrico ferric
  glycerophosphate
glicerofosfato potásico potassium
  glycerophosphate
gliceroilo glyceroyl
glicerol glycerol, glycerin
glicerosa glycerose
glicida glycide
glicidilo glycidyl
glicidol glycidol
glicilglicina glycylglycine
glicilo glycyl
glicina glycine
glicinato de cobre copper glycinate
glicinio glycinium
glico glyco
glicogenasa glycogenase
glicogénesis glycogenesis
glicógeno glycogen
glicol glycol
glicolal glycolal
glicolaldehído glycolaldehyde
glicolato cálcico calcium glycolate
glicolato de etilo ethyl glycolate
glicolato de zirconio zirconium glycolate
glicolato de zirconio sódico sodium
  zirconium glycolate
glicolato sódico sodium glycolate
glicólido glycolide
glicolilo glycolyl
glicolípido glycolipid
glicólisis glycolysis
glicoloilo glycoloyl
glicolonitrilo glycolonitrile
gliconitrilo glyconitrile
glicoproteína glycoprotein
glicosaminoglicano glycosaminoglycan
glicosidasa glycosidase
glicosídico glycosidic
glicósido glycoside
glicosilación glycosylation
gliftal glyphtal
glime glyme
glioxal glyoxal
glioxalasa glyoxalase
glioxalina glyoxaline
glioxilasa glyoxylase
glioxima glyoxime

globulina globulin
glucagón glucagon
glucano glucan
glucasa glucase
glucinio glucinium
glucógeno glycogen
glucoheptonato sódico sodium
  glucoheptonate
gluconato gluconate
gluconato alumínico aluminum gluconate
gluconato amónico ammonium gluconate
gluconato cálcico calcium gluconate
gluconato de cinc zinc gluconate
gluconato de cobre copper gluconate
gluconato de manganeso manganese
  gluconate
gluconato ferroso ferrous gluconate
gluconato magnésico magnesium
  gluconate
gluconato potásico potassium gluconate
gluconato sódico sodium gluconate
glucopiranosa glucopyranose
glucosa glucose
glucosamina glucosamine
glucosazona glucosazone
glucosidasa glucosidase
glucósido glucoside, glycoside
glucosina glucosin
glucuronida glucuronide
glucuronidasa glucuronidase
glucuronolactona glucuronolactone
glusida gluside
glutamato amónico ammonium
  glutamate
glutamato cálcico calcium glutamate
glutamato monosódico monosodium
  glutamate
glutamato potásico potassium glutamate
glutamato sódico sodium glutamate
glutamilcisteinilglicina
  glutamylcysteinylglycine
glutamilo glutamyl
glutamina glutamine
glutamoilo glutamoyl
glutaraldehído glutaraldehyde
glutárico glutaric
glutaronitrilo glutaronitrile
glutationa glutathione
glutelina glutelin

gluten gluten
glutenina glutenin
glutetimida glutethimide
glutosa glutose
gmelinita gmelinite
gneis gneiss
gnoscopina gnoscopine
goetita goethite
golpeteo knock
goma arábiga arabic gum
goma de acacia acacia gum
goma de celulosa cellulose gum
goma de éster ester gum
goma de tragacanto tragacanth gum
goma de xantano xanthan gum
goma de yaca yacca gum
goma guar guar gum
goma karaya karaya gum
goma laca shellac
goma soluble en agua water-soluble gum
goma vegetal vegetable gum
gonadotropina coriónica chorionic
    gonadotropin
gosipol gossypol
goslarita goslarite
gradación grading
gradiente gradient
gradiente de concentración
    concentration gradient
gradiente osmótico osmotic gradient
grado degree, grade
grado de Brix Brix degree
grado de dilución degree of dilution
grado de disociación degree of
    dissociation
grado de dureza degree of hardness
grado de hidrólisis degree of hydrolysis
grado de ionización ionization degree
grado de polimerización degree of
    polymerization
grado de substitución degree of
    substitution
grado de temperatura degree of
    temperature
grado de vacío degree of vacuum
grados de libertad degrees of freedom
graduación graduation
graduado graduated
gradual gradual

grafitización graphitization
grafito graphite
gramicidina gramicidin
gramina gramine
gramnegativo gram-negative
gramo gram
grampositivo gram-positive
granate garnet
granito granite
granulación granulation
granular granular
gránulo granule
granulosa granulose
granulosidad granularity
grasa fat, grease
grasa de lana wool fat
grasa mineral mineral fat
grasa poliinsaturada polyunsaturated fat
grasa saturada saturated fat
gravedad específica specific gravity
gravilla de ebullición boiling chips
gravimetría gravimetry
gravimétrico gravimetric
gray gray
greenockita greenockite
grifita griphite
griolita gryolite
grosularita grossularite
grunerita grunerite
grupo group
grupo acídico acidic group
grupo ácido acid group
grupo acilo acyl group
grupo activo active group
grupo alifático aliphatic group
grupo alilo allyl group
grupo amida amide group
grupo amino amino group
grupo atómico atomic group
grupo básico basic group
grupo carbonilo carbonyl group
grupo carboxilo carboxyl group
grupo cuántico quantum group
grupo de elementos group of elements
grupo funcional functional group
grupo negativo negative group
grupo positivo positive group
grupo prostético prosthetic group
grupo protector protecting group

grupo substituyente substituent
guaiol guaiol
guanacilo guanazyl
guanamina guanamine
guanasa guanase
guanidina guanidine
guanidinio guanidinium
guanidino guanidino
guanido guanido

guanilato sódico sodium guanylate
guanina guanine
guano guano
guanosina guanosine
guantes protectores protective gloves
guayacol guaiacol
guerra química chemical warfare
gusto taste
gutapercha gutta-percha

H

hábito habit
hachís hashish
hafnio hafnium
halazona halazone
halita halite
halo halo
haloalcano haloalkane
haloamina haloamine
halocarburo halocarbon
halocromismo halochromism
haloformo haloform
halogenación halogenation
halogenado halogenated
halogenar halogenate
halogeniuro halogenide
halógeno halogen
halohidrina halohydrin
haloide haloid
halonio halonium
haloperidol haloperidol
halotano halothane
haluro halide
haluro ácido acid halide
haluro de acilo acyl halide
haluro de alquilo alkyl halide
haluro de cadmio cadmium halide
haluro de carbonilo carbonyl halide
haluro de cloro chlorine halide
haluro de cobalto cobalt halide
haluro de erbio erbium halide
haluro de flúor fluorine halide
haluro de galio gallium halide
haluro de germanio germanium halide
haluro de litio lithium halide
haluro de manganeso manganese halide
haluro de oro gold halide
haluro de plomo lead halide
haluro magnésico magnesium halide
haluro sódico sodium halide
halloysita halloysite
hamartita hamartite
hanksita hanksite
hapto hapto

harina de aceite de linaza linseed oil meal
harina de aceite de maní peanut oil meal
harina de madera wood flour, wood meal
harina de pizarra slate flour
harina de semilla de algodón cottonseed meal
harmina harmine
harmotoma harmotome
hartita hartite
hausmannita hausmannite
hauyna hauyne
haz electrónico electron beam
haz iónico molecular molecular ion beam
haz molecular molecular beam
haz neutrónico neutron beam
hectorita hectorite
hedenbergita hedenbergite
hedonal hedonal
helcosol helcosol
helenina helenine
helenita helenite
hélice helix
hélice alfa alpha helix
helicoidal helical
helio helium
helvita helvite
hem heme
hemateína hematein
hematina hematin
hematita hematite
hematoxilina hematoxylin
hemel hemel
hemi hemi
hemiacetal hemiacetal
hemicelulosa hemicellulose
hemihedral hemihedral
hemimeliteno hemimellitene
hemina hemin
hemiquinonoide hemiquinonoid
hemoglobina hemoglobin

hempa hempa
hendecano hendecane
hendidura cleavage
heneicosano heneicosane
heneicosanoato de metilo methyl
    heneicosanoate
hentriacontano hentriacontane
heparina heparin
hepta hepta
heptabarbital heptabarbital
heptaclorepóxido heptachlorepoxide
heptacloro heptachlor
heptacosano heptacosane
heptadecano heptadecane
heptadecanoato de metilo methyl
    heptadecanoate
heptadecanol heptadecanol
heptadecanona heptadecanone
heptadecilglioxalidina
    heptadecylglyoxalidine
heptadieno heptadiene
heptahidrato heptahydrate
heptaldehído heptaldehyde
heptaleno heptalene
heptametildisilazano
    heptamethyldisilazane
heptametileno heptamethylene
heptametilnonano heptamethylnonane
heptanal heptanal
heptano heptane
heptanoato heptanoate
heptanoato de etilo ethyl heptanoate
heptanodioilo heptanedioyl
heptanoilo heptanoyl
heptanol heptanol
heptanona heptanone
heptasulfuro de fósforo phosphorus
    heptasulfide
heptasulfuro de renio rhenium
    heptasulfide
heptavalente heptavalent
heptenilo heptenyl
hepteno heptene
heptilamina heptylamine
heptileno heptylene
heptilo heptyl
heptino heptyne
heptoato de heptilo heptyl heptoate
heptosa heptose

heptóxido heptoxide
heptóxido de cloro chlorine heptoxide
heptóxido de manganeso manganese
    heptoxide
heptóxido de renio rhenium heptoxide
herapatita herapathite
herbicida herbicide
hercinita hercynite
herderita herderite
hermético hermetic
hermético al agua watertight
heroína heroin
herrerita herrerite
herrumbre rust
hervir boil
hesperidina hesperidin
hessita hessite
hessonita hessonite
hetero hetero
heteroaromático heteroaromatic
heteroatómico heteroatomic
heteroauxina heteroauxin
heteroazeótropo heteroazeotrope
heterobárico heterobaric
heterocíclico heterocyclic
heterogéneo heterogeneous
heteroléptico heteroleptic
heterólisis heterolysis
heterolítico heterolytic
heterometría heterometry
heteromolibdato heteromolybdate
heteronuclear heteronuclear
heteropolar heteropolar
heteropoli heteropoly
heteropolímero heteropolymer
heterotipo heterotype
heterotópico heterotopic
heterótopo heterotope
heulandita heulandite
hexa hexa
hexaamina hexaamine
hexaborano hexaborane
hexabromo hexabromo
hexabromoetano hexabromoethane
hexabromoplatinato amónico
    ammonium hexabromoplatinate
hexabromuro de benceno benzene
    hexabromide
hexacarbonilo de cromo chromium

hexacarbonyl
hexacarbonilo de molibdeno
molybdenum hexacarbonyl
hexacarbonilo de tungsteno tungsten
hexacarbonyl
hexacarbonilo de vanadio vanadium
hexacarbonyl
hexacarbonilo potásico potassium
hexacarbonyl
hexaciano hexacyano
hexacianoferrato amónico ammonium
hexacyanoferrate
hexacianoferrato bárico barium
hexacyanoferrate
hexacianoferrato de cinc zinc
hexacyanoferrate
hexacianoferrato de cobre copper
hexacyanoferrate
hexacianoferrato de hierro iron
hexacyanoferrate
hexacianoferrato de manganeso
manganese hexacyanoferrate
hexacianoferrato de mercurio mercury
hexacyanoferrate
hexacianoferrato de plata silver
hexacyanoferrate
hexacianoferrato de plomo lead
hexacyanoferrate
hexacianoferrato de uranilo uranyl
hexacyanoferrate
hexacianoferrato potásico potassium
hexacyanoferrate
hexacianoferrato sódico sodium
hexacyanoferrate
hexacianógeno hexacyanogen
hexacloro hexachloro
hexacloroacetona hexachloroacetone
hexaclorobenceno hexachlorobenzene
hexaclorobutadieno hexachlorobutadiene
hexaclorociclohexano
hexachlorocyclohexane
hexaclorociclopentadieno
hexachlorocyclopentadiene
hexacloroestannato amónico ammonium
hexachlorostannate
hexacloroetano hexachloroethane
hexaclorofeno hexachlorophene
hexaclorometilcarbonato
hexachloromethylcarbonate

hexacloronaftaleno
hexachloronaphthalene
hexacloroosmiato sódico sodium
hexachloroosmate
hexacloroplatinato hexachloroplatinate
hexacloroplatinato amónico ammonium
hexachloroplatinate
hexacloroplatinato bárico barium
hexachloroplatinate
hexacloroplatinato de cesio cesium
hexachloroplatinate
hexacloroplatinato de litio lithium
hexachloroplatinate
hexacloroplatinato de rubidio rubidium
hexachloroplatinate
hexacloroplatinato potásico potassium
hexachloroplatinate
hexacloroplatinato sódico sodium
hexachloroplatinate
hexacloropropano hexachloropropane
hexacloropropileno hexachloropropylene
hexacloruro de benceno benzene
hexachloride
hexacloruro de carbono carbon
hexachloride
hexacloruro de tungsteno tungsten
hexachloride
hexacontano hexacontane
hexacosano hexacosane
hexacosanol hexacosanol
hexacosilo hexacosyl
hexadecanal hexadecanal
hexadecano hexadecane
hexadecanoato hexadecanoate
hexadecanoilo hexadecanoyl
hexadecanol hexadecanol
hexadeceno hexadecene
hexadecenolida hexadecenolide
hexadecilmercaptano hexadecyl
mercaptan
hexadecilo hexadecyl
hexadeciltriclorosilano
hexadecyltrichlorosilane
hexadecino hexadecyne
hexadieno hexadiene
hexadiíno hexadiyne
hexafenildisilano hexaphenyldisilane
hexafeniletano hexaphenylethane
hexafluoroacetona hexafluoroacetone

hexafluoroaluminato amónico
ammonium hexafluoroaluminate
hexafluoroaluminato sódico sodium
hexafluoroaluminate
hexafluorobenceno hexafluorobenzene
hexafluorodisilano hexafluorodisilane
hexafluoroetano hexafluoroethane
hexafluorofosfato amónico ammonium
hexafluorophosphate
hexafluorofosfato potásico potassium
hexafluorophosphate
hexafluorogermantano bárico barium
hexafluorogermantane
hexafluoropropeno hexafluoropropene
hexafluoropropileno
hexafluoropropylene
hexafluorosilicato hexafluorosilicate
hexafluorosilicato amónico ammonium
hexafluorosilicate
hexafluorosilicato cálcico calcium
hexafluorosilicate
hexafluorosilicato cobaltoso cobaltous
hexafluorosilicate
hexafluorosilicato cúprico cupric
hexafluorosilicate
hexafluorosilicato cuproso cuprous
hexafluorosilicate
hexafluorosilicato de cesio cesium
hexafluorosilicate
hexafluorosilicato de cinc zinc
hexafluorosilicate
hexafluorosilicato de estaño tin
hexafluorosilicate
hexafluorosilicato de estroncio
strontium hexafluorosilicate
hexafluorosilicato de hidrógeno
hydrogen hexafluorosilicate
hexafluorosilicato de hierro iron
hexafluorosilicate
hexafluorosilicato de litio lithium
hexafluorosilicate
hexafluorosilicato de manganeso
manganese hexafluorosilicate
hexafluorosilicato de mercurio mercury
hexafluorosilicate
hexafluorosilicato de níquel nickel
hexafluorosilicate
hexafluorosilicato de plata silver
hexafluorosilicate

hexafluorosilicato de plomo lead
hexafluorosilicate
hexafluorosilicato de rubidio rubidium
hexafluorosilicate
hexafluorosilicato de talio thallium
hexafluorosilicate
hexafluorosilicato magnésico
magnesium hexafluorosilicate
hexafluorosilicato sódico sodium
hexafluorosilicate
hexafluoruro de azufre sulfur
hexafluoride
hexafluoruro de telurio tellurium
hexafluoride
hexafluoruro de uranio uranium
hexafluoride
hexafluorurosilicato de cadmio
cadmium hexafluorosilicate
hexaglicerol hexaglycerol
hexagonita hexagonite
hexágono hexagon
hexahelicina hexahelicine
hexahidrato hexahydrate
hexahidrato de piperazina piperazine
hexahydrate
hexahídrico hexahydric
hexahidro hexahydro
hexahidroanilina hexahydroaniline
hexahidroantraceno
hexahydroanthracene
hexahidrobenceno hexahydrobenzene
hexahidrocresol hexahydrocresol
hexahidrofenol hexahydrophenol
hexahidrometilo hexahydromethyl
hexahidropiridina hexahydropyridine
hexahidrotolueno hexahydrotoluene
hexahidroxi hexahydroxy
hexahidroxiciclohexano
hexahydroxycyclohexane
hexahidroxileno hexahydroxylene
hexaldehído hexaldehyde
hexalina hexalin
hexametafosfato hexametaphosphate
hexametafosfato sódico sodium
hexametaphosphate
hexametilbenceno hexamethylbenzene
hexametildisilano hexamethyldisilane
hexametildisilazano
hexamethyldisilazane

hexametilendiamina
hexamethylenediamine
hexametilenglicol hexamethylene glycol
hexametilenimina hexamethyleneimine
hexametileno hexamethylene
hexametilentetramina
hexamethylenetetramine
hexametilmelamina
hexamethylmelamine
hexamina hexamine
hexanafteno hexanaphthene
hexanal hexanal
hexanitrato de manitol mannitol
hexanitrate
hexanitro hexanitro
hexanitrocobaltato potásico potassium
hexanitrocobaltate
hexanitrocobaltato sódico sodium
hexanitrocobaltate
hexanitruro bárico barium hexanitride
hexano hexane
hexanoato hexanoate
hexanoato de etilo ethyl hexanoate
hexanoato de propenilo propenyl
hexanoate
hexanodiamida hexanediamide
hexanodinitrilo hexanedinitrile
hexanodioato hexanedioate
hexanodiol hexanediol
hexanodiona hexanedione
hexanoilo hexanoyl
hexanol hexanol
hexanolactama hexanelactam
hexanona hexanone
hexanotriol hexanetriol
hexaprismo hexaprismo
hexatriacontano hexatriacontane
hexayodo hexaiodo
hexayoduro de azufre sulfur hexaiodide
hexenal hexenal
hexénico hexenic
hexenilo hexenyl
hexeno hexene
hexenol hexenol
hexestrol hexestrol
hexetidina hexetidine
hexilamina hexylamine
hexilenglicol hexylene glycol
hexileno hexylene

hexilfenol hexylphenol
hexilmercaptano hexyl mercaptan
hexilo hexyl
hexilresorcina hexylresorcinol
hexiltriclorosilano hexyltrichlorosilane
hexino hexyne
hexinol hexynol
hexobarbital hexobarbital
hexobarbitona hexobarbitone
hexoico hexoic
hexona hexone
hexoquinasa hexokinase
hexosa hexose
hexotriosa hexotriose
hialofano hyalophane
hialosiderita hyalosiderite
hialuronidasa hyaluronidase
hibridización hybridization
hibridización de enlace bond
hybridization
hidantoína hydantoin
hiddenita hiddenite
hidracilo hydrazyl
hidramina hydramine
hidrargillita hydrargillite
hidrasa hydrase
hidrastina hydrastine
hidratación hydration
hidratado hydrated
hidrato hydrate
hidrato alumínico aluminum hydrate
hidrato amónico ammonium hydrate
hidrato bárico barium hydrate
hidrato cálcico calcium hydrate
hidrato de amileno amylene hydrate
hidrato de amilo amyl hydrate
hidrato de berilio beryllium hydrate
hidrato de bismuto bismuth hydrate
hidrato de carbono carbohydrate
hidrato de cerio cerium hydrate
hidrato de cesio cesium hydrate
hidrato de cloral chloral hydrate
hidrato de cloro chlorine hydrate
hidrato de cobalto cobalt hydrate
hidrato de cromo chromium hydrate
hidrato de diamida diamide hydrate
hidrato de etano ethane hydrate
hidrato de etileno ethylene hydrate
hidrato de fenilo phenyl hydrate

hidrato de gas gas hydrate
hidrato de hidrazina hydrazine hydrate
hidrato de manganeso manganese hydrate
hidrato de pentilo pentyl hydrate
hidrato de propano propane hydrate
hidrato de terpina terpin hydrate
hidrato férrico ferric hydrate
hidrato potásico potassium hydrate
hidrato sódico sodium hydrate
hidrazi hydrazi
hidrazida hydrazide
hidrazida maleica maleic hydrazide
hidrazidina hydrazidine
hidrazido hydrazido
hidrazina hydrazine
hidrazina de etanol ethanol hydrazine
hidrazina de metilo methyl hydrazine
hidrazinio hydrazinium
hidrazino hydrazino
hidrazinobenceno hydrazinobenzene
hidrazo hydrazo
hidrazobenceno hydrazobenzene
hidrazona hydrazone
hidrazono hydrazono
hidrazotolueno hydrazotoluene
hidrdato de butilcloral butyl chloral hydrate
hidrido hydrido
hidrina hydrin
hidro hydro
hidrobenzoína hydrobenzoin
hidrobiotita hydrobiotite
hidroboración hydroboration
hidrocarbonilo de cobalto cobalt hydrocarbonyl
hidrocarburo hydrocarbon
hidrocarburo alifático aliphatic hydrocarbon
hidrocarburo aromático aromatic hydrocarbon
hidrocarburo cíclico cyclic hydrocarbon
hidrocarburo clorado chlorinated hydrocarbon
hidrocarburo de acetileno acetylene hydrocarbon
hidrocarburo de cadena recta straight-chain hydrocarbon
hidrocarburo insaturado unsaturated hydrocarbon
hidrocarburo ligero light hydrocarbon
hidrocarburo normal normal hydrocarbon
hidrocarburo policíclico polycyclic hydrocarbon
hidrocarburo policíclico alifático aliphatic polycyclic hydrocarbon
hidrocarburo polinuclear polynuclear hydrocarbon
hidrocarburo saturado saturated hydrocarbon
hidrocelulosa hydrocellulose
hidrocianuro hydrocyanide
hidrocincita hydrozincite
hidrocoloide hydrocolloid
hidrocortisona hydrocortisone
hidrocraqueo hydrocracking
hidrodesalquilación hydrodealkylation
hidrodestilación hydrodistillation
hidroextractor hydroextractor
hidrofílico hydrophilic
hidrofilita hydrophilite
hidrófilo hydrophile
hidroflumetiazida hydroflumethiazide
hidrofóbico hydrophobic
hidroformación hydroforming
hidroformilación hydroformylation
hidrofuramida hydrofuramide
hidrogasificación hydrogasification
hidrogel hydrogel
hidrogenación hydrogenation
hidrogenado hydrogenated
hidrogenar hydrogenate
hidrógeno hydrogen
hidrógeno acídico acidic hydrogen
hidrógeno atómico atomic hydrogen
hidrógeno ligero light hydrogen
hidrógeno nasciente nascent hydrogen
hidrógeno pesado heavy hydrogen
hidrógenofosfato disódico disodium hydrogenphosphate
hidrogenólisis hydrogenolysis
hidrogenosufuro de indio indium hydrogensulfide
hidrogenosulfato de cesio cesium hydrogensulfate
hidrohematita hydrohematite
hidrol hydrol

hidrolasa hydrolase
hidrolicuefacción hydroliquefaction
hidrólisis hydrolysis
hidrólisis de Donnan Donnan hydrolysis
hidrólisis por membrana membrane
   hydrolysis
hidrolítico hydrolytic
hidrólito hydrolyte
hidrolizado hydrolized
hidrolizar hydrolyze
hidromagnesita hydromagnesite
hidrómetro hydrometer
hidrona hydrone
hidroperóxido hydroperoxide
hidroperóxido de cumeno cumene
   hydroperoxide
hidropolisulfuro hydropolysulfide
hidropónica hydroponics
hidroquinol hydroquinol
hidroquinona hydroquinone
hidrorrefinación hydrofining
hidroseleno hydroseleno
hidrosol hydrosol
hidrosoluble water soluble
hidrosolvatación hydrosolvation
hidrostático hydrostatic
hidrosulfito hydrosulfite
hidrosulfito de cinc zinc hydrosulfite
hidrosulfito-formaldehído hydrosulfite-
   formaldehyde
hidrosulfito sódico sodium hydrosulfite
hidrosulfuro hydrosulfide
hidrosulfuro amónico ammonium
   hydrosulfide
hidrosulfuro bárico barium hydrosulfide
hidrosulfuro cálcico calcium
   hydrosulfide
hidrosulfuro de etilo ethyl hydrosulfide
hidrosulfuro potásico potassium
   hydrosulfide
hidrosulfuro sódico sodium hydrosulfide
hidrotetrazona hydrotetrazone
hidrótropo hydrotrope
hidroxi hydroxy
hidroxiacetal hydroxyacetal
hidroxiacetanilida hydroxyacetanilide
hidroxiacetofenona
   hydroxyacetophenone
hidroxiacetona hydroxyacetone

hidroxiadipaldehído
   hydroxyadipaldehyde
hidroxialanina hydroxyalanine
hidroxiamino hydroxyamino
hidroxianilina hydroxyaniline
hidroxianisol hydroxyanisole
hidroxianisol butilado butylated
   hydroxyanisole
hidroxiantraceno hydroxyanthracene
hidroxiapatita hydroxyapatite
hidroxibenceno hydroxybenzene
hidroxibenzaldehído
   hydroxybenzaldehyde
hidroxibenzamida hydroxybenzamide
hidroxibenzofenona
   hydroxybenzophenone
hidroxibutanal hydroxybutanal
hidroxibutiraldehído
   hydroxybutyraldehyde
hidroxibutiranilida hydroxybutyranilide
hidroxicanfano hydroxycamphane
hidroxicaroteno hydroxycarotene
hidroxicerusita hydroxycerussite
hidroxicitronelal hydroxycitronellal
hidroxicloruro de zirconilo zirconyl
   hyroxychloride
hidroxicobalamina hydroxycobalamin
hidroxicolestano hydroxycholestane
hidroxicolina hydroxycholine
hidroxicorticosterona
   hydroxycorticosterone
hidroxidifenilamina
   hydroxydiphenylamine
hidroxidifenilo hydroxydiphenyl
hidroxidimetilbenceno
   hydroxydimethylbenzene
hidróxido hydroxide
hidróxido alumínico aluminum
   hydroxide
hidróxido amónico ammonium
   hydroxide
hidróxido anfótero amphoteric
   hydroxide
hidróxido áurico auric hydroxide
hidróxido bárico barium hydroxide
hidróxido cálcico calcium hydroxide
hidróxido cérico ceric hydroxide
hidróxido ceroso cerous hydroxide
hidróxido cobáltico cobaltic hydroxide

hidróxido cobaltoso cobaltous hydroxide
hidróxido crómico chromic hydroxide
hidróxido cromoso chromous hydroxide
hidróxido cúprico cupric hydroxide
hidróxido cuproso cuprous hydroxide
hidróxido de bencenodiazonio benzenediazonium hydroxide
hidróxido de berilio beryllium hydroxide
hidróxido de bismutilo bismuthyl hydroxide
hidróxido de bismuto bismuth hydroxide
hidróxido de boro boron hydroxide
hidróxido de cadmio cadmium hydroxide
hidróxido de cesio cesium hydroxide
hidróxido de cinc zinc hydroxide
hidróxido de cobalto cobalt hydroxide
hidróxido de cobre copper hydroxide
hidróxido de cromo chromium hydroxide
hidróxido de diazobenceno diazobenzene hydroxide
hidróxido de escandio scandium hydroxide
hidróxido de estaño tin hydroxide
hidróxido de estroncio strontium hydroxide
hidróxido de fenilo phenyl hydroxide
hidróxido de fosfonio phosphonium hydroxide
hidróxido de gadolinio gadolinium hydroxide
hidróxido de galio gallium hydroxide
hidróxido de germanio germanium hydroxide
hidróxido de hafnio hafnium hydroxide
hidróxido de hierro iron hydroxide
hidróxido de indio indium hydroxide
hidróxido de itrio yttrium hydroxide
hidróxido de litio lithium hydroxide
hidróxido de manganeso manganese hydroxide
hidróxido de molibdeno molybdenum hydroxide
hidróxido de neodimio neodymium hydroxide
hidróxido de oro gold hydroxide
hidróxido de paladio palladium hydroxide

hidróxido de plomo lead hydroxide
hidróxido de rodio rhodium hydroxide
hidróxido de rubidio rubidium hydroxide
hidróxido de rutenio ruthenium hydroxide
hidróxido de talio thallium hydroxide
hidróxido de telurio tellurium hydroxide
hidróxido de terbio terbium hydroxide
hidróxido de tetraetanolamonio tetraethanolammonium hydroxide
hidróxido de titanio titanium hydroxide
hidróxido de torio thorium hydroxide
hidróxido de trifenilestaño triphenyltin hydroxide
hidróxido de uranilo uranyl hydroxide
hidróxido de uranio uranium hydroxide
hidróxido de vanadio vanadium hydroxide
hidróxido de zirconilo zirconyl hydroxide
hidróxido de zirconio zirconium hydroxide
hidróxido estánnico stannic hydroxide
hidróxido estannoso stannous hydroxide
hidroxido fenilmercúrico phenylmercuric hydroxide
hidróxido férrico ferric hydroxide
hidróxido ferroso ferrous hydroxide
hidróxido gálico gallic hydroxide
hidróxido magnésico magnesium hydroxide
hidróxido mangánico manganic hydroxide
hidróxido manganoso manganous hydroxide
hidróxido mercúrico mercuric hydroxide
hidróxido niquélico nickelic hydroxide
hidróxido niqueloso nickelous hydroxide
hidróxido potásico potassium hydroxide
hidróxido samárico samaric hydroxide
hidróxido sódico sodium hydroxide
hidróxido talioso thallous hydroxide
hidróxido titánico titanic hydroxide
hidróxido vanadioso vanadous hydroxide
hidroxietilacetamida hydroxyethylacetamide
hidroxietilamina hydroxyethylamine
hidroxietilcelulosa hydroxyethylcellulose

hidroxietilcelulosa de etilo ethyl
hydroxyethylcellulose
hidroxietiletilendiamina
hydroxyethylethylenediamine
hidroxietilhidrazina
hydroxyethylhydrazine
hidroxifencano hydroxyfenchane
hidroxifenilalanina
hydroxyphenylalanine
hidroxifenilo hydroxyphenyl
hidroxifenol hydroxyphenol
hidroxihidrazida hydroxyhydrazide
hidroxiimino hydroxyimino
hidroxilación hydroxylation
hidroxilación de Woodward Woodward
hydroxylation
hidroxilamina hydroxylamine
hidroxilo hydroxyl
hidroximercuricresol
hydroxymercuricresol
hidroximesitileno hydroxymesitylene
hidroximetano hydroxymethane
hidroximetileno hydroxymethylene
hidroximetilfuraldehído
hydroxymethylfuraldehyde
hidroximetilo hydroxymethyl
hidroxinaftaleno hydroxynaphthalene
hidroxinaftoquinona
hydroxynaphthoquinone
hidroxipiperidina hydroxypiperidine
hidroxiprogesterona
hydroxyprogesterone
hidroxiprolina hydroxyproline
hidroxipropanona hydroxypropanone
hidroxipropilamina hydroxypropylamine
hidroxipropilcelulosa hydroxypropyl
cellulose
hidroxipropilmetilcelulosa
hydroxypropyl methylcellulose
hidroxipropionitrilo
hydroxypropionitrile
hidroxiquinolina hydroxyquinoline
hidroxitolueno hydroxytoluene
hidroxitolueno butilado butylated
hydroxytoluene
hidroxitriacontano hydroxytriacontane
hidroxitriptamina hydroxytryptamine
hidroxo hydroxo
hidroxonio hydroxonium

hidruro hydride
hidruro alumínico aluminum hydride
hidruro alumínico cálcico calcium
aluminum hydride
hidruro alumínico de litio lithium
aluminum hydride
hidruro alumínico sódico sodium
aluminum hydride
hidruro antimonioso antimonous hydride
hidruro arsenioso arsenous hydride
hidruro bárico barium hydride
hidruro cálcico calcium hydride
hidruro cálcico alumínico aluminum
calcium hydride
hidruro de amilo amyl hydride
hidruro de antimonio antimony hydride
hidruro de arsénico arsenic hydride
hidruro de boro boron hydride
hidruro de bromo bromine hydride
hidruro de carbonilo carbonyl hydride
hidruro de cinamilo cinnamyl hydride
hidruro de decilo decyl hydride
hidruro de estaño tin hydride
hidruro de etileno ethylene hydride
hidruro de flúor fluorine hydride
hidruro de formacilo formazyl hydride
hidruro de formoxilo formoxyl hydride
hidruro de fósforo phosphorus hydride
hidruro de galio gallium hydride
hidruro de germanio germanium
hydride
hidruro de litio lithium hydride
hidruro de litio alumínico aluminum
lithium hydride
hidruro de metilo methyl hydride
hidruro de nitrógeno nitrogen hydride
hidruro de nonilo nonyl hydride
hidruro de paladio palladium hydride
hidruro de pentilo pentyl hydride
hidruro de rubidio rubidium hydride
hidruro de silicio silicon hydride
hidruro de titanio titanium hydride
hidruro de uranio uranium hydride
hidruro de zirconio zirconium hydride
hidruro magnésico magnesium hydride
hidruro potásico potassium hydride
hidruro sódico sodium hydride
hielmita hielmite
hielo ice

hielo seco dry ice
hierro iron
hierro amoniacal ammoniated iron
hierro beta beta iron
hierro de lingote ingot iron
hierro delta delta iron
hierro electrolítico electrolytic iron
hierro en lingotes pig iron
hierro fundido cast iron
hierro galvanizado galvanized iron
hierro gamma gamma iron
hierro pasivo passive iron
higrométrico hygrometric
higrómetro hygrometer
higromicina hygromycin
higroscópico hygroscopic
hiosciamina hyoscyamine
hioscina hyoscine
hiper hyper
hiperconjugación hyperconjugation
hiperconjugación isovalente isovalent
   hyperconjugation
hipercrómico hyperchromic
hiperón hyperon
hiperóxido hyperoxide
hipersorción hypersorption
hipertónico hypertonic
hipo hypo
hipoalergénico hypoallergenic
hipobromato hypobromate
hipobromito hypobromite
hipobromito sódico sodium hypobromite
hipoclorito hypochlorite
hipoclorito cálcico calcium hypochlorite
hipoclorito de litio lithium hypochlorite
hipoclorito de plata silver hypochlorite
hipoclorito potásico potassium
   hypochlorite
hipoclorito sódico sodium hypochlorite
hipocrómico hypochromic
hipofamina hypophamine
hipofluorita hypofluorite
hipofosfato hypophosphate
hipofosfato amónico ammonium
   hypophosphate
hipofosfato bárico barium
   hypophosphate
hipofosfato de plata silver
   hypophosphate

hipofosfito hypophosphite
hipofosfito bárico barium hypophosphite
hipofosfito cálcico calcium
   hypophosphite
hipofosfito de cinc zinc hypophosphite
hipofosfito de manganeso manganese
   hypophosphite
hipofosfito férrico ferric hypophosphite
hipofosfito ferroso ferrous
   hypophosphite
hipofosfito potásico potassium
   hypophosphite
hipofosfito sódico sodium hypophosphite
hiponitrilo hyponitrile
hiponitrito sódico sodium hyponitrite
hiposulfato hyposulfate
hiposulfato bárico barium hyposulfate
hiposulfato ferroso ferrous hyposulfate
hiposulfato potásico potassium
   hyposulfate
hiposulfato sódico sodium hyposulfate
hiposulfito hyposulfite
hiposulfito bárico barium hyposulfite
hiposulfito de estroncio strontium
   hyposulfite
hiposulfito de plomo lead hyposulfite
hiposulfito de plomo sódico sodium lead
   hyposulfite
hiposulfito magnésico magnesium
   hyposulfite
hiposulfito potásico potassium
   hyposulfite
hiposulfito sódico sodium hyposulfite
hipótesis hypothesis
hipótesis de Avogadro Avogadro
   hypothesis
hipotónico hypotonic
hipoxantina hypoxanthine
hipoyodato hypoiodate
hipurato hippurate
hipurato potásico potassium hippurate
hipuroilo hippuroyl
histamina histamine
histaminasa histaminase
histéresis hysteresis
histidina histidine
histona histone
histoquímica histochemistry
hogbomita hogbomite

hoja leaf, foil
hoja de oro gold foil, gold leaf
hoja metálica metal foil
hojalata tin, tinplate
hojuela foil
holmio holmium
holocelulosa holocellulose
holografía holography
holohedral holohedral
hollín soot
homatropina homatropine
homilita homilite
homo homo
homocéntrico homocentric
homocíclico homocyclic
homocrómico homochromic
homogéneo homogeneous
homogenización homogenization
homoisohídrico homoisohydric
homoléptico homoleptic
homólisis homolysis
homomorfo homomorph
homonuclear homonuclear
homopolar homopolar
homopolímero homopolymer
homosalato homosalate
homótopo homotope
hongo fungus
hopcalita hopcalite
hormona hormone

hormona corticoide corticoid hormone
hormona tirotrópica thyrotropic
hormone
hornablenda hornblende
horno oven, furnace, kiln
horno de combustión combustion
furnace
horno de coque coke oven
horno de Fletcher Fletcher furnace
horno eléctrico electric furnace
horno solar solar furnace
horsfordita horsfordite
howlita howlite
hubnerita hübnerite
hueco hole
hulla coal
humeante fuming
humectante humectant
humedad humidity, moisture
humedad absoluta absolute humidity
humedad atmosférica atmospheric
humidity
humedad crítica critical humidity
humedad relativa relative humidity
humero flue
humidificación humidification
humidímetro humidity meter
humo smoke, fume
humo químico chemical smoke
humus humus

I

ibogaína ibogaine
ibuprofeno ibuprofen
ico ic
icosa icosa
icosaedro icosahedron
icosano icosane
icosanol icosanol
icosilo icosyl
ictamol ichthammol
ictiocola ichthyocolla
ictiolato ichthyolate
ideno idene
idino idyne
idiocromático idiochromatic
ido ide
idosa idose
ignición ignition
ignición espontánea spontaneous ignition
ignífugo flame-resistant
ilideno ylidene
ilidino ylidyne
ilinio illinium
ilmenita ilmenite
ilo yl
iluro ylide
ilvaíta ilvaite
imagen de espejo mirror-image
imagen electrónica electron image
imazalil imazalil
imbibición imbibition
imida imide
imida de alcanfor camphor imide
imidazol imidazole
imidazolidona imidazolidone
imidazolilo imidazolyl
imidazolio imidazolium
imidazoltriona imidazoletrione
imidina imidine
imido imido
imina imine
iminio iminium
imino imino
iminoácido imino acid

iminobispropilamina
  iminobispropylamine
iminodiacetonitrilo iminodiacetonitrile
iminourea iminourea
imodoilo imidoyl
impacto impact
impacto electrónico electron impact
impalpable impalpable
impedimento estérico steric hindrance
impermeable waterproof
implosión implosion
impregnación impregnation, permeation
impregnar impregnate
impulso pulse
impulsor impeller
impureza impurity
in vitro in vitro
in vivo in vivo
ina ine
inactivo inactive
incandescencia incandescence
incandescente incandescent
incendiario incendiary
incidencia incidence
incineración incineration, ashing
incombustible flameproof
incompatibilidad química chemical
  incompatibility
incompatible incompatible
incompresible incompressible
incrustación de caldera boiler
  incrustation
incubación incubation
indamina indamine
indandiona indandione
indanilo indanyl
indano indan
indantreno indanthrene
indantrona indanthrone
indazol indazole
indenilo indenyl
indeno indene
indenona indenone

414

independencia de carga charge independence
indeterminado indeterminate
indicador indicator
indicador ácido-base acid-base indicator
indicador acromático achromatic indicator
indicador de absorción absorption indicator
indicador de adsorción adsorption indicator
indicador de color color indicator
indicador de Degener Degener indicator
indicador de fenantrolina phenanthroline indicator
indicador de fluorescencia fluorescence indicator
indicador de Formanek Formanek indicator
indicador de humedad humidity indicator
indicador de ion hidrógeno hydrogen-ion indicator
indicador de oxidación-reducción oxidation-reduction indicator
indicador de pH pH indicator
indicador de precipitación precipitation indicator
indicador de reacción reaction indicator
indicador de Sörensen Sörensen indicator
indicador isotópico isotopic indicator
indicador metalocrómico metallochromic indicator
indicador mixto mixed indicator
indicador químico chemical indicator
indicador radiactivo radioactive indicator, tracer
indicador redox redox indicator
indicador universal universal indicator
indicador visual display
indicadores de Clark-Lubs Clark-Lubs indicators
indicán indican
índice index
índice amortiguador buffer index
índice de actividad óptica optical-activity index
índice de Bellier Bellier index

índice de excitación excitation index
índice de octano octane rating, octane number
índice de refracción refractive index
índice de retención retention index
índice de viscosidad viscosity index
índice térmico heat index
índigo indigo
indigotina indigotin
indilo indyl
indio indium
indofenol indophenol
indógeno indogen
indogenuro indogenide
indol indole
indolilo indolyl
indolizina indolizine
indolol indolol
indometacina indomethacin
indoxilo indoxyl
inducción induction
inducción asimétrica asymmetric induction
inducción fotoquímica photochemical induction
inducido por neutrones neutron-induced
inducido por radiación radiation-induced
indurita indurite
industrial industrial
inelástico inelastic
inerte inert
inestable unstable
inflamabilidad inflammability, flammability
inflamable inflammable, flammable
inflamador inflamer
inflamar inflame
infrarrojo infrared
infrarrojo con transformación de Fourier Fourier transform infrared
infrarrojo de cromatografía de gases gas chromatography infrared
infusión infusion
ingeniería de seguridad safety engineering
ingeniería química chemical engineering
inhibición inhibition
inhibición no competitiva

noncompetitive inhibition
inhibidor inhibitor
inhibidor de colinesterasa cholinesterase
   inhibitor
inhibidor de reacción reaction inhibitor
inhibidor químico chemical inhibitor
iniciador initiator
inicio de cadena chain initiation
injerto grafting
inmersión immersion
inmiscible immiscible
inmunoglobulina immunoglobulin
inmunoquímica immunochemistry
ino yne, ino
inorgánico inorganic
inosamina inosamine
inosina inosine
inosinato sódico sodium inosinate
inositol inositol
insaturación unsaturation
insaturado unsaturated
insecticida insecticide
insolubilidad insolubility
insoluble insoluble
instrumentación instrumentation
instrumento instrument
instrumento espectrométrico
   spectrometric instrument
insulina insulin
intensidad intensity, strength
intensidad de campo field intensity
intensidad de olor odor intensity
intensidad de radiación radiation
   intensity
intensidad térmica thermal intensity
interacción de Gamow-Teller Gamow-
   Teller interaction
interacción espín-espín spin-spin
   interaction
interacción espín-orbital spin-orbit
   interaction
interacción nuclear nuclear interaction
interatómico interatomic
intercambiador térmico heat exchanger
intercambio exchange
intercambio de cationes cation exchange
intercambio electrónico electron
   exchange
intercambio iónico ion exchange

intercambio químico chemical exchange
intercambio térmico heat exchange
interdifusión interdiffusion
interelectródico interelectrode
interfacial interfacial
interfase interface
interferencia interference
interferón interferon
interhalógeno interhalogen
interiónico interionic
intermetálico intermetallic
intermolecular intermolecular
interno internal
intersticial interstitial
intersticio interstice
intervalo amortiguador buffer range
intervalo de indicador indicator range
intervalo de puntos de congelación
   freezing-point range
intervalo de puntos de ebullición
   boiling-point range
intervalo de transición transition
   interval
intoxicación intoxication
intraanular intraannular
intraatómico intraatomic
intramolecular intramolecular
intranuclear intranuclear
intrínseco intrinsic
introfacción introfaction
intrón intron
inulina inulin
inulinasa inulinase
inundación química chemical flooding
invención invention
inversión inversion, reversal
inversión de fase phase reversal
inversión de Walden Walden inversion
invertasa invertase
investigación research
investigación aplicada applied research
investigación fundamental fundamental
   research
investigación química chemical research
investigación y desarrollo research and
   development
iolita iolite
ion ion
ion amónico ammonium ion

ion carbenio carbenium ion
ion carbonio carbonium ion
ion cloronio chloronium ion
ion complejo complex ion
ion dipolar dipolar ion
ion excitado excited ion
ion-gramo gram-ion
ion hidrógeno hydrogen ion
ion hidronio hydronium ion
ion hidroxilo hydroxyl ion
ion libre free ion
ion metaestable metastable ion
ion negativo negative ion
ion oxonio oxonium ion
ion positivo positive ion
ion primario primary ion
iónico ionic
ionización ionization
ionización de campo field ionization
ionización de gas gas ionization
ionización de Penning Penning ionization
ionización electrolítica electrolytic ionization
ionización por colisión collision ionization
ionización por radiación radiation ionization
ionización primaria primary ionization
ionización proporcional proportional ionization
ionización térmica thermal ionization
ionizado ionized
ionizante ionizing
ionizar ionize
ionogénico ionogenic
ionógeno ionogen
ionografía ionography
ionol ionol
ionómero ionomer
ionona ionone
iotalamato sódico sodium iothalamate
iperita yperite
iprodiona iprodione
iridio iridium
irido irido
iridosmina iridosmine
irona irone
irradiación irradiation

irradiante irradiating
irreversible irreversible
isatina isatin
isazol isazol
iso iso
isoamildicloroarsina isoamyldichloroarsine
isoamileno isoamylene
isoamilmercaptano isoamyl mercaptan
isoamilo isoamyl
isoascorbato sódico sodium isoascorbate
isóbaro isobar
isóbaro estable stable isobar
isoborneol isoborneol
isobornilo isobornyl
isobutano isobutane
isobutanol isobutanol
isobutanolamina isobutanolamine
isobuteno isobutene
isobutilamina isobutylamine
isobutilaminobenzoato isobutylaminobenzoate
isobutilbenceno isobutylbenzene
isobutilcarbinol isobutyl carbinol
isobutileno isobutylene
isobutilmercaptano isobutyl mercaptan
isobutilo isobutyl
isobutilxantato sódico sodium isobutylxanthate
isobutiraldehído isobutyraldehyde
isobutirato cálcico calcium isobutyrate
isobutirato de etilo ethyl isobutyrate
isobutirato de fenetilo phenethyl isobutyrate
isobutirato de isobutilo isobutyl isobutyrate
isobutirato de linalilo linalyl isobutyrate
isobutirato de tolilo tolyl isobutyrate
isobutírico isobutyric
isobutirilo isobutyryl
isobutiroilo isobutyroyl
isobutironitrilo isobutyronitrile
isobutoxi isobutoxy
isocianato isocyanate
isocianato de alilo allyl isocyanate
isocianato de ciclohexilo cyclohexyl isocyanate
isocianato de clorofenilo chlorophenyl isocyanate

isocianato de etilo ethyl isocyanate
isocianato de fenilo phenyl isocyanate
isocianato de metilo methyl isocyanate
isocianato de octadecilo octadecyl isocyanate
isociano isocyano
isocianurato isocyanurate
isocianuro isocyanide
isocianuro de etilo ethyl isocyanide
isocianuro de fenilo phenyl isocyanide
isocianuro de pentilo pentyl isocyanide
isocíclico isocyclic
isócora isochore
isócora de van't Hoff van't Hoff isochore
isodecaldehído isodecaldehyde
isodecano isodecane
isodecanol isodecanol
isodecilo isodecyl
isodinámico isodynamic
isodipersión isodispersion
isodrina isodrin
isodureno isodurene
isoeléctrico isoelectric
isoelectrónico isoelectronic
isoenzima isoenzyme
isoestructural isostructural
isoeugenol isoeugenol
isoforona isophorone
isoftalato de dialilo diallyl isophthalate
isoftalato de etilo ethyl isophthalate
isoheptano isoheptane
isohexano isohexane
isoléptico isoleptic
isoleucina isoleucine
isólogo isolog
isomerasa isomerase
isomerasa de glucosa glucose isomerase
isomérico isomeric
isomerismo isomerism
isomerismo cis-trans cis-trans isomerism
isomerismo de anillo ring isomerism
isomerismo de enlace doble double-bond isomerism
isomerismo de saturación saturation isomerism
isomerismo dinámico dynamic isomerism
isomerismo estructural structural isomerism
isomerismo geométrico geometrical isomerism
isomerismo nuclear nuclear isomerism
isomerismo por coordinación coordination isomerism
isomerización isomerization
isomerización de cadena chain isomerization
isomerización de valencia valence isomerization
isómero isomer
isómero cis cis-isomer
isómero geométrico geometric isomer
isómero nuclear nuclear isomer
isómero óptico optical isomer
isómero trans trans-isomer
isométrico isometric
isomolécula isomolecule
isomórfico isomorphic
isomorfismo isomorphism
isomorfo isomorph
isonicotinato de metilo methyl isonicotinate
isonitrilo isonitrile
isonitro isonitro
isonitroso isonitroso
isonitrosocetona isonitrosoketone
isooctano isooctane
isoocteno isooctene
isoparafina isoparaffin
isopentaldehído isopentaldehyde
isopentano isopentane
isopoliácido isopolyacid
isopoliéster isopolyester
isopolimolibdato isopolymolybdate
isopolimorfismo isopolymorphism
isopolitungstato isopolytungstate
isopreno isoprene
isoprenoide isoprenoid
isopropanol isopropanol
isopropanolamina isopropanolamine
isopropilacetona isopropylacetone
isopropilamina isopropylamine
isopropilaminoetanol isopropylaminoethanol
isopropilanilina isopropylaniline
isopropilato alumínico aluminum isopropylate

isopropilato de titanio titanium isopropylate

isopropilbenceno isopropylbenzene

isopropilbencilo isopropylbenzyl

isopropilcarbinol isopropylcarbinol

isopropilcresol isopropylcresol

isopropiletileno isopropylethylene

isopropilfenol isopropylphenol

isopropilideno isopropylidene

isopropilmercaptano isopropyl mercaptan

isopropilmetilbenceno isopropylmethylbenzene

isopropilnaftaleno isopropylnaphthalene

isopropilo isopropyl

isopropiltolueno isopropyltoluene

isopropóxido alumínico aluminum isopropoxide

isopropoxietanol isopropoxyethanol

isopulegol isopulegol

isoquinolina isoquinoline

isosafrol isosafrole

isostérico isosteric

isosterismo isosterism

isosulfocianato de alilo allyl isosulfocyanate

isotáctico isotactic

isoterma isotherm

isoterma de adsorción de Gibbs Gibbs adsorption isotherm

isoterma de adsorción de Langmuir Langmuir adsorption isotherm

isoterma de Freundlich Freundlich isotherm

isoterma de van't Hoff van't Hoff isotherm

isotiocianato isothiocyanate

isotiocianato de etoxicarbonilo ethoxycarbonyl isothiocyanate

isotiocianato de fenilo phenyl isothiocyanate

isotiocianato de metilo methyl isothiocyanate

isotiocianato de propenilo propenyl isothiocyanate

isotiocinato de alilo allyl isothiocyanate

isotónico isotonic

isótono isotone

isotópico isotopic

isótopo isotope

isótopo estable stable isotope

isótopo inestable unstable isotope

isótopo pesado heavy isotope

isótopo radiactivo radioactive isotope

isotrón isotron

isotrópico isotropic

isovalente isovalent

isovaleraldehído isovaleraldehyde

isovalerato de bornilo bornyl isovalerate

isovalerato de etilo ethyl isovalerate

isovalerato de isoamilo isoamyl isovalerate

isovalerato sódico sodium isovalerate

isoxazolilo isoxazolyl

iterbia ytterbia

iterbio ytterbium

itria yttria

itrialita yttrialite

itrio yttrium

itrocerita yttrocerite

itrotantalita yttrotantalite

iuro ide

J

jabón soap
jabón de alquitrán tar soap
jabón metálico metallic soap
jaborina jaborine
jade jade
jadeíta jadeite
jalea jelly
jalea mineral mineral jelly
jarabe syrup

jarabe de almidón starch syrup
jarabe de cacao cacao syrup
jarabe de maíz corn syrup
jardín químico chemical garden
jaspe jasper
jazmona jasmone
johimbina johimbine
joseíta joseite
jugo gástrico gastric juice

K

kanamicina kanamycin
kappa kappa
kernita kernite
kerógeno kerogen
kieselguhr kieselguhr
kieserita kieserite
kilogramo kilogram

kimberlita kimberlite
kleinita kleinite
koppita koppite
kovar kovar
kremersita kremersite
krugita krügite
kunzita kunzite

L

lábil labile
laboratorio laboratory
laboratorio de investigación research
  laboratory
labradorita labradorite
laca lacquer, lake
laca amarilla yellow lake
laca de tungsteno tungsten lake
lacasa laccase
lacatato de plata silver lactate
lacmoide lacmoid
lacrimógeno lachrymatory
lactalbúmina lactalbumin
lactama lactam
lactamida lactamide
lactasa lactase
lactato lactate
lactato alumínico aluminum lactate
lactato amónico ammonium lactate
lactato bárico barium lactate
lactato cálcico calcium lactate
lactato ceroso cerous lactate
lactato de antimonio antimony lactate
lactato de butilo butyl lactate
lactato de cadmio cadmium lactate
lactato de cinc zinc lactate
lactato de cobre copper lactate
lactato de estroncio strontium lactate
lactato de etilo ethyl lactate
lactato de litio lithium lactate
lactato de manganeso manganese lactate
lactato de metilo methyl lactate
lactato de plomo lead lactate
lactato de zirconio zirconium lactate
lactato de zirconio sódico sodium
  zirconium lactate
lactato fenilmercúrico phenylmercuric
  lactate
lactato ferroso ferrous lactate
lactato magnésico magnesium lactate
lactato mercúrico mercuric lactate
lactato sódico sodium lactate
lactida lactide

lactilo lactyl
lactima lactim
lactobiosa lactobiose
lactogénico lactogenic
lactoglobulina lactoglobulin
lactoilo lactoyl
lactol lactol
lactolida lactolide
lactona lactone
lactona beta beta lactone
lactona de angélica angelica lactone
lactona de nonilo nonyl lactone
lactonitrilo lactonitrile
lactopreno lactoprene
lactosa lactose
lagoriolita lagoriolite
lambda lambda
lambert lambert
lámina beta beta sheet
laminado laminated
laminarina laminarin
lámpara de halógeno halogen lamp
lámpara de neón neon lamp
lámpara de sodio sodium lamp
lana wool
lana de vidrio glass wool
lanarkita lanarkite
landsbergita landsbergite
langbeinita langbeinite
lanolina lanolin
lanosterol lanosterol
lansfordita lansfordite
lantana lanthana
lantánido lanthanide
lantano lanthanum
lantionina lanthionine
lapislázuli lapis lazuli
larvicida larvicide
láser laser
laterita laterite
látex latex
latón brass
latón amarillo yellow brass

latón de aluminio aluminum brass
latón duro hard brass
latón rojo red brass
laudanidina laudanidine
laudanina laudanine
láudano laudanum
laudanosina laudanosine
laumonita laumonite
lauraldehído lauraldehyde
laurato amónico ammonium laurate
laurato de butoxietilo butoxyethyl
    laurate
laurato de cinc zinc laurate
laurato de diglicol diglycol laurate
laurato de isocetilo isocetyl laurate
laurato de metilo methyl laurate
laurato de plata silver laurate
laurato de plomo lead laurate
laurato de tetrahidrofurfurilo
    tetrahydrofurfuryl laurate
laurato magnésico magnesium laurate
laurato potásico potassium laurate
laurencio lawrencium
laurilacetato lauryl acetate
laurilbromuro lauryl bromide
laurilcloruro lauryl chloride
laurilmercaptano lauryl mercaptan
laurilmetacrilato lauryl methacrylate
laurilo lauryl
laurilsulfato lauryl sulfate
laurilsulfato de trietanolamina
    triethanolamine lauryl sulfate
laurilsulfato magnésico magnesium
    lauryl sulfate
laurilsulfato sódico sodium lauryl sulfate
laurita laurite
lauroilo lauroyl
lauroilsarcosina lauroylsarcosine
lauroleato de metilo methyl lauroleate
laurona laurone
lautal lautal
lavado washing
lavado cáustico caustic wash
lavador de ojos eye-wash
lazulita lazulite
lazurita lazurite
lecitina lecithin
lectina lectin
lechada de cal whitewash

leche de magnesia milk of magnesia
lejía lye
lenacil lenacil
leno lene
leonardita leonardite
leonita leonite
lepidina lepidine
lepidolita lepidolite
lepidona lepidone
leptón lepton
letal lethal
leucilo leucyl
leucina leucine
leucita leucite
leuco leuco
leucoanilina leucoaniline
leucolina leucoline
leuconita leukonite
leucovorina leucovorin
levadura yeast
levadura de cerveza brewer's yeast
levadura de tórula torula yeast
levo levo
levorrotatorio levorotatory
levulosa levulose
lewisita lewisite
ley cero de termodinámica zeroth law of
    thermodynamics
ley de acción de masas mass action law
ley de alternación alternation law
ley de Avogadro Avogadro law
ley de Babo Babo law
ley de Beer Beer law
ley de Beer-Lambert Beer-Lambert law
ley de Blagden Blagden law
ley de Boltzmann Boltzmann law
ley de Bouguer-Lambert Bouguer-
    Lambert law
ley de Boyle Boyle law
ley de Bragg Bragg law
ley de Bunsen-Kirchhoff Bunsen-
    Kirchhoff law
ley de Clausius Clausius law
ley de Clausius-Mosotti Clausius-
    Mosotti law
ley de Coehn Coehn law
ley de composición definida definite
    composition law
ley de conservación conservation law

ley de conservación de masa mass conservation law
ley de Coppet Coppet law
ley de Curie Curie law
ley de Curie-Weiss Curie-Weiss law
ley de Charles Charles law
ley de Dalton Dalton law
ley de desplazamiento displacement law
ley de desplazamiento de Soddy Soddy displacement law
ley de desplazamiento radiactivo radioactive displacement law
ley de Despretz Despretz law
ley de difusión diffusion law
ley de difusión de Fick Fick diffusion law
ley de dilución dilution law
ley de dilución de Ostwald Ostwald dilution law
ley de distribución distribution law
ley de Draper Draper law
ley de Einstein Einstein law
ley de equilibrio equilibrium law
ley de equivalente fotoquímico photochemical equivalent law
ley de esterificación esterification law
ley de Fajans-Soddy Fajans-Soddy law
ley de Faraday Faraday law
ley de Gay-Lussac Gay-Lussac law
ley de Goldschmidt Goldschmidt law
ley de Graham Graham law
ley de Grottius-Draper Grottius-Draper law
ley de Henry Henry law
ley de Hess Hess law
ley de Joule Joule law
ley de Kopp Kopp law
ley de Lambert Lambert law
ley de Lambert-Beer Lambert-Beer law
ley de Mendeleev Mendeleev law
ley de Meyer Meyer law
ley de Moseley Moseley law
ley de Nernst Nernst law
ley de Oudeman Oudeman law
ley de partición partition law
ley de Pascal Pascal law
ley de proporciones definidas definite proportions law
ley de proporciones múltiples multiple proportions law
ley de proporciones recíprocas reciprocal proportions law
ley de Ramsay-Young Ramsay-Young law
ley de Raoult Raoult law
ley de reacción reaction law
ley de Retger Retger law
ley de Richter Richter law
ley de Sievert Sievert law
ley de Stark-Einstein Stark-Einstein law
ley de Stokes Stokes law
ley de termodinámica thermodynamics law
ley de van't Hoff van't Hoff law
ley de Wullner Wullner law
ley periódica periodic law
leyes de gases gas laws
leyes de termodinámica laws of thermodynamics
leyes químicas chemical laws
liberación liberation, release
licopeno lycopene
licopodio lycopodium
licor liquor
licuación liquation
licuado liquefied
licuefacción liquefaction
lidocaína lidocaine
ligador binder
ligadura binding
ligando ligand
ligando cuadridentado quadridentate ligand
ligasa ligase
lignina lignin
lignita lignite
lignocaína lignocaine
lignocerato de metilo methyl lignocerate
lignosulfonato lignosulfonate
ligroína ligroin
límite de absorción absorption limit
límite de concentración concentration limit
límite de convergencia convergence limit
límite de identificación identification limit
límite de inflamabilidad flammability

limit
limoneno limonene
limonita limonite
limpieza de gas gas cleanup
limpieza electrolítica electrolytic
cleaning
linalilo linalyl
linalol linalool
linamarina linamarin
linarita linarite
lincomicina lincomycin
lindano lindane
lindgrenita lindgrenite
línea de absorción absorption line
línea de Becke Becke line
línea de hidrógeno hydrogen line
línea de isoactividad isoactivity line
línea de Rayleigh Rayleigh line
línea de Stokes Stokes line
línea espectral spectral line, spectrum
line
línea prohibida forbidden line
línea realzada enhanced line
lineal linear
líneas de Balmer Balmer lines
líneas de emisión emission lines
líneas de Fraunhofer Fraunhofer lines
líneas de Raman Raman lines
líneas de Schmidt Schmidt lines
linnaeíta linnaeite
linoleato linoleate
linoleato amónico ammonium linoleate
linoleato cobaltoso cobaltous linoleate
linoleato de cinc zinc linoleate
linoleato de cobalto cobalt linoleate
linoleato de manganeso manganese
linoleate
linoleato de metilo methyl linoleate
linoleato de plomo lead linoleate
linoleato potásico potassium linoleate
linolenato de metilo methyl linolenate
liofílico lyophilic
liofilización lyophilization, freeze drying
liofóbico lyophobic
liogel lyogel
lionita lionite
liotópico lyotopic
liótropo lyotrope
lipasa lipase

lípido lipid
lipofílico lipophilic
lipólisis lipolysis
lipoproteína lipoprotein
lipoxidasa lipoxidase
lipoxigenasa lipoxygenase
liquen islándico Iceland moss
líquido liquid
líquido polar polar liquid
líquido saturado saturated liquid
líquido subenfriado supercooled liquid
líquidos asociados associated liquids
liroconita liroconite
lisidina lysidine
lisina lysine
lisis lysis
lisozima lysozyme
litarge litharge
litidionita litidionite
litina lithia
litio lithium
litro liter, litre
livingstonita livingstoneite
lixiviación lixiviation, leaching
lixosa lyxose
lobelina lobeline
localización localization
loción blanca white lotion
lodo anódico anodic mud, anodic slime
longitud de onda de de Broglie de
Broglie wavelength
lorandita lorandite
lorazepam lorazepam
lote matriz master batch
lubricación lubrication
lubricante lubricant
luciferina luciferin
ludwigita ludwigite
lugar activo active site
lugar de unión binding site
luminancia luminance
luminífero luminiferous
luminiscencia luminescence, glow
luminiscencia de resonancia resonance
luminescence
luminiscencia persistente afterglow
luminiscencia residual afterglow
luminiscente luminescent
luminol luminol

luminometría luminometry
luminosidad luminosity
luminoso luminous
lupinidina lupinidine
lúpulo hops
lutecia lutetia
lutecio lutetium
luteína lutein
luteotropina luteotropin

lutidina lutidine
luz coherente coherent light
luz fría cold light
luz no polarizada unpolarized light
luz polarizada polarized light
luz polarizada en un plano plane-polarized light
luz química chemical light
luz ultravioleta ultraviolet light

LL

llama flame
llama de Bunsen Bunsen flame
llama oxiacetilénica oxyacetylene flame
llama oxidante oxidizing flame

llama reductora reducing flame
lluvia ácida acid rain
lluvia amarilla yellow rain

M

maclado twinning
macro macro
macroanálisis macroanalysis
macrocíclico macrocyclic
macrólido macrolide
macromolécula macromolecule
macromolecular macromolecular
macroquímica macrochemistry
macrosa macrose
macroscópico macroscopic
maduración ripening
maduración de Ostwald Ostwald
    ripening
magenta magenta
magnalio magnalium
magnalita magnalite
magnesia magnesia
magnesio magnesium
magnesioferrita magnesioferrite
magnesita magnesite
magnético magnetic
magnetismo magnetism
magnetita magnetite
magnetización magnetization
magnetohidrodinámica
    magnetohydrodynamics
magnetón magneton
magnetón de Bohr Bohr magneton
magnetón nuclear nuclear magneton
magnetoquímica magnetochemistry
magnolio magnolium
magnolita magnolite
malamida malamide
malaquita malachite
malatión malathion
malato malate
malato bárico barium malate
malato cálcico calcium malate
malato de cinc zinc malate
malato de dietilo diethyl malate
malato de etilo ethyl malate
malato de plomo lead malate
maldonita maldonite

maleabilidad malleability
maleable malleable
maleato maleate
maleato de dialilo diallyl maleate
maleato de dibutilo dibutyl maleate
maleato de dietilo diethyl maleate
maleato de feniramina pheniramine
    maleate
maleato de plomo lead maleate
malenoide malenoid
maleoilo maleoyl
maloilo maloyl
malonamida malonamide
malonato malonate
malonato bárico barium malonate
malonato de dimetilo dimethyl malonate
malonato de etilo ethyl malonate
malonilo malonyl
malonilurea malonyl urea
malononitrilo malononitrile
malquita malchite
malta malt
maltasa maltase
malteno malthene
maltol maltol
maltosa maltose
mallardita mallardite
maná manna
mandelonitrilo mandelonitrile
maneb maneb
manejo de materiales materials handling
manejo remoto remote handling
manganato manganate
manganato bárico barium manganate
manganato potásico potassium
    manganate
manganato sódico sodium manganate
manganeso manganese
manganeso-boro manganese-boron
manganeso rojo red manganese
manganeso-titanio manganese-titanium
mangánico manganic
manganina manganin

manganita manganite
manganoestilbita manganostilbite
manganosita manganosite
manganoso manganous
manita mannite
manitol mannitol
manitosa mannitose
mano de mortero pestle
manómetro manometer
manosa mannose
manteca lard
manteca de cacao cocoa butter, cacao butter
mantequilla butter
maquinaria de laboratorio laboratory machinery
marcado tagged, labeled
marcador tag, label
marcasita marcasite
margarina margarine
margarita margarite
marialita marialite
mariguana marijuana
mármol marble
martensita martensite
martita martite
martonita martonite
masa mass
masa activa active mass
masa atómica atomic mass
masa atómica relativa relative atomic mass
masa crítica critical mass
masa de electrón electron mass
masa de partícula particle mass
masa en reposo de electrón electron rest mass
masa en reposo de neutrón neutron rest mass
masa isotópica isotopic mass
masa mesónica meson mass
masa molecular molecular mass
masa molecular relativa relative molecular mass
masa nuclear nuclear mass
masa reducida reduced mass
mascagnita mascagnite
máser maser
masicote massicot

masilla putty
masticación mastication
mata matte
materia matter
materia colorante dyestuff
materia prima raw material
material material
material a prueba de fuego fireproof material
material ablativo ablative material
material antideflagrante flameproof material
material combustible combustible material
material corrosivo corrosive material
material de acabado finishing material
material de blanqueo bleaching material
material fértil fertile material
material incombustible flameproof material
material inflamable flammable material
material no combustible noncombustible material
material oxidante oxidizing material
material peligroso hazardous material
material pirofórico pyrophoric material
material radiactivo radioactive material
material reductor reducing material
materialización materialization
matildita matildite
matlockita matlockite
matraz flask
matraz de base plana flat-bottom flask
matraz de Claisen Claisen flask
matraz de cuello largo long-necked flask
matraz de destilación distillation flask
matraz de Erlenmeyer Erlenmeyer flask
matraz de filtración filter flask
matraz de Giles Giles flask
matraz de Kjeldahl Kjeldahl flask
matraz de medida measuring flask
matraz de tres cuellos three-necked flask
matraz de Wurtz Wurtz flask
matraz volumétrico volumetric flask
matriz matrix
maxivalencia maxivalence
meaformaldehído metaformaldehyde
mecánica cuántica quantum mechanics
mecánica estadística statistical

mechanics
mecanismo mechanism
mecanismo de reacción reaction
   mechanism
mecanismo de reacción orgánica
   organic reaction mechanism
mecanismo en cadena chain mechanism
mecanoquímica mechanochemistry
meconina meconin
mechero burner
mechero de Boyce Boyce burner
mechero de Bunsen Bunsen burner
mechero de Teclu Teclu burner
media reacción half-reaction
media vida half-life
media vida radiactiva radioactive half-
   life
medida measurement
medida cuantitativa quantitative
   measurement
medida electroquímica electrochemical
   measurement
medidor meter, gage
medidor de ion hidrógeno hydrogen-ion
   meter
medidor de pH pH meter
medio medium
medio de dispersión dispersion medium
medio para filtración filter medium
megnesilo magnesyl
meimacita meymacite
melamina melamine
melanilina melaniline
melanina melanin
melaza molasses
melfalano melphalan
melibiosa melibiose
melitato mellitate
melonita melonite
membrana membrane
membrana parcialmente permeable
   partially permeable membrane
membrana permeable permeable
   membrane
membrana semipermeable
   semipermeable membrane
mena ore
menacanita menachanite
menadiona menadione

mendelevio mendelevium
mendocita mendozite
meneghinita meneghinite
menisco meniscus
mensuración mensuration
menta verde spearmint
mentadieno menthadiene
mentalcanfor menthacamphor
mentano menthane
mentanodiamina menthanediamine
mentanol menthanol
mentanona menthanone
mentenilo menthenyl
menteno menthene
mentenol menthenol
mentilo menthyl
mentol menthol
mentona menthone
mentonafteno menthonaphthene
meprobamato meprobamate
mercaptal mercaptal
mercaptamina mercaptamine
mercaptano mercaptan
mercapto mercapto
mercaptoacetato sódico sodium
   mercaptoacetate
mercaptobenzotiazol
   mercaptobenzothiazole
mercaptobenzotiazol sódico sodium
   mercaptobenzothiazole
mercaptoetanol mercaptoethanol
mercaptofenilo mercaptophenyl
mercaptol mercaptol
mercaptotiazolina mercaptothiazoline
mercuración mercuration
mercurial mercurial
mercúrico mercuric
mercurificación mercurification
mercurio mercury
mercurio amoniacal ammoniated
   mercury
mercurioso mercurous
mercurización mercurization
mercururo mercuride
meridional meridional
merwinita merwinite
mescalina mescaline
mesilato de benztropina benztropine
   mesylate

mesilo mesyl
mesitileno mesitylene
mesitilo mesityl
mesitita mesitite
mesitol mesitol
meso meso
mesocoloide mesocolloid
mesoiónico mesoionic
mesomería mesomerism
mesomérico mesomeric
mesómero mesomer
mesomórfico mesomorphic
mesón meson
mesónico mesonic
mesoscópico mesoscopic
mesotomía mesotomy
mesotorio mesothorium
mesoxalilo mesoxalyl
mesoxalo mesoxalo
meta meta
metaarsenito de cinc zinc metaarsenite
metaarsenito potásico potassium metaarsenite
metabisulfito potásico potassium metabisulfite
metabisulfito sódico sodium metabisulfite
metabólico metabolic
metabolismo metabolism
metabolito metabolite
metaborato metaborate
metaborato de cobre copper metaborate
metaborato de litio lithium metaborate
metaborato sódico sodium metaborate
metacetina methacetin
metacrilamida methacrylamide
metacrilato methacrylate
metacrilato de estearilo stearyl methacrylate
metacrilato de etilo ethyl methacrylate
metacrilato de hexilo hexyl methacrylate
metacrilato de hidroxietilo hydroxyethyl methacrylate
metacrilato de isobutilo isobutyl methacrylate
metacrilato de metilo methyl methacrylate
metacrilato sódico sodium methacrylate
metacrilonitrilo methacrylonitrile

metacroleína methacrolein
metacromasia metachromasia
metacromatismo metachromatism
metadona methadone
metaestable metastable
metaestructura metastructure
metafenileno metaphenylene
metafosfato metaphosphate
metafosfato alumínico aluminum metaphosphate
metafosfato amónico ammonium metaphosphate
metafosfato bárico barium metaphosphate
metafosfato cálcico calcium metaphosphate
metafosfato de berilio beryllium metaphosphate
metafosfato potásico potassium metaphosphate
metafosfato sódico sodium metaphosphate
metal metal
metal alcalino alkali metal
metal alcalinotérreo alkaline-earth metal
metal blanco white metal
metal coloidal colloidal metal
metal de Aich Aich metal
metal de Babbitt Babbitt metal
metal de base base metal
metal de Borcher Borcher metal
metal de Devarda Devarda metal
metal de Frary Frary metal
metal de Lewis Lewis metal
metal de tierras raras rare-earth metal
metal ligero light metal
metal Muntz Muntz metal
metal nativo native metal
metal noble noble metal
metal pasivo passive metal
metal pesado heavy metal
metal precioso precious metal
metal radiactivo radioactive metal
metal raro rare metal
metal rojo red metal
metalación metalation
metaldehído metaldehyde
metálico metallic
metalificar metallify

metalilo methallyl
metalización metallization, metallizing, plating
metalización al vacío vacuum metalizing
metalización electrolítica electrolytic metallization
metalización galvánica galvanic metallization
metalizado metallized, metal-coated, plated
metalizar metallize
metaloceno metallocene
metalocrómico metallochromic
metalografía metallography
metaloide metalloid
metalorgánico metalorganic
metalurgia metallurgy
metamerismo metamerism
metámero metamer
metanación methanation
metanal methanal
metanamida methanamide
metanilato sódico sodium metanilate
metanloato sódico sodium methanolate
metano methane
metanoato methanoate
metanoato de etilo ethyl methanoate
metanoato sódico sodium methanoate
metanoilo methanoyl
metanol methanol
metanolato methanolate
metanosulfonato methanesulfonate
metanotiol methanethiol
metaperyodato sódico sodium metaperiodate
metasilicato bárico barium metasilicate
metasilicato cálcico calcium metasilicate
metasilicato de litio lithium metasilicate
metasilicato de plomo lead metasilicate
metasilicato sódico sodium metasilicate
metasomatosis metasomatosis
metástasis metastasis
metavanadato amónico ammonium metavanadate
metavanadato sódico sodium metavanadate
metenamina methenamine
metenilo methenyl
meteno methene

meteórico meteoric
metepa metepa
meticilina methicillin
metida methide
metilacetileno methylacetylene
metilacetofenona methylacetophenone
metilacetona methyl acetone
metilación de Haworth Haworth methylation
metilación de Purdie Purdie methylation
metilacrilamida methylacrylamide
metilado methylated
metilal methylal
metilalanina methylalanine
metilaluminio methylaluminum
metilamilcetona methylamyl ketone
metilamilo methylamyl
metilamina methylamine
metilamino methylamino
metilaminofenol methylaminophenol
metilanilina methylaniline
metilanisol methylanisole
metilantraceno methylanthracene
metilantranilato methylanthranilate
metilantraquinona methylanthraquinone
metilar methylate
metilato de litio lithium methylate
metilato magnésico magnesium methylate
metilato sódico sodium methylate
metilbenceno methylbenzene
metilbencilamina methylbenzylamine
metilbencilo methylbenzyl
metilbutano methylbutane
metilbutanol methylbutanol
metilbuteno methylbutene
metilbutilamina methylbutylamine
metilbutilcetona methylbutyl ketone
metilbutinol methylbutynol
metilcelulosa methylcellulose
metilciclohexanilo methylcyclohexanyl
metilciclohexano methylcyclohexane
metilciclohexanol methylcyclohexanol
metilciclohexanona methylcyclohexanone
metilciclohexilamina methylcyclohexylamine
metilciclopentano methylcyclopentane
metilcolantreno methylcholanthrene

metildicloroarsina methyldichloroarsine
metildiclorosilano methyldichlorosilane
metildietanolamina
    methyldiethanolamine
metildifenilamina methyldiphenylamine
metildioxolano methyldioxolane
metildipropilmetano
    methyldipropylmethane
metileno methylene
metilenociclopentadieno
    methylenecyclopentadiene
metilestireno methylstyrene
metiletilcelulosa methylethylcellulose
metiletilcetona methylethyl ketone
metilfenilcetona methylphenyl ketone
metilfenildiclorosilano
    methylphenyldichlorosilane
metilformanilida methylformanilide
metilfurano methylfuran
metilfurfurilamina methylfurfurylamine
metilglicol methyl glycol
metilheptano methylheptane
metilheptenona methylheptenone
metilheptilamina methylheptylamine
metilhexano methylhexane
metilhexilcetona methylhexyl ketone
metilhidrazina methylhydrazine
metilhidrazona methylhydrazone
metilhidroxibutanona
    methylhydroxybutanone
metílico methylic
metilidino methylidyne
metilindol methylindole
metilisoamilcetona methylisoamyl
    ketone
metilisobutilcarbinol methylisobutyl
    carbinol
metilisobutilcetona methylisobutyl
    ketone
metilmetano methylmethane
metilnaftaleno methylnaphthalene
metilnitrobenceno methylnitrobenzene
metilo methyl
metilol methylol
metilolurea methylolurea
metilparabeno methylparaben
metilpentadieno methylpentadiene
metilpentaldehído methylpentaldehyde
metilpentano methylpentane

metilpentanol methylpentanol
metilpenteno methylpentene
metilpentosa methylpentose
metilpiperazina methylpiperazine
metilpirrol methylpyrrole
metilpirrolidina methylpyrrolidine
metilpropano methylpropane
metilpropeno methylpropene
metilpropilcetona methylpropyl ketone
metilquinolina methylquinoline
metilsulfato bárico barium methylsulfate
metilsulfonilo methylsulfonyl
metiltaurina methyltaurine
metiltriclorosilano methyltrichlorosilane
metiltrinitrobenceno
    methyltrinitrobenzene
metilundecanoato methylundecanoate
metina methine
metino methyne
metiodal sódico sodium methiodal
metionilo methionyl
metionina methionine
método científico scientific method
método conductométrico conductometric
    method
método de acetilación de Barnett
    Barnett acetylation method
método de Boeseken Boeseken method
método de Born-Oppenheimer Born-
    Oppenheimer method
método de Bouveault-Blanc Bouveault-
    Blanc method
método de Bragg Bragg method
método de Carius Carius method
método de Craig Craig method
método de Debye-Scherrer Debye-
    Scherrer method
método de Dumas Dumas method
método de Frankland Frankland method
método de Freund Freund method
método de Helferich Helferich method
método de Kjeldahl Kjeldahl method
método de Kolbe Kolbe method
método de Körner Körner method
método de Mohr Mohr method
método de Rast Rast method
método de Regnault Regnault method
método de Sonn-Muller Sonn-Muller
    method

método de Van Slyke Van Slyke method
método de Volhard Volhard method
método de Ziegler Ziegler method
método gasométrico gasometric method
metomilo methomyl
metosa methose
metotrexato methotrexate
metoxalilo methoxalyl
metoxi methoxy
metoxiacetaldehído
  methoxyacetaldehyde
metoxiacetanilida methoxyacetanilide
metoxiacetofenona
  methoxyacetophenone
metoxiamina methoxyamine
metoxibenceno methoxybenzene
metoxibenzaldehído
  methoxybenzaldehyde
metoxibutanol methoxybutanol
metóxido methoxide
metóxido de benciltrimetilamonio
  benzyltrimethylammonium methoxide
metóxido de litio lithium methoxide
metóxido magnésico magnesium
  methoxide
metóxido potásico potassium methoxide
metóxido sódico sodium methoxide
metoxietilo methoxyethyl
metoxifenilo methoxyphenyl
metoxifenol methoxyphenol
metoxilo methoxyl
metoxinaftaleno methoxynaphthalene
metoxipropanol methoxypropanol
metoxipropilamina methoxypropylamine
metro meter, metre
mezcla mixture, blend
mezcla azeotrópica azeotropic mixture
mezcla binaria binary mixture
mezcla de ebullición constante constant-
  boiling mixture
mezcla de Eschka Eschka mixture
mezcla distéctica dystectic mixture
mezcla efervescente effervescent mixture
mezcla estequiométrica stoichiometric
  mixture
mezcla eutéctica eutectic mixture
mezcla frigorífica frigorific mixture
mezcla gaseosa gaseous mixture
mezcla racémica racemic mixture

mezcla rica rich mixture
mezclador de Banbury Banbury mixer
miargirita miargyrite
miazina miazine
mica mica, isinglass
micela micelle
micotoxina mycotoxin
micrilo micril
micro micro
microanálisis microanalysis
microanálisis con rayos X X-ray
  microanalysis
microanálisis cuantitativo quantitative
  microanalysis
microanalizador microanalyzer
microanalizador electrónico electron
  microanalyzer
microbalanza microbalance
microbicida microbicide
microbio microbe
microcápsula microcapsule
microcristal microcrystal
microcristalino microcrystalline
microcuerpo microbody
microcurie microcurie
microdensitómetro microdensitometer
microdifusión microdiffusion
microelectrodo microelectrode
microelectroforesis microelectrophoresis
microelectrólisis microelectrolysis
microelemento trace element
microencapsulación microencapsulation
microespectrofotómetro
  microspectrophotometer
microespectroscopio microspectroscope
microfotómetro microphotometer
microgalvanómetro microgalvanometer
microgamilo microgammil
micrografía micrography
microgramo microgram
microincineración microincineration
microlina microline
micrómetro micrometer
micrón micron
micronutriente micronutrient
microonda microwave
microorganismo microorganism
micropipeta micropipet
microporoso microporous

microquímica microchemistry
microquímico microchemical
microrreacción microreaction
microscopia microscopy
microscopia de fluorescencia
    fluorescence microscopy
microscopia electrónica electron
    microscopy
microscopia electrónica de barrido
    scanning electron microscopy
microscopia química chemical
    microscopy
microscópico microscopic
microscopio microscope
microscopio de emisión de campo field-
    emission microscope
microscopio de fuerza atómica atomic
    force microscope
microscopio de infrarrojo infrared
    microscope
microscopio electrónico electron
    microscope
microscopio electrónico de barrido
    scanning electron microscope
microscopio iónico de campo field-ion
    microscope
microscopio óptico optical microscope
microscopio protónico proton
    microscope
microsegundo microsecond
microsonda microprobe
microsonda electrónica electron
    microprobe
miel honey
miemita miemite
migración migration
migración atómica atomic migration
migración de plata silver migration
migración electroquímica
    electrochemical migration
migración iónica ion migration
mili milli
milibarra millibar
milicurie millicurie
miliequivalente milliequivalent
miligramo milligram
mililitro milliliter
milímetro millimeter
milímetro de mercurio millimeter of

mercury
milimol millimole
milinormal millinormal
mimetita mimetite
mineral mineral, ore
mineral anaranjado orange mineral
mineral de carbonato carbonate mineral
mineral de hierro iron ore
mineral de plata silver ore
mineral de sulfato sulfate mineral
mineral de titanio titanium ore
mineral de uranio uranium ore
mineral radiactivo radioactive mineral
mineralización mineralization
mineralogía mineralogy
minio minium
minivalencia minivalence
minulita minulite
mioglobina myoglobin
mioquinasa myokinase
miosina myosin
mirabilita mirabilite
mirceno myrcene
miricilo myricyl
miristato de butilo butyl myristate
miristato de etilo ethyl myristate
miristato de isocetilo isocetyl myristate
miristato de litio lithium myristate
miristato de metilo methyl myristate
miristato de plata silver myristate
miristato magnésico magnesium
    myristate
miristoilo myristoyl
mirra myrrh
miscibilidad miscibility
miscible miscible
mischmetal mischmetal
mispiquel mispickel
mitomicina mitomycin
mixina myxin
mixita mixite
mixo mixo
moción electrónica electron motion
modelo model
modelo atómico atomic model
modelo de Frenkel-Kontorowa Frenkel-
    Kontorowa model
modelo de Gouy-Chapman Gouy-
    Chapman model

modelo de Helmholtz Helmholtz model
modelo de Ising Ising model
modelo de Thomas-Fermi Thomas-Fermi model
modelo vectorial de estructura atómica vector model of atomic structure
moderador moderator
modificación modification
modificación de Eschweiler-Clarke Eschweiler-Clarke modification
modificación química chemical modification
modo mode
modo de desintegración decay mode
modo de desintegración nuclear nuclear-decay mode
módulo de elasticidad modulus of elasticity
moho mildew, mold, rust
mol mole, mol
molal molal
molalidad molality
molar molar
molaridad molarity
molécula molecule
molécula activada activated molecule
molécula compuesta compound molecule
molécula de van der Waals van der Waals molecule
molécula diatómica diatomic molecule
molécula elemental elementary molecule
molécula-gramo gram-molecule
molécula heteronuclear heteronuclear molecule
molécula homonuclear homonuclear molecule
molécula libre free molecule
molécula lineal linear molecule
molécula marcada tagged molecule
molécula monoatómica monoatomic molecule
molécula neutra neutral molecule
molécula no lineal nonlinear molecule
molécula no polar nonpolar molecule
molécula polar polar molecule
molécula poliatómica polyatomic molecule
molécula triatómica triatomic molecule
molecular molecular

molecularidad molecularity
moler grind
molibdato molybdate
molibdato amónico ammonium molybdate
molibdato bárico barium molybdate
molibdato cálcico calcium molybdate
molibdato de cinc zinc molybdate
molibdato de cobalto cobalt molybdate
molibdato de cobre copper molybdate
molibdato de estroncio strontium molybdate
molibdato de litio lithium molybdate
molibdato de plomo lead molybdate
molibdato magnésico magnesium molybdate
molibdato potásico potassium molybdate
molibdato sódico sodium molybdate
molibdenilo molybdenyl
molibdenita molybdenite
molibdeno molybdenum
molíbdico molybdic
molibdita molybdite
molibdofosfato sódico sodium molybdophosphate
molibdosilicato sódico sodium molybdosilicate
molido grinding
molienda química chemical milling
molino mill
molino coloidal colloid mill
molino de bolas ball mill
molisita molysite
molybdato de cadmio cadmium molybdate
momento angular nuclear nuclear angular momentum
momento angular orbital orbital angular momentum
momento cuadripolar nuclear nuclear quadripole moment
momento dipolar dipole moment
momento dipolar electrónico electron dipole moment
momento magnético magnetic moment
momento magnético atómico atomic magnetic moment
momento magnético de espín spin magnetic moment

momento magnético electrónico electron magnetic moment

momento magnético neutrónico neutron magnetic moment

momento magnético nuclear nuclear magnetic moment

momento nuclear nuclear moment

momento orbital orbital moment

monacita monazite

mónada monad

monetita monetite

monístico monistic

mono mono

monoacetato monoacetate

monoacetato de celulosa cellulose monoacetate

monoacetato de etilenglicol ethylene glycol monoacetate

monoacetato de glicerilo glyceryl monoacetate

monoacetato de glicerol glycerol monoacetate

monoacetato de glicol glycol monoacetate

monoacetato de resorcina resorcinol monoacetate

monoacetina monoacetin

monoácido monoacid

monoamina monoamine

monoaminoácido monoamino acid

monoamonio monoammonium

monoatómico monoatomic

monobásico monobasic

monobenzoato de resorcina resorcinol monobenzoate

monobromado monobromated

monobromo monobromo

monobromuro de antipireno antipyrene monobromide

monobromuro de hidrazina hydrazine monobromide

monobromuro de oro gold monobromide

monobromuro de selenio selenium monobromide

monobromuro de yodo iodine monobromide

monobromuro sulfúrico sulfuric monobromide

monocapa monolayer

monocarburo de uranio uranium monocarbide

monocloro monochloro

monocloroacetona monochloroacetone

monoclorobenceno monochlorobenzene

monocloroetano monochloroethane

monoclorofenol monochlorophenol

monoclorohidrina de glicerol glycerol monochlorohydrin

monoclorometano monochloromethane

monocloruro de azufre sulfur monochloride

monocloruro de hidrazina hydrazine monochloride

monocloruro de oro gold monochloride

monocloruro de selenio selenium monochloride

monocloruro de yodo iodine monochloride

monocloruro sulfúrico sulfuric monochloride

monocromático monochromatic

monodisperso monodisperse

monoestearato alumínico aluminum monostearate

monoestearato de glicerol glycerol monostearate

monoestearato de polioxietileno polyoxyethylene monostearate

monoestearato de propilenglicol propylene glycol monostearate

monoestearato de sucrosa sucrose monostearate

monoestearato de tetraetilenglicol tetraethylene glycol monostearate

monoéster monoester

monoetanolamina monoethanolamine

monoetilamina monoethylamine

monofluoruro de azufre sulfur monofluoride

monofluoruro de cloro chlorine monofluoride

monofosfato monophosphate

monofosfato bárico barium monophosphate

monofosfato de adenosina adenosine monophosphate

monofosfato de adenosina cíclico cyclic

adenosine monophosphate

monofosfato de guanosina guanosine
monophosphate

monofosfato de uridina uridine
monophosphate

monofosfato potásico potassium
monophosphate

monofuncional monofunctional

monoglicérido monoglyceride

monoglime monoglyme

monohidrato monohydrate

monohidrato bárico barium
monohydrate

monohidrato de carbonato sódico
sodium carbonate monohydrate

monohidrato de perborato sódico
sodium perborate monohydrate

monohidrato de sulfato de cinc zinc
sulfate monohydrate

monohidrato de sulfato sódico sodium
sulfate monohydrate

monohídrico monohydric

monolaurato de glicerilo glyceryl
monolaurate

monolaurato de glicerol glycerol
monolaurate

monómero monomer

monómero de estireno styrene monomer

monometilamina monomethylamine

monométrico monometric

monomolecular monomolecular

monomorfo monomorphous

mononitrilo malónico malonic
mononitrile

mononitrorresorcinato de plomo lead
mononitroresorcinate

mononuclear mononuclear

monooleato de glicerol glycerol
monooleate

monorrefringente monorefringent

monorricinoleato de propilenglicol
propylene glycol monoricinoleate

monosacárido monosaccharide

monosulfito de paladio palladium
monosulfite

monosulfuro monosulfide

monosulfuro bárico barium monosulfide

monosulfuro de carbono carbon
monosulfide

monosulfuro de estaño tin monosulfide

monosulfuro de estroncio strontium
monosulfide

monosulfuro de platino platinum
monosulfide

monosulfuro de plomo lead monosulfide

monosulfuro potásico potassium
monosulfide

monosulfuro sódico sodium monosulfide

monotartrato de etilo ethyl monotartrate

monoterpeno monoterpene

monotropía monotropy

monotrópico monotropic

monovalente monovalent

monoxicloro monoxychlor

monóxido monoxide

monóxido bárico barium monoxide

monóxido de azufre sulfur monoxide

monóxido de carbono carbon monoxide

monóxido de cloro chlorine monoxide

monóxido de cobalto cobalt monoxide

monóxido de cobre copper monoxide

monóxido de dipenteno dipentene
monoxide

monóxido de estroncio strontium
monoxide

monóxido de galio gallium monoxide

monóxido de germanio germanium
monoxide

monóxido de hierro iron monoxide

monóxido de manganeso manganese
monoxide

monóxido de nitrógeno nitrogen
monoxide

monóxido de osmio osmium monoxide

monóxido de paladio palladium
monoxide

monóxido de plomo lead monoxide

monóxido de silicio silicon monoxide

monóxido de talio thallium monoxide

monóxido de telurio tellurium monoxide

monóxido de titanio titanium monoxide

monóxido de vanadio vanadium
monoxide

monóxido de vinilciclohexeno
vinylcyclohexene monoxide

monóxido sódico sodium monoxide

monóxido talioso thallous monoxide

monoxima de benzoquinona
   benzoquinone monoxime
monoyoduro de oro gold monoiodide
montanita montanite
monticelita monticellite
montmorillonita montmorillonite
montroidita montroydite
mordiente mordant
morenosita morenosite
morfina morphine
morfolina morpholine
morfología morphology
morfosán morphosan
morina morin
mortero mortar
mostasa de nitrógeno nitrogen mustard
movilidad mobility
movimiento browniano Brownian
   motion
mu mu
mucopéptido mucopeptide
mucopolisacárido mucopolysaccharide
muestra sample, sampling

muestreo sampling
multi multi
multiplete multiplet
multiplicación electrónica electron
   multiplication
multiplicación neutrónica neutron
   multiplication
multiplicador electrónico electron
   multiplier
multiplicidad electrónica electron
   multiplicity
multivalente multivalent
mullita mullite
muon muon
murexida murexide
muscarina muscarine
muscovita muscovite
musgo de Irlanda Irish moss
mutamerismo mutamerism
mutámero mutamer
mutarrotación mutarotation
mutasa mutase
mutualidad de fases mutuality of phases

N

nabam nabam
nácar nacre
nacrita nacrite
nadorita nadorite
nafta naphtha
nafta de petróleo petroleum naphtha
naftaceno naphthacene
naftacridina naphthacridine
naftal naphthal
naftaldehído naphthaldehyde
naftalendiamina naphthalenediamine
naftalendiol naphthalenediol
naftalenilo naphthalenyl
naftaleno naphthalene
naftaleno clorado chlorinated
    naphthalene
naftalenosulfonato de plomo lead
    naphthalenesulfonate
naftalenosulfonato sódico sodium
    naphthalenesulfonate
naftalentiol naphthalenethiol
naftalida naphthalide
naftalimido naphthalimido
naftalina naphthalin
naftamida naphthamide
naftamina naphthamine
naftano naphthane
naftecetol naphthacetol
naftenato cálcico calcium naphthenate
naftenato cobaltoso cobaltous
    naphthenate
naftenato de cinc zinc naphthenate
naftenato de cobalto cobalt naphthenate
naftenato de cobre copper naphthenate
naftenato de cromo chromium
    naphthenate
naftenato de manganeso manganese
    naphthenate
naftenato de plomo lead naphthenate
naftenato de zirconio zirconium
    naphthenate
naftenato fenilmercúrico
    phenylmercuric naphthenate

naftenato ferroso ferrous naphthenate
naftenato potásico potassium
    naphthenate
naftenato sódico sodium naphthenate
naftenilo naphthenyl
nafteno naphthene
naftieno naphthieno
naftilamina naphthylamine
naftilbenceno naphthylbenzene
naftilbenzoato naphthylbenzoate
naftilendiamina naphthylenediamine
naftileno naphthylene
naftiletiléter naphthylethyl ether
naftilmercúrico naphthylmercuric
naftilmetanol naphthylmethanol
naftilmetileno naphthylmethylene
naftilmetilo naphthylmethyl
naftilnaftilo naphthylnaphthyl
naftilo naphthyl
naftiloxi naphthyloxy
naftiltiourea naphthylthiourea
naftindeno naphthindene
naftionato sódico sodium naphthionate
naftiridina naphthyridine
nafto naphtho
naftodiantreno naphthodianthrene
naftoilo naphthoyl
naftol naphthol
naftolato naphtholate
naftolén naftolen
naftonitrilo naphthonitrile
naftoquinona naphthoquinone
naftorresorcina naphthoresorcinol
naftotiazol naphthothiazole
naftoxi naphthoxy
naled naled
nalorfina nalorphine
naloxona naloxone
nanómetro nanometer
nanotecnología nanotechnology
nantoquita nantokite
napalina napalin
napalm napalm

naproxén naproxen
naptalam naptalam
narceína narceine
narcótico narcotic
naringina naringin
nasciente nascent
natrolita natrolite
natrón natron
natural natural
nectar nectar
nefelita nephelite
nefelometría nephelometry
nefelómtero nephelometer
nefrita nephrite
negativo negative
negro de acetileno acetylene black
negro de alizarina alizarin black
negro de anilina aniline black
negro de antimonio antimony black
negro de carbón carbon black
negro de cobalto cobalt black
negro de gas gas black
negro de hierro iron black
negro de horno furnace black
negro de huesos bone black
negro de humo lampblack
negro de manganeso manganese black
negro de paladio palladium black
negro de petróleo oil black
negro de platino platinum black
negro de renio rhenium black
negro de rodio rhodium black
nemático nematic
neo neo
neocianita neocianite
neodimio neodymium
neohexano neohexane
neolita neolite
neomicina neomycin
neón neon
neopentano neopentane
neopentilglicol neopentyl glycol
neopentilo neopentyl
neopreno neoprene
neptunilo neptunyl
neptunio neptunium
nerilo neryl
nerol nerol
nerolidol nerolidol

nervona nervone
nesquehonita nesquehonite
neurina neurine
neutralidad neutrality
neutralización neutralization
neutralizado neutralized
neutralizante neutralizing
neutralizar neutralize
neutrino neutrino
neutro neutral
neutrón neutron
neutrón de alta energía high-energy neutron
neutrón de baja energía low-energy neutron
neutrón de fisión fission neutron
neutrón de resonancia resonance neutron
neutrón lento slow neutron
neutrón nuclear nuclear neutron
neutrón rápido fast neutron
neutrón retardado delayed neutron
neutrón térmico thermal neutron
neutron termonuclear thermonuclear neutron
neutrónica neutronics
neutrónico neutronic
niacina niacin
niacinamida niacinamide
nialamida nialamide
niccolita niccolite
nicotina nicotine
nicotinamida nicotinamide
nicotinoilo nicotinoyl
nicromo nichrome
nido nido
niebla fog
niebla contaminante smog
nieve de dióxido de carbono carbon dioxide snow
nilón nylon
ninhidrina ninhydrin
niobato niobate
niobato de litio lithium niobate
niobato sódico sodium niobate
nióbico niobic
niobilo niobyl
niobio niobium
niobio-estaño niobium-tin

niobio-titanio niobium-titanium
niobio-uranio niobium-uranium
niobita niobite
niobo niobus
nioboxi nioboxy
nionel nionel
níquel nickel
níquel-cadmio nickel-cadmium
níquel de Raney Raney nickel
níquel-plata nickel-silver
níquel-rodio nickel-rhodium
niquélico nickelic
niquelina nickeline
niqueloceno nickelocene
niqueloso nickelous
niranio niranium
nisina nisin
nitobifenilo nitrobiphenyl
nitracidio nitracidium
nitración nitration
nitración de Zincke Zincke nitration
nitralina nitralin
nitramida nitramide
nitramina nitramine
nitranilato sódico sodium nitranilate
nitranilida nitranilide
nitranilina nitraniline
nitrato nitrate
nitrato alumínico aluminum nitrate
nitrato alumínico cálcico calcium aluminum nitrate
nitrato amónico ammonium nitrate
nitrato amónico cérico ceric ammonium nitrate
nitrato amónico de lantano lanthanum ammonium nitrate
nitrato amónico de neodimio neodymium ammonium nitrate
nitrato bárico barium nitrate
nitrato cálcico calcium nitrate
nitrato cérico ceric nitrate
nitrato ceroso cerous nitrate
nitrato cobaltoso cobaltous nitrate
nitrato crómico chromic nitrate
nitrato cúprico cupric nitrate
nitrato de acetamida acetamide nitrate
nitrato de acetilo acetyl nitrate
nitrato de amilo amyl nitrate
nitrato de antimonilo sódico sodium

antimonyl nitrate
nitrato de atropina atropine nitrate
nitrato de berilio beryllium nitrate
nitrato de bismutilo bismuthyl nitrate
nitrato de bismuto bismuth nitrate
nitrato de brucina brucine nitrate
nitrato de butilo butyl nitrate
nitrato de cadmio cadmium nitrate
nitrato de celulosa cellulose nitrate
nitrato de cerio cerium nitrate
nitrato de cesio cesium nitrate
nitrato de cinc zinc nitrate
nitrato de cinconina cinchonine nitrate
nitrato de cobalto cobalt nitrate
nitrato de cobre copper nitrate
nitrato de cocaína cocaine nitrate
nitrato de cromo chromium nitrate
nitrato de diazobenceno diazobenzene nitrate
nitrato de didimio didymium nitrate
nitrato de disprosio dysprosium nitrate
nitrato de erbio erbium nitrate
nitrato de estroncio strontium nitrate
nitrato de etileno ethylene nitrate
nitrato de etilo ethyl nitrate
nitrato de europio europium nitrate
nitrato de flúor fluorine nitrate
nitrato de gadolinio gadolinium nitrate
nitrato de galio gallium nitrate
nitrato de guanidina guanidine nitrate
nitrato de hidrazina hydrazine nitrate
nitrato de hierro iron nitrate
nitrato de indio indium nitrate
nitrato de isopropilo isopropyl nitrate
nitrato de itrio yttrium nitrate
nitrato de lantano lanthanum nitrate
nitrato de litio lithium nitrate
nitrato de lutecio lutetium nitrate
nitrato de manganeso manganese nitrate
nitrato de mercurio mercury nitrate
nitrato de metilo methyl nitrate
nitrato de miconazol miconazole nitrate
nitrato de neodimio neodymium nitrate
nitrato de níquel nickel nitrate
nitrato de paladio palladium nitrate
nitrato de pentilo pentyl nitrate
nitrato de peroxibenzoilo peroxybenzoyl nitrate
nitrato de plata silver nitrate

nitrato de plomo   lead nitrate
nitrato de propanol   propanol nitrate
nitrato de propilo   propyl nitrate
nitrato de rodio   rhodium nitrate
nitrato de rubidio   rubidium nitrate
nitrato de talio   thallium nitrate
nitrato de telurio   tellurium nitrate
nitrato de terbio   terbium nitrate
nitrato de titanio   titanium nitrate
nitrato de torio   thorium nitrate
nitrato de uranilo   uranyl nitrate
nitrato de uranio   uranium nitrate
nitrato de urea   urea nitrate
nitrato de zirconilo   zirconyl nitrate
nitrato de zirconio   zirconium nitrate
nitrato fenilmercúrico   phenylmercuric nitrate
nitrato férrico   ferric nitrate
nitrato ferroso   ferrous nitrate
nitrato magnésico   magnesium nitrate
nitrato manganoso   manganous nitrate
nitrato mercúrico   mercuric nitrate
nitrato mercurioso   mercurous nitrate
nitrato niqueloso   nickelous nitrate
nitrato paladioso   palladous nitrate
nitrato potásico   potassium nitrate
nitrato potásico de cobalto   cobalt potassium nitrate
nitrato samárico   samaric nitrate
nitrato sódico   sodium nitrate
nitrato talioso   thallous nitrate
nitreno   nitrene
nitridación   nitridation
nitrido   nitrido
nitrificación   nitrification
nitrilasa   nitrilase
nitrilo   nitrile
nitrilo de bencilo   benzyl nitrile
nitrilo fórmico   formic nitrile
nitrilo graso   fatty nitrile
nitrilotriacetato sódico   sodium nitrilotriacetate
nitrilotriacetonitrilo   nitrilotriacetonitrile
nitrito   nitrite
nitrito amónico   ammonium nitrite
nitrito amónico de cinc   zinc ammonium nitrite
nitrito bárico   barium nitrite
nitrito cálcico   calcium nitrite

nitrito cúprico   cupric nitrite
nitrito de alquilo   alkyl nitrite
nitrito de amilo   amyl nitrite
nitrito de butilo   butyl nitrite
nitrito de cobre   copper nitrite
nitrito de estroncio   strontium nitrite
nitrito de etilo   ethyl nitrite
nitrito de isoamilo   isoamyl nitrite
nitrito de metilo   methyl nitrite
nitrito de plata   silver nitrite
nitrito de plomo   lead nitrite
nitrito de propilo   propyl nitrite
nitrito magnésico   magnesium nitrite
nitrito potásico   potassium nitrite
nitrito potásico cobáltico   cobaltic potassium nitrite
nitrito sódico   sodium nitrite
nitro   nitro
nitro de sosa   soda niter
nitroacetanilida   nitroacetanilide
nitroalmidón   nitrostarch
nitroamina   nitroamine
nitroamino   nitroamino
nitroanilina   nitroaniline
nitroanisol   nitroanisole
nitroantraceno   nitroanthracene
nitroantraquinona   nitroanthraquinone
nitroaromático   nitroaromatic
nitrobarita   nitrobarite
nitrobenceno   nitrobenzene
nitrobencilo   nitrobenzyl
nitrobenzaldehído   nitrobenzaldehyde
nitrobenzamida   nitrobenzamide
nitrobenzoilo   nitrobenzoyl
nitrobenzonitrilo   nitrobenzonitrile
nitrobenzotrifluoruro   nitrobenzotrifluoride
nitrobromoformo   nitrobromoform
nitrobutanol   nitrobutanol
nitrocelulosa   nitrocellulose
nitroclorobenceno   nitrochlorobenzene
nitrocloroformo   nitrochloroform
nitrocobalamina   nitrocobalamin
nitrocresol   nitrocresol
nitrodifenilamina   nitrodiphenylamine
nitrodifenilo   nitrodiphenyl
nitroestireno   nitrostyrene
nitroetano   nitroethane
nitroetilpropanodiol

nitroethylpropanediol

nitrofenetol nitrophenetole

nitrofenida nitrophenide

nitrofenilhidrazina nitrophenylhydrazine

nitrofenol nitrophenol

nitrofenolato sódico sodium nitrophenolate

nitroferricianuro sódico sodium nitroferricyanide

nitrofosfato nitrophosphate

nitrofurano nitrofuran

nitrofurantoína nitrofurantoin

nitrogenado nitrogenated

nitrogenasa nitrogenase

nitrógeno nitrogen

nitrógeno amoniacal ammonia nitrogen

nitrógeno líquido liquid nitrogen

nitroglicerina nitroglycerin

nitroguanidina nitroguanidine

nitrol nitrol

nitromagnesita nitromagnesite

nitromanita nitromannite

nitromanitol nitromannitol

nitromersol nitromersol

nitrometano nitromethane

nitrometilpropanodiol nitromethylpropanediol

nitrometoxianilina nitromethoxyaniline

nitrón nitron

nitrona nitrone

nitronaftaleno nitronaphthalene

nitronaftilamina nitronaphthylamine

nitronaftol nitronaphthol

nitronio nitronium

nitroparafina nitroparaffin

nitropropano nitropropane

nitroquinolina nitroquinoline

nitrosilo nitrosyl

nitrosita de estireno styrene nitrosite

nitroso nitroso

nitrosoamina nitrosoamine

nitrosobenceno nitrosobenzene

nitrosodimetilamina nitrosodimethylamine

nitrosodimetilanilina nitrosodimethylaniline

nitrosoetano nitrosoethane

nitrosofenol nitrosophenol

nitrosoguanidina nitrosoguanidine

nitrosometilurea nitrosomethylurea

nitrosonaftilamina nitrosonaphthylamine

nitrosonaftol nitrosonaphthol

nitrosonio nitrosonium

nitrosonitrato de rutenio ruthenium nitrosonitrate

nitrosooxi nitrosooxy

nitrosotolueno nitrosotoluene

nitrosulfatiazol nitrosulfathiazole

nitrotiofeno nitrothiophene

nitrotolueno nitrotoluene

nitrotoluidina nitrotoluidine

nitrotriclorometano nitrotrichloromethane

nitrotrifluorometilclorobenceno nitrotrifluoromethyl chlorobenzene

nitrourea nitrourea

nitroxileno nitroxylene

nitroxilo nitroxyl

nitruro nitride

nitruro alumínico aluminum nitride

nitruro cálcico calcium nitride

nitruro de berilio beryllium nitride

nitruro de boro boron nitride

nitruro de carbono carbon nitride

nitruro de cinc zinc nitride

nitruro de cobalto cobalt nitride

nitruro de fosforilo phosphoryl nitride

nitruro de fósforo phosphorus nitride

nitruro de hafnio hafnium nitride

nitruro de litio lithium nitride

nitruro de plata silver nitride

nitruro de selenio selenium nitride

nitruro de silicio silicon nitride

nitruro de tántalo tantalum nitride

nitruro de titanio titanium nitride

nitruro de vanadio vanadium nitride

nitruro de zirconio zirconium nitride

nitruro magnésico magnesium nitride

nitruro sódico sodium nitride

nivel atómico atomic level

nivel cuántico quantum level

nivel de energía energy level

nivel de energía atómica atomic energy level

nivel de energía electrónica electron energy level

nivel de energía molecular molecular energy level

nivel de energía nuclear nuclear energy level
nivel de Fermi Fermi level
nivel de saturación saturation level
nivel molecular molecular level
nivel superior higher level
nivel vibracional vibrational level
nivel virtual virtual level
niveles de Landau Landau levels
nivenita nivenite
no acuoso nonaqueous
no combustible noncombustible
no destructivo nondestructive
no electrólito nonelectrolyte
no estequiométrico nonstoichiometric
no ferroso nonferrous
no ideal nonideal
no inflamable nonflammable
no iónico nonionic
no ionizante nonionizing
no metal nonmetal
no metálico nonmetallic
no polar nonpolar
no saturación unsaturation
no saturado unsaturated
nobelio nobelium
noble noble
nocivo noxious
nódulo nodule
nombre aprobado approved name
nombre genérico generic name
nombre sistemático systematic name
nombre trivial trivial name
nomenclatura nomenclature
nomenclatura conjuntiva conjuctive nomenclature
nomenclatura química chemical nomenclature
nomenclatura radicofuncional radicofunctional nomenclature
non non
nona nona
nonacíclico nonacyclic
nonacosano nonacosane
nonacosanol nonacosanol
nonadecano nonadecane
nonadecanol nonadecanol
nonadecanona nonadecanone
nonahidrato nonahydrate

nonalactona nonalactone
nonanal nonanal
nonano nonane
nonanoato nonanoate
nonanoato de butilo butyl nonanoate
nonanoato de etilo ethyl nonanoate
nonanoato de nonilo nonyl nonanoate
nonanodiol nonanediol
nonanoilo nonanoyl
nonanol nonanol
nonanona nonanone
nonanonitrilo nonanenitrile
noneno nonene
nonilamina nonylamine
nonilbenceno nonylbenzene
nonilcarbinol nonylcarbinol
nonileno nonylene
nonilfenol nonylphenol
nonilo nonyl
noniltriclorosilano nonyltrichlorosilane
nonino nonyne
nopineno nopinene
nor nor
noradrenalina noradrenaline
norbornadieno norbornadiene
norborneno norbornene
norbornenometanol norbornenemethanol
norepinefrina norepinephrine
norgestrel norgestrel
norleucina norleucine
normal normal
normalidad normality
normas de almacenamiento storage guidelines
nornicotina nornicotine
noscapina noscapine
noselita noselite
notación notation
notación de Dyson Dyson notation
notación de Frankland Frankland notation
notación de Wiswesser Wiswesser notation
notación lineal de Wiswesser Wiswesser line notation
novobiocina novobiocin
novobiocina sódica sodium novobiocin
novolaca novolak
nube electrónica electron cloud

nube iónica ion cloud
nucleación nucleation
nucleado nucleate
nuclear nuclear
nucleasa nuclease
nucleido nuclide
nucleido radiactivo radioactive nuclide
nucleidos de Wigner Wigner nuclides
núcleo nucleus
núcleo aromático aromatic nucleus
núcleo atómico atomic nucleus
núcleo compuesto compound nucleus
núcleo de recristalización
    recrystallization nucleus
núcleo estable stable nucleus
núcleo excitado excited nucleus
núcleo impar-impar odd-odd nucleus
núcleo impar-par odd-even nucleus
núcleo inestable unstable nucleus
núcleo par-impar even-odd nucleus
núcleo par-par even-even nucleus
núcleo pesado heavy nucleus
núcleo primario primary nucleus
núcleo radiactivo radioactive nucleus
nucleofílico nucleophilic
nucleófilo nucleophile
nucleogénesis nucleogenesis
nucleón nucleon
nucleónico nucleonic
nucleoproteína nucleoprotein
núcleos equivalentes equivalent nuclei
núcleos espejos mirror nuclei
nucleósido nucleoside
nucleótido nucleotide
numeración bicíclica bicyclo numbering
número atómico atomic number
número atómico efectivo effective
    atomic number
número cuántico quantum number
número cuántico acimutal azimuthal
    quantum number
número cuántico de espín spin quantum
    number
número cuántico magnético magnetic
    quantum number
número cuántico orbital orbital quantum
    number
número cuántico principal principal

quantum number
número de acetona acetone number
número de ácido acid number
número de Avogadro Avogadro number
número de base base number
número de Bellier Bellier number
número de cetano cetane number
número de cobre copper number
número de combinación combining
    number
número de coordinación coordination
    number
número de Drew Drew number
número de entalpía de reacción reaction
    enthalpy number
número de éster ester number
número de Hehner Hehner number
número de Loschmidt Loschmidt
    number
número de masa mass number
número de masa atómica atomic mass
    number
número de neutralización neutralization
    number
número de neutrones neutron number
número de oro gold number
número de oxidación oxidation number
número de peróxido peroxide number
número de Polenske Polenske number
número de Prandtl Prandtl number
número de precipitación precipitation
    number
número de saponificación saponification
    number
número de transferencia transference
    number
número de transporte transport number
número de valencia valence number
número de yodo iodine number
número isotópico isotopic number
número molecular molecular number
número nuclear nuclear number
número protónico proton number
números mágicos magic numbers
nutrición nutrition
nutriente nutrient
nutrificación nutrification

O

ocimeno ocimene
oclusión occlusion
ocre ocher
ocre rojo red ocher
octa octa
octaacetato de sucrosa sucrose
    octaacetate
octabenzona octabenzone
octacloronaftaleno
    octachloronaphthalene
octacloruro de silicio silicon
    octachloride
octacosano octacosane
octadeca octadeca
octadecanal octadecanal
octadecano octadecane
octadecanoato de butilo butyl
    octadecanoate
octadecanol octadecanol
octadeceno octadecene
octadecenol octadecenol
octadecilo octadecyl
octadeciltriclorosilano
    octadecyltrichlorosilane
octadieno octadiene
octafluorobuteno octafluorobutene
octafluorociclobutano
    octafluorocyclobutane
octafluoropropano octafluoropropane
octahedral octahedral
octahedro octahedron
octahidrato octahydrate
octahidrato bárico barium octahydrate
octametiltrisiloxano
    octamethyltrisiloxane
octanal octanal
octano octane
octanoato octanoate
octanoato de cinc zinc octanoate
octanoato de etilo ethyl octanoate
octanoilo octanoyl
octanol octanol
octanona octanone

octavalente octavalent
octeno octene
octeto octet
octeto electrónico electron octet
octilamina octylamine
octilcarbinol octyl carbinol
octileno octylene
octilfenilo octylphenyl
octilfenol octylphenol
octilmercaptano octyl mercaptan
octilo octyl
octilsulfato sódico sodium octyl sulfate
octiltriclorosilano octyltrichlorosilane
octino octyne
octoato cálcico calcium octoate
octoato de cinc zinc octoate
octoato de etilo ethyl octoate
octoato de hierro iron octoate
octoato de manganeso manganese
    octoate
octoato de plomo lead octoate
octoato estannoso stannous octoate
octoato férrico ferric octoate
octoato ferroso ferrous octoate
octometileno octomethylene
octóxido de triuranio triuranium
    octoxide
odorante odorant
odorífero odoriferous
odorimetría odorimetry
odorómetro odorometer
oico oic
oide oid
okonita okonite
oleamida oleamide
oleato oleate
oleato alumínico aluminum oleate
oleato amónico ammonium oleate
oleato bárico barium oleate
oleato cálcico calcium oleate
oleato cobaltoso cobaltous oleate
oleato de bismuto bismuth oleate
oleato de butilo butyl oleate

447

oleato de butoxietilo butoxyethyl oleate
oleato de cinc zinc oleate
oleato de cobalto cobalt oleate
oleato de cobre copper oleate
oleato de diglicol diglycol oleate
oleato de etilo ethyl oleate
oleato de isocetilo isocetyl oleate
oleato de manganeso manganese oleate
oleato de metilo methyl oleate
oleato de plomo lead oleate
oleato estannoso stannous oleate
oleato fenilmercúrico phenylmercuric
　oleate
oleato férrico ferric oleate
oleato magnésico magnesium oleate
oleato mercúrico mercuric oleate
oleato potásico potassium oleate
oleato sódico sodium oleate
olefina olefin
oleilo oleyl
oleína olein
oleo oleo
oleoilo oleoyl
oleoilsarcosina oleoylsarcosine
oleorresina oleoresin
oligo oligo
oligoelemento trace element
oligómero oligomer
oligonita oligonite
oligonucleótido oligonucleotide
oligopéptido oligopeptide
oligosacárido oligosaccharide
olivenita olivenite
olivetol olivetol
olivina olivine
olor odor
olor de advertencia warning odor
omega omega
ona one
oncógeno oncogen
onda wave
onio onium
opacidad opacity
opaco opaque
ópalo opal
operación operation
opiato opiate
opio opium
opioide opioid

ópticamente optically
ópticamente activo optically active
órbita orbit
órbita atómica atomic orbit
órbita de Bohr Bohr orbit
órbita electrónica electron orbit
orbital orbital
orbital antienlazante antibonding orbital
orbital atómico atomic orbital
orbital de antienlazado antibonding
　orbital
orbital de enlazado bonding orbital
orbital degenerado degenerate orbital
orbital delta delta orbital
orbital electrónico electron orbital
orbital enlazante bonding orbital
orbital exterior outer orbital
orbital fronterizo frontier orbital
orbital híbrido hybrid orbital
orbital molecular molecular orbital
**orbital molecular de máxima energía
　ocupado** highest occupied molecular
　orbital
**orbital molecular de mínima energía
　ocupado** lowest occupied molecular
　orbital
orbital molecular desocupado
　unoccupied molecular orbital
orbital molecular ocupado occupied
　molecular orbital
orbitales solapados overlapping orbitals
orcina orcin
orden order
órden de Irving-Williams Irving-
　Williams order
orden de reacción reaction order
orgánico organic
organoarcilla organoclay
organoborano organoborane
organogel organogel
organometálico organometallic
organometaloide organometalloid
organosilano organosilane
organosilíceo organosilicon
organosol organosol
orientación superficial surface
　orientation
ornitina ornithine
oro gold

oro blanco white gold
oro de mosaico mosaic gold
oro musivo mosaic gold
oropimente orpiment
ortamina orthamine
ortita orthite
orto ortho
ortoarseniato de cinc zinc orthoarsenate
ortoclasa orthoclase
ortofosfato orthophosphate
ortofosfato alumínico aluminum orthophosphate
ortofosfato bárico barium orthophosphate
ortofosfato cálcico calcium orthophosphate
ortofosfato de cinc zinc orthophosphate
ortofosfato de etilo ethyl orthophosphate
ortofosfato de litio lithium orthophosphate
ortofosfato de plata silver orthophosphate
ortofosfato de plomo lead orthophosphate
ortofosfato de urea-amonio urea-ammonium orthophosphate
ortofosfato de zirconio zirconium orthophosphate
ortofosfato dicálcico dicalcium orthophosphate
ortofosfato magnésico magnesium orthophosphate
ortofosfato manganoso manganous orthophosphate
ortofosfato potásico potassium orthophosphate
ortofosfato sódico sodium orthophosphate
ortofosfato tricálcico tricalcium orthophosphate
ortofosfato trimercúrico trimercuric orthophosphate
ortofosfato trimercurioso trimercurous orthophosphate
ortofosfato tripotásico tripotassium orthophosphate
ortofosfato trisódico trisodium orthophosphate
ortohidrógeno orthohydrogen

ortonitrógeno orthonitrogen
ortoperyodato bárico barium orthoperiodate
ortopropionato de etilo ethyl orthopropionate
ortosilicato orthosilicate
ortosilicato de etilo ethyl orthosilicate
ortosilicato sódico sodium orthosilicate
ortotungstato cálcico calcium orthotungstate
ortovanadato sódico sodium orthovanadate
osamina osamine
osazona osazone
oscilación oscillation
oscilación atómica atomic oscillation
oscilador armónico harmonic oscillator
oscilador electrónico electron oscillator
osmiato osmate
osmiato potásico potassium osmate
ósmica osmics
osmio osmium
osmiridio osmiridium
osmoceno osmocene
osmol osmole
osmolalidad osmolality
osmolaridad osmolarity
osmometría osmometry
osmómetro osmometer
osmómetro de presión de vapor vapor-pressure osmometer
ósmosis osmosis
ósmosis inversa reverse osmosis
osmoso osmous
osmótico osmotic
osono osone
osotriazol osotriazole
ovaleno ovalene
ovicida ovicide
oxa oxa
oxalato oxalate
oxalato ácido potásico potassium acid oxalate
oxalato alumínico aluminum oxalate
oxalato amónico ammonium oxalate
oxalato amónico de titanio titanium ammonium oxalate
oxalato amónico férrico ferric ammonium oxalate

oxalato antimonioso antimonous oxalate
oxalato bárico barium oxalate
oxalato cálcico calcium oxalate
oxalato ceroso cerous oxalate
oxalato cobaltoso cobaltous oxalate
oxalato cromoso chromous oxalate
oxalato cúprico cupric oxalate
oxalato de alilo allyl oxalate
oxalato de anilina aniline oxalate
oxalato de berilio beryllium oxalate
oxalato de bismuto bismuth oxalate
oxalato de cadmio cadmium oxalate
oxalato de cerio cerium oxalate
oxalato de cesio cesium oxalate
oxalato de cinc zinc oxalate
oxalato de cobre copper oxalate
oxalato de dibutilo dibutyl oxalate
oxalato de dietilo diethyl oxalate
oxalato de erbio erbium oxalate
oxalato de escandio scandium oxalate
oxalato de estroncio strontium oxalate
oxalato de etilo ethyl oxalate
oxalato de europio europium oxalate
oxalato de gadolinio gadolinium oxalate
oxalato de hierro iron oxalate
oxalato de hierro amónico ammonium
   iron oxalate
oxalato de holmio holmium oxalate
oxalato de iterbio ytterbium oxalate
oxalato de itrio yttrium oxalate
oxalato de lantano lanthanum oxalate
oxalato de litio lithium oxalate
oxalato de manganeso manganese
   oxalate
oxalato de neodimio neodymium oxalate
oxalato de niobio niobium oxalate
oxalato de níquel nickel oxalate
oxalato de plata silver oxalate
oxalato de plomo lead oxalate
oxalato de praseodimio praseodymium
   oxalate
oxalato de titanio titanium oxalate
oxalato de torio thorium oxalate
oxalato de tulio thulium oxalate
oxalato de urea urea oxalate
oxalato estannoso stannous oxalate
oxalato férrico ferric oxalate
oxalato ferroso ferrous oxalate
oxalato magnésico magnesium oxalate

oxalato mercúrico mercuric oxalate
oxalato mercurioso mercurous oxalate
oxalato niqueloso nickelous oxalate
oxalato potásico potassium oxalate
oxalato potásico crómico chromium
   potassium oxalate
oxalato potásico de antimonio antimony
   potassium oxalate
oxalato samárico samaric oxalate
oxalato sódico sodium oxalate
oxalato sódico férrico ferric sodium
   oxalate
oxalato titanoso titanous oxalate
oxalilo oxalyl
oxalilurea oxalylurea
oxalo oxalo
oxamato amónico ammonium oxamate
oxamida oxamide
oxamido oxamido
oxamoilo oxamoyl
oxantrol oxanthrol
oxatilo oxatyl
oxazol oxazole
óxdo niqueloso nickelous oxide
oxetano oxetane
oxetilo oxethyl
oxetona oxetone
oxi oxy
oxiácido oxyacid
oxiácido de azufre sulfur oxyacid
oxiácido de fósforo phosphorus oxyacid
oxiácido de selenio selenium oxyacid
oxiamida oxyamide
oxianión oxyanion
oxibromuro de bismuto bismuth
   oxybromide
oxibromuro de carbono carbon
   oxybromide
oxibromuro de fósforo phosphorus
   oxybromide
oxibromuro de zirconio zirconium
   oxybromide
oxicarbonato de bismuto bismuth
   oxycarbonate
oxicarbonilo oxycarbonyl
oxicianógeno oxycyanogen
oxicianuro de carbono carbon
   oxycyanide
oxicianuro mercúrico mercuric

oxycyanide

oxicloruro antimonioso antimonous oxychloride

oxicloruro arsenioso arsenous oxychloride

oxicloruro cálcico calcium oxychloride

oxicloruro crómico chromic oxychloride

oxicloruro de antimonio antimony oxychloride

oxicloruro de azufre sulfur oxychloride

oxicloruro de bismuto bismuth oxychloride

oxicloruro de carbono carbon oxychloride

oxicloruro de cinc zinc oxychloride

oxicloruro de cobre copper oxychloride

oxicloruro de cromo chromium oxychloride

oxicloruro de fósforo phosphorus oxychloride

oxicloruro de rutenio ruthenium oxychloride

oxicloruro de selenio selenium oxychloride

oxicloruro de silicio silicon oxychloride

oxicloruro de telurio tellurium oxychloride

oxicloruro de tungsteno tungsten oxychloride

oxicloruro de uranio uranium oxychloride

oxicloruro de vanadio vanadium oxychloride

oxicloruro de zirconio zirconium oxychloride

oxicloruro sulfúrico sulfuric oxychloride

oxicloruro sulfuroso sulfurous oxychloride

oxicorte flame cutting

oxidación oxidation

oxidación anódica anodic oxidation

oxidación de Jones Jones oxidation

oxidación de Oppenauer Oppenauer oxidation

oxidación de Pfitzner-Moffatt Pfitzner-Moffatt oxidation

oxidación de Riley Riley oxidation

oxidación de Sarett Sarett oxidation

oxidación electrolítica electrolytic

oxidation

oxidación electroquímica electrochemical oxidation

oxidación por aire air oxidation

oxidación-reducción oxidation-reduction

oxidante oxidant

oxidante fotoquímico photochemical oxidant

oxidar oxidize

oxidasa oxidase

oxidasa de ácido ascórbico ascorbic acid oxidase

oxidasa de glucosa glucose oxidase

oxidasa de xantina xanthine oxidase

oxidativo oxidative

oxidicloruro de vanadio vanadium oxydichloride

oxidimetría oxidimetry

oxidipropionitrilo oxydipropionitrile

óxido oxide

óxido acético acetic oxide

óxido ácido acid oxide

óxido alcalinotérreo alkaline-earth oxide

óxido alumínico aluminum oxide

óxido alumínico hidratado hydrated aluminum oxide

óxido anfótero amphoteric oxide

óxido antimonioso antimonous oxide

óxido argéntico argentic oxide

óxido arsenioso arsenous oxide

óxido áurico auric oxide

óxido bárico barium oxide

óxido básico basic oxide

óxido bórico boric oxide

óxido cálcico calcium oxide

óxido cérico ceric oxide

óxido cobáltico cobaltic oxide

óxido cobaltoso cobaltous oxide

óxido colbatosocobáltico cobaltouscobaltic oxide

óxido crómico chromic oxide

óxido cromoso chromous oxide

óxido cúprico cupric oxide

óxido cuproso cuprous oxide

óxido de acetilo acetyl oxide

óxido de alquilo alkyl oxide

óxido de antimonio antimony oxide

óxido de arsénico arsenic oxide

óxido de azufre sulfur oxide

óxido de benzofenona benzophenone oxide

óxido de benzoilo benzoyl oxide

óxido de berilio beryllium oxide

óxido de bifenileno biphenylene oxide

óxido de bismuto bismuth oxide

óxido de boro boron oxide

óxido de butileno butylene oxide

óxido de cacodilo cacodyl oxide

óxido de cadmio cadmium oxide

óxido de carbono carbon oxide

óxido de cerio cerium oxide

óxido de ciclohexeno cyclohexene oxide

óxido de cinc zinc oxide

óxido de cloro chlorine oxide

óxido de cobalto cobalt oxide

óxido de cromo chromium oxide

óxido de deuterio deuterium oxide

óxido de dibutilestaño dibutyltin oxide

óxido de dicloro dichlorine oxide

óxido de dicloroetilo dichloroethyl oxide

óxido de didimio didymium oxide

oxido de dietilo diethyl oxide

óxido de difenilo diphenyl oxide

óxido de difenilo clorometilado chloromethylated diphenyl oxide

óxido de dinitrógeno dinitrogen oxide

óxido de disprosio dysprosium oxide

óxido de divinilo divinyl oxide

óxido de erbio erbium oxide

óxido de escandio scandium oxide

óxido de estaño tin oxide

óxido de estireno styrene oxide

óxido de estroncio strontium oxide

óxido de etileno ethylene oxide

óxido de etilo ethyl oxide

óxido de europio europium oxide

óxido de fenoxipropileno phenoxypropylene oxide

óxido de flúor fluorine oxide

óxido de fósforo phosphorus oxide

óxido de gadolinio gadolinium oxide

óxido de galio gallium oxide

óxido de germanio germanium oxide

óxido de hafnio hafnium oxide

óxido de hexaclorodifenilo hexachlorodiphenyl oxide

óxido de hidrógeno hydrogen oxide

óxido de hierro iron oxide

óxido de holmio holmium oxide

óxido de indio indium oxide

óxido de iridio iridium oxide

óxido de iterbio ytterbium oxide

óxido de itrio yttrium oxide

óxido de lantano lanthanum oxide

óxido de litio lithium oxide

óxido de lutecio lutetium oxide

óxido de manganeso manganese oxide

óxido de mercurio mercury oxide

óxido de mercurio rojo red mercury oxide

óxido de mesitilo mesityl oxide

óxido de metileno methylene oxide

óxido de metilo methyl oxide

óxido de molibdeno molybdenum oxide

óxido de neodimio neodymium oxide

óxido de niobio niobium oxide

óxido de níquel nickel oxide

óxido de nitrógeno nitrogen oxide

óxido de octileno octylene oxide

óxido de oro gold oxide

óxido de osmio osmium oxide

óxido de paladio palladium oxide

óxido de pentametileno pentamethylene oxide

óxido de pentilo pentyl oxide

óxido de piridina pyridine oxide

óxido de plata silver oxide

óxido de platino platinum oxide

óxido de plomo lead oxide

óxido de plutonio plutonium oxide

óxido de polietileno polyethylene oxide

óxido de polifenileno polyphenylene oxide

óxido de polipropileno polypropylene oxide

óxido de praseodimio praseodymium oxide

óxido de propileno propylene oxide

óxido de renio rhenium oxide

óxido de rodio rhodium oxide

óxido de rubidio rubidium oxide

óxido de rutenio ruthenium oxide

óxido de samario samarium oxide

óxido de selenio selenium oxide

óxido de silicio silicon oxide

óxido de succinilo succinyl oxide

óxido de talio thallium oxide

óxido de tántalo   tantalum oxide
óxido de telurio   tellurium oxide
óxido de terbio   terbium oxide
óxido de titanio   titanium oxide
óxido de torio   thorium oxide
óxido de tributilestaño   tributyltin oxide
óxido de triclorobutileno   trichlorobutylene oxide
óxido de trifluoroamina   trifluoroamine oxide
óxido de trimetilamina   trimethylamine oxide
óxido de trimetileno   trimethylene oxide
óxido de tulio   thulium oxide
óxido de tungsteno   tungsten oxide
óxido de uranilo   uranyl oxide
óxido de uranio   uranium oxide
óxido de vanadio   vanadium oxide
óxido de vinilo   vinyl oxide
óxido de yodo   iodine oxide
óxido de zirconio   zirconium oxide
óxido estánnico   stannic oxide
óxido estannoso   stannous oxide
óxido etilarsenoiso   ethylarsenious oxide
óxido férrico   ferric oxide
óxido férrico gamma   gamma ferric oxide
óxido ferromagnético   ferromagnetic oxide
óxido ferroso   ferrous oxide
óxido ferrosoférrico   ferrosoferric oxide
óxido fosfórico   phosphoric oxide
óxido gálico   gallic oxide
óxido galoso   gallous oxide
óxido grafítico   graphitic oxide
óxido magnésico   magnesium oxide
óxido mangánico   manganic oxide
óxido manganoso   manganous oxide
óxido mercúrico   mercuric oxide
óxido mercurioso   mercurous oxide
óxido negro de cobre   copper oxide black
óxido niquélico   nickelic oxide
óxido nítrico   nitric oxide
óxido nitroso   nitrous oxide
óxido platínico   platinic oxide
óxido plumboso   plumbous oxide
óxido potásico   potassium oxide
óxido rojo   red oxide
óxido selenioso   selenious oxide

óxido sódico   sodium oxide
óxido talioso   thallous oxide
óxido trioctilfosfínico   trioctylphosphinic oxide
óxido túngstico   tungstic oxide
óxido uránico   uranic oxide
óxido uranoso   uranous oxide
oxidorreductasa   oxidoreductase
oxifluoruro de carbono   carbon oxyfluoride
oxifluoruro de cromo   chromium oxyfluoride
oxifluoruro potásico de niobio   niobium potassium oxyfluoride
oxifosforano   oxyphosphorane
oxigenación   oxygenation
oxigenado   oxygenated
oxigenar   oxygenate
oxígeno   oxygen
oxígeno consumido   oxygen consumed
oxígeno disuelto   dissolved oxygen
oxígeno líquido   liquid oxygen
oxígeno pesado   heavy oxygen
oxihidrato de bismuto   bismuth oxyhydrate
oxihidrógeno   oxyhydrogen
oxiluminiscencia   oxyluminescence
oxima   oxime
oxima de acetofenona   acetophenone oxime
oxima de alcanfor   camphor oxime
oxima de benzamida   benzamide oxime
oxima de benzoína   benzoin oxime
oximetileno   oxymethylene
oximeturea   oxymethurea
oximido   oximido
oxina   oxine
oxinitrato de bismuto   bismuth oxynitrate
oxiquinolina   oxyquinoline
oxiquinona   oxyquinone
oxirano   oxirane
oxireno   oxirene
oxisulfuro antimonioso   antimonous oxysulfide
oxisulfuro de carbono   carbon oxysulfide
oxisulfuro de talio   thallium oxysulfide
oxitetraciclina   oxytetracycline
oxitoxina   oxytoxin
oxitricloruro de vanadio   vanadium

oxytrichloride

oxo oxo

oxoácido oxo acid

oxobromuro de bismuto bismuth oxobromide

oxocloruro de bismuto bismuth oxochloride

oxofluoruro de bismuto bismuth oxofluoride

oxoproceso oxo process

oxorreacción oxo reaction

oxosilano oxosilane

oxoyoduro de bismuto bismuth oxoiodide

ozocerita ozocerite, earth wax

ozonación ozonation

ozonidación ozonidation

ozónido ozonide

ozonización ozonization

ozonizador ozonator

ozono ozone

ozonólisis ozonolysis

ozoquerita ozokerite

P

pacnolita pachnolite
pagodita pagodite
paladato palladate
paládico palladic
paladio palladium
paladioso palladous
palamina pallamine
palas pallas
palconato sódico sodium palconate
paligorskita paligorskite
pálmico palmic
palmierita palmierite
palmitato palmitate
palmitato alumínico aluminum palmitate
palmitato amónico ammonium palmitate
palmitato cálcico calcium palmitate
palmitato de cetilo cetyl palmitate
palmitato de cinc zinc palmitate
palmitato de etilo ethyl palmitate
palmitato de isooctilo isooctyl palmitate
palmitato de litio lithium palmitate
palmitato de metilo methyl palmitate
palmitato de plata silver palmitate
palmitato de plomo lead palmitate
palmitato magnésico magnesium
    palmitate
palmitato sódico sodium palmitate
palmitina palmitin
palmitoilo palmitoyl
palmitonitrilo palmitonitrile
palo de campeche logwood
pan de oro gold leaf
panclastita panclastite
pancreatina pancreatin
pancromático panchromatic
pandermita pandermite
pantacromático pantachromatic
pantalla screen
pantenol panthenol
pantetina pantethine
pantocaína pantocaine
pantolactona pantolactone
pantomorfismo pantomorphism

pantotenato cálcico calcium pantothenate
pantotenol pantothenol
papaína papain
papaverina papaverine
papel paper
papel de absorción absorption paper
papel de aluminio aluminum foil
papel de filtro filter paper
papel de filtro plegado fluted filter paper
papel de prueba test paper
papel de tornasol litmus paper
papel indicador indicator paper
par ácido-base acid-base pair
par de electrones electron pair
par donador-aceptor donor-acceptor
    pair
par iónico ion pair
para para
parabeno paraben
parabituminoso parabituminous
paracaseína paracasein
paracetaldehído paracetaldehyde
paracetamol paracetamol
paracianógeno paracyanogen
paraconina paraconine
paradiazina paradiazine
parafina paraffin
parafina clorada chlorinated paraffin
paraformaldehído paraformaldehyde
paragonita paragonite
parahelio parahelium
parahidrógeno parahydrogen
paralaje parallax
paraldehído paraldehyde
paraldol paraldol
paralelosterismo parallelosterism
paralisol paralysol
paramagnético paramagnetic
paramagnetismo paramagnetism
paramagnetismo molecular molecular
    paramagnetism
paramagnetismo nuclear nuclear
    paramagnetism

parámetro parameter
parámetros de Racah Racah parameters
paramomicina paromomycin
paramorfismo paramorphism
paramorfo paramorph
paranitranilina paranitraniline
paraperyodato sódico sodium paraperiodate
paraquat paraquat
pararrosanilina pararosaniline
parasiticida parasiticide
parásito parasite
paratión parathion
parawolframato parawolframate
paraxantina paraxanthine
paraxileno paraxylene
parcial partial
parileno parylene
parisita parisite
paroxazina paroxazine
partenina parthenine
partición partition
partícula particle
partícula alfa alpha particle
partícula atómica atomic particle
partícula beta beta particle
partícula cargada charged particle
partícula de alta energía high-energy particle
partícula de baja energía low-energy particle
partícula de desintegración decay particle
partícula elemental elementary particle
partícula fundamental fundamental particle
partícula nuclear nuclear particle
partícula relativista relativistic particle
partícula sigma sigma particle
partícula subatómica subatomic particle
partícula subnuclear subnuclear particle
partícula tau tau particle
partícula virtual virtual particle
partícula xi xi particle
particulado particulate
partinio partinium
parvolina parvoline
parvulina parvuline
pasivación passivation

pasivación electroquímica electrochemical passivation
pasivación química chemical passivation
pasivador passivator
pasividad passivity
pasividad química chemical passivity
pasivo passive
paso step
paso de reacción reaction step
paso determinante de velocidad rate-determining step
pasta paste
pasteurización pasteurization
patentabilidad patentability
patogénico pathogenic
patoquímica pathochemistry
patrón standard
patrón de color color standard
patrón de difracción diffraction pattern
patrón de laboratorio laboratory standard
patrón de Laue Laue pattern
patrón de radiactividad radioactivity standard
patrón químico chemical standard
patrón termoquímico thermochemical standard
patronita patronite
paucina paucine
pebulato pebulate
pecblenda pitchblende
pectato pectate
péctico pectic
pectina pectin
pectinasa pectinase
pectización pectization
pectolita pectolite
pectosa pectose
pedernal flint
pedesis pedesis
pegmatita pegmatite
pelargonato de bencilo benzyl pelargonate
pelargonato de butilo butyl pelargonate
pelargonato de etilo ethyl pelargonate
pelargonato de heptilo heptyl pelargonate
pelargonato de isoamilo isoamyl pelargonate

pelargonato de propilo propyl
    pelargonate
pelargonilo pelargonyl
pelargonina pelargonin
pelargonitrilo pelargonitrile
peletierina pelletierine
película film
película anódica anodic film
película de acetato acetate film
película de Langmuir-Blodgett
    Langmuir-Blodgett film
película de plástico plastic film
película fina thin film
película gruesa thick film
película interfacial interfacial film
película metálica metal film
película molecular molecular film
película monomolecular monomolecular
    film
película poliéster polyester film
película unimolecular unimolecular film
peligro de radiación radiation danger
peligroso dangerous, hazardous
pelotina pellotine
peltre pewter
penetración penetration
penetrante penetrant
penicilamina penicillamine
penicilanasa penicillanase
penicilina penicillin
penicilina potásica potassium penicillin
penta penta
pentaamino pentaamino
pentabásico pentabasic
pentaborano pentaborane
pentaborato amónico ammonium
    pentaborate
pentaborato sódico sodium pentaborate
pentabromo pentabromo
pentabromobenceno pentabromobenzene
pentabromuro pentabromide
pentabromuro de fósforo phosphorus
    pentabromide
pentabromuro de tántalo tantalum
    pentabromide
pentabromuro de yodo iodine
    pentabromide
pentacarbonilo pentacarbonyl
pentacarbonilo de hierro iron

pentacarbonyl
pentacarboxílico pentacarboxylic
pentaceno pentacene
pentacetato pentacetate
pentacíclico pentacyclic
pentacloro pentachloro
pentacloroanilina pentachloroaniline
pentaclorobenceno pentachlorobenzene
pentacloroetano pentachloroethane
pentaclorofenato sódico sodium
    pentachlorophenate
pentaclorofenol pentachlorophenol
pentacloronaftaleno
    pentachloronaphthalene
pentacloronitrobenceno
    pentachloronitrobenzene
pentaclorotiofenol pentachlorothiophenol
pentacloruro pentachloride
pentacloruro de antimonio antimony
    pentachloride
pentacloruro de azufre sulfur
    pentachloride
pentacloruro de fósforo phosphorus
    pentachloride
pentacloruro de molibdeno molybdenum
    pentachloride
pentacloruro de niobio niobium
    pentachloride
pentacloruro de renio rhenium
    pentachloride
pentacloruro de tántalo tantalum
    pentachloride
pentacloruro de tungsteno tungsten
    pentachloride
pentacosano pentacosane
pentada pentad
pentadecano pentadecane
pentadecanoato de metilo methyl
    pentadecanoate
pentadecanol pentadecanol
pentadecanolida pentadecanolide
pentadecanona pentadecanone
pentadecenilo pentadecenyl
pentadecilo pentadecyl
pentadieno pentadiene
pentaeritritol pentaerythritol
pentaetilbenceno pentaethylbenzene
pentafluoruro pentafluoride
pentafluoruro de antimonio antimony

pentafluoride

pentafluoruro de arsénico arsenic
pentafluoride

pentafluoruro de bismuto bismuth
pentafluoride

pentafluoruro de bromo bromine
pentafluoride

pentafluoruro de fósforo phosphorus
pentafluoride

pentafluoruro de oro gold pentafluoride

pentafluoruro de tántalo tantalum
pentafluoride

pentafluoruro de yodo iodine
pentafluoride

pentagaloilo pentagalloyl

pentaglicerina pentaglycerin

pentaglicol pentaglycol

pentaglucosa pentaglucose

pentahidrato pentahydrate

pentahidrato bárico barium
pentahydrate

pentahidro pentahydro

pentahidroxi pentahydroxy

pentahidroxiciclohexano
pentahydroxycyclohexane

pentalina pentalin

pentametilbenceno pentamethylbenzene

pentametilenamina
pentamethyleneamine

pentametilendiamina
pentamethylenediamine

pentametilenglicol pentamethylene
glycol

pentametileno pentamethylene

pentametilfenol pentamethylphenol

pentametilo pentamethyl

pentanal pentanal

pentano pentane

pentanodiol pentanediol

pentanodiona pentanedione

pentanol pentanol

pentanona pentanone

pentanotiol pentanethiol

pentaóxido pentaoxide

pentaprismo pentaprismo

pentaseleniuro de fósforo phosphorus
pentaselenide

pentasulfuro pentasulfide

pentasulfuro de antimonio antimony

pentasulfide

pentasulfuro de arsénico arsenic
pentasulfide

pentasulfuro de fósforo phosphorus
pentasulfide

pentasulfuro de plomo lead pentasulfide

pentasulfuro de vanadio vanadium
pentasulfide

pentasulfuro potásico potassium
pentasulfide

pentatriacontano pentatriacontane

pentavalente pentavalent

pentayodo pentaiodo

pentazdieno pentazdiene

pentazocina pentazocine

pentazolilo pentazolyl

pentel pentel

pentenilo pentenyl

penteno pentene

pentenol pentenol

pentilamina pentylamine

pentilbenceno pentylbenzene

pentilbenzoato pentylbenzoate

pentilcetona pentyl ketone

pentilcinamaldehído
pentylcinnamaldehyde

pentileno pentylene

pentilfenilcetona pentyl phenyl ketone

pentilfenol pentylphenol

pentilideno pentylidene

pentilidino pentylidyne

pentilo pentyl

pentiloxi pentyloxy

pentiloxifenol pentyloxyphenol

pentiloxihidrato pentyloxyhydrate

pentilsulfato bárico barium pentylsulfate

pentilurea pentylurea

pentiluretano pentylurethane

pentino pentyne

pentita pentite

pentitol pentitol

pentlandita pentlandite

pentobarbital pentobarbital

pentobarbital sódico sodium
pentobarbital

pentol pentol

pentolita pentolite

pentosa pentose

pentosana pentosan

pentósido pentoside

pentóxido pentoxide

pentóxido de antimonio antimony pentoxide

pentóxido de arsénico arsenic pentoxide

pentóxido de bismuto bismuth pentoxide

pentóxido de fósforo phosphorus pentoxide

pentóxido de niobio niobium pentoxide

pentóxido de nitrógeno nitrogen pentoxide

pentóxido de tántalo tantalum pentoxide

pentóxido de tungsteno tungsten pentoxide

pentóxido de vanadio vanadium pentoxide

pentóxido de yodo iodine pentoxide

pentoxifenol pentoxyphenol

pentrita pentrite

pepsina pepsin

pepsinógeno pepsinogen

peptidasa peptidase

péptido peptide

peptización peptization

peptona peptone

per per

peracetato de butilo butyl peracetate

peracidez peracidity

perácido peracid

peralcohol peralcohol

perbenzoato de butilo butyl perbenzoate

perborato perborate

perborato cálcico calcium perborate

perborato de cinc zinc perborate

perborato magnésico magnesium perborate

perborato sódico sodium perborate

perbromato perbromate

perbromo perbromo

perbromuro fosfórico phosphoric perbromide

percarbamida percarbamide

percarbonato percarbonate

percarbonato potásico potassium percarbonate

percarbonato sódico sodium percarbonate

percarburo percarbide

perclorato perchlorate

perclorato amónico ammonium perchlorate

perclorato bárico barium perchlorate

perclorato cálcico calcium perchlorate

perclorato cobaltoso cobaltous perchlorate

perclorato de cesio cesium perchlorate

perclorato de cinc zinc perchlorate

perclorato de estroncio strontium perchlorate

perclorato de etilo ethyl perchlorate

perclorato de hidrazina hydrazine perchlorate

perclorato de hierro iron perchlorate

perclorato de litio lithium perchlorate

perclorato de níquel nickel perchlorate

perclorato de nitronio nitronium perchlorate

perclorato de nitrosilo nitrosyl perchlorate

perclorato de plata silver perchlorate

perclorato de plomo lead perchlorate

perclorato de rubidio rubidium perchlorate

perclorato férrico ferric perchlorate

perclorato ferroso ferrous perchlorate

perclorato magnésico magnesium perchlorate

perclorato potásico potassium perchlorate

perclorato sódico sodium perchlorate

perclorilo perchloryl

percloro perchloro

perclorobenceno perchlorobenzene

perclorociclopentadieno perchlorocyclopentadiene

percloroetano perchloroethane

percloroéter perchloroether

percloroetileno perchloroethylene

perclorometano perchloromethane

perclorometilmercaptano perchloromethyl mercaptan

perclorometilo perchloromethyl

percloropropileno perchloropropylene

percloruro perchloride

percloruro de estaño tin perchloride

percloruro férrico ferric perchloride

percloruro fosfórico phosphoric perchloride

percolación percolation
percolador percolator
percolar percolate
percristalización percrystallization
percromato perchromate
percromato amónico ammonium
    perchromate
perdestilación perdistillation
perfenazina perphenazine
perfluoro perfluoro
perfluorobuteno perfluorobutene
perfluorocarburo perfluorocarbon
perfluorociclobutano
    perfluorocyclobutane
perfluoroetileno perfluoroethylene
perfluoropropano perfluoropropane
perfluoropropeno perfluoropropene
perfume perfume
perfusión perfusion
pergenol pergenol
perhidrato perhydrate
perhidro perhydro
perhidrol perhydrol
perhidronaftaleno perhydronaphthalene
peri peri
pericíclico pericyclic
periciclo pericyclo
pericinético perkinetic
periclasa periclase
periclina pericline
perileno perylene
perimidina perimidine
perimidinilo perimidinyl
perimorfo perimorph
periodicidad periodicity
periodo period
periodo de identidad identity period
periodo de inducción induction period
peritéctico peritectic
peritectoide peritectoid
perjudicial hazardous
perlas de vidrio glass beads
perlita perlite
permaleación permalloy
permanencia de olor odor permanence
permanganato permanganate
permanganato amónico ammonium
    permanganate
permanganato bárico barium

permanganate
permanganato cálcico calcium
    permanganate
permanganato de bismuto bismuth
    permanganate
permanganato de cadmio cadmium
    permanganate
permanganato de cesio cesium
    permanganate
permanganato de cinc zinc
    permanganate
permanganato de plata silver
    permanganate
permanganato magnésico magnesium
    permanganate
permanganato potásico potassium
    permanganate
permanganato sódico sodium
    permanganate
permanganilo permanganyl
permeabilidad permeability
permeabilidad efectiva effective
    permeability
permeable permeable
permeametría permeametry
permetrina permethrin
permisible allowable
permitido allowed
permutación permutation
permutita permutite
perovskita perovskite
peroxi peroxy
peroxiacetato de butilo butyl
    peroxyacetate
peroxibenzoato de butilo butyl
    peroxybenzoate
peroxicromo peroxychromium
peroxidación peroxidation
peroxidasa peroxidase
peroxidisulfato potásico potassium
    peroxydisulfate
peroxidisulfato sódico sodium
    peroxydisulfate
peróxido peroxide
peróxido bárico barium peroxide
peróxido cálcico calcium peroxide
peróxido de acetilo acetyl peroxide
peróxido de ácido succínico succinic
    acid peroxide

peróxido de benzoilo benzoyl peroxide
peróxido de bismuto bismuth peroxide
peróxido de butilo butyl peroxide
peróxido de carbamida carbamide peroxide
peróxido de carbonato sódico sodium carbonate peroxide
peróxido de cesio cesium peroxide
peróxido de ciclohexanona cyclohexanone peroxide
peróxido de cinc zinc peroxide
peróxido de clorobenzoilo chlorobenzoyl peroxide
peróxido de decanoilo decanoyl peroxide
peróxido de dibenzoilo dibenzoyl peroxide
peroxido de dietilo diethyl peroxide
peróxido de estaño tin peroxide
peróxido de estroncio strontium peroxide
peróxido de etilo ethyl peroxide
peróxido de hidrógeno hydrogen peroxide
peróxido de hidrógeno sódico sodium hydrogen peroxide
peróxido de lauroilo lauroyl peroxide
peróxido de litio lithium peroxide
peróxido de manganeso manganese peroxide
peróxido de mercurio mercury peroxide
peróxido de miristoilo myristoyl peroxide
peróxido de níquel nickel peroxide
peróxido de nitrógeno nitrogen peroxide
peróxido de octilo octyl peroxide
peróxido de pelargonilo pelargonyl peroxide
peróxido de pirofosfato sódico sodium pyrophosphate peroxide
peróxido de plata silver peroxide
peróxido de plomo lead peroxide
peróxido de propionilo propionyl peroxide
peróxido de rubidio rubidium peroxide
peróxido de titanio titanium peroxide
peróxido de urea urea peroxide
peróxido magnésico magnesium peroxide
peróxido potásico potassium peroxide

peróxido sódico sodium peroxide
peroxidol peroxydol
peroxiisobutirato de butilo butyl peroxyisobutyrate
peroxo peroxo
peroxoborato amónico ammonium peroxoborate
peroxoborato de cinc zinc peroxoborate
peroxoborato potásico potassium peroxoborate
peroxodicarbonato sódico sodium peroxodicarbonate
peroxodisulfato amónico ammonium peroxodisulfate
peroxodisulfato bárico barium peroxodisulfate
peroxohidrato de carbonato sódico sodium carbonate peroxohydrate
peroxosulfato potásico potassium peroxosulfate
perrenato perrhenate
persorción persorption
persulfato persulfate
persulfato amónico ammonium persulfate
persulfato potásico potassium persulfate
persulfato sódico sodium persulfate
persulfuro persulfide
persulfuro de antimonio antimony persulfide
persulfuro de fósforo phosphorus persulfide
persulfuro de hidrógeno hydrogen persulfide
pertio perthio
pertiocarbonato perthiocarbonate
pervaporación pervaporation
peryodato periodate
peryodato bárico barium periodate
peryodato de cesio cesium periodate
peryodato potásico potassium periodate
peryodato sódico sodium periodate
pesado heavy
pesaje doble double weighing
peso weight
peso atómico atomic weight
peso atómico-gramo gram-atomic weight
peso de combinación combining weight
peso de fórmula formula weight

peso equivalente equivalent weight
peso equivalente-gramo gram-equivalent
    weight
peso específico specific weight
peso isotópico isotopic weight
peso molar molar weight
peso molecular molecular weight,
    formula weight
peso molecular-gramo gram-molecular
    weight
peso molecular medio average molecular
    weight
peso relativo relative weight
pesticida pesticide
petalita petalite
petrolato petrolatum, petroleum jelly
petróleo petroleum, oil
petróleo bruto crude petroleum, crude
    oil
petroquímico petrochemical
petroxolina petroxolin
petzita petzite
pez pitch
pH pH
phi phi
pi pi
piceno picene
pico peak
pico de absorción absorption peak
pico de Bragg Bragg peak
picocurie picocurie
picolilamina picolylamine
picolilo picolyl
picolina picoline
picotita picotite
picramato sódico sodium picramate
picramida picramide
picrato picrate
picrato amónico ammonium picrate
picrato de plata silver picrate
picrato de plomo lead picrate
picrato de torio thorium picrate
picrato potásico potassium picrate
picrilo picryl
picro picro
picromerita picromerite
picrotoxina picrotoxin
piedmontita piedmontite
piedra de azufre brimstone

piedra de toque touchstone
piedra preciosa precious stone,
    gemstone, gem
piezocristalización piezocrystallization
piezoelectricidad piezoelectricity
piezoquímica piezochemistry
pigmento pigment
pigmento de cadmio cadmium pigment
pigmento de cromo chrome pigment
pigmento fluorescente fluorescent
    pigment
pigmento mineral mineral pigment
pigmento orgánico organic pigment
pila cell
pila de Bunsen Bunsen cell
pila de cadmio cadmium cell
pila de Castner-Kellner Castner-Kellner
    cell
pila de Clark Clark cell
pila de combustible fuel cell
pila de Daniell Daniell cell
pila de Edison Edison cell
pila de gas gas cell
pila de Leclanché Leclanché cell
pila de mercurio mercury cell
pila de Weston Weston cell
pila galvánica galvanic cell
pila patrón standard cell
pila seca dry cell
pila secundaria secondary cell
pila voltaica voltaic cell
pilocarpina pilocarpine
pimaricina pimaricin
pimelita pimelite
pinacoide pinacoid
pinacol pinacol
pinacolona pinacolone
pinacona pinacone
pinanilo pinanyl
pinano pinane
pineno pinene
pinocarveol pinocarveol
pintura paint
pipecolina pipecoline
piperalina piperaline
piperazina piperazine
piperazina de hidroxietilo hydroxyethyl
    piperazine
piperidilo piperidyl

piperidina piperidine
piperidinio piperidinium
piperidino piperidino
piperidinoetanol piperidinoethanol
piperileno piperylene
piperilo piperyl
piperina piperine
piperitol piperitol
piperitona piperitone
piperonal piperonal
piperonilideno piperonylidene
piperonilo piperonyl
piperoniloilo piperonyloyl
pipeta pipet
pipeta automática automatic pipet
pipeta capilar capillary pipet
pipeta graduada graduated pipet
pipeta volumétrica volumetric pipet
piranilo pyranyl
pirano pyran
piranona pyranone
piranosa pyranose
piranosida pyranoside
pirantreno pyranthrene
pirazina pyrazine
pirazol pyrazole
pirazolidina pyrazolidine
pirazolilo pyrazolyl
pirazolina pyrazoline
pirazolona pyrazolone
pirenilo pyrenyl
pireno pyrene
piretrina pyrethrin
piretro pyrethrum
piretroide pyrethroid
piretrolona pyrethrolone
piridazina pyridazine
piridilamina pyridylamine
piridilcarbinol pyridylcarbinol
piridilo pyridyl
piridina pyridine
piridinilo pyridinyl
piridinio pyridinium
piridona pyridone
piridoxal pyridoxal
piridoxina pyridoxine
pirimidina pyrimidine
pirimidinilo pyrimidinyl
pirimitato pyrimithate

pirita pyrite
pirita de hierro iron pyrite
piritiamina pyrithiamine
piritiona de cinc zinc pyrithione
piro pyro
piroantimoniato sódico sodium
    pyroantimonate
piroborato potásico potassium
    pyroborate
piroborato sódico sodium pyroborate
pirocatecol pyrocatechol
pirocelulosa pyrocellulose
pirocloro pyrochlore
pirocroíta pyrochroite
pirofilita pyrophyllite
pirofórico pyrophoric
pirofosfato pyrophosphate
pirofosfato ácido sódico sodium acid
    pyrophosphate
pirofosfato bárico barium pyrophosphate
pirofosfato cálcico calcium
    pyrophosphate
pirofosfato de cinc zinc pyrophosphate
pirofosfato de etilo ethyl pyrophosphate
pirofosfato de hierro iron pyrophosphate
pirofosfato de hierro sódico sodium iron
    pyrophosphate
pirofosfato de plomo lead pyrophosphate
pirofosfato de tiamina thiamine
    pyrophosphate
pirofosfato de zirconio zirconium
    pyrophosphate
pirofosfato estannoso stannous
    pyrophosphate
pirofosfato férrico ferric pyrophosphate
pirofosfato magnésico magnesium
    pyrophosphate
pirofosfato manganoso manganous
    pyrophosphate
pirofosfato potásico potassium
    pyrophosphate
pirofosfato sódico sodium pyrophosphate
pirofosfato tetrapotásico tetrapotassium
    pyrophosphate
pirofosfato tetrasódico tetrasodium
    pyrophosphate
pirofosforilo pyrophosphoryl
pirogalato pyrogallate
pirogalol pyrogallol

pirolán pyrolan
pirolignita de hierro iron pyrolignite
pirólisis pyrolysis
pirolusita pyrolusite
pirometría pyrometry
pirómetro pyrometer
piromorfita pyromorphite
pirona pyrone
pirorracemato sódico sodium
　　pyroracemate
pirosina pyrosin
pirosulfato pyrosulfate
pirosulfato potásico potassium
　　pyrosulfate
pirosulfato sódico sodium pyrosulfate
pirosulfito pyrosulfite
pirosulfito potásico potassium
　　pyrosulfite
pirosulfito sódico sodium pyrosulfite
pirosulfurilo pyrosulfuryl
pirotecnia pyrotechnics
pirovanadato sódico sodium
　　pyrovanadate
piroxeno pyroxene
piroxilina pyroxylin
pirrilo pyrryl
pirrocolina pyrrocoline
pirroilo pyrroyl
pirrol pyrrole
pirrolidina pyrrolidine
pirrolidinilo pyrrolidinyl
pirrolidona pyrrolidone
pirrolilo pyrrolyl
pirrolina pyrroline
pirrona pyrrone
pirrotita pyrrhotite
piruvaldehído pyruvaldehyde
piruvato de etilo ethyl pyruvate
piruvato de metilo methyl pyruvate
piruvato sódico sodium pyruvate
piruvonitrilo pyruvonitrile
pivaldehído pivaldehyde
pizarra slate
placa plate
placa perforada perforated plate
placa teorética theoretical plate
plano atómico atomic plane
plano de polarización plane of
　　polarization

plano de simetría plane of symmetry
planta plant
planta piloto pilot plant
plasma plasma
plásmido plasmid
plasmina plasmin
plasmógeno plasmogen
plasticidad plasticity
plasticizador plasticizer
plástico plastic
plástico alílico allyl plastic
plástico bituminoso bituminous plastic
plástico celular cellular plastic
plástico celulósico cellulosic plastic
plástico de acrilato acrylate plastic
plástico de estireno styrene plastic
plástico de halocarburo halocarbon
　　plastic
plástico reforzado reinforced plastic
plastisol plastisol
plastoquinona plastoquinone
plata silver
plata esterlina sterling silver
plateado silver-plating
platinado platinization
platinicloruro sódico sodium
　　platinichloride
platínico platinic
platino platinum
platino-litio platinum-lithium
platinocianuro platinocyanide
platinocloruro platinochloride
platinocloruro sódico sodium
　　platinochloride
platinoso platinous
plato de evaporación evaporation dish
plazolita plazolite
pleocroismo pleochroism
plomado lead-coating
plomo lead
plomo azul blue lead
plomo blanco white lead
plomo negro black lead
plomo rojo red lead
plomo telúrico telluric lead
plumbago plumbago
plumbato plumbate
plumbato cálcico calcium plumbate
plumbato sódico sodium plumbate

plúmbico plumbic
plumbito sódico sodium plumbite
plumboso plumbous
plutonio plutonium
poder de absorción absorptive power
poder de resolución resolving power
poder rotatorio rotatory power
polar polar
polaridad polarity
polaridad química chemical polarity
polarimetría polarimetry
polarímetro polarimeter
polarizabilidad polarizability
polarizabilidad molecular molecular polarizability
polarizable polarizable
polarización polarization
polarización anódica anodic polarization
polarización atómica atomic polarization
polarización catódica cathodic polarization
polarización circular circular polarization
polarización electrolítica electrolytic polarization
polarización electroquímica electrochemical polarization
polarización nuclear nuclear polarization
polarizado polarized
polarizado en un plano plane-polarized
polarizador polarizer
polarizante polarizing
polarizar polarize
polarografía polarography
polarograma polarogram
poli poly
poliacetal polyacetal
poliacetaldehído polyacetaldehyde
poliacetileno polyacetylene
poliacrilamida polyacrylamide
poliacrilato polyacrylate
poliacrilonitrilo polyacrylonitrile
polialcano polyalkane
polialcohol polyalcohol
polialómero polyallomer
polialqueno polyalkene
polialquilideno polyalkylidene
poliamida polyamide
poliamina polyamine

poliaminotriazol polyaminotriazole
poliatómico polyatomic
polibásico polybasic
polibencimidazol polybenzimidazole
polibutadieno polybutadiene
polibuteno polybutene
polibutileno polybutylene
policarbonato polycarbonate
policarboxílico polycarboxylic
policíclico polycyclic
policloral polychloral
policloro polychlor
policloropreno polychloroprene
policlorotrifluoroetano polychlorotrifluoroethane
policondensación polycondensation
policromático polychromatic
polidimetilsiloxano polydimethylsiloxane
polidispersión polydispersion
polielectrólito polyelectrolyte
polieno polyene
poliespiro polyspiro
poliéster polyester
poliéster de sucrosa sucrose polyester
poliestireno polystyrene
polietenoide polyethenoid
poliéter polyether
poliéter clorado chlorinated polyether
polieterglicól polyether glycol
polietilenglicol polyethylene glycol
polietilenimina polyethyleneimine
polietileno polyethylene
polifásico polyphase
polifenilo polyphenyl
poliformación polyforming
poliformaldehído polyformaldehyde
polifosfato polyphosphate
polifosfato amónico ammonium polyphosphate
polifosfato sódico sodium polyphosphate
polifuncional polyfunctional
polígeno polygen
poliglicerol polyglycerol
poliglicol polyglycol
polihalita polyhalite
polihexafluoropropeno polyhexafluoropropene
polihidrato polyhydrate
polihídrico polyhydric

poliimida polyimide
poliinsaturado polyunsaturated
poliisobuteno polyisobutene
poliisobutileno polyisobutylene
poliisocianurato polyisocyanurate
poliisopreno polyisoprene
polimérico polymeric
polimerismo polymerism
polimerización polymerization
polimerización aniónica anionic
    polymerization
polimerización catalítica catalytic
    polymerization
polimerización catiónica cationic
    polymerization
polimerización de vinilo vinyl
    polymerization
polimerización de Ziegler-Natta
    Ziegler-Natta polymerization
polimerización en emulsión emulsion
    polymerization
polimerización iónica ionic
    polymerization
polimerizar polymerize
polímero polymer
polímero acrílico acrylic polymer
polímero alternante alternating polymer
polímero amorfo amorphous polymer
polímero atáctico atactic polymer
polímero con flúor fluorine polymer
polímero con injerto graft polymer
polímero de ácido acrílico acrylic acid
    polymer
polímero de acrilamida acrylamide
    polymer
polímero de acrilonitrilo acrylonitrile
    polymer
polímero de acroleína acrolein polymer
polímero de adición addition polymer
polímero de aldehído aldehyde polymer
polímero de alilo allyl polymer
polímero de cadena chain polymer
polímero de clorotrifluoroetileno
    chlorotrifluoroethylene polymer
polímero de cloruro de vinilideno
    vinylidene chloride polymer
polímero de condensación condensation
    polymer
polímero de coordinación coordination

polymer
polímero de estereobloques stereoblock
    polymer
polímero de estireno styrene polymer
polímero de eteno ethene polymer
polímero de fluorocarburo fluorocarbon
    polymer
polímero de isopreno isoprene polymer
polímero de piridina pyridine polymer
polímero de propeno propene polymer
polímero de tetrahidrofurano
    tetrahydrofuran polymer
polímero de vinilo vinyl polymer
polímero en escalera ladder polymer
polímero epóxido epoxy polymer
polímero estereoespecífico stereospecific
    polymer
polímero estereorregular stereoregular
    polymer
polímero inorgánico inorganic polymer
polímero lineal linear polymer
polímero nitroso nitroso polymer
polímero regular regular polymer
polímero táctico tactic polymer
polimetafosfato potásico potassium
    polymetaphosphate
polimetanal polymethanal
polimetilbenceno polymethylbenzene
polimetilenglicol polymethylene glycol
polimetilpenteno polymethylpentene
polimezcla polyblend
polimixina polymyxin
polimorfismo polymorphism
polinuclear polynuclear
polinuclídico polynuclidic
poliol polyol
poliolefina polyolefin
poliolefina clorada chlorinated
    polyolefin
polioxadiazol polyoxadiazole
polioxamida polyoxamide
polioxi polyoxy
polioxietileno polyoxyethylene
polioxietilenoxipropileno
    polyoxyethyleneoxypropylene
polioximetileno polyoxymethylene
polioxipropilendiamina
    polyoxypropylenediamine
polipéptido polypeptide

polipirrolidina polypyrrolidine
poliprenol polyprenol
polipropilenbenceno
　　polypropylenebenzene
polipropilenglicol polypropylene glycol
polipropileno polypropylene
polirrotaxama polyrotaxame
polisacárido polysaccharide
polisilicato cálcico calcium polysilicate
polisiloxano polysiloxane
polisorbato polysorbate
polisulfona polysulfone
polisulfuro polysulfide
polisulfuro amónico ammonium
　　polysulfide
polisulfuro potásico potassium
　　polysulfide
polisulfuro sódico sodium polysulfide
politeno polythene
politetrafluoroeteno
　　polytetrafluoroethene
politetrafluoroetileno
　　polytetrafluoroethylene
politiacilo polythiazyl
politiadazol polythiadazole
politionato polythionate
poliuretano polyurethane
polivalencia polyvalence
polivalente polyvalent
polivinilacetal polyvinyl acetal
polivinilbutiral polyvinyl butyral
polivinilcarbazol polyvinyl carbazole
polivinilideno polyvinylidene
polivinilo polyvinyl
polivinilpirrolidina polyvinylpyrrolidine
polixematileno polyhexamethylene
poliyoduro polyiodide
polonio polonium
polucita pollucite
polvo powder, dust
polvo azul blue powder
polvo de blanqueo bleaching powder
polvo de cinc zinc dust
polvo de hierro iron powder
polvo de hornear baking powder
polvo negro black powder
pólvora gunpowder
pólvora de algodón guncotton
pólvora sin humo smokeless powder

pómez pumice
porcelana porcelain
porcentaje percentage
porfina porphin
porfirina porphyrin
porfirinógeno porphyrinogen
poro pore
porométrico porometric
porosidad porosity
porosímetro porosimeter
poroso porous
portador carrier
portador de catalizador catalyst carrier
portador de oxígeno oxygen carrier
portadora de hidrógeno hydrogen
　　carrier
posición adyacente adjacent position
posición alfa alpha position
posición beta beta position
positivo positive
positrón positron
positronio positronium
posprecipitación postprecipitation
potasa potash
potasa cáustica caustic potash
potasán potasan
potásico potassic
potasio potassium
potencial potential
potencial atómico atomic potential
potencial crítico critical potential
potencial de contacto contact potential
potencial de deposición deposition
　　potential
potencial de electrodo electrode
　　potential
potencial de equilibrio equilibrium
　　potential
potencial de ionización ionization
　　potential, ionizing potential
potencial de Morse Morse potential
potencial de Nernst Nernst potential
potencial de oxidación oxidation
　　potential
potencial de oxidación-reducción
　　oxidation-reduction potential
potencial de polarización polarization
　　potential
potencial de redox redox potential

potencial de reducción reduction
potential
potencial de sedimentación
sedimentation potential
potencial electrocinético electrokinetic
potential
potencial electrolítico electrolytic
potential
potencial electronegativo electronegative
potential
potencial electropositivo electropositive
potential
potencial electroquímico electrochemical
potential
potencial noble noble potential
potencial químico chemical potential
potencial termodinámico
thermodynamic potential
potencial zeta zeta potential
potenciometría redox redox
potentiometry
potenciométrico potentiometric
praseodimio praseodymium
precauciones de almacenamiento
storage precautions
precipitabilidad precipitability
precipitable precipitable
precipitación precipitation
precipitación ácida acid precipitation
precipitación fraccionaria fractional
precipitation
precipitación isoeléctrica isoelectric
precipitation
precipitación rítmica rhythmic
precipitation
precipitado precipitate, deposit
precipitado amarillo yellow precipitate
precipitado blanco white precipitate
precipitado coherente coherent
precipitate
precipitado rojo red precipitate
precipitador precipitator
precipitador electrostático electrostatic
precipitator
precipitante precipitant
precursor precursor
predisociación predissociation
prednisolona prednisolone
prednisona prednisone

preferencial preferential
pregananodiol pregnanediol
pregnenodiona pregnenedione
pregnenolona pregnenolone
prehnita prehnite
prenilo prenyl
preparación preparation
prepolímero prepolymer
preservativo preservative
presión pressure
presión ambiente ambient pressure
presión atmosférica atmospheric
pressure
presión barométrica barometric pressure
presión constante constant pressure
presión crítica critical pressure
presión de disociación dissociation
pressure
presión de disolución solution pressure
presión de gas gas pressure
presión de vapor vapor pressure
presión normal normal pressure,
standard pressure
presión osmótica osmotic pressure
presión parcial partial pressure
presión superficial surface pressure
primario primary
primera ley de termodinámica first law
of thermodynamics
principal principal
principio amargo bitter principle
principio de correspondencia
correspondence principle
principio de distribución distribution
principle
principio de exclusión exclusion
principle
principio de exclusión de Pauli Pauli
exclusion principle
principio de exclusión de Pauli-Fermi
Pauli-Fermi exclusion principle
principio de Franck-Condon Franck-
Condon principle
principio de incertidumbre uncertainty
principle
**principio de incertidumbre de
Heisenberg** Heisenberg uncertainty
principle
principio de Le Chatelier Le Chatelier

principle
principio de Pauli Pauli principle
prisma prism
prisma de Nicol Nicol prism
pristano pristane
probeta test tube
procedimento de Pregl Pregl procedure
procedimiento procedure
procedimiento de Darzens Darzens
　　procedure
procedimiento de Wilzbach Wilzbach
　　procedure
procesamiento processing
procesamiento químico chemical
　　processing
proceso process
proceso adiabático adiabatic process
proceso cíclico cyclic process
proceso de Acker Acker process
proceso de Acheson Acheson process
proceso de amalgamación amalgamation
　　process
proceso de Barff Barff process
proceso de Bayer Bayer process
proceso de Benfield Benfield process
proceso de Bergius Bergius process
proceso de Bessemer Bessemer process
proceso de Betterton-Kroll Betterton-
　　Kroll process
proceso de Betts Betts process
proceso de Birkeland-Eyde Birkeland-
　　Eyde process
proceso de Borcher Borcher process
proceso de Bosch Bosch process
proceso de Brewster Brewster process
proceso de Brin Brin process
proceso de Castner Castner process
proceso de cianamida cyanamide
　　process
proceso de cianuro cyanide process
proceso de Clark Clark process
proceso de Claude Claude process
proceso de Coahran Coahran process
proceso de contacto contact process
proceso de conversión conversion
　　process
proceso de copelación cupellation
　　process
proceso de Creighton Creighton process

proceso de cumeno cumene process
proceso de Chance-Claus Chance-Claus
　　process
proceso de Deacon Deacon process
proceso de Downs Downs process
proceso de Edeleanu Edeleanu process
proceso de Fament Fament process
proceso de Farrar Farrar process
proceso de fenolato phenolate process
proceso de Fischer-Tropsch Fischer-
　　Tropsch process
proceso de Frasch Frasch process
proceso de Goldschmidt Goldschmidt
　　process
proceso de Haber Haber process
proceso de Hall Hall process
proceso de Hargreaves Hargreaves
　　process
proceso de Houdriflow Houdriflow
　　process
proceso de Houdry Houdry process
proceso de Hunter Hunter process
proceso de Keyes Keyes process
proceso de Kroll Kroll process
proceso de Lebedev Lebedev process
proceso de Linde Linde process
proceso de Lyovac Lyovac process
proceso de Marathon-Howard
　　Marathon-Howard process
proceso de Mond Mond process
proceso de oxidación oxidation process
proceso de Parkes Parkes process
proceso de Pattinson Pattinson process
proceso de Raschig Raschig process
proceso de recombinación
　　recombination process
proceso de Reich Reich process
proceso de Reppe Reppe process
proceso de Roesler Roesler process
proceso de Rosenstein Rosenstein
　　process
proceso de Siemens Siemens process
proceso de Siemens-Halske Siemens-
　　Halske process
proceso de Simons Simons process
proceso de Sperry Sperry process
proceso de Steffen Steffen process
proceso de Stengel Stengel process
proceso de Thermofor Thermofor

process
proceso de Toth Toth process
proceso de Trona Trona process
proceso de Twitchell Twitchell process
proceso de Wacker Wacker process
proceso de Weldon Weldon process
proceso de Wohlwill Wohlwill process
proceso de Wulff Wulff process
proceso de Ziegler Ziegler process
proceso de Ziervogel Ziervogel process
proceso electrolítico electrolytic process
proceso electroquímico electrochemical
    process
proceso elemental elementary process
proceso en cascada cascade process
proceso estocástico stochastic process
proceso fotoquímico photochemical
    process
proceso irreversible irreversible process
proceso isentrópico isentropic process
proceso isotérmico isothermal process
proceso nuclear nuclear process
proceso químico chemical process
proceso radioquímico radiochemical
    process
proceso reversible reversible process
proceso termatómico thermatomic
    process
producción production
producción quimionuclear
    chemonuclear production
producto product
producto de desintegración decay
    product
producto de desintegración radiactiva
    radioactive decay product
producto de fisión fission product
producto de fusión fusion product
producto de reacción reaction product
producto de solubilidad solubility
    product
producto de volatilidad volatility
    product
producto iónico ionic product
producto químico chemical product,
    chemical
producto químico agrícola agricultural
    chemical
producto químico aleopático

allelopathic chemical
producto químico fino fine chemical
producto químico nocivo noxious
    chemical
producto químico peligroso hazardous
    chemical
producto químico pesado heavy
    chemical
producto químico seco dry chemical
producto radiactivo radioactive product
producto secundario by-product
productor de neutrones neutron
    producer
progesterona progesterone
proguanil proguanil
prolactina prolactin
prolamina prolamine
prolilo prolyl
prolina proline
promazina promazine
prometazina promethazine
prometio promethium
promotor promoter
promotor de catalizador catalyst
    promotor
promotor de reacción reaction promotor
propadieno propadiene
propagación propagation
propanal propanal
propanamida propanamide
propanil propanil
propano propane
propanodiamina propanediamine
propanodiol propanediol
propanoilo propanoyl
propanol propanol
propanolamina propanolamine
propanolpiridina propanolpyridine
propanona propanone
propanonitrilo propanenitrile
propanosultona propane sultone
propanotiol propanethiol
propargilo propargyl
propargita propargite
propelente propellant
propenal propenal
propenilamina propenylamine
propenilanisol propenylanisole
propenileno propenylene

propenilguaetol propenyl guaethol
propenilideno propenylidene
propenilo propenyl
propeniltiourea propenylthiourea
propeno propene
propenol propenol
propenonitrilo propenenitrile
propenotiol propenethiol
propidedad periódica periodic property
propiedad property
propiedad coligativa colligative property
propiedad física physical property
propiedad química chemical property
propiedades atómicas atomic properties
propiedades de grupo group properties
propiedades intensivas intensive
    properties
propiedades seudocríticas pseudocritical
    properties
propiedades seudorreducidas
    pseudoreduced properties
propiedades tóxicas toxic properties
propilacetona propylacetone
propilamina propylamine
propilanilina propylaniline
propilbenceno propylbenzene
propilendiamina propylenediamine
propilenglicol propylene glycol
propilenimina propyleneimine
propileno propylene
propilhidroxilamina
    propylhydroxylamine
propilideno propylidene
propilmercaptano propyl mercaptan
propilo propyl
propilparabeno propylparaben
propilpiperidina propylpiperidine
propilpiridina propylpyridine
propiltriclorosilano propyltrichlorosilane
propilurea propylurea
propinal propynal
propinilo propynyl
propino propyne
propinol propynol
propiofenona propiophenone
propiolactona propiolactone
propiolato de etilo ethyl propiolate
propioloilo propioloyl
propiona propione

propionaldehído propionaldehyde
propionato propionate
propionato bárico barium propionate
propionato cálcico calcium propionate
propionato cobaltoso cobaltous
    propionate
propionato de amilo amyl propionate
propionato de bencilo benzyl propionate
propionato de bismuto bismuth
    propionate
propionato de butilo butyl propionate
propionato de cadmio cadmium
    propionate
propionato de celulosa cellulose
    propionate
propionato de cinc zinc propionate
propionato de etilo ethyl propionate
propionato de geranilo geranyl
    propionate
propionato de glicol glycol propionate
propionato de isoamilo isoamyl
    propionate
propionato de isobutilo isobutyl
    propionate
propionato de linalilo linalyl propionate
propionato de metilo methyl propionate
propionato de pentilo pentyl propionate
propionato de plomo lead propionate
propionato de propilo propyl propionate
propionato de vinilo vinyl propionate
propionato fenilmercúrico
    phenylmercuric propionate
propionato magnésico magnesium
    propionate
propionato sódico sodium propionate
propionilbenceno propionylbenzene
propionilo propionyl
propionitrilo propionitrile
propiono propiono
proporción proportion
proporción constante constant
    proportion
proporción fija fixed proportion
proporcional proportional
proporciones definidas definite
    proportions
proporciones equivalentes equivalent
    proportions
propoxi propoxy

propóxido alumínico aluminum propoxide

propoxifeno propoxyphene

propoxipropanol propoxypropanol

propranonol propranonol

proquiral prochiral

prostaglandina prostaglandin

protactinio protactinium

protamina protamine

proteasa protease

protección ambiental environmental protection

protección catódica cathodic protection

protección contra fuego fire protection

protección contra humedad humidity protection

protección sacrificial sacrificial protection

protector químico chemical protector

proteína protein

proteína hidrolizada hydrolized protein

proteólisis proteolysis

protilo protyl

protio protium

proto proto

protocloruro de estaño tin protochloride

protocloruro de hierro iron protochloride

protofílico protophilic

protogénico protogenic

protólisis protolysis

protón proton

protón nuclear nuclear proton

protón retardado delayed proton

protonación protonation

protoplasma protoplasm

protosulfato protosulfate

protosulfuro de estaño tin protosulfide

protosulfuro de hierro iron protosulfide

prototrópico prototropic

protóxido bárico barium protoxide

protóxido de estaño tin protoxide

protóxido de manganeso manganese protoxide

protóxido de níquel nickel protoxide

protóxido de plomo lead protoxide

protrombina prothrombin

provitamina provitamin

proyección projection

proyección de Fischer Fischer projection

proyección de Haworth Haworth projection

proyección de Newman Newman projection

proyección en cabrilla sawhorse projection

prueba test, proof

prueba con llama flame test

prueba cualitativa qualitative test

prueba cuantitativa quantitative test

prueba de acroleína acrolein test

prueba de Almen Almen test

prueba de anillo ring test

prueba de anillo pardo brown-ring test

prueba de azufre sulfur test

prueba de Babcock Babcock test

prueba de banco bench test

prueba de Barfoed Barfoed test

prueba de Beilstein Beilstein test

prueba de bencidina benzidine test

prueba de Benedict Benedict test

prueba de Bettendorf Bettendorf test

prueba de biuret biuret test

prueba de Boettger Boettger test

prueba de Brown-Boverti Brown-Boverti test

prueba de color de Rosenheim Rosenheim color test

prueba de dureza de Brinell Brinell hardness test

prueba de Fehling Fehling test

prueba de Foulger Foulger test

prueba de Hinsberg Hinsberg test

prueba de Kjeldahl Kjeldahl test

prueba de Lassaigne Lassaigne test

prueba de Marsh Marsh test

prueba de Millon Millon test

prueba de Molisch Molisch test

prueba de perla bead test

prueba de perla con bórax borax-bead test

prueba de Reinsch Reinsch test

prueba de Riegler Riegler test

prueba de Schiff Schiff test

prueba de solubilidad solubility test

prueba de Tauber Tauber test

prueba de Tollens Tollens test

prueba de yodo iodine test

prueba física physical test
prueba no destructiva nondestructive test
prueba ordinal ordinal test
prueba química chemical test
prusiato amarillo de potasa yellow prussiate of potash
prusiato amarillo de sosa yellow prussiate of soda
prusiato potásico potassium prussiate
prusiato sódico sodium prussiate
pseudo pseudo
psi psi
psilocibina psilocybin
psilocina psilocin
psilomelano psilomelane
pteridina pteridine
pterina pterin
ptialina ptyalin
ptomaína ptomaine
puente bridge
puente hidrídico hydridic bridge
puente protónico protonic bridge
puente salino salt bridge
pulegona pulegone
pulpa pulp
pulpa de madera wood pulp
pulpa de sosa soda pulp
pulpa de sulfito sulfite pulp
pululanasa pullulanase
pulverización pulverization, comminution
pulverización catódica cathodic sputtering, sputtering
punta peak
punto crítico critical point
punto cuádruple quadruple point
punto de anilina aniline point
punto de autoignición autoignition point
punto de condensación al vacío vacuum condensing point
punto de congelación freezing point, ice point
punto de Curie Curie point
punto de descomposición decomposition point
punto de ebullición boiling point

punto de ebullición absoluto absolute boiling point
punto de ebullición inicial initial boiling point
punto de equilibrio equilibrium point
punto de equivalencia equivalence point
punto de fluidez pour point
punto de flujo pour point
punto de fusión melting point, fusion point
punto de fusión mixto mixed melting point
punto de ignición ignition point
punto de inflamabilidad flash point
punto de inversión inversion point
punto de rocío dew point
punto de saturación saturation point
punto de solidificación setting point
punto de sublimación sublimation point
punto de transición transition point
punto de turbidez turbidity point
punto de vapor steam point
punto eutéctico eutectic point
punto final end point
punto invariante invariant point
punto isobéstico isobestic point
punto isoeléctrico isoelectric point
punto lambda lambda point
punto peritéctico peritectic point
punto seco dry point
punto triple triple point
pureza purity, fineness
pureza colorimétrica colorimetric purity
pureza óptica optical purity
pureza química chemical purity
purificación purification
purificación de agua water purification
purina purine
puro pure
puromicina puromycin
púrpura de bromocresol bromocresol purple
púrpura de cresol cresol purple
púrpura de Tiro Tyrian purple
purpurina purpurin
putrefacción putrefaction
putrescina putrescine

Q

quark quark
quebracho quebracho
quebradizo brittle
queiramidina cheiramidine
quelación chelation
quelato chelate
quelato de titanio titanium chelate
queleritrina chelerythrine
quemado burnt
quemadura química chemical burn
queratina keratin
queratinasa keratinase
quercitol quercitol
quercitrina quercitrin
queroseno kerosene
quevkinita chevkinite
quianita kyanite
quilate karat, carat
química chemistry
química ambiental environmental
    chemistry
química analítica analytical chemistry
química aplicada applied chemistry
química bioinorgánica bioinorganic
    chemistry
química biomimética biomimetic
    chemistry
química clínica clinical chemistry
química computacional computational
    chemistry
química cuántica quantum chemistry
química de estado sólido solid-state
    chemistry
química de radiación radiation
    chemistry
química física physical chemistry
química forense forensic chemistry
química industrial industrial chemistry
química inorgánica inorganic chemistry
química legal legal chemistry
química medicinal medicinal chemistry
química nuclear nuclear chemistry
química orgánica organic chemistry

química superficial surface chemistry
química supramolecular supramolecular
    chemistry
químicamente inducido chemically
    induced
químicamente puro chemically pure
químico chemist
quimiodinámica chemodynamics
quimiólisis chemolysis
quimioluminiscencia chemiluminescence
quimiometría chemometrics
quimionuclear chemonuclear
quimiosíntesis chemosynthesis
quimiósmosis chemosmosis
quimiosmótico chemosmotic
quimiosorción chemosorption
quimióstato chemostat
quimiotaxis chemotaxis
quimiotaxonomía chemotaxonomy
quimioterapia chemotherapy
quimiotropismo chemotropism
quimisorción chemisorption
quimosina chymosin
quimotripsina chymotrypsin
quimotripsinógeno chymotrypsinogen
quina cinchona bark
quinacrina quinacrine
quinaftol quinaphthol
quinaldina quinaldine
quinamina quinamine
quinazolina quinazoline
quinhidrona quinhydrone
quinidina quinidine
quinina quinine
quininona quininone
quinitol quinitol
quinizarina quinizarin
quinógeno quinogen
quinoidina chinoidine
quinoilo quinoyl
quinol quinol
quinolilo quinolyl
quinolina quinoline

quinolinio quinolinium
quinolinol quinolinol
quinolinona quinolinone
quinolizina quinolizine
quinolona quinolone
quinona quinone
quinonadioxima quinonedioxime
quinonilo quinonyl
quinosol quinosol
quinoxalina quinoxaline
quinoxalinilo quinoxalinyl
quinquevalente quinquevalent

quinteto quintet
quintoceno quintozene
quinuclidina quinuclidine
quinurenina kynurenine
quiral chiral
quiralidad chirality, handedness
quiroóptico chirooptical
quitapintura paint remover
quitina chitin
quitinasa chitinase
quitosana chitosan

R

racema raceme
racemación racemation
racemasa racemase
racemato racemate
racémico racemic
racemización racemization
racemización parcial partial
 racemization
rad rad
radiación radiation
radiación actínica actinic radiation
radiación beta beta radiation
radiación dañina harmful radiation
radiación de fondo background radiation
radiación de fuga leakage radiation
radiación de ionización ionizing
 radiation
radiación de resonancia resonance
 radiation
radiación dipolar dipole radiation
radiación electromagnética
 electromagnetic radiation
radiación fluorescente fluorescent
 radiation
radiación gamma gamma radiation
radiación infrarroja infrared radiation
radiación nuclear nuclear radiation
radiación primaria primary radiation
radiación secundaria secondary
 radiation
radiación ultravioleta ultraviolet
 radiation
radiactividad radioactivity
radiactividad artificial artificial
 radioactivity
radiactivo radioactive
radiado radiated
radiador radiator
radiador alfa alpha radiator
radiancia radiance
radiante radiant, radiating
radiar radiate
radiativo radiative

radical radical
radical de ácido acid radical
radical electronegativo electronegative
 radical
radical enmascarado masked radical
radical libre free radical
radical negativo negative radical
radical orgánico organic radical
radical positivo positive radical
radio radium, radius
radio atómico atomic radius
radio covalente covalent radius
radio de Bohr Bohr radius
radio de electrón clásico classical
 electron radius
radio de van der Waals van der Waals
 radius
radio hidrodinámico hydrodynamic
 radius
radio iónico ionic radius
radio nuclear nuclear radius
radioactinio radioactinium
radioanálisis radioassay
radiobario radiobarium
radiobiología radiobiology
radiocarbono radiocarbon
radiocesio radiocesium
radiocristalografía radiocrystallography
radiocromatografía
 radiochromatography
radioelemento radioelement
radiogénico radiogenic
radiografía radiography
radioinmunoanálisis radioimmunoassay
radioisótopo radioisotope
radiólisis radiolysis
radiolita radiolite
radiología radiology
radiológico radiological
radioluminiscencia radioluminescence
radiometalografía radiometallography
radiometría radiometry
radiométrico radiometric

radiómetro radiometer
radiomimético radiomimetic
radión radion
radionitrógeno radionitrogen
radionucleido radionuclide
radioplomo radiolead
radioquímica radiochemistry
radioquímico radiochemical
radiosodio radiosodium
radiotelurio radiotellurium
radiotor radiothor
radiotorio radiothorium
radioyodo radioiodine
radón radon
rafinato raffinate
rafinosa raffinose
ralstonita ralstonite
ramificación branching
ramificado branched
ramio ramie
ramnitol rhamnitol
ramnosa rhamnose
rancio rancid, sour
rankinita rankinite
ranksita ranksite
rápido fast
rarefacción rarefaction
rarefactor getter
rareficar rarefy
rasorita rasorite
rauwolfia rauwolfia
rayo ray
rayo beta beta ray
rayo delta delta ray
rayón rayon
rayos alfa alpha rays
rayos canales canal rays
rayos catódicos cathodic rays
rayos de Becquerel Becquerel rays
rayos gamma gamma rays
rayos moleculares molecular rays
razón ratio
razón de abundancia abundance ratio
razón de carbono carbon ratio
razón de carga a masa charge-mass ratio
razón de conversión conversion ratio
razón de dilución dilution ratio
razón de equilibrio equilibrium ratio

razón de masa a carga mass-to-charge ratio
razón de mezcla de gases gas admixture ratio
razón de neutralización neutralization ratio
razón de radios radius ratio
razón giromagnética gyromagnetic ratio
razón isotópica isotopic ratio
reacción reaction
reacción ácida acid reaction
reacción alcalina alkaline reaction
reacción analítica analytical reaction
reacción balanceada balanced reaction
reacción bimolecular bimolecular reaction
reacción caótica chaotic reaction
reacción compleja complex reaction
reacción con llama flame reaction
reacción cualitativa qualitative reaction
reacción cuantitativa quantitative reaction
reacción de Acree-Rosenheim Acree-Rosenheim reaction
reacción de Adamkiewicz Adamkiewicz reaction
reacción de adición addition reaction
reacción de Adkins-Peterson Adkins-Peterson reaction
reacción de Allan-Robinson Allan-Robinson reaction
reacción de anillo ring reaction
reacción de Bart Bart reaction
reacción de Barton Barton reaction
reacción de Baudisch Baudisch reaction
reacción de Béchamp Béchamp reaction
reacción de Beckmann Beckmann reaction
reacción de Belousov-Zhabotinskii Belousov-Zhabotinskii reaction
reacción de Benary Benary reaction
reacción de Berthelot Berthelot reaction
reacción de Betti Betti reaction
reacción de Biginelli Biginelli reaction
reacción de biuret biuret reaction
reacción de Blaise Blaise reaction
reacción de Blanc Blanc reaction
reacción de Bodroux Bodroux reaction
reacción de Bohn-Schmidt Bohn-

Schmidt reaction

reacción de Bradsher Bradsher reaction

reacción de Bucherer Bucherer reaction

reacción de Bucherer-Bergs Bucherer-
Bergs reaction

reacción de cadena ramificada
branched-chain reaction

reacción de Cannizzaro Cannizzaro
reaction

reacción de Carroll Carroll reaction

reacción de ciclización de Nazarov
Nazarov cyclization reaction

reacción de Claisen Claisen reaction

reacción de condensación condensation
reaction

reacción de condensación de Mannich
Mannich condensation reaction

reacción de Conrad-Limpach Conrad-
Limpach reaction

reacción de Criegee Criegee reaction

reacción de Curtius Curtius reaction

reacción de Chichibabin Chichibabin
reaction

reacción de Chugaev Chugaev reaction

reacción de Dakin Dakin reaction

reacción de Darzens Darzens reaction

reacción de De Mayo De Mayo reaction

reacción de Delepine Delepine reaction

reacción de desplazamiento
displacement reaction

reacción de Dieckmann Dieckmann
reaction

reacción de Diels-Alder Diels-Alder
reaction

reacción de Dische Dische reaction

reacción de disociación dissociation
reaction

reacción de Doebner Doebner reaction

reacción de Doebner-Miller Doebner-
Miller reaction

reacción de Dragendorff Dragendorff
reaction

reacción de Duff Duff reaction

reacción de Dutt-Wormall Dutt-
Wormall reaction

reacción de Ehrlich-Sachs Ehrlich-Sachs
reaction

reacción de Einhorn-Brunner Einhorn-
Brunner reaction

reacción de Elbs Elbs reaction

reacción de eliminación elimination
reaction

reacción de eliminación de Cope Cope
elimination reaction

reacción de Eltekoff Eltekoff reaction

reacción de Emmert Emmert reaction

reacción de enaminas de Stork Stork
enamine reaction

reacción de Étard Étard reaction

reacción de Fenton Fenton reaction

reacción de Ferrario Ferrario reaction

reacción de Feulgen Feulgen reaction

reacción de Finkelstein Finkelstein
reaction

reacción de Fischer-Tropsch Fischer-
Tropsch reaction

reacción de Fittig Fittig reaction

reacción de Flood Flood reaction

reacción de Forster Forster reaction

reacción de fragmentación de Grob
Grob fragmentation reaction

reacción de Franchimont Franchimont
reaction

reacción de Frankland Frankland
reaction

reacción de Frankland-Duppa
Frankland-Duppa reaction

reacción de Friedel-Crafts Friedel-
Crafts reaction

reacción de Fujimoto-Belleau Fujimoto-
Belleau reaction

reacción de Fujiwara Fujiwara reaction

reacción de Gabriel Gabriel reaction

reacción de Gattermann Gattermann
reaction

reacción de Gattermann-Koch
Gattermann-Koch reaction

reacción de Goldschmidt Goldschmidt
reaction

reacción de Gomberg Gomberg reaction

reacción de Gomberg-Bachmann
Gomberg-Bachmann reaction

reacción de Gould-Jacobs Gould-Jacobs
reaction

reacción de Graebe-Ullman Graebe-
Ullman reaction

reacción de Griess Griess reaction

reacción de Grignard Grignard reaction

reacción de Guerbet Guerbet reaction
reacción de haces cruzadas crossed-beam reaction
reacción de Haller-Bauer Haller-Bauer reaction
reacción de Hammick Hammick reaction
reacción de Hell-Volhard-Zelinsky Hell-Volhard-Zelinsky reaction
reacción de Henkel Henkel reaction
reacción de Henry Henry reaction
reacción de Herz Herz reaction
reacción de Hilbert-Johnson Hilbert-Johnson reaction
reacción de Hill Hill reaction
reacción de Hinsberg Hinsberg reaction
reacción de Hofmann Hofmann reaction
reacción de Hofmann-Sand Hofmann-Sand reaction
reacción de Hooker Hooker reaction
reacción de Hopkins-Cole Hopkins-Cole reaction
reacción de Houben-Hoesch Houben-Hoesch reaction
reacción de Hunsdiecker Hunsdiecker reaction
reacción de indofenina indophenine reaction
reacción de inserción insertion reaction
reacción de intercambio exchange reaction
reacción de Janovsky Janovsky reaction
reacción de Japp-Klingemann Japp-Klingemann reaction
reacción de Kendall-Mattox Kendall-Mattox reaction
reacción de Knoevenagel Knoevenagel reaction
reacción de Kochi Kochi reaction
reacción de Kolbe Kolbe reaction
reacción de Komarowsky Komarowsky reaction
reacción de Kucherov Kucherov reaction
reacción de Kuhn-Winterstein Kuhn-Winterstein reaction
reacción de Lehmsted-Tanasescu Lehmsted-Tanasescu reaction
reacción de Leuckart Leuckart reaction

reacción de Lieben Lieben reaction
reacción de Liebermann Liebermann reaction
reacción de Maillard Maillard reaction
reacción de Malaprade Malaprade reaction
reacción de Mannich Mannich reaction
reacción de McFayden-Stevens McFayden-Stevens reaction
reacción de Mendius Mendius reaction
reacción de Menschutkin Menschutkin reaction
reacción de Meyer Meyer reaction
reacción de Michael Michael reaction
reacción de Michaelis-Arbuzov Michaelis-Arbuzov reaction
reacción de Mignonac Mignonac reaction
reacción de Mitsunobu Mitsunobu reaction
reacción de Nef Nef reaction
reacción de Nencki Nencki reaction
reacción de Nierenstein Nierenstein reaction
reacción de orden cero zero-order reaction
reacción de oxidación oxidation reaction
reacción de oxidación-reducción oxidation-reduction reaction
reacción de ozónidos de Harries Harries ozonide reaction
reacción de Passerini Passerini reaction
reacción de Pasteur Pasteur reaction
reacción de Paterno-Buchi Paterno-Buchi reaction
reacción de Pellizzari Pellizzari reaction
reacción de Perkin Perkin reaction
reacción de Perkow Perkow reaction
reacción de Peterson Peterson reaction
reacción de Pfitzinger Pfitzinger reaction
reacción de Pictet-Hubert Pictet-Hubert reaction
reacción de Pinner Pinner reaction
reacción de Piria Piria reaction
reacción de Polonovski Polonovski reaction
reacción de Pomeranz-Fritsch Pomeranz-Fritsch reaction

reacción de Ponzio Ponzio reaction

reacción de Prevost Prevost reaction

reacción de primer orden first-order reaction

reacción de Prins Prins reaction

reacción de Pschorr Pschorr reaction

reacción de Quelet Quelet reaction

reacción de Ramberg-Backlund Ramberg-Backlund reaction

reacción de redox redox reaction

reacción de Reed Reed reaction

reacción de Reformatsky Reformatsky reaction

reacción de Reimer Reimer reaction

reacción de Reimer-Tiemann Reimer-Tiemann reaction

reacción de Reissert Reissert reaction

reacción de reordenamiento rearrangement reaction

reacción de Reverdin Reverdin reaction

reacción de Ritter Ritter reaction

reacción de Robinson Robinson reaction

reacción de Rosenmund Rosenmund reaction

reacción de Rothemund Rothemund reaction

reacción de Sachse Sachse reaction

reacción de Sandmeyer Sandmeyer reaction

reacción de Schiemann Schiemann reaction

reacción de Schmidt Schmidt reaction

reacción de Scholl Scholl reaction

reacción de Schorigin Schorigin reaction

reacción de Schotten-Baumann Schotten-Baumann reaction

reacción de segundo orden second-order reaction

reacción de Semmler-Wolff Semmler-Wolff reaction

reacción de Serini Serini reaction

reacción de Sharpless Sharpless reaction

reacción de Simmons-Smith Simmons-Smith reaction

reacción de Simonini Simonini reaction

reacción de Sommelet Sommelet reaction

reacción de Staudinger Staudinger reaction

reacción de Stobbe Stobbe reaction

reacción de Strecker Strecker reaction

reacción de substitución substitution reaction

reacción de Sullivan Sullivan reaction

reacción de Swarts Swarts reaction

reacción de telomerización telomerization reaction

reacción de tercer orden third-order reaction

reacción de Thiele Thiele reaction

reacción de Thorpe Thorpe reaction

reacción de Tishchenko Tishchenko reaction

reacción de Tscherniac-Einhorn Tscherniac-Einhorn reaction

reacción de Ullman Ullman reaction

reacción de Varrentrapp Varrentrapp reaction

reacción de Vilsmeier-Haack Vilsmeier-Haack reaction

reacción de von Braun von Braun reaction

reacción de von Richter von Richter reaction

reacción de Wackenroder Wackenroder reaction

reacción de Wacker Wacker reaction

reacción de Wagner-Jauregg Wagner-Jauregg reaction

reacción de Wharton Wharton reaction

reacción de Whiting Whiting reaction

reacción de Willgerodt Willgerodt reaction

reacción de Williamson Williamson reaction

reacción de Wittig Wittig reaction

reacción de Wohl Wohl reaction

reacción de Wohl-Ziegler Wohl-Ziegler reaction

reacción de Wolff-Kishner Wolff-Kishner reaction

reacción de Wolffenstein-Boters Wolffenstein-Boters reaction

reacción de Wurtz Wurtz reaction

reacción de Wurtz-Fittig Wurtz-Fittig reaction

reacción de Ziesel Ziesel reaction

reacción de Zimmermann Zimmermann

reaction

reacción de Zincke-Suhl Zincke-Suhl reaction

reacción electrocíclica electrocyclic reaction

reacción electroorgánica electroorganic reaction

reacción en cadena chain reaction

reacción en cadena nuclear nuclear chain reaction

reacción endotérmica endothermic reaction

reacción espontánea spontaneous reaction

reacción estereoespecífica stereospecific reaction

reacción exotérmica exothermic reaction

reacción explosiva explosive reaction

reacción fotonuclear photonuclear reaction

reacción fotoquímica photochemical reaction

reacción heterogénea heterogeneous reaction

reacción heterolítica heterolytic reaction

reacción homogénea homogeneous reaction

reacción homolítica homolytic reaction

reacción incompleta incomplete reaction

reacción inducida induced reaction

reacción inducida por neutrones neutron-induced reaction

reacción interna internal reaction

reacción intramolecular intramolecular reaction

reacción irreversible irreversible reaction

reacción isotópica isotopic reaction

reacción lenta slow reaction

reacción metatética metathetical reaction

reacción neutra neutral reaction

reacción nuclear nuclear reaction

reacción nuclear directa direct nuclear reaction

reacción obscura dark reaction

reacción oscilante oscillating reaction

reacción oxidante oxidizing reaction

reacción pericíclica pericyclic reaction

reacción primaria primary reaction

reacción química chemical reaction

reacción química lenta slow chemical reaction

reacción química rápida fast chemical reaction

reacción química reversible reversible chemical reaction

reacción rápida fast reaction

reacción reductora reducing reaction

reacción reversible reversible reaction

reacción secundaria secondary reaction, side reaction

reacción sigmatrópica sigmatropic reaction

reacción simultánea simultaneous reaction

reacción subatómica subatomic reaction

reacción superficial surface reaction

reacción termonuclear thermonuclear reaction

reacción topoquímica topochemical reaction

reacción unimolecular unimolecular reaction

reacciones de Baeyer-Villiger Baeyer-Villiger reactions

reactante reactant

reactivación reactivation

reactividad reactivity

reactividad química chemical reactivity

reactivo reagent

reactivo catiónico cationic reagent

reactivo de Abel Abel reagent

reactivo de Agulhon Agulhon reagent

reactivo de Barfoed Barfoed reagent

reactivo de Bessonoff Bessonoff reagent

reactivo de Bettendorf Bettendorf reagent

reactivo de Bolton Bolton reagent

reactivo de Brady Brady reagent

reactivo de Carnot Carnot reagent

reactivo de Caro Caro reagent

reactivo de Cleland Cleland reagent

reactivo de Collman Collman reagent

reactivo de Denigè Denigè reagent

reactivo de Dobbin Dobbin reagent

reactivo de Dragendorff Dragendorff reagent

reactivo de Erdmann Erdmann reagent

reactivo de Fehling Fehling reagent
reactivo de Fenton Fenton reagent
reactivo de Fischer Fischer reagent
reactivo de Fraude Fraude reagent
reactivo de Froehde Froehde reagent
reactivo de Gerard Gerard reagent
reactivo de Girard Girard reagent
reactivo de Griess Griess reagent
reactivo de Grignard Grignard reagent
reactivo de Günzberg Günzberg reagent
reactivo de Hanus Hanus reagent
reactivo de Huber Huber reagent
reactivo de Hübl Hübl reagent
reactivo de Ivanov Ivanov reagent
reactivo de Jacquemart Jacquemart
  reagent
reactivo de Jones Jones reageant
reactivo de Karl Fischer Karl Fischer
  reagent
reactivo de Klein Klein reagent
reactivo de Mandelin Mandelin reagent
reactivo de Marme Marme reagent
reactivo de Mayer Mayer reagent
reactivo de Millon Millon reagent
reactivo de Nessler Nessler reagent
reactivo de Normant Normant reagent
reactivo de Nylander Nylander reagent
reactivo de Obermayer Obermayer
  reagent
reactivo de Petroff Petroff reagent
reactivo de Scheibler Scheibler reagent
reactivo de Schiff Schiff reagent
reactivo de Schweitzer Schweitzer
  reagent
reactivo de Simmons-Smith Simmons-
  Smith reagent
reactivo de Stead Stead reagent
reactivo de Tollens Tollens reagent
reactivo de Twitchell Twitchell reagent
reactivo de Vilsmeier Vilsmeier reagent
reactivo de Wagner Wagner reagent
reactivo de Zerewitinoff Zerewitinoff
  reagent
reactivo electrofílico electrophilic
  reagent
reactivo polimérico polymeric reagent
reactor reactor
reactor de producción production
  reactor

reactor de torio thorium reactor
reactor de uranio uranium reactor
reactor engendrador breeder
reactor nuclear nuclear reactor
reactor térmico thermal reactor
recalescencia recalescence
receptor receptor, receiver
receptor de Bruehl Bruehl receiver
reciclado recycling
recipiente de Dewar Dewar flask
recocido annealing
recolección electrónica electron
  collection
reconstitución reconstitution
recorrido libre free path
recorrido libre medio mean free path
recristalización recrystallization
recristalizar recrystallize
rectificación rectification
rectivo de Halphen Halphen reagent
rectorita rectorite
recuperación recovery
red lattice
red centrada en el cuerpo body-centered
  lattice
red centrada en las caras face-centered
  lattice
red cristalina crystal lattice
red molecular molecular lattice
redestilación redistillation
redingita reddingite
redisposición rearrangement
redox redox
reducción reduction
reducción de Béchamp Béchamp
  reduction
reduccion de Bouveault-Blanc
  Bouveault-Blanc reduction
reducción de Clemmensen Clemmensen
  reduction
reducción de punto de congelación
  freezing-point reduction
reducción de punto de ebullición
  boiling-point reduction
reducción de Rosenmund Rosenmund
  reduction
reducción de Sabatier-Senderens
  Sabatier-Senderens reduction
reducción de Wolff-Kishner Wolff-

Kishner reduction
reducción diferencial differential reduction
reducción electrolítica electrolytic reduction
reducción electroquímica electrochemical reduction
reducción química chemical reduction
reducido reduced
reducir reduce
reductasa reductase
reductona reductone
reductor reducer, reductor
reductor de Jones Jones reductor
reemplazo replacement
reemplazo doble double replacement
referencia reference
refinación refining
refinación de petróleo petroleum refining
refinación por solvente solvent refining
refinado refined
refinamiento refinement
refinamiento por zonas zone refining
refinar refine
refinería refinery
reflujo reflux, refluxing
reflujo total total reflux
reformación reforming
reformación al vapor steam reforming
reformación catalítica catalytic reforming
reformación térmica thermal reforming
refracción refraction
refracción atómica atomic refraction
refracción doble double refraction
refractario refractory
refractividad refractivity
refractividad específica specific refractivity
refractivo refractive
refractómetro refractometer
refrigeración refrigeration
refrigerante refrigerant, coolant
regelación regelation
regeneración regeneration
regenerativo regenerative
región de Geiger-Müller Geiger-Müller region

región espectral spectral region
regioselectividad regioselectivity
registrador de pH pH recorder
regla de Abbegg Abbegg rule
regla de Alder-Rickert Alder-Rickert rule
regla de Aston Aston rule
regla de Badger Badger rule
regla de Barlow Barlow rule
regla de Blanc Blanc rule
regla de Bragg Bragg rule
regla de Bredt Bredt rule
regla de Compton Compton rule
regla de Cram Cram rule
regla de Dieterici Dieterici rule
regla de dilución dilution rule
regla de Dühring Dühring rule
regla de Erlenmeyer Erlenmeyer rule
regla de exclusión mutua mutual exclusion rule
regla de fases phase rule
regla de fases de Gibbs Gibbs phase rule
regla de Fries Fries rule
regla de Geiger-Nutall Geiger-Nutall rule
regla de Gibbs Gibbs rule
regla de Guldberg Guldberg rule
regla de Hofmann Hofmann rule
regla de Hückel Hückel rule
regla de Hume-Rothery Hume-Rothery rule
regla de intervalos de Landé Landé interval rule
regla de Konowaloff Konowaloff rule
regla de Kopp Kopp rule
regla de Kundt Kundt rule
regla de lactonas de Hudson Hudson lactone rule
regla de Leduc Leduc rule
regla de Magnus Magnus rule
regla de Markovnikoff Markovnikoff rule
regla de Markovnikov Markovnikov rule
regla de Nernst-Thomson Nernst-Thomson rule
regla de Ostwald Ostwald rule
regla de Paneth Paneth rule
regla de Pauli Pauli rule

regla de Perrin Perrin rule
regla de selección de Laporte Laporte selection rule
regla de Stuffer Stuffer rule
regla de Teller-Redlich Teller-Redlich rule
regla de Welter Welter rule
reglas de Alder-Stein Alder-Stein rules
reglas de Baldwin Baldwin rules
reglas de clorinación de Hass Hass chlorination rules
reglas de Fajans Fajans rules
reglas de Hammick-Illingworth Hammick-Illingworth rules
reglas de Hund Hund rules
reglas de isorrotación de Hudson Hudson isorotation rules
reglas de Pascal Pascal rules
reglas de secuencia sequence rules
reglas de selección selection rules
reglas de Woodward-Hoffmann Woodward-Hoffmann rules
regulador regulator
regulador de aire air regulator
regulador de gas gas regulator
regulador de temperatura temperature regulator
rehidratación rehydration
reinita reinite
rejalgar realgar
rejilla grating, grid
rejilla de difracción diffraction grating
rejilla escalonada echelon grating
relación de Keesom Keesom relationship
relación diagonal diagonal relationship
relajación relaxation
relajación molecular molecular relaxation
relatividad relativity
relativista relativistic
relativo relative
reloj radiactivo radioactive clock
relleno filler
rem rem
renato rhenate
renaturación renaturation
rendimiento yield
rendimiento cuántico quantum yield
rendimiento de fisión fission yield

rendimiento de fluorescencia fluorescence yield
rendimiento fotoquímico photochemical yield
renina rennin, renin
renio rhenium
reología rheology
reómetro rheometer
reordenamiento rearrangement
reordenamiento alílico allylic rearrangement
reordenamiento de Amadori Amadori rearrangement
reordenamiento de Beckmann Beckmann rearrangement
reordenamiento de Ciamician-Dennstedt Ciamician-Dennstedt rearrangement
reordenamiento de Claisen Claisen rearrangement
reordenamiento de Cope Cope rearrangement
reordenamiento de Curtius Curtius rearrangement
reordenamiento de Chapman Chapman rearrangement
reordenamiento de Demjanov Demjanov rearrangement
reordenamiento de dienona-fenol dienone-phenol rearrangement
reordenamiento de éteres de Wittig Wittig ether rearrangement
reordenamiento de Favorskii Favorskii rearrangement
reordenamiento de Fischer-Hepp Fischer-Hepp rearrangement
reordenamiento de Fries Fries rearrangement
reordenamiento de Gabriel-Colman Gabriel-Colman rearrangement
reordenamiento de Hayasi Hayashi rearrangement
reordenamiento de Hofmann-Martius Hofmann-Martius rearrangement
reordenamiento de Jacobsen Jacobsen rearrangement
reordenamiento de Ladenburg Ladenburg rearrangement
reordenamiento de Lossen Lossen rearrangement

reordenamiento de McLafferty McLafferty rearrangement

reordenamiento de Meisenheimer Meisenheimer rearrangement

reordenamiento de Neber Neber rearrangement

reordenamiento de Perkin Perkin rearrangement

reordenamiento de pinacol pinacol rearrangement

reordenamiento de Pummerer Pummerer rearrangement

reordenamiento de retropinacol retropinacol rearrangement

reordenamiento de Rowe Rowe rearrangement

reordenamiento de Smiles Smiles rearrangement

reordenamiento de Sommelet-Hauser Sommelet-Hauser rearrangement

reordenamiento de Stevens Stevens rearrangement

reordenamiento de Stieglitz Stieglitz rearrangement

reordenamiento de Tafel Tafel rearrangement

reordenamiento de Tiemann Tiemann rearrangement

reordenamiento de Wagner-Meerwein Wagner-Meerwein rearrangement

reordenamiento de Wallach Wallach rearrangement

reordenamiento de Wessely-Moser Wessely-Moser rearrangement

reordenamiento de Westphalen-Lettre Westphalen-Lettre rearrangement

reordenamiento de Wittig Wittig rearrangement

reordenamiento de Wolff Wolff rearrangement

reordenamiento molecular molecular rearrangement

repelencia repellency

repelente repellent

repercolación repercolation

repetible repeatable

reprocesamiento reprocessing

reproducibilidad reproducibility

reproducible reproducible

reptación reptation

repulsión repulsion

repulsión molecular molecular repulsion

repulsión nuclear nuclear repulsion

requisitos ambientales environmental requirements

resbenzofenona resbenzophenone

reseno resene

reserpina reserpine

reserva reserve

residual residual

residuo residue

residuo de destilación distillation residue

residuos anódicos anodic mud, anodic slime

resina resin, rosin

resina acetálica acetal resin

resina acrílica acrylic resin

resina alílica allyl resin

resina alquídica alkyd resin

resina amarilla yellow resin

resina aniónica anionic resin

resina celulósica cellulosic resin

resina de acetato de vinilo vinyl acetate resin

resina de acrilato acrylate resin

resina de adición addition resin

resina de cloruro de vinilo vinyl chloride resin

resina de contacto contact resin

resina de copaiba copaiba resin

resina de estireno-acrilonitrilo styrene-acrylonitrile resin

resina de fenol-formaldehído phenol-formaldehyde resin

resina de fenol-furfural phenol-furfural resin

resina de fluorocarburo fluorocarbon resin

resina de halocarburo halocarbon resin

resina de hidrocarburos hydrocarbon resin

resina de intercambio iónico ion-exchange resin

resina de ionómero ionomer resin

resina de isocianato isocyanate resin

resina de Manila Manila resin

resina de melamina melamine resin

resina de poliamida polyamide resin

resina de poliamina-metileno
  polyamine-methylene resin
resina de polietileno polyethylene resin
resina de poliimida polyimide resin
resina de polimetacrilato
  polymethacrylate resin
resina de polioxialquileno
  polyoxyalkylene resin
resina de politerpeno polyterpene resin
resina de politetrafluoroetileno
  polytetrafluoroethylene resin
resina de poliuretano polyurethane resin
resina de polivinilideno polyvinylidene
  resin
resina de polivinilo polyvinyl resin
resina de resorcina-formaldehído
  resorcinol-formaldehyde resin
resina de triazona triazone resin
resina de urea-formaldehído urea-
  formaldehyde resin
resina de uretano urethane resin
resina de vinilacetal vinyl acetal resin
resina de viniléter vinyl ether resin
resina de vinilideno vinylidene resin
resina de vinilo vinyl resin
resina en polvo powdered resin
resina epóxica epoxy resin
resina etilénica ethylene resin
resina fenólica phenolic resin
resina fenoxi phenoxy resin
resina olefínica olefin resin
resina poliéster polyester resin
resina poliéter polyether resin
resina sintética synthetic resin
resina soluble en agua water-soluble
  resin
resina termoplástica thermoplastic resin
resinamina resinamine
resinato resinate
resinato alumínico aluminum resinate
resinato cálcico calcium resinate
resinato cobaltoso cobaltous resinate
resinato de cinc zinc resinate
resinato de cobalto cobalt resinate
resinato de cobre copper resinate
resinato de estaño tin resinate
resinato de hierro iron resinate
resinato de manganeso manganese
  resinate

resinato de plomo lead resinate
resinato férrico ferric resinate
resinato sódico sodium resinate
resinificación resinification
resinografía resinography
resinoide resinoid
resinol resinol
resinoso resinous
resistencia al fuego fire resistance
resistente al ácido acid resistant
resistente al fuego fire resistant
resita resite
resitol resitol
resocianina resocyanin
resol resol
resolución resolution
resonancia resonance
resonancia atómica atomic resonance
resonancia cuadripolar nuclear nuclear
  quadripole resonance
resonancia cuádruple quadruple
  resonance
resonancia de espín spin resonance
resonancia de espín electrónico
  electron-spin resonance
resonancia de Fermi Fermi resonance
resonancia electrónica electron
  resonance
resonancia magnética magnetic
  resonance
resonancia magnética electrónica
  electron magnetic resonance
resonancia magnética nuclear nuclear
  magnetic resonance
resonancia nuclear nuclear resonance
resonancia paramagnética paramagnetic
  resonance
resonancia paramagnética electrónica
  electron paramagnetic resonance
resonante resonant
resorcilo resorcyl
resorcina resorcin
resorcinol resorcinol
restitución restitution
resultante resultant
retardador retarder
retardante de llama flame-retarding
retardo retardation
retardo iónico ion retardation

retención retention
reteno retene
reticulado reticulated
retículo grating, lattice
retinal retinal
retineno retinene
retinita retinite
retinol retinol
retorta retort
retro retro
retroalimentación feedback
retrodispersión backscatter
retrodonación back donation
retroextracción back-extraction, stripping
retrogradación retrogradation
retrorcina retrorsine
retrosíntesis retrosynthesis
retrotitulación back titration
retrovaloración back titration
revelador developer
reversible reversible
reversión reversion
revertosa revertose
revestido coated
revestimiento coating, cladding
revestimiento anódico anodic coating
revestimiento básico basic lining
revestimiento catódico cathodic coating
revestimiento de oro gold coating
revestimiento protector protective coating
revivir revive
rho rho
riboflavina riboflavin
ribofuranosiladenina ribofuranosyladenine
ribonucleasa ribonuclease
ribonucleoproteína ribonucleoprotein
ribosa ribose
ribosida riboside
ribosilo ribosyl
ribosoma ribosome
ribulosa ribulose
ricina ricin
ricinina ricinine
ricinoleato ricinoleate
ricinoleato amónico ammonium ricinoleate

ricinoleato cálcico calcium ricinoleate
ricinoleato de butilo butyl ricinoleate
ricinoleato de cadmio cadmium ricinoleate
ricinoleato de cinc zinc ricinoleate
ricinoleato de litio lithium ricinoleate
ricinoleato de metilo methyl ricinoleate
ricinoleato potásico potassium ricinoleate
ricinoleato sódico sodium ricinoleate
ricinoleína ricinolein
rickardita rickardite
richmondita richmondite
riesgo aceptable acceptable risk
rigoleno rhigolene
ripidolita ripidolite
ristocetina ristocetin
roca carbonatada carbonate rock
roca de carbonato carbonate rock
roca de fosfato phosphate rock
rodalina rhodalline
rodamina rhodamine
rodanato rhodanate
rodanato sódico sodium rhodanate
rodanida rhodanide
rodanina rhodanine
rodanometría rhodanometry
rodenticida rodenticide
rodinilo rhodinyl
rodinol rhodinol, rodinol
rodio rhodium
rodita rhodite
rodizita rhodizite
rodocrosita rhodochrosite
rodolita rhodolite
rodonita rhodonite
rodopsina rhodopsin
rodoxantina rhodoxanthin
roentgen roentgen, röntgen
roentgenograma roentgenogram
rojo congo congo red
rojo de alizarina alizarin red
rojo de antimonio antimony red
rojo de bromofenol bromophenol red
rojo de cadmio cadmium red
rojo de clorofenol chlorophenol red
rojo de cresol cresol red
rojo de cromo chrome red
rojo de índigo indigo red

rojo de la India Indian red
rojo de metilo methyl red
rojo de óxido férrico ferric oxide red
rojo de paranitranilina paranitraniline red
rojo de París Paris red
rojo de Persia Persian red
rojo de quinolina quinoline red
rojo de rutenio ruthenium red
rojo de toluidina toluidine red
rojo fenol phenol red
rojo neutro neutral red
rojo para para red
rojo turco Turkey red
rojo veneciano Venetian red
rómbico rhombic
rombohedral rhombohedral
rombohedro rhombohedron
rompedor de emulsión emulsion breaker
rompimiento de anillo ring breakage
rompimiento de emulsión emulsion breaking
rosa de Bengala Rose Bengal
rosanilina rosaniline
roscoelita roscoelite
roselita roselite
rosinol rosinol
rotación rotation
rotación específica specific rotation
rotación interna restringida restricted internal rotation
rotación molecular molecular rotation
rotación óptica optical rotation
rotacional rotational
rotámero rotamer
rotámetro rotameter
rotanio rhotanium
rotatorio rotatory
rotaversión rotaversion
rotenona rotenone
rotulación labeling
rotulado labeled
rowlandita rowlandite
rubeno rubene
ruberita ruberite
rubí ruby
rubí espinela ruby spinel
rubiceno rubicene
rubidina rubidine
rubidio rubidium
rubreno rubrene
rutefordita rutherfordite
rutenato ruthenate
ruténico ruthenic
rutenio ruthenium
rutenioso ruthenious
rutenoceno ruthenocene
rutilo rutile
rutina rutin
rydberg rydberg

S

sabineno sabinene
sabinol sabinol
sabor taste, flavor
sacarasa saccharase
sacarato saccharate
sacarato ácido potásico potassium acid
  saccharate
sacarato de estroncio strontium
  saccharate
sacárico saccharic
sacárido saccharide
sacarificación saccharification
sacarificar saccharify
sacarímetro saccharimeter
sacarina saccharin
sacarosa saccharose, sucrose
saccarina sódica sodium saccharin
safranina safranine
safrol safrole
sahlita sahlite
sal salt
sal ácida acid salt
sal amarilla yellow salt
sal amoníaco sal ammoniac
sal amónica ammonium salt
sal amónica cuaternaria quaternary
  ammonium salt
sal amortiguadora buffer salt
sal azul blue salt
sal básica basic salt
sal biliar bile salt
sal binaria binary salt
sal compleja complex salt
sal común common salt
sal cuaternaria quaternary salt
sal de anilina aniline salt
sal de antimonio antimony salt
sal de diazonio diazonium salt
sal de Engel Engel salt
sal de Erdmann Erdmann salt
sal de estaño tin salt
sal de Étard Étard salt
sal de Fischer Fischer salt

sal de fosfonio phosphonium salt
sal de Fremy Fremy salt
sal de Glauber Glauber salt
sal de Graham Graham salt
sal de Lemery salt of Lemery
sal de Macquer Macquer salt
sal de Maddrell Maddrell salt
sal de Magnus Magnus salt
sal de mesa table salt
sal de Mohr Mohr salt
sal de oro gold salt
sal de Pelilgot Peligot salt
sal de plata silver salt
sal de Reinecke Reinecke salt
sal de roca rock salt
sal de Rochelle Rochelle salt
sal de Schäffer Schäffer salt
sal de Seidlitz Seidlitz salt
sal de tártaro salt of tartar
sal de Vauquelin Vauquelin salt
sal de Wurster Wurster salt
sal de Zeise Zeise salt
sal débil weak salt
sal doble double salt
sal esencial essential salt
sal fundida molten salt, fused salt
sal mixta mixed salt
sal neutra neutral salt
sal normal normal salt
sal orgánica organic salt
sal simple simple salt
sal sosa sal soda
sal verde green salt
sala limpia clean room
salamida salamide
sales de Epsom Epsom salts
sales de tierras raras rare-earth salts
salicilado salicylated
salicilal salicylal
salicilaldehído salicylaldehyde
salicilamida salicylamide
salicilanilida salicylanilide
salicilato salicylate

salicilato alumínico   aluminum salicylate
salicilato amónico   ammonium salicylate
salicilato bárico   barium salicylate
salicilato de amilo   amyl salicylate
salicilato de atropina   atropine salicylate
salicilato de bencilo   benzyl salicylate
salicilato de bornilo   bornyl salicylate
salicilato de cafeína   caffeine salicylate
salicilato de cesio   cesium salicylate
salicilato de cinc   zinc salicylate
salicilato de codeína   codeine salicylate
salicilato de estroncio   strontium salicylate
salicilato de etilo   ethyl salicylate
salicilato de feniletilo   phenylethyl salicylate
salicilato de fenilo   phenyl salicylate
salicilato de isoamilo   isoamyl salicylate
salicilato de isobornilo   isobornyl salicylate
salicilato de isobutilo   isobutyl salicylate
salicilato de litio   lithium salicylate
salicilato de mentilo   menthyl salicylate
salicilato de metilo   methyl salicylate
salicilato de octilfenilo   octylphenyl salicylate
salicilato de pentilo   pentyl salicylate
salicilato de plata   silver salicylate
salicilato de plomo   lead salicylate
salicilato fenilmercúrico   phenylmercuric salicylate
salicilato magnésico   magnesium salicylate
salicilato mercúrico   mercuric salicylate
salicilato potásico   potassium salicylate
salicilato sódico   sodium salicylate
salicilideno   salicylidene
salicilo   salicyl
saliciloilo   salicyloyl
salicilonitrilo   salicylonitrile
salicina   salicin
sálico   salic
salicoilo   salicoyl
salida de gas   gas outlet
saligenina   saligenin
saligenol   saligenol
salímetro   salimeter
salinidad   salinity
salino   saline

salinómetro   salinometer
salitre   saltpeter
salmina   salmine
salmuera   brine
salol   salol
salto cuántico   quantum jump
samárico   samaric
samario   samarium
samaroso   samarous
samarskita   samarskite
sandaraca   sandarac
sangrar   bleed
sanguinaria   sanguinaria
santalilo   santalyl
santalol   santalol
saponificación   saponification
saponina   saponin
saponita   saponite
sarcolactato   sarcolactate
sarcolisina   sarcolysin
sarcolita   sarcolite
sarcosina   sarcosine
sarcosinato sódico   sodium sarcosinate
sardinianita   sardinianite
sardónica   sardonyx
sarina   sarin
saturable   saturable
saturación   saturation
saturación con agua   water saturation
saturado   saturated
saturador   saturator
saturante   saturating
saturar   saturate
saxitoxina   saxitoxin
scheelita   scheelite
schonita   schönite
sebacato de butilo   butyl sebacate
sebacato de dimetilo   dimethyl sebacate
sebacato de dioctilo   dioctyl sebacate
sebacoilo   sebacoyl
sebaconitrilo   sebaconitrile
sebo   tallow
sec   sec
secado   drying
secado instantáneo   flash drying
secado por solvente   solvent drying
secador   dryer
sección transversal   cross-section
sección transversal de fisión   fission

cross-section
sección transversal gamma gamma
　cross-section
sección transversal nuclear nuclear
　cross-section
seco dry
secobarbital secobarbital
secobarbital sódico sodium secobarbital
secuencia isoelectrónica isoelectronic
　sequence
secuestrante sequestering
secuestro sequestration
secundario secondary
seda silk
sedante sedative
sedimentación sedimentation
sedimento sediment, sludge
sedimento químico chemical sediment
sedimentos de aceite foots
segregación segregation
segunda ley de termodinámica second
　law of thermodynamics
segundo second
seguridad safety
selectividad selectivity
selectividad de catalizador catalyst
　selectivity
selectividad de filtro filter selectivity
selectivo selective
selena selena
seleniato selenate
seleniato amónico ammonium selenate
seleniato bárico barium selenate
seleniato cálcico calcium selenate
seleniato cúprico cupric selenate
seleniato de cadmio cadmium selenate
seleniato de plomo lead selenate
seleniato sódico sodium selenate
selénico selenic
selenilo selenyl
seleninilo seleninyl
selenino selenino
selenio selenium
selenioso selenious
selenita selenite
selenito amónico ammonium selenite
selenito de cobalto cobalt selenite
selenito sódico sodium selenite
seleniuro selenide

seleniuro arsenioso arsenous selenide
seleniuro bárico barium selenide
seleniuro cuproso cuprous selenide
seleniuro de arsénico arsenic selenide
seleniuro de bismuto bismuth selenide
seleniuro de cadmio cadmium selenide
seleniuro de cinc zinc selenide
seleniuro de dietilo diethyl selenide
seleniuro de dimetilo dimethyl selenide
seleniuro de hidrógeno hydrogen
　selenide
seleniuro de mercurio mercury selenide
seleniuro de plata silver selenide
seleniuro de plomo lead selenide
seleniuro de tetrafósforo
　tetraphosphorus selenide
seleniuro ferroso ferrous selenide
seleno seleno
selenocianato selenocyanate
selenofurano selenofuran
selenol selenol
selenomio selenomium
selenona selenone
selenonilo selenonyl
selenonio selenonium
selenono selenono
selenourea selenourea
selenóxido selenoxide
sellado sealed
sellador sealant
sellaíta sellaite
semi semi
semicarbazida semicarbazide
semicarbazido semicarbazido
semicarbazino semicarbazino
semicarbazona semicarbazone
semicarbazona de acetona acetone
　semicarbazone
semicarbazono semicarbazono
semicelda half-cell
semiconductor semiconductor
semidina semidine
semiempírico semiempirical
semilla seed
semilla de algodón cottonseed
semimetal semimetal
semimetálico semimetallic
semimicroanálisis semimicroanalysis
semimicroquímica semimicrochemistry

semipermeable semipermeable
semipila half-cell
semiprecioso semiprecious
semisintético semisynthetic
semisólido semisolid
semivalencia semivalence
senarmontita senarmontite
sensibilizador sensitizer
sensible sensitive
sensible a la luz light-sensitive
sensible a la temperatura temperature-sensitive
sensible a radiación radiation-sensitive
sensor sensor
sensor de humedad humidity sensor
separación separation
separación centrífuga centrifugal separation
separación de aminas de Hofmann Hofmann amine separation
separación de isótopos isotope separation
separación electrolítica electrolytic separation
separación electromagnética electromagnetic separation
separación magnética magnetic separation
separador separator
separador de isótopos isotope separator
sepia sepia
sepiolita sepiolite
septifeno septiphene
serendipismo serendipity
serie series
serie actínida actinide series
serie alifática aliphatic series
serie aromática aromatic series
serie colateral collateral series
serie de actinio actinium series
serie de actividad activity series
serie de alquenos alkene series
serie de Balmer Balmer series
serie de benceno benzene series
serie de Brackett Brackett series
serie de desintegración decay series
serie de desintegración de uranio uranium decay series
serie de desintegración radiactiva radioactive decay series
serie de desplazamiento displacement series
serie de hidrocarburos hydrocarbon series
serie de Hofmeister Hofmeister series
serie de líneas series of lines
serie de Lyman Lyman series
serie de Paschen Paschen series
serie de redox redox series
serie de torio thorium series
serie de uranio uranium series
serie de uranio-radio uranium-radium series
serie de Volta Volta series
serie electromotriz electromotive series
serie electroquímica electrochemical series
serie espectral spectral series, spectrum series
serie espectroquímica spectrochemical series
serie eutrópica eutropic series
serie fundamental fundamental series
serie galvánica galvanic series
serie homogénea homogeneous series
serie homóloga homologous series
serie lantánida lanthanide series
serie liotrópica lyotropic series
serie perilógica perilogic series
serie periódica periodic series
serie principal principal series
serie química chemical series
serie radiactiva radioactive series
serina serine
serotonina serotonin
serpentina serpentine
sesamina sesamin
sesamol sesamol
sesamolina sesamolin
sesona sesone
sesqui sesqui
sesquibromuro férrico ferric sesquibromide
sesquicarbonato sesquicarbonate
sesquicarbonato sódico sodium sesquicarbonate
sesquicloruro sesquichloride
sesquicloruro de cromo chromium

sesquichloride
sesquicloruro de etilaluminio ethylaluminum sesquichloride
sesquicloruro de rutenio ruthenium sesquichloride
sesquicloruro de talio thallium sesquichloride
sesquicloruro talioso thallous sesquichloride
sesquióxido sesquioxide
sesquióxido de galio gallium sesquioxide
sesquióxido de iridio iridium sesquioxide
sesquióxido de lantano lanthanum sesquioxide
sesquióxido de manganeso manganese sesquioxide
sesquióxido de molibdeno molybdenum sesquioxide
sesquióxido de níquel nickel sesquioxide
sesquióxido de plomo lead sesquioxide
sesquióxido de rutenio ruthenium sesquioxide
sesquióxido de vanadio vanadium sesquioxide
sesquisal sesquisalt
sesquisilicato sódico sodium sesquisilicate
sesquisosa sesquisoda
sesquisulfato de titanio titanium sesquisulfate
sesquisulfuro de fósforo phosphorus sesquisulfide
sesquiterpeno sesquiterpene
seudo pseudo
seudoácido pseudo acid
seudoaromático pseudoaromatic
seudoasimétrico pseudoasymmetric
seudobase pseudo base
seudocumeno pseudocumene
seudocumidina pseudocumidine
seudocumilo pseudocumyl
seudocumol pseudocumol
seudohalógeno pseudohalogen
seudoisomerismo pseudoisomerism
seudoisótopo pseudoisotope
seudomórfico pseudomorphic
sexivalente sexivalent
sexteto sextet
siderita siderite

siderotilato siderotilate
sidnona sydnone
siembra seeding
siena sienna
sievert sievert
sigma sigma
sigmatrópico sigmatropic
siicato de zirconio zirconium silicate
silano silane
silanodiilo silanediyl
silanol silanol
silanotriilo silanetriyl
silatrano silatrane
silazano silazane
silicano silicane
silicato silicate
silicato alumínico aluminum silicate
silicato bárico barium silicate
silicato cálcico calcium silicate
silicato cobaltoso cobaltous silicate
silicato de berilio beryllium silicate
silicato de cinc zinc silicate
silicato de cobre copper silicate
silicato de etilbutilo ethylbutyl silicate
silicato de etilo ethyl silicate
silicato de litio lithium silicate
silicato de manganeso manganese silicate
silicato de plomo lead silicate
silicato de sosa silicate of soda
silicato magnésico magnesium silicate
silicato manganoso manganous silicate
silicato potásico potassium silicate
silicato sódico sodium silicate
silicato tricálcico tricalcium silicate
sílice silica
sílice fundida fused silica
sílice hidratada hydrated silica
silíceo siliceous
silícico silicic
silicileno silicylene
silicilo silicyl
silicio silicon
silicio-cobre silicon-copper
siliciuro silicide
siliciuro bárico barium silicide
siliciuro cálcico calcium silicide
siliciuro crómico chromic silicide
siliciuro de cobalto cobalt silicide

siliciuro de niobio niobium silicide
siliciuro de rutenio ruthenium silicide
siliciuro de tungsteno tungsten silicide
siliciuro de zirconio zirconium silicide
siliciuro magnésico magnesium silicide
silico silico
silicoaluminato sódico sodium
  silicoaluminate
silicobromoformo silicobromoform
silicocloroformo silicochloroform
silicoetano silicoethane
silicofluoruro silicofluoride
silicofluoruro alumínico sódico sodium
  aluminum silicofluoride
silicofluoruro bárico barium
  silicofluoride
silicofluoruro cálcico calcium
  silicofluoride
silicofluoruro de plomo lead
  silicofluoride
silicofluoruro magnésico magnesium
  silicofluoride
silicofluoruro potásico potassium
  silicofluoride
silicofluoruro sódico sodium
  silicofluoride
silicoheptano silicoheptane
silicol silicol
silicomanganeso silicomanganese
silicometano silicomethane
silicomolibdato sódico sodium
  silicomolybdate
silicona silicone
siliconato de metilo sódico sodium
  methyl siliconate
silicotungstato sódico sodium
  silicotungstate
silicoyodoformo silicoiodoform
sililación silylation
silileno silylene
sililo sylyl
silimanita sillimanite
siloxano siloxane
siloxi siloxy
silumina silumin
silvanita sylvanite
silvano sylvan
silvestreno silvestrene
silvex silvex

silvina sylvine
silvinita sylvinite
silviquímico silvichemical
silvita sylvite
simazina simazine
símbolo symbol
símbolo atómico atomic symbol
símbolo químico chemical symbol
simetría symmetry
simetría de cristal crystal symmetry
simetría interna internal symmetry
simetría orbital orbital symmetry
simétrico symmetrical
sin solvente solventless
sindiotáctico syndiotactic
sinéresis syneresis
sinergismo synergism
sinergista synergist
singulete singlet
sinigrina sinigrin
sintáctico syntactic
sínter sinter
sinterización sintering
sinterizar sinterize
síntesis synthesis
síntesis alicíclica de Perkin Perkin
  alicyclic synthesis
síntesis asimétrica asymmetric synthesis
síntesis de aldehídos de Bouveault
  Bouveault aldehyde synthesis
síntesis de aldehídos de Gattermann
  Gattermann aldehyde synthesis
síntesis de aldehídos de Kröhnke
  Kröhnke aldehyde synthesis
síntesis de Arndt-Eistert Arndt-Eistert
  synthesis
síntesis de Bogert-Cook Bogert-Cook
  synthesis
síntesis de Campbell Campbell synthesis
síntesis de cetenas de Schmidlin
  Schmidlin ketene synthesis
síntesis de ciclopropanos de Kishner
  Kishner cyclopropane synthesis
síntesis de Darzens Darzens synthesis
síntesis de Erlenmeyer Erlenmeyer
  synthesis
síntesis de ésteres malónicos malonic
  ester synthesis
síntesis de éteres de Williamson

Williamson ether synthesis

síntesis de Favorskii-Babayan
Favorskii-Babayan synthesis

síntesis de Feist-Benary Feist-Benary
synthesis

síntesis de fenilhidrazinas de Fischer
Fischer phenylhydrazine synthesis

síntesis de Fischer-Tropsch Fischer-
Tropsch synthesis

síntesis de Fittig Fittig synthesis

síntesis de Frankland Frankland
synthesis

síntesis de Freund Freund synthesis

síntesis de Friedlander Friedlander
synthesis

síntesis de Gabriel Gabriel synthesis

síntesis de Gastaldi Gastaldi synthesis

síntesis de Gattermann Gattermann
synthesis

síntesis de Gattermann-Koch
Gattermann-Koch synthesis

síntesis de Gogte Gogte synthesis

síntesis de Graebe-Ullman Graebe-
Ullman synthesis

síntesis de Grignard Grignard synthesis

síntesis de Grove Grove synthesis

síntesis de Grundmann Grundmann
synthesis

síntesis de Haworth Haworth synthesis

síntesis de Heumann-Pfleger Heumann-
Pfleger synthesis

síntesis de Hinsberg Hinsberg synthesis

síntesis de Hoch-Campbell Hoch-
Campbell synthesis

síntesis de Hoesch Hoesch synthesis

síntesis de Houben-Fischer Houben-
Fischer synthesis

síntesis de Houben-Hoesch Houben-
Hoesch synthesis

síntesis de indoles de Fischer Fischer
indole synthesis

síntesis de indoles de Nenitzescu
Nenitzescu indole synthesis

síntesis de indoles de Reissert Reissert
indole synthesis

síntesis de isonitrilos de Hofmann
Hofmann isonitrile synthesis

síntesis de Kiliani-Fischer Kiliani-
Fischer synthesis

síntesis de Kolbe Kolbe synthesis

síntesis de Letts Letts synthesis

síntesis de Madelung Madelung
synthesis

síntesis de Martinet Martinet synthesis

síntesis de Meyer Meyer synthesis

síntesis de Nef Nef synthesis

síntesis de oxazoles de Fischer Fischer
oxazole synthesis

síntesis de Paal-Knorr Paal-Knorr
synthesis

síntesis de Patterson Patterson synthesis

síntesis de Pelouze Pelouze synthesis

síntesis de Piloty-Robinson Piloty-
Robinson synthesis

síntesis de Pinner Pinner synthesis

**síntesis de pirazinas de Staedel-
Rugheimer** Staedel-Rugheimer
pyrazine synthesis

síntesis de pirazoles de Knorr Knorr
pyrazole synthesis

síntesis de pirazoles de Pechmann
Pechmann pyrazole synthesis

síntesis de piridinas de Chichibabin
Chichibabin pyridine synthesis

síntesis de piridinas de Hantzsch
Hantzsch pyridine synthesis

síntesis de pirroles de Hantach Hantach
pyrrole synthesis

síntesis de pirroles de Knorr Knorr
pyrrole synthesis

síntesis de polipéptidos de Fischer
Fischer polypeptide synthesis

síntesis de purinas de Traube Traube
purine synthesis

**síntesis de quinazolinas de
Niementowski** Niementowski
quinazoline synthesis

síntesis de quinolinas de Camps Camps
quinoline synthesis

síntesis de quinolinas de Combes
Combes quinoline synthesis

síntesis de quinolinas de Knorr Knorr
quinoline synthesis

síntesis de quinolinas de Niementowski
Niementowski quinoline synthesis

síntesis de quinolinas de Riehm Riehm
quinoline synthesis

síntesis de Riemschneider

Riemschneider synthesis

síntesis de Rosenmund-von Braun Rosenmund-von Braun synthesis

síntesis de Sandmeyer Sandmeyer synthesis

síntesis de Skraup Skraup synthesis

síntesis de Stolle Stolle synthesis

síntesis de Strecker Strecker synthesis

síntesis de Urech Urech synthesis

síntesis de von Richter von Richter synthesis

síntesis de Widman-Stoermer Widman-Stoermer synthesis

síntesis de Williamson Williamson synthesis

síntesis de Wöhler Wöhler synthesis

síntesis de Wurtz Wurtz synthesis

síntesis estereoespecífica stereospecific synthesis

síntesis orgánica organic synthesis

síntesis química chemical synthesis

sintético synthetic

sintetizar synthesize

sintol synthol

sirosingopina syrosingopine

sisal sisal

sistema absoluto absolute system

sistema binario binary system

sistema coloidal colloidal system

sistema conjugado conjugated system

sistema cristalográfico crystal system

sistema cuantificado quantized system

sistema cuaternario quaternary system

sistema cúbico cubic system

sistema de anillo ring system

sistema de Claude Claude system

sistema de Ginebra Geneva system

sistema de Hermann-Mauguin Hermann-Mauguin system

sistema de redox redox system

sistema de Schoenflies Schoenflies system

sistema de Stock Stock system

sistema degenerado degenerate system

sistema dinámico dynamic system

sistema disipativo dissipative system

sistema disperso dispersed system

sistema espiro spiro system

sistema estable stable system

sistema eutéctico eutectic system

sistema extintor de fuego fire-extinguisher system

sistema heterogéneo heterogeneous system

sistema hexagonal hexagonal system

sistema homogéneo homogeneous system

sistema internacional international system

sistema monodisperso monodisperse system

sistema ortorrómbico orthorhombic system

sistema periódico periodic system

sistema peritéctico peritectic system

sistema polifuncional polyfunctional system

sistema regular regular system

sistema rómbico rhombic system

sistema rombohedral rhombohedral system

sistema ternario ternary system

sistema tetragonal tetragonal system

sistema triclínico triclinic system

sistema trigonal trigonal system

sistémico systemic

sitosterol sitosterol

sklodowskita sklodowskite

skutterudita skutterudite

smog smog

sobrenadante supernatant

sobrepotencial overpotential

sobretensión overvoltage

soda soda

sodalita sodalite

sodamida sodamide

sodilo sodyl

sodio sodium

sodioacetato sódico sodium sodioacetate

soja soybean

sol sol

solación solation

solana solan

solanina solanine

solato solate

soldadura welding

soldadura con hidrógeno atómico atomic hydrogen welding

soldadura fuerte brazing

soldante solder
solidificación solidification, setting
solidificar solidify
sólido solid
sólidos suspendidos suspended solids
solidus solidus
solubilidad solubility
solubilidad de equilibrio equilibrium solubility
solubilidad de gas gas solubility
solubilizar solubilize
soluble soluble
soluble en agua water soluble
soluble en grasa fat soluble
solución solution
solución ácida acid solution
solución acuosa aqueous solution
solución alcalina alkaline solution
solución amortiguadora buffer solution
solución anisotónica anisotonic solution
solución anódica anodic solution
solución básica basic solution
solución centinormal centinormal solution
solución coloidal colloidal solution
solución concentrada concentrated solution
solución de Benedict Benedict solution
solución de Bouchardat Bouchardat solution
solución de Brodie Brodie solution
solución de Carl Carl solution
solución de Channing Channing solution
solución de Dakin Dakin solution
solución de Eder Eder solution
solución de Fehling Fehling solution
solución de Feiser Feiser solution
solución de Fischer Fischer solution
solución de Folin Folin solution
solución de Gabbet Gabbet solution
solución de Hanus Hanus solution
solución de Hübl Hübl solution
solución de Lugol Lugol solution
solución de nutrientes nutrient solution
solución de Pavy Pavy solution
solución de Pearson Pearson solution
solución de reactivo reagent solution
solución de Ringer Ringer solution
solución de Rohrbach Rohrbach solution

solución de sal de Pasteur Pasteur salt solution
solución de Schiff Schiff solution
solución de titulación titrant
solución de valoración titrant
solución de Volhard Volhard solution
solución de Wagner Wagner solution
solución de Wijs Wijs solution
solución decimolar decimolar solution
solución decinormal decinormal solution, tenth-normal solution
solución diluida dilute solution
solución electrolítica electrolytic solution
solución física physical solution
solución hipertónica hypertonic solution
solución hipotónica hypotonic solution
solución ideal ideal solution
solución iónica ionic solution
solución isotónica isotonic solution
solución molal molal solution
solución molar molar solution
solución molecular molecular solution
solución molecular-gramo gram-molecular solution
solución neutra neutral solution
solución no acuosa nonaqueous solution
solución no saturada unsaturated solution
solución normal normal solution, standard solution
solución perfecta perfect solution
solución química chemical solution
solución saturada saturated solution
solución sólida solid solution
solución supersaturada supersaturated solution
solución volumétrica volumetric solution
soluto solute
solutropo solutrope
solvatación solvation
solvato solvate
solvatocromismo solvatochromism
solvente solvent
solvente ácido acid solvent
solvente acuoso aqueous solvent
solvente aprótico aprotic solvent
solvente básico basic solvent
solvente clorado chlorinated solvent
solvente de ionización ionizing solvent

solvente de Stoddard Stoddard solvent
solvente físico physical solvent
solvente latente latent solvent
solvente no acuosa nonaqueous solvent
solvente no ionizante nonionizing solvent
solvente no polar nonpolar solvent
solvente orgánico organic solvent
solvente polar polar solvent
solvente químico chemical solvent
solvente volátil volatile solvent
solvólisis solvolysis
solvolítico solvolytic
solvus solvus
sonda probe
sonda electrónica electron probe
sonólisis sonolysis
sonoquímica sonochemistry
sorbato cálcico calcium sorbate
sorbato de butilo butyl sorbate
sorbato potásico potassium sorbate
sorbato sódico sodium sorbate
sorbente sorbent
sórbido sorbide
sorbitán sorbitan
sorbitol sorbitol
sorbosa sorbose
sorción sorption
sosa soda
sosa cáustica caustic soda
sosa de lavar washing soda
sosoloide sosoloid
sub sub
subacetato subacetate
subacetato de plomo lead subacetate
subatómica subatomics
subatómico subatomic
subcapa subshell
subcarbonato subcarbonate
subcarbonato cúprico cupric subcarbonate
subcarbonato de bismuto bismuth subcarbonate
subcarbonato de plomo lead subcarbonate
subcarbonato férrico ferric subcarbonate
subcloruro de azufre sulfur subchloride
subcloruro de bismuto bismuth subchloride

subenfriado undercooling, supercooling
suberano suberane
suberilo suberyl
suberona suberone
subestructura substructure
subgalato de bismuto bismuth subgallate
sublación sublation
subletal sublethal
sublimación sublimation
sublimador sublimator
sublimador de Hortvet Hortvet sublimator
sublimar sublimate
subnitrato subnitrate
subnitrato de bismuto bismuth subnitrate
subnivel sublevel
subnuclear subnuclear
subóxido suboxide
subóxido de carbono carbon suboxide
subóxido de plata silver suboxide
subóxido de plomo lead suboxide
subóxido de vanadio vanadium suboxide
subsalicilato de bismuto bismuth subsalicylate
substancia substance
substancia aprótica aprotic substance
substancia controlada controlled substance
substancia corrosiva corrosive substance
substancia dañina harmful substance
substancia de contacto contact substance
substancia delicuescente deliquescent substance
substancia explosiva explosive substance
substancia fotosensible photosensitive substance
substancia inerte inert substance
substancia inflamable flammable substance
substancia nociva noxious substance
substancia peligrosa hazardous substance
substancia química chemical substance
substancia racémica racemic substance
substancia radiomimética radiomimetic substance
substancia tóxica toxic substance
substitución substitution

substitución electrofílica electrophilic substitution

substitución nucleofílica nucleophilic substitution

substituir substitute

substituto de azúcar sugar substitute

substrato substrate

subsulfato de paladio palladium subsulfate

subsulfato férrico ferric subsulfate

subsulfito sódico sodium subsulfite

subtilina subtilin

succinaldehído succinaldehyde

succinamilo succinamyl

succinamoilo succinamoyl

succinato succinate

succinato amónico ammonium succinate

succinato bárico barium succinate

succinato cálcico calcium succinate

succinato cobaltoso cobaltous succinate

succinato de bencilo benzyl succinate

succinato de bencilo sódico sodium benzyl succinate

succinato de cadmio cadmium succinate

succinato de cinc zinc succinate

succinato de dietilo diethyl succinate

succinato de etilo ethyl succinate

succinato magnésico magnesium succinate

succinato mercúrico mercuric succinate

succinato sódico sodium succinate

succinilo succinyl

succinimida succinimide

succinimido succinimido

succiniodimida succiniodimide

succinonitrilo succinonitrile

sucrasa sucrase

sucrato sucrate

sucrol sucrol

sucrosa sucrose

suelo soil

suero serum

suero de la leche whey

suero de la verdad truth serum

sujetador de crisol crucible holder

sulfacetamida sulfacetamide

sulfadimidina sulfadimidine

sulfadiazina sulfadiazine

sulfaguanidina sulfaguanidine

sulfaldehído sulfaldehyde

sulfalizarato sódico sodium sulfalizarate

sulfamato sulfamate

sulfamato amónico ammonium sulfamate

sulfamato cálcico calcium sulfamate

sulfamerazina sulfamerazine

sulfametizol sulfamethizole

sulfámico sulfamic

sulfamida sulfamide

sulfamilo sulfamyl

sulfamoilo sulfamoyl

sulfanilamida sulfanilamide

sulfanilato sulfanilate

sulfanilato sódico sodium sulfanilate

sulfano sulfane

sulfantimoniato sódico sodium sulfantimonate

sulfarseniato sulfarsenate

sulfatación sulfation

sulfatasa sulfatase

sulfatiazol sulfathiazole

sulfato sulfate

sulfato ácido de hidroxilamina hydroxylamine acid sulfate

sulfato ácido de metilo acid methyl sulfate

sulfato ácido potásico potassium acid sulfate

sulfato ácido sódico sodium acid sulfate

sulfato alumínico aluminum sulfate

sulfato alumínico de cesio cesium aluminum sulfate

sulfato alumínico potásico potassium aluminum sulfate

sulfato alumínico sódico sodium aluminum sulfate

sulfato amónico ammonium sulfate

sulfato amónico ácido acid ammonium sulfate

sulfato amónico alumínico aluminum ammonium sulfate

sulfato amónico cobaltoso cobaltous ammonium sulfate

sulfato amónico de cobalto cobalt ammonium sulfate

sulfato amónico de cromo chromium ammonium sulfate

sulfato amónico de hierro iron ammonium sulfate

sulfato amónico de níquel nickel ammonium sulfate

sulfato amónico férrico ferric ammonium sulfate

sulfato amónico ferroso ferrous ammonium sulfate

sulfato amónico sódico sodium ammonium sulfate

sulfato antimonioso antimonous sulfate

sulfato áurico auric sulfate

sulfato bárico barium sulfate

sulfato cálcico calcium sulfate

sulfato cérico ceric sulfate

sulfato ceroso cerous sulfate

sulfato cobáltico cobaltic sulfate

sulfato cobaltoso cobaltous sulfate

sulfato crómico chromic sulfate

sulfato cromoso chromous sulfate

sulfato cúprico cupric sulfate

sulfato cuproso cuprous sulfate

sulfato de aluminio amónico ammonium aluminum sulfate

sulfato de anilina aniline sulfate

sulfato de antimonio antimony sulfate

sulfato de atropina atropine sulfate

sulfato de bencidina benzidine sulfate

sulfato de berilio beryllium sulfate

sulfato de bismuto bismuth sulfate

sulfato de brucina brucine sulfate

sulfato de cadmio cadmium sulfate

sulfato de cerio cerium sulfate

sulfato de cesio cesium sulfate

sulfato de cinc zinc sulfate

sulfato de cinc alumínico aluminum zinc sulfate

sulfato de cobalto cobalt sulfate

sulfato de cobre copper sulfate

sulfato de cocaína cocaine sulfate

sulfato de codeína codeine sulfate

sulfato de cromo chromium sulfate

sulfato de cromo amónico ammonium chromium sulfate

sulfato de dextrano sódico sodium dextran sulfate

sulfato de didimio didymium sulfate

sulfato de dietilo diethyl sulfate

sulfato de dihidrazina dihydrazine sulfate

sulfato de dimetilo dimethyl sulfate

sulfato de disprosio dysprosium sulfate

sulfato de efedrina ephedrine sulfate

sulfato de erbio erbium sulfate

sulfato de escandio scandium sulfate

sulfato de estaño tin sulfate

sulfato de estroncio strontium sulfate

sulfato de etilo ethyl sulfate

sulfato de europio europium sulfate

sulfato de fenilo phenyl sulfate

sulfato de fisostigmina physostigmine sulfate

sulfato de gadolinio gadolinium sulfate

sulfato de galio gallium sulfate

sulfato de guanilurea guanylurea sulfate

sulfato de hafnio hafnium sulfate

sulfato de hidrazina hydrazine sulfate

sulfato de hidrógeno hydrogen sulfate

sulfato de hidrógeno amónico ammonium hydrogen sulfate

sulfato de hidrógeno potásico potassium hydrogen sulfate

sulfato de hidrógeno sódico sodium hydrogen sulfate

sulfato de hidroxilamina hydroxylamine sulfate

sulfato de hidroxiquinolina hydroxyquinoline sulfate

sulfato de hierro iron sulfate

sulfato de hierro amónico ammonium iron sulfate

sulfato de indio indium sulfate

sulfato de iterbio ytterbium sulfate

sulfato de itrio yttrium sulfate

sulfato de kanamicina kanamycin sulfate

sulfato de lantano lanthanum sulfate

sulfato de litio lithium sulfate

sulfato de lutecio lutetium sulfate

sulfato de manganeso manganese sulfate

sulfato de manganeso amónico ammonium manganese sulfate

sulfato de metilo methyl sulfate

sulfato de neodimio neodymium sulfate

sulfato de níquel nickel sulfate

sulfato de níquel amónico ammonium nickel sulfate

sulfato de paladio palladium sulfate

sulfato de paramomicina paromomycin sulfate

sulfato de pentilo pentyl sulfate

sulfato de plata silver sulfate
sulfato de platino platinum sulfate
sulfato de plomo lead sulfate
sulfato de praseodimio praseodymium sulfate
sulfato de quinina quinine sulfate
sulfato de radio radium sulfate
sulfato de rodio rhodium sulfate
sulfato de rubidio rubidium sulfate
sulfato de rubidio alumínico aluminum rubidium sulfate
sulfato de talio thallium sulfate
sulfato de talio alumínico aluminum thallium sulfate
sulfato de telurio tellurium sulfate
sulfato de terbio terbium sulfate
sulfato de titanilo titanyl sulfate
sulfato de titanio titanium sulfate
sulfato de torio thorium sulfate
sulfato de uranilo uranyl sulfate
sulfato de uranio uranium sulfate
sulfato de vanadilo vanadyl sulfate
sulfato de vanadio vanadium sulfate
sulfato de yodo iodine sulfate
sulfato de zirconilo zirconyl sulfate
sulfato de zirconio zirconium sulfate
sulfato de zirconio sódico sodium zirconium sulfate
sulfato estánnico stannic sulfate
sulfato estannoso stannous sulfate
sulfato férrico ferric sulfate
sulfato ferroso ferrous sulfate
sulfato magnésico magnesium sulfate
sulfato magnésico amónico ammonium magnesium sulfate
sulfato manganoso manganous sulfate
sulfato mercúrico mercuric sulfate
sulfato mercurioso mercurous sulfate
sulfato niqueloso nickelous sulfate
sulfato platínico platinic sulfate
sulfato potásico potassium sulfate
sulfato potásico alumínico aluminum potassium sulfate
sulfato potásico berílico beryllium potassium sulfate
sulfato potásico crómico chromium potassium sulfate
sulfato potásico de cobalto cobalt potassium sulfate

sulfato potásico de manganeso manganese potassium sulfate
sulfato potásico de níquel nickel potassium sulfate
sulfato samárico samaric sulfate
sulfato samaroso samarous sulfate
sulfato sódico sodium sulfate
sulfato sódico alumínico aluminum sodium sulfate
sulfato sódico amónico ammonium sodium sulfate
sulfato sódico anhidro anhydrous sodium sulfate
sulfato talioso thallous sulfate
sulfato titánico titanic sulfate
sulfato titanoso titanous sulfate
sulfato vanádico vanadic sulfate
sulfazida sulfazide
sulfenamida sulfenamide
sulfénico sulfenic
sulfeno sulfeno
sulfenona sulfenone
sulfhidrato cálcico calcium sulfhydrate
sulfhidrato de etilo ethyl sulfhydrate
sulfhidrato de glicol glycol sulfhydrate
sulfhidrato sódico sodium sulfhydrate
sulfhidrilo sulfhydryl
sulfima sulfime
sulfimida sulfimide
sulfimido benzoico benzoic sulfimide
sulfina sulfine
sulfinato sulfinate
sulfinilimida sulfinylimide
sulfinilo sulfinyl
sulfino sulfino
sulfito sulfite
sulfito ácido potásico potassium acid sulfite
sulfito ácido sódico sodium acid sulfite
sulfito amónico ammonium sulfite
sulfito bárico barium sulfite
sulfito cálcico calcium sulfite
sulfito cúprico cupric sulfite
sulfito cuproso cuprous sulfite
sulfito de bismuto bismuth sulfite
sulfito de cesio cesium sulfite
sulfito de cinc zinc sulfite
sulfito de dietilo diethyl sulfite
sulfito de escandio scandium sulfite

sulfito de estroncio strontium sulfite
sulfito de hidrógeno cálcico calcium
   hydrogen sulfite
sulfito de hidrógeno sódico sodium
   hydrogen sulfite
sulfito de hierro iron sulfite
sulfito de litio lithium sulfite
sulfito de manganeso manganese sulfite
sulfito de plata silver sulfite
sulfito de plomo lead sulfite
sulfito de rubidio rubidium sulfite
sulfito magnésico magnesium sulfite
sulfito manganoso manganous sulfite
sulfito potásico potassium sulfite
sulfito sódico sodium sulfite
sulfo sulfo
sulfoamino sulfoamino
sulfobromoftaleína sódica sodium
   sulfobromophthalein
sulfocarbanilida sulfocarbanilide
sulfocarbimida sulfocarbimide
sulfocarbolato sódico sodium
   sulfocarbolate
sulfocarbolida sulfocarbolide
sulfocarbonato potásico potassium
   sulfocarbonate
sulfocarbonato sódico sodium
   sulfocarbonate
sulfocianato sulfocyanate
sulfocianato cálcico calcium sulfocyanate
sulfocianato mercúrico mercuric
   sulfocyanate
sulfocianato potásico potassium
   sulfocyanate
sulfocianato sódico sodium sulfocyanate
sulfocianuro sulfocyanide
sulfocianuro bárico barium sulfocyanide
sulfocianuro de plomo lead sulfocyanide
sulfocianuro potásico potassium
   sulfocyanide
sulfocianuro sódico sodium sulfocyanide
sulfohidrato sulfohydrate
sulfolano sulfolane
sulfona sulfone
sulfonación sulfonation
sulfonamida sulfonamide
sulfonato sulfonate
sulfonato de alizarina alizarin sulfonate
sulfonato de alquilbenceno alkylbenzene

sulfonate
sulfonato sódico sodium sulfonate
sulfonildifenol sulfonyldiphenol
sulfonilo sulfonyl
sulfonio sulfonium
sulfonóxido sulfinoxide
sulfopropionitrilo sódico sodium
   sulfopropionitrile
sulforricinoleato sódico sodium
   sulforicinoleate
sulfoseleniuro sulfoselenide
sulfóxido sulfoxide
sulfóxido de dimetilo dimethyl sulfoxide
sulfóxido de talio thallium sulfoxide
sulfoxilato sulfoxylate
sulfoxilato de cinc zinc sulfoxylate
sulfoxilato sódico sodium sulfoxylate
sulfuración sulfuration
sulfurado sulfurated
sulfúrico sulfuric
sulfurilo sulfuryl
sulfuro sulfide
sulfuro alumínico aluminum sulfide
sulfuro amónico ammonium sulfide
sulfuro antimónico antimonic sulfide
sulfuro antimonioso antimonous sulfide
sulfuro arsenioso arsenous sulfide
sulfuro áurico auric sulfide
sulfuro bárico barium sulfide
sulfuro cálcico calcium sulfide
sulfuro cérico ceric sulfide
sulfuro cobáltico cobaltic sulfide
sulfuro cobaltoso cobaltous sulfide
sulfuro cúprico cupric sulfide
sulfuro cuproso cuprous sulfide
sulfuro de alilo allyl sulfide
sulfuro de amilo amyl sulfide
sulfuro de antimonio antimony sulfide
sulfuro de arsénico arsenic sulfide
sulfuro de bencilo benzyl sulfide
sulfuro de benzofenona benzophenone
   sulfide
sulfuro de benzoilo benzoyl sulfide
sulfuro de bismuto bismuth sulfide
sulfuro de boro boron sulfide
sulfuro de butilo butyl sulfide
sulfuro de cadmio cadmium sulfide
sulfuro de carbonilo carbonyl sulfide
sulfuro de caucho rubber sulfide

sulfuro de cianógeno cyanogen sulfide
sulfuro de cinc zinc sulfide
sulfuro de cobre copper sulfide
sulfuro de dialilo diallyl sulfide
sulfuro de diamilo diamyl sulfide
sulfuro de dibromodietilo
dibromodiethyl sulfide
sulfuro de dibutilestaño dibutyltin
sulfide
sulfuro de diclorodietilo dichlorodiethyl
sulfide
sulfuro de didecilo didecyl sulfide
sulfuro de diectilo diectyl sulfide
sulfuro de diestearilo distearyl sulfide
sulfuro de dietilo diethyl sulfide
sulfuro de dimetilo dimethyl sulfide
sulfuro de dimiristilo dimyristyl sulfide
sulfuro de dipentilo dipentyl sulfide
sulfuro de dipropilo dipropyl sulfide
sulfuro de divinilo divinyl sulfide
sulfuro de estaño tin sulfide
sulfuro de estroncio strontium sulfide
sulfuro de etilo ethyl sulfide
sulfuro de fósforo phosphorus sulfide
sulfuro de galio gallium sulfide
sulfuro de germanio germanium sulfide
sulfuro de hidrógeno hydrogen sulfide
sulfuro de hidrógeno amónico
ammonium hydrogen sulfide
sulfuro de hidrógeno sódico sodium
hydrogen sulfide
sulfuro de hierro iron sulfide
sulfuro de indio indium sulfide
sulfuro de itrio yttrium sulfide
sulfuro de lantano lanthanum sulfide
sulfuro de litio lithium sulfide
sulfuro de manganeso manganese sulfide
sulfuro de metilo methyl sulfide
sulfuro de molibdeno molybdenum
sulfide
sulfuro de neodimio neodymium sulfide
sulfuro de nitrógeno nitrogen sulfide
sulfuro de oro gold sulfide
sulfuro de plata silver sulfide
sulfuro de plomo lead sulfide
sulfuro de renio rhenium sulfide
sulfuro de rodio rhodium sulfide
sulfuro de rubidio rubidium sulfide
sulfuro de rutenio ruthenium sulfide

sulfuro de selenio selenium sulfide
sulfuro de silicio silicon sulfide
sulfuro de talio thallium sulfide
sulfuro de telurio tellurium sulfide
sulfuro de uranilo uranyl sulfide
sulfuro de vanadio vanadium sulfide
sulfuro de viniletoxietilo
vinylethoxyethyl sulfide
sulfuro de vinilo vinyl sulfide
sulfuro estánnico stannic sulfide
sulfuro estannoso stannous sulfide
sulfuro férrico ferric sulfide
sulfuro ferroso ferrous sulfide
sulfuro fosfórico phosphoric sulfide
sulfuro magnésico magnesium sulfide
sulfuro manganoso manganous sulfide
sulfuro mercúrico mercuric sulfide
sulfuro mercurioso mercurous sulfide
sulfuro niquélico nickelic sulfide
sulfuro plumboso plumbous sulfide
sulfuro potásico potassium sulfide
sulfuro samárico samaric sulfide
sulfuro sódico sodium sulfide
sulfuro talioso thallous sulfide
sulfuro vanádico vanadic sulfide
sulfuroso sulfurous
sultama sultam
sultona sultone
sulvanita sulvanite
superácido superacid
superactínido superactinide
superaleación superalloy
supercalentamiento superheating
supercarbonato supercarbonate
superconductividad superconductivity
superconductor superconductor
supercrítico supercritical
superficie surface
superficie de Fermi Fermi surface
superficie molar molar surface
superfluidez superfluidity
superfluido superfluid
superfosfato superphosphate
superfosfato amoniacal ammoniated
superphosphate
superfosfato cálcico calcium
superphosphate
superfosfato triple triple superphosphate
supernormal supernormal

superóxido superoxide
superóxido bárico barium superoxide
superóxido cálcico calcium superoxide
superóxido de plomo lead superoxide
superóxido potásico potassium
    superoxide
superóxido sódico sodium superoxide
superpalita superpalite
superpesado superheavy
superplasticidad superplasticity
superpolímero superpolymer
supersaturación supersaturation
supersaturado supersaturated
supersolubilidad supersolubility

supra supra
surfactante surfactant
surfusión surfusion
surrosión surrosion
susceptibilidad atómica atomic
    susceptibility
susceptibilidad magnética magnetic
    susceptibility
susceptibilidad molar molar
    susceptibility
suspensión suspension
suspensión acuosa slurry
suspensión coloidal colloidal suspension
suspensoide suspensoid

T

tabaco tobacco

tabla periódica periodic table

tabún tabun

taconita taconite

tacticidad tacticity

tactosol tactosol

taenita taenite

talato thallate

talato de manganeso manganese tallate

talato de plomo lead tallate

talato férrico ferric tallate

talco talc

tálico thallic

talidomida thalidomide

talio thallium

talioso thallous

talosa talose

tamaño de partícula particle size

tamiz screen

tamizado screening

tampón buffer

tanasa tannase

tanato cálcico calcium tannate

tanato de bismuto bismuth tannate

tanato férrico ferric tannate

tanilo tannyl

tanino tannin

tanque solar solar pond

tantalato tantalate

tantalato ferroso ferrous tantalate

tantalio tantalum

tantalita tantalite

tántalo tantalum

tantaloso tantalous

tantirón tantiron

tapa lid

tapioca tapioca

tapolita tapiolite

tapón stopper

tapón con llave stopper with tap

tapón de vidrio glass stopper

taquiol tachiol

tara tare

tártaro emético tartar emetic

tartaroilo tartaroyl

tartrato tartrate

tartrato ácido amónico ammonium acid tartrate

tartrato ácido potásico potassium acid tartrate

tartrato ácido sódico sodium acid tartrate

tartrato alumínico aluminum tartrate

tartrato amónico ammonium tartrate

tartrato amónico ácido acid ammonium tartrate

tartrato bárico barium tartrate

tartrato cálcico calcium tartrate

tartrato cobaltoso cobaltous tartrate

tartrato crómico chromic tartrate

tartrato cúprico cupric tartrate

tartrato de cadmio cadmium tartrate

tartrato de cinc zinc tartrate

tartrato de dibutilo dibutyl tartrate

tartrato de dietilo diethyl tartrate

tartrato de estroncio strontium tartrate

tartrato de hidrazina hydrazine tartrate

tartrato de hidrógeno amónico ammonium hydrogen tartrate

tartrato de hidrógeno potásico potassium hydrogen tartrate

tartrato de hidrógeno sódico sodium hydrogen tartrate

tartrato de hierro amónico ammonium iron tartrate

tartrato de plata silver tartrate

tartrato de plomo lead tartrate

tartrato de rubidio rubidium tartrate

tartrato estannoso stannous tartrate

tartrato ferroso ferrous tartrate

tartrato magnésico magnesium tartrate

tartrato manganoso manganous tartrate

tartrato mercurioso mercurous tartrate

tartrato potásico potassium tartrate

tartrato potásico alumínico aluminum potassium tartrate

tartrato potásico de antimonio antimony potassium tartrate
tartrato potásico de bismuto bismuth potassium tartrate
tartrato potásico sódico sodium potassium tartrate
tartrato sódico sodium tartrate
tartrato sódico de antimonio antimony sodium tartrate
tartrato sódico potásico potassium sodium tartrate
tartrazina tartrazine
tartronoilo tartronoyl
tau tau
taurilo tauryl
taurina taurine
tautomerasa tautomerase
tautomería tautomerism
tautomérico tautomeric
tautomerismo de valencia valence tautomerism
tautómero tautomer
tautourea tautourea
taxol taxol
tebaína thebaine
tebenidina thebenidine
tecnecio technetium
técnica de datación dating technique
técnica de Paneth Paneth technique
técnico technical
tecnología química chemical technology
tectita tektite
tefroíta tephroite
teína theine
tela de alambre wire cloth
telodrina telodrin
telomerización telomerization
teluluro de hidrógeno hydrogen telluride
teluluro de indio indium telluride
telurato tellurate
telurato amónico ammonium tellurate
telurato de hidrógeno hydrogen tellurate
telurato sódico sodium tellurate
telurilo telluryl
telurinilo tellurinyl
telurio tellurium
telurita tellurite
telurita potásica potassium tellurite
telurita sódica sodium tellurite

telurocetona telluroketone
teluronio telluronium
teluroso tellurous
telururo telluride
telururo de bismuto bismuth telluride
telururo de cadmio cadmium telluride
telururo de cinc zinc telluride
telururo de dimetilo dimethyl telluride
telururo de germanio germanium telluride
telururo de mercurio mercury telluride
telururo de plata silver telluride
telururo de plomo lead telluride
temperatura temperature
temperatura absoluta absolute temperature
temperatura ambiente ambient temperature
temperatura ambiente room temperature
temperatura característica de Debye Debye characteristic temperature
temperatura constante constant temperature
temperatura correspondiente corresponding temperature
temperatura crítica critical temperature
temperatura de cero absoluto absolute zero temperature
temperatura de condensación condensation temperature
temperatura de Curie Curie temperature
temperatura de Flory Flory temperature
temperatura de ignición ignition temperature
temperatura de inversión inversion temperature
temperatura de Néel Néel temperature
temperatura de sala room temperature
temperatura de solución crítica critical solution temperature
temperatura de transición transition temperature
temperatura de transición de vidrio glass-transition temperature
temperatura efectiva effective temperature
temperatura eutéctica eutectic temperature

temperatura normal standard temperature, normal temperature
temperatura peritéctica peritectic temperature
temperatura y presión normales standard temperature and pressure
templa mash
templar temper
tenazas tongs
tenazas de crisol crucible tongs
tenildiamina thenyldiamine
tenilo thenyl
tennantita tennantite
tenorita tenorite
tensión acuosa aqueous tension
tensión de electrodo electrode voltage
tensión de vapor vapor tension
tensión superficial surface tension
teobromina theobromine
teofilina theophylline
teorema de Bloch Bloch theorem
teorema de Kramers Kramers theorem
teorético theoretical
teoría atómica atomic theory
teoría atómica de Dalton Dalton atomic theory
teoría cinética kinetic theory
teoría cuántica quantum theory
teoría cuántica relativista relativistic quantum theory
teoría de Avogadro Avogadro theory
teoría de Bohr Bohr theory
teoría de Bohr-Sommerfeld Bohr-Sommerfeld theory
teoría de Brönsted Brönsted theory
teoría de Brönsted-Lowry Brönsted-Lowry theory
teoría de campo cristalino crystal-field theory
teoría de campo de ligandos ligand-field theory
teoría de Debye-Hückel Debye-Hückel theory
teoría de deformación strain theory
teoría de deformación de Baeyer Baeyer strain theory
teoría de Langmuir Langmuir theory
teoría de Lewis Lewis theory
teoría de Lowry-Brønsted Lowry-Brønsted theory
teoría de Lucas Lucas theory
teoría de McMillan-Mayer McMillan-Mayer theory
teoría de octeto octet theory
teoría de valencia valence theory
teoría de van't Hoff van't Hoff theory
teoría electrónica de Lewis Lewis electron theory
teoría nuclear nuclear theory
teoría orbital orbital theory
ter ter
teratógeno teratogen
terbia terbia
terbio terbium
tercera ley de termodinámica third law of thermodynamics
terciario tertiary
terebeno terebene
tereftalaldehído terephthalaldehyde
tereftalato de dimetilo dimethyl terephthalate
tereftalonitrilo terephthalonitrile
terfenilo terphenyl
termal thermal, thermic
termalizar thermalize
térmico thermic, thermal
terminación de cadena chain termination
terminología terminology
termita thermite
termo thermo
termoaislamiento heat insulation
termoanálisis thermoanalysis
termobalanza thermobalance
termocatalítico thermocatalytic
termocroico thermocroic
termodifusión thermodiffusion
termodinámica thermodynamics
termodinámica química chemical thermodynamics
termodinámico thermodynamic
termoelectricidad thermoelectricity
termoeléctrico thermoelectric
termoestable thermostable
termofisión thermofission
termofusión thermofusion
termogénico thermogenic
termógrafo thermograph
termólisis thermolysis

termolítico thermolytic
termoluminiscencia thermoluminescence
termoluminiscente thermoluminescent
termometamorfismo
    thermometamorphism
termométrico thermometric
termómetro thermometer
termómetro armado armored
    thermometer
termómetro de ampolleta húmeda wet-
    bulb thermometer
termómetro de Beckmann Beckmann
    thermometer
termómetro de Einchluss Einchluss
    thermometer
termómetro de gas gas thermometer
termómetro de pentano pentane
    thermometer
termonatrita thermonatrite
termoneutralidad thermoneutrality
termonuclear thermonuclear
termopar thermocouple
termoplástico thermoplastic
termoquímica thermochemistry
termoquímico thermochemical
termorregulador thermoregulator
termostático thermostatic
termostato thermostat
termotrópico thermotropic
ternario ternary
teroría de Werner Werner theory
teróxido teroxide
terpadieno terpadiene
terpeno terpene
terpenoide terpenoid
terpenol terpenol
terpilenol terpilenol
terpineno terpinene
terpineol terpineol
terpinoleno terpinolene
terpolímero terpolymer
tervalente tervalent
testosterona testosterone
tetina thetine
tetra tetra
tetraacetato de pentaeritritol
    pentaerythritol tetraacetate
tetraacetato de plomo lead tetraacetate
tetraacetilacetonato de zirconio

zirconium tetraacetylacetonate
tetrabásico tetrabasic
tetrabencilo tetrabenzyl
tetraborano tetraborane
tetraborato de cinc zinc tetraborate
tetraborato de litio lithium tetraborate
tetraborato de mercurio mercury
    tetraborate
tetraborato potásico potassium
    tetraborate
tetraborato sódico sodium tetraborate
tetrabromo tetrabromo
tetrabromoaurato sódico sodium
    tetrabromoaurate
tetrabromobenceno tetrabromobenzene
tetrabromoetano tetrabromoethane
tetrabromoetileno tetrabromoethylene
tetrabromofluoresceína
    tetrabromofluorescein
tetrabromometano tetrabromomethane
tetrabromosilano tetrabromosilane
tetrabromuro de acetileno acetylene
    tetrabromide
tetrabromuro de carbono carbon
    tetrabromide
tetrabromuro de estaño tin tetrabromide
tetrabromuro de etileno ethylene
    tetrabromide
tetrabromuro de germanio germanium
    tetrabromide
tetrabromuro de iridio iridium
    tetrabromide
tetrabromuro de selenio selenium
    tetrabromide
tetrabromuro de silicio silicon
    tetrabromide
tetrabromuro de telurio tellurium
    tetrabromide
tetrabromuro de uranio uranium
    tetrabromide
tetrabutilestaño tetrabutyltin
tetracaína tetracaine
tetracarbonilo de cobalto cobalt
    tetracarbonyl
tetracarbonilo de níquel nickel
    tetracarbonyl
tetracarboxibutano tetracarboxybutane
tetraceno tetrazene
tetracetona tetraketone

tetraciano tetracyano

tetracianoaurato potásico potassium
tetracyanoaurate

tetracianoetileno tetracyanoethylene

tetracianoquinonadimetano
tetracyanoquinonedimethane

tetraciclina tetracycline

tetraciclona tetracyclone

tetracloro tetrachloro

tetracloroaluminato sódico sodium
tetrachloroaluminate

tetracloroanilina tetrachloroaniline

tetracloroaurato sódico sodium
tetrachloroaurate

tetraclorobenceno tetrachlorobenzene

tetraclorodifeniletano
tetrachlorodiphenylethane

tetracloroetano tetrachloroethane

tetracloroetileno tetrachloroethylene

tetraclorofenato sódico sodium
tetrachlorophenate

tetraclorofenol tetrachlorophenol

tetraclorometano tetrachloromethane

tetracloronaftaleno
tetrachloronaphthalene

tetracloropaladato potásico potassium
tetrachloropalladate

tetracloropaladato sódico sodium
tetrachloropalladate

tetracloroplatinato sódico sodium
tetrachloroplatinate

tetraclorosalicilanilida
tetrachlorosalicylanilide

tetraclorosilano tetrachlorosilane

tetracloruro tetrachloride

tetracloruro de azufre sulfur
tetrachloride

tetracloruro de carbono carbon
tetrachloride

tetracloruro de estaño tin tetrachloride

tetracloruro de etileno ethylene
tetrachloride

tetracloruro de germanio germanium
tetrachloride

tetracloruro de iridio iridium
tetrachloride

tetracloruro de manganeso manganese
tetrachloride

tetracloruro de osmio osmium
tetrachloride

tetracloruro de platino platinum
tetrachloride

tetracloruro de plomo lead tetrachloride

tetracloruro de silicio silicon
tetrachloride

tetracloruro de titanio titanium
tetrachloride

tetracloruro de torio thorium
tetrachloride

tetracloruro de uranio uranium
tetrachloride

tetracloruro de vanadio vanadium
tetrachloride

tetracloruro de zirconio zirconium
tetrachloride

tetracosano tetracosane

tetracosilo tetracosyl

tetrada tetrad

tetradecano tetradecane

tetradecanol tetradecanol

tetradeceno tetradecene

tetradecilamina tetradecylamine

tetradecileno tetradecylene

tetradecilo tetradecyl

tetradecilsulfato sódico sodium
tetradecyl sulfate

tetradeciltiol tetradecyl thiol

tetradifón tetradifon

tetraédrico tetrahedronal

tetraedro tetrahedron

tetraetilbenceno tetraethylbenzene

tetraetilenglicol tetraethylene glycol

tetraetilenopentamina
tetraethylenepentamine

tetraetilestaño tetraethyltin

tetraetilgermanio tetraethylgermanium

tetraetilo tetraethyl

tetraetilplomo tetraethyllead

tetraetoxipropano tetraethoxypropane

tetrafenilborato sódico sodium
tetraphenylborate

tetrafenilbutadieno tetraphenylbutadiene

tetrafenilciclopentadienona
tetraphenylcyclopentadienone

tetrafenileno tetraphenylene

tetrafenilestaño tetraphenyltin

tetrafeniletano tetraphenylethane

tetrafenilmetano tetraphenylmethane

tetrafenilo tetraphenyl
tetrafenilsilano tetraphenylsilane
tetrafenilurea tetraphenylurea
tetrafluoroborato tetrafluoroborate
tetrafluoroborato potásico potassium
    tetrafluoroborate
tetrafluorodicloroetano
    tetrafluorodichloroethane
tetrafluoroetileno tetrafluoroethylene
tetrafluorohidrazina
    tetrafluorohydrazine
tetrafluorometano tetrafluoromethane
tetrafluorosilano tetrafluorosilane
tetrafluoruro de azufre sulfur
    tetrafluoride
tetrafluoruro de carbono carbon
    tetrafluoride
tetrafluoruro de germanio germanium
    tetrafluoride
tetrafluoruro de manganeso manganese
    tetrafluoride
tetrafluoruro de selenio selenium
    tetrafluoride
tetrafluoruro de silicio silicon
    tetrafluoride
tetrafluoruro de uranio uranium
    tetrafluoride
tetrafluoruro de xenón xenon
    tetrafluoride
tetrafluoruro de zirconio zirconium
    tetrafluoride
tetrafosfato de hexaetilo hexaethyl
    tetraphosphate
tetrafosfato sódico sodium
    tetraphosphate
tetraftalato de polietileno polyethylene
    tetraphthalate
tetragaloilo tetragalloyl
tetragonal tetragonal
tetrahidrato tetrahydrate
tetrahidrato de perborato sódico
    sodium perborate tetrahydrate
tetrahidro tetrahydro
tetrahidroaluminato de litio lithium
    tetrahydroaluminate
tetrahidrobenceno tetrahydrobenzene
tetrahidroborato potásico potassium
    tetrahydroborate
tetrahidroborato sódico sodium

tetrahydroborate
tetrahidrocannibol tetrahydrocannibol
tetrahidrofenol tetrahydrophenol
tetrahidrofurano tetrahydrofuran
tetrahidrofuranodimetanol
    tetrahydrofurandimethanol
tetrahidrofurfurilamina
    tetrahydrofurfurylamine
tetrahidronaftaleno
    tetrahydronaphthalene
tetrahidropiranometanol
    tetrahydropyranmethanol
tetrahidropiridina tetrahydropyridine
tetrahidrotiofeno tetrahydrothiophene
tetrahidroxi tetrahydroxy
tetrahidroxibenceno
    tetrahydroxybenzene
tetrahidroxibutano tetrahydroxybutane
tetrahidruro de germanio germanium
    tetrahydride
tetrahidruro de silicio silicon
    tetrahydride
tetrakis tetrakis
tetralina tetralin
tetralita tetralite
tetralona tetralone
tetrámero tetramer
tetrametilbenceno tetramethylbenzene
tetrametilbencidina tetramethylbenzidine
tetrametilciclobutanodiol
    tetramethylcyclobutanediol
tetrametildiaminobenzhidrol
    tetramethyldiaminobenzhydrol
tetrametildiaminobenzofenona
    tetramethyldiaminobenzophenone
tetrametildiaminodifenilsulfona
    tetramethyldiaminodiphenylsulfone
tetrametilendiamina
    tetramethylenediamine
tetrametileno tetramethylene
tetrametilestaño tetramethyltin
tetrametiletilendiamina
    tetramethylethylenediamine
tetrametilmetano tetramethylmethane
tetrametilo tetramethyl
tetrametilplomo tetramethyllead
tetrametilsilano tetramethylsilane
tetrametilurea tetramethylurea
tetrametoxipropano

tetramethoxypropane
tetramina tetramine
tetramolecular tetramolecular
tetramorfismo tetramorphism
tetranitrato tetranitrate
tetranitrato de pentaeritritol
pentaerythritol tetranitrate
tetranitro tetranitro
tetranitroanilina tetranitroaniline
tetranitrobifenilo tetranitrobiphenyl
tetranitrofenol tetranitrophenol
tetranitrometano tetranitromethane
tetraóxido tetraoxide
tetrapropileno tetrapropylene
tetrasilano tetrasilane
tetrasulfuro bárico barium tetrasulfide
tetrasulfuro de arsénico arsenic
tetrasulfide
tetrasulfuro de cromo chromium
tetrasulfide
tetrasulfuro de molibdeno molybdenum
tetrasulfide
tetrasulfuro sódico sodium tetrasulfide
tetratioarseniato tetrathioarsenate
tetratioarsenito tetrathioarsenite
tetrationato tetrathionate
tetravalente tetravalent
tetrayodo tetraiodo
tetrayodoaurato potásico potassium
tetraiodoaurate
tetrayoduro de carbono carbon
tetraiodide
tetrayoduro de estaño tin tetraiodide
tetrayoduro de germanio germanium
tetraiodide
tetrayoduro de iridio iridium tetraiodide
tetrayoduro de silicio silicon tetraiodide
tetrazol tetrazole
tetrazolilo tetrazolyl
tetrazolio tetrazolium
tetrazona tetrazone
tetrel tetrel
tetrilo tetryl
tetritol tetritol
tetrol tetrol
tetrona tetrone
tetrosa tetrose
tetroxalato potásico potassium
tetroxalate

tetróxido tetroxide
tetróxido de bismuto bismuth tetraoxide
tetróxido de cesio cesium tetraoxide
tetróxido de dinitrógeno dinitrogen
tetraoxide
tetróxido de manganeso manganese
tetraoxide
tetróxido de nitrógeno nitrogen
tetraoxide
tetróxido de osmio osmium tetraoxide
tetróxido de plomo lead tetraoxide
tetróxido de rutenio ruthenium
tetraoxide
tetróxido de vanadio vanadium
tetraoxide
texafirina texaphyrin
texilo thexyl
textura texture
thalenita thalenite
theta theta
thomsonita thomsonite
thortveitita thortveitite
tia thia
tiabendazol thiabendazole
tiacilo thiazyl
tial thial
tiamina thiamine
tiantreno thianthrene
tiazina thiazine
tiazol thiazole
tiazolilo thiazolyl
tiempo de generación generation time
tiempo de reacción reaction time
tiempo de relajación relaxation time
tiempo de retención retention time
tiempo de transición transition time
tiempo libre medio mean free time
tienilo thienyl
tierra earth
tierra alcalina alkaline earth
tierra de batán fuller's earth
tierra de diatomeas diatomaceous earth
tierra de pipas pipe clay
tierra de sombra umber
tierra rara rare earth
timerosal thimerosal
timidina thymidine
timilo thymyl
timina thymine

timol thymol
timolftaleína thymolphthalein
timolol timolol
timolsulfonaftaleína
    thymolsulfonephthalein
tincal tincal
tinta ink
tinte dye, colorant
tintómetro tintometer
tintura tincture, stain
tintura de yodo iodine tincture
tio thio
tioacetal thioacetal
tioacetaldehído thioacetaldehyde
tioacetamida thioacetamide
tioacetanilida thioacetanilide
tioacetato amónico ammonium
    thioacetate
tioácido thioacid
tioalcohol thioalcohol
tioaldehído thioaldehyde
tioanhídrido thioanhydride
tioantimoniato sódico sodium
    thioantimonate
tioarseniato thioarsenate
tioarseniato de arsénico arsenic
    thioarsenate
tioato thioate
tioaurato sódico sodium thioaurate
tiobenzaldehído thiobenzaldehyde
tiobenzamida thiobenzamide
tiobenzofenona thiobenzophenone
tiocarbamoilo thiocarbamoyl
tiocarbanilida thiocarbanilide
tiocarbonato thiocarbonate
tiocarbonato celulósico cellulosic
    thiocarbonate
tiocarbonato sódico sodium
    thiocarbonate
tiocarbonilo thiocarbonyl
tiocarboxi thiocarboxy
tiocetona thioketone
tiocianato thiocyanate
tiocianato alumínico aluminum
    thiocyanate
tiocianato amónico ammonium
    thiocyanate
tiocianato bárico barium thiocyanate
tiocianato cálcico calcium thiocyanate

tiocianato cuproso cuprous thiocyanate
tiocianato de arsénico arsenic
    thiocyanate
tiocianato de bencilo benzyl thiocyanate
tiocianato de cinc zinc thiocyanate
tiocianato de etilo ethyl thiocyanate
tiocianato de guanidina guanidine
    thiocyanate
tiocianato de hierro iron thiocyanate
tiocianato de litio lithium thiocyanate
tiocianato de nonilo nonyl thiocyanate
tiocianato de pentilo pentyl thiocyanate
tiocianato de plata silver thiocyanate
tiocianato de plomo lead thiocyanate
tiocianato férrico ferric thiocyanate
tiocianato ferroso ferrous thiocyanate
tiocianato magnésico magnesium
    thiocyanate
tiocianato mercúrico mercuric
    thiocyanate
tiocianato potásico potassium
    thiocyanate
tiocianato sódico sodium thiocyanate
tiocianiato de butilo butyl thiocyanate
tiocianoacetato de isobornilo isobornyl
    thiocyanoacetate
tiocianógeno thiocyanogen
tiocianuro thiocyanide
tiocloruro de fósforo phosphorus
    thiochloride
tiocromo thiochrome
tiodifenol thiodiphenol
tiodiglicol thiodiglycol
tiodipropionitrilo thiodipropionitrile
tioestannato amónico ammonium
    thiostanate
tioéster thioester
tioetano de etilo ethyl thioethane
tioetanol de etilo ethyl thioethanol
tioéter thioether
tioéter alílico allyl thioether
tiofano thiophane
tiofenaldehído thiophenealdehyde
tiofenato de cinc zinc thiophenate
tiofenilo thiophenyl
tiofeno thiophene
tiofenol thiophenol
tioflavina thioflavine
tiofosforilo thiophosphoryl

tiofosgeno thiophosgene
tiofurano thiofuran
tioglicerol thioglycerol
tioglicolato de isooctilo isooctyl thioglycolate
tioglicolato sódico sodium thioglycolate
tiohidantoína thiohydantoin
tiohidroxi thiohydroxy
tiol thiol
tiolato thiolate
tiomolibdato thiomolybdate
tionato thionate
tionilo thionyl
tiopental sódico thiopental sodium
tioridazina thioridazine
tiosemicarbazida thiosemicarbazide
tiosorbitol thiosorbitol
tiosulfato thiosulfate
tiosulfato amónico ammonium thiosulfate
tiosulfato bárico barium thiosulfate
tiosulfato de estroncio strontium thiosulfate
tiosulfato de hierro iron thiosulfate
tiosulfato de plata silver thiosulfate
tiosulfato de plata sódico sodium silver thiosulfate
tiosulfato de plomo lead thiosulfate
tiosulfato de plomo sódico sodium lead thiosulfate
tiosulfato magnésico magnesium thiosulfate
tiosulfato potásico potassium thiosulfate
tiosulfato sódico sodium thiosulfate
tiosulfato sódico de plata silver sodium thiosulfate
tiotungstato thiotungstate
tiouracil thiouracil
tiourea thiourea
tiourea de etileno ethylene thiourea
tioxanteno thioxanthene
tioxilenol thioxylenol
tioxo thioxo
típico typical
tipo type
tiram thiram
tiramina tyramine
tirocidina tyrocidine
tironina thyronine

tirosilo tyrosyl
tirosina tyrosine
tirosinasa tyrosinase
tirotricina thyrothricin
tiroxina thyroxine
tisonita tysonite
titanato titanate
titanato bárico barium titanate
titanato cálcico calcium titanate
titanato de butilo butyl titanate
titanato de estroncio strontium titanate
titanato de litio lithium titanate
titanato de níquel nickel titanate
titanato de plomo lead titanate
titanato de tetrabutilo tetrabutyl titanate
titanato de tetraetilhexilo tetraethylhexyl titanate
titanato de tetraisopropilo tetraisopropyl titanate
titanato ferroso ferrous titanate
titanato potásico potassium titanate
titanato sódico sodium titanate
titania titania
titánico titanic
titanilo titanyl
titanio titanium
titanoso titanous
titulación titration
titulación ácido-base acid-base titration
titulación amperométrica amperometric titration
titulación automática automatic titration
titulación complejométrica complexometric titration
titulación con permanganato permanganate titration
titulación conductiométrica conductometric titration
titulación de Mohr Mohr titration
titulación de Volhard Volhard titration
titulación diferencial differential titration
titulación electrométrica electrometric titration
titulación entálpica enthalpy titration
titulación fotométrica photometric titration
titulación magnetométrica magnetometric titration

titulación oscilométrica oscillometric titration
titulación potenciométrica potentiometric titration
titulación radiométrica radiometric titration
titulación redox redox titration
titulación térmica thermal titration
titulación termométrica thermometric titration
título titer
tixotropía thixotropy
tocoferol tocopherol
tocofersolán tocophersolan
tolano tolan
tolidina tolidine
tolilaldehído tolylaldehyde
tolildietanolamina tolyldiethanolamine
tolilendiamina tolylenediamine
tolileno tolylene
tolilideno tolylidene
tolilo tolyl
tolnaftato tolnaftate
tolualdehído tolualdehyde
toluato sódico sodium toluate
toluendiamina toluenediamine
toluendiisocianato toluenediisocyanate
toluenilo toluenyl
tolueno toluene
toluenosulfonato sódico sodium toluenesulfonate
toluensulfamina toluenesulfamine
toluensulfanilida toluenesulfanilide
toluensulfonamida toluenesulfonamide
toluensulfonilo toluenesulfonyl
toluentiol toluenethiol
toluidina toluidine
toluidino toluidino
toluilendiamina toluylenediamine
toluileno toluylene
toluilo toluyl
toluoilo toluoyl
toluquinona toluquinone
toma de gas gas outlet
tomatina tomatine
tonca tonka
topacio topaz
topoquímica topochemistry
topotáctico topotactic

toria thoria
toriado thoriated
torianita thorianite
torina thorin
torio thorium
toriode toroid
torita thorite
tornasol litmus
torón thoron
torr torr
torre tower
torre de absorción absorption tower
torre de destilación distillation tower
torre de Gay-Lussac Gay-Lussac tower
torre empaquetada packed tower
torta amarilla yellow cake
torta de sal salt cake
tosilación tosylation
tosilo tosyl
toxafeno toxaphene
toxicidad toxicity
tóxico toxic
toxicología toxicology
toxina toxin
trans trans
transactínido transactinide
transalquilación transalkylation
transaminación transamination
transaminasa transaminase
transductor transducer
transesterificación transesterification
transferasa transferase
transferecia térmica heat transfer
transferencia de cadena chain transfer
transferencia de carga charge transfer
transferencia de fase phase transfer
transferencia electrónica electron transfer
transferencia radiativa radiative transfer
transformación transformation
transformación atómica atomic transformation
transformación de Curtius Curtius transformation
transformación de Fourier Fourier transform
transformación de Hofmann Hofmann transformation
transformación nuclear nuclear

transformation
transformación radiactiva radioactive
transformation
transformación termonuclear
thermonuclear transformation
transición transition
transición cuántica quantum transition
transición de fase phase transition
transición de Fermi Fermi transition
transición electrónica electronic
transition
transición gamma gamma transition
transición isomérica isomeric transition
transición permitida allowed transition
transición prohibida forbidden transition
transicional transitional
transmitancia transmittance
transmutación transmutation
transmutación atómica atomic
transmutation
transoide transoid
transparencia transparency
transporte transport
transporte activo active transport
transposición transposition,
rearrangement
transuránico transuranic
travertino travertine
trayectoria electrónica electron
trajectory
trayectoria libre molecular molecular
free path
traza trace
trazador tracer
trazador isotópico isotopic tracer
trazador radiactivo radioactive tracer
trementina turpentine
trementina de madera wood turpentine
tremolita tremolite
trenardita threnardite
treo threo
treonina threonine
treosa threose
tretamina tretamine
tri tri
triaceno triazene
triacetato triacetate
triacetato alumínico aluminum triacetate
triacetato de celulosa cellulose triacetate

triacetato de glicerilo glyceryl triacetate
triacetilgalato de etilo ethyl
triacetylgallate
triacetina triacetin
triacetonamina triacetoneamine
triacontano triacontane
triacontanol triacontanol
triacontilo triacontyl
tríada triad
trialilamina triallylamine
trialilo triallyl
trialquilsilanol trialkylsilanol
triamcinolona triamcinolone
triamilamina triamylamine
triamilbenceno triamylbenzene
triamina triamine
triaminobenceno triaminobenzene
triaminotolueno triaminotoluene
triángulo de crisol crucible triangle
triatómico triatomic
triazano triazane
triazina triazine
triazinilo triazinyl
triazol triazole
tribásico tribasic
tribencilamina tribenzylamine
tribencilo tribenzyl
triboluminiscencia triboluminescence
tribromo tribromo
tribromoacetaldehído
tribromoacetaldehyde
tribromoanilina tribromoaniline
tribromobenceno tribromobenzene
tribromoetanal tribromoethanal
tribromoetano tribromoethane
tribromoetanol tribromoethanol
tribromofenol tribromophenol
tribromometano tribromomethane
tribromopropano tribromopropane
tribromosalicilanilida
tribromosalicylanilide
tribromuro tribromide
tribromuro de alilo allyl tribromide
tribromuro de antimonio antimony
tribromide
tribromuro de arsénico arsenic
tribromide
tribromuro de bismuto bismuth
tribromide

tribromuro de boro boron tribromide
tribromuro de fósforo phosphorus tribromide
tribromuro de iridio iridium tribromide
tribromuro de nitrógeno nitrogen tribromide
tribromuro de oro gold tribromide
tribromuro férrico ferric tribromide
tributilaluminio tributylaluminum
tributilamina tributylamine
tributilborano tributylborane
tributilestaño tributyltin
tributilo tributyl
tricarbimida tricarbimide
tricetona triketone
triciano tricyano
tricianuro de oro gold tricyanide
tricíclico tricyclic
triciclodecano tricyclodecane
triclorfón trichlorfon
tricloro trichloro
tricloroacetaldehído trichloroacetaldehyde
tricloroacetamida trichloroacetamide
tricloroacetato sódico sodium trichloroacetate
tricloroanisol trichloroanisole
triclorobenceno trichlorobenzene
tricloroborazol trichloroborazole
triclorobromometano trichlorobromomethane
tricloroetanal trichloroethanal
tricloroetano trichloroethane
tricloroetanol trichloroethanol
tricloroeteno trichloroethene
tricloroetileno trichloroethylene
triclorofenato potásico potassium trichlorophenate
triclorofenato sódico sodium trichlorophenate
triclorofenol trichlorophenol
triclorofluorometano trichlorofluoromethane
triclorometano trichloromethane
triclorometilo trichloromethyl
tricloronaftaleno trichloronaphthalene
tricloronitrosometano trichloronitrosomethane
tricloropropano trichloropropane

triclorosilano trichlorosilane
triclorosilano de ciclohexilo cyclohexyl trichlorosilane
triclorotolueno trichlorotoluene
triclorotrifluoroacetona trichlorotrifluoroacetone
triclorotrifluoroetano trichlorotrifluoroethane
tricloruro trichloride
tricloruro benzoico benzoic trichloride
tricloruro de antimonio antimony trichloride
tricloruro de arsénico arsenic trichloride
tricloruro de bencenilo benzenyl trichloride
tricloruro de bismuto bismuth trichloride
tricloruro de boro boron trichloride
tricloruro de butilina butylin trichloride
tricloruro de carbono carbon trichloride
tricloruro de cromo chromium trichloride
tricloruro de etileno ethylene trichloride
tricloruro de formilo formyl trichloride
tricloruro de fósforo phosphorus trichloride
tricloruro de galio gallium trichloride
tricloruro de hierro iron trichloride
tricloruro de indio indium trichloride
tricloruro de iridio iridium trichloride
tricloruro de nitrógeno nitrogen trichloride
tricloruro de oro gold trichloride
tricloruro de renio rhenium trichloride
tricloruro de rodio rhodium trichloride
tricloruro de rutenio ruthenium trichloride
tricloruro de titanio titanium trichloride
tricloruro de tolueno toluene trichloride
tricloruro de trimesoilo trimesoyl trichloride
tricloruro de vanadio vanadium trichloride
tricloruro de vinilo vinyl trichloride
tricloruro de yodo iodine trichloride
tricloruro férrico ferric trichloride
tricosano tricosane
tricosanol tricosanol
tricresilo tricresyl

tridecano tridecane
tridecanoato de metilo methyl tridecanoate
tridecanol tridecanol
trideceno tridecene
tridecilbenceno tridecylbenzene
tridecilo tridecyl
tridimita tridymite
tridodecilamina tridodecyl amine
triestearato de glicerilo glyceryl tristearate
triestearato de glicerol glycerol tristearate
trietanolamina triethanolamine
trietilaluminio triethylaluminum
trietilamina triethylamine
trietilbenceno triethylbenzene
trietilborano triethylborane
trietilendiamina triethylenediamine
trietilenfosforamida triethylenephosphoramide
trietilenglicol triethylene glycol
trietilenmelamina triethylenemelamine
trietileno triethylene
trietilentetramina triethylenetetramine
trietilentriamina triethylenetriamine
trietilmetano triethylmethane
trietilo triethyl
trietilortoformiato triethylorthoformate
trietoxi triethoxy
trietoxihexano triethoxyhexane
trietoximetano triethoxymethane
trietoximetoxipropano triethoxymethoxypropane
trifásico triphasic
trifenilantimonio triphenylantimony
trifenilbenceno triphenylbenzene
trifenilboro triphenylboron
trifenilcarbinol triphenylcarbinol
trifenileno triphenylene
trifenilestaño triphenyltin
trifenilfosfina triphenylphosphine
trifenilfósforo triphenylphosphorus
trifenilguanidina triphenylguanidine
trifenilmetilo triphenylmethyl
trifenilo triphenyl
trifenol triphenol
triflato triflate
trifluoro trifluoro

trifluoroacetato de talio thallium trifluoroacetate
trifluoroacetilacetonato de cobre copper trifluoroacetylacetonate
trifluorobromometano trifluorobromomethane
trifluoroclorometano trifluorochloromethane
trifluoroestireno trifluorostyrene
trifluorometilo trifluoromethyl
trifluoronitrosometano trifluoronitrosomethane
trifluorotricloroetano trifluorotrichloroethane
trifluoroyodometano trifluoroiodomethane
trifluoruro de antimonio antimony trifluoride
trifluoruro de arsénico arsenic trifluoride
trifluoruro de boro boron trifluoride
trifluoruro de bromo bromine trifluoride
trifluoruro de cloro chlorine trifluoride
trifluoruro de cobalto cobalt trifluoride
trifluoruro de cromo chromium trifluoride
trifluoruro de nitrógeno nitrogen trifluoride
trifluoruro de rutenio ruthenium trifluoride
trifluoruro de tolueno toluene trifluoride
trifosfato bárico barium triphosphate
trifosfato de adenosina adenosine triphosphate
trifosfato pentasódico pentasodium triphosphate
trifosfato sódico sodium triphosphate
trifosgeno triphosgene
trigaloilo trigalloyl
triglicérido triglyceride
triglicerol triglycerol
triglicina triglycine
triglime triglyme
trigol trigol
trigonal trigonal
trigonelina trigonelline
trihexilo trihexyl
trihidrato trihydrate
trihidrato de alúmina alumina trihydrate

trihidrato de bismuto bismuth trihydrate
trihídrico trihydric
trihidroxi trihydroxy
trihidroxiantraquinona
    trihydroxyanthraquinone
trihidroxibenceno trihydroxybenzene
trihidróxido de bismuto bismuth
    trihydroxide
trihidroxipropano trihydroxypropane
triisobutilaluminio triisobutylaluminum
triisobutileno triisobutylene
triisopropanolamina triisopropanolamine
trilaurilamina trilaurylamine
trilaurilfosfito trilauryl phosphite
trilaurilo trilauryl
trilauriltritiofosfito trilauryl
    trithiophosphite
trímero trimer
trimetadiona trimethadione
trimetálico trimetallic
trimetano trimethano
trimetilaluminio trimethylaluminum
trimetilamina trimethylamine
trimetilanilina trimethylaniline
trimetilbenceno trimethylbenzene
trimetilbutano trimethylbutane
trimetilciclododecatrieno
    trimethylcyclododecatriene
trimetilciclohexano
    trimethylcyclohexane
trimetilciclohexanol
    trimethylcyclohexanol
trimetilclorosilano trimethylchlorosilane
trimetilenglicol trimethylene glycol
trimetileno trimethylene
trimetilestaño trimethyltin
trimetilhexano trimethylhexane
trimetilhexanol trimethylhexanol
trimetilmetano trimethylmethane
trimetilnonanona trimethylnonanone
trimetilo trimethyl
trimetiloletano trimethylolethane
trimetilolpropano trimethylolpropane
trimetilpentano trimethylpentane
trimetilpentanodiol trimethylpentanediol
trimetilpenteno trimethylpentene
trimetilpiridina trimethylpyridine
trimetilsililo trimethylsilyl
trimetoxi trimethoxy

trimetoxiboroxina trimethoxyboroxine
trimetoximetano trimethoxymethane
trimolecular trimolecular
trimorfismo trimorphism
trinitrato trinitrate
trinitrato de bismuto bismuth trinitrate
trinitrato de celulosa cellulose trinitrate
trinitrato de fenol phenol trinitrate
trinitrato de glicerilo glyceryl trinitrate
trinitrato de glicerol glycerol trinitrate
trinitro trinitro
trinitroanilina trinitroaniline
trinitroanisol trinitroanisole
trinitrobenceno trinitrobenzene
trinitrofenol trinitrophenol
trinitrofenolato amónico ammonium
    trinitrophenolate
trinitroglicerina trinitroglycerin
trinitrometano trinitromethane
trinitronaftaleno trinitronaphthalene
trinitrorresorcina trinitroresorcinol
trinitrorresorcinato de plomo lead
    trinitroresorcinate
trinitrotolueno trinitrotoluene
trinor trinor
triol triol
triona trione
triosa triose
trioxa trioxa
trioxano trioxane
trioxicloruro de renio rhenium
    trioxychloride
trióxido trioxide
trióxido de antimonio antimony trioxide
trióxido de arsénico arsenic trioxide
trióxido de azufre sulfur trioxide
trióxido de bismuto bismuth trioxide
trióxido de cesio cesium trioxide
trióxido de cromo chromium trioxide
trióxido de hierro iron trioxide
trióxido de iridio iridium trioxide
trióxido de lantano lanthanum trioxide
trióxido de manganeso manganese
    trioxide
trióxido de molibdeno molybdenum
    trioxide
trióxido de nitrógeno nitrogen trioxide
trióxido de oro gold trioxide
trióxido de samario samarium trioxide

trióxido de telurio tellurium trioxide
trióxido de titanio titanium trioxide
trióxido de tungsteno tungsten trioxide
trióxido de uranio uranium trioxide
trióxido de vanadio vanadium trioxide
trióxido férrico ferric trioxide
trioxígeno trioxygen
trioxima trioxime
trioximetileno trioxymethylene
trioxina trioxin
tripalmitato de glicerilo glyceryl tripalmitate
tripalmitina tripalmitin
triplete triplet
triplita triplite
trípode tripod
trípoli tripoli
tripolifosfato potásico potassium tripolyphosphate
tripolifosfato sódico sodium tripolyphosphate
tripolita tripolite
tripropilaluminio tripropylaluminum
tripropilamina tripropylamine
tripropilenglicol tripropylene glycol
tripropileno tripropylene
tripropilo tripropyl
tripsina trypsin
tripsinógeno trypsinogen
triptano triptane
tripticeno triptycene
triptófano tryptophan
triptofila tryptophyl
triquitas whiskers
tris tris
trishidroxifenilpropano trishydroxyphenylpropane
trisilano trisilane
trisilicato magnésico magnesium trisilicate
trisulfato de antimonio antimony trisulfate
trisulfato férrico ferric trisulfate
trisulfuro trisulfide
trisulfuro bárico barium trisulfide
trisulfuro de antimonio antimony trisulfide
trisulfuro de arsénico arsenic trisulfide
trisulfuro de bismuto bismuth trisulfide

trisulfuro de boro boron trisulfide
trisulfuro de cesio cesium trisulfide
trisulfuro de fósforo phosphorus trisulfide
trisulfuro de molibdeno molybdenum trisulfide
trisulfuro de oro gold trisulfide
trisulfuro de vanadio vanadium trisulfide
tritelururo de bismuto bismuth tritelluride
triterpeno triterpene
tritilo trityl
tritio tritium
tritioacetaldehído trithioacetaldehyde
tritiocarbonato trithiocarbonate
tritiocarbonato sódico sodium trithiocarbonate
tritionato bárico barium trithionate
trititanato sódico sodium trititanate
tritón triton
triturar triturate
trivalente trivalent
triyodo triiodo
triyodometano triiodomethane
triyodotironina triiodothyronine
triyoduro de antimonio antimony triiodide
triyoduro de bismuto bismuth triiodide
triyoduro de formilo formyl triiodide
triyoduro de fósforo phosphorus triiodide
triyoduro de nitrógeno nitrogen triiodide
triyoduro de oro gold triiodide
triyoduro potásico potassium triiodide
trombina thrombin
trona trona
tropanol tropanol
tropilidina tropilidine
tropilio tropylium
tropina tropine
tropoilo tropoyl
tropolona tropolone
tubería de vidrio glass tubing
tubo con embudo funnel tube
tubo de absorción absorption tube
tubo de absorción de Babo Babo absorption tube
tubo de Arndt Arndt tube

tubo de Bowen Bowen tube
tubo de Carius Carius tube
tubo de combustión combustion tube
tubo de condensación fraccionaria
  fractional condensation tube
tubo de Coolidge Coolidge tube
tubo de Crookes Crookes tube
tubo de destilación fraccionaria
  fractional distillation tube
tubo de ensayo test tube
tubo de Faraday Faraday tube
tubo de fermentación fermentation tube
tubo de filtro filter tube
tubo de Fleming Fleming tube
tubo de fusión fusion tube
tubo de LaBel LaBel tube
tubo de plástico plastic pipe
tubo de rayos X X-ray tube
tubo de vidrio glass tube
tubo en U U-tube
tubo espectral spectral tube
tubo secante drying tube
tubos de Nessler Nessler tubes
tulia thulia
tulio thulium
tungstato tungstate
tungstato amónico ammonium tungstate
tungstato cálcico calcium tungstate
tungstato cobaltoso cobaltous tungstate
tungstato de bismuto bismuth tungstate

tungstato de cadmio cadmium tungstate
tungstato de cobalto cobalt tungstate
tungstato de cobre copper tungstate
tungstato de litio lithium tungstate
tungstato de manganeso manganese
  tungstate
tungstato de plomo lead tungstate
tungstato ferroso ferrous tungstate
tungstato magnésico magnesium
  tungstate
tungstato potásico potassium tungstate
tungstato sódico sodium tungstate
tungsteno tungsten
túngstico tungstic
tungstita tungstite
tungstofosfato tungstophosphate
tungstofosfato sódico sodium
  tungstophosphate
tungstosilicato tungstosilicate
tungstosilicato sódico sodium
  tungstosilicate
tunsgstato bárico barium tungstate
turba peat
turbidez turbidity
turbidimetría turbidimetry
turbio cloudy
turmalina tourmaline
turquesa turquoise
tuyeno thujene
tuyona thujone

U

ubiquinona ubiquinone
ulexina ulexine
ulexita ulexite
ulmina ulmin
ultra ultra
ultracentrífuga ultracentrifuge
ultrafiltración ultrafiltration
ultramicroscópico ultramicroscopic
ultramicroscopio ultramicroscope
ultrapuro ultrapure
ultrarrápido ultrarapid
ultrasensible ultrasensitive
ultrasónica ultrasonics
ultrasónico ultrasonic
ultravioleta ultraviolet
umangita umangite
umbeliferona umbelliferone
umbelulona umbellulone
umbral threshold
umbral de detección threshold of
    detection
umbral de fisión fission threshold
unario unary
undeca undeca
undecalactona undecalactone
undecanal undecanal
undecano undecane
undecanol undecanol
undecanona undecanone
undecenilo undecenyl
undeceno undecene
undecenol undecenol
undecilamina undecylamine
undecilenato cálcico calcium
    undecylenate
undecilenato de cinc zinc undecylenate
undecilenato potásico potassium
    undecylenate
undecilenato sódico sodium undecylenate
undecileno undecylene
undecilo undecyl
ungüento ointment
uni uni

uniaxial uniaxial
unidad unit
unidad absoluta absolute unit
unidad angstrom angstrom unit
unidad atómica atomic unit
unidad básica base unit
unidad cuántica quantum unit
unidad de masa atómica atomic mass
    unit
unidad de peso atómico atomic weight
    unit
unidad de radiación radiation unit
unidad de radio radium unit
unidad de sacarosa saccharose unit
unidad derivada derived unit
unidad fundamental fundamental unit
unidad internacional international unit
unidad repetitiva repeating unit
unimolecular unimolecular
unión union
unión espiro spiro union
univalente univalent
ur ur
uracil uracil
uralita uralite
uramido uramido
uramina uramine
uramino uramino
uranato uranate
uranato sódico sodium uranate
urania urania
uránico uranic
uranilo uranyl
uranina uranine
uraninita uraninite
uranio uranium
uranio agotado depleted uranium
uranio enriquecido enriched uranium
uranita uranite
uranoceno uranocene
uranocircita uranocircite
uranofano uranophane
uranosferita uranospherite

uranoso uranous
uranotalita uranothallite
urato magnésico magnesium urate
urea urea
urea de etileno ethylene urea
ureasa urease
ureido ureido

uretano urethane
uridina uridine
urilón urylon
uronio uronium
urotropina urotropin
ursina ursin
uvarovita uwarowite

V

vacante vacancy
vacío vacuum
vainillilideno vanillylidene
vainillilo vanillyl
vainillina vanillin
vainilloilo vanilloyl
valacidina valacidin
valencia valence, valency
valencia principal principal valency
valentinita valentinite
valeral valeral
valeraldehído valeraldehyde
valeramida valeramide
valerato valerate
valerato amónico ammonium valerate
valerato cálcico calcium valerate
valerato ceroso cerous valerate
valerato de atropina atropine valerate
valerato de butilo butyl valerate
valerato de cadmio cadmium valerate
valerato de etilo ethyl valerate
valerato de fenilo phenyl valerate
valerato de isoamilo isoamyl valerate
valerato de isobutilo isobutyl valerate
valerato de mentol menthol valerate
valerato de pentilo pentyl valerate
valerilo valeryl
valerolactona valerolactone
valilo valyl
valina valine
valor calorífico calorific value
valor de acetilo acetyl value
valor de ácido acid value
valor de amortiguador buffer value
valor de bario barium value
valor de peróxido peroxide value
valor de pH pH value
valor de saponificación saponification value
valor de yodo iodine value
valor delta delta value
valor sigma sigma value
valor tau tau value

valor umbral threshold value
valor umbral límite threshold limit value
valoración titration
valoración ácido-base acid-base titration
valoración amperométrica amperometric titration
valoración automática automatic titration
valoración complejométrica complexometric titration
valoración con permanganato permanganate titration
valoración conductométrica conductometric titration
valoración de Mohr Mohr titration
valoración de Volhard Volhard titration
valoración diferencial differential titration
valoración electrométrica electrometric titration
valoración entálpica enthalpy titration
valoración fotométrica photometric titration
valoración magnetométrica magnetometric titration
valoración oscilométrica oscillometric titration
valoración potenciométrica potentiometric titration
valoración radiométrica radiometric titration
valoración redox redox titration
valoración térmica thermal titration
valoración termométrica thermometric titration
valproato sódico sodium valproate
válvula de admisión de gas gas-admittance valve
válvula de entrada de gas gas-inlet valve
vanadato vanadate
vanadato amónico ammonium vanadate
vanadato de itrio yttrium vanadate
vanadato de litio lithium vanadate

vanadato de plata silver vanadate
vanadato de plomo lead vanadate
vanadato férrico ferric vanadate
vanadato sódico sodium vanadate
vanadílico vanadylic
vanadilo vanadyl
vanadiloso vanadylous
vanadinita vanadinite
vanadio vanadium
vanadioso vanadous
vanadita vanadite
vanadofosfato sódico sodium vanadophosphate
vanadol vanadol
vanilato de etilo ethyl vanillate
vanilina de etilo ethyl vanillin
vapor vapor
vapor de agua steam
vapor saturado saturated vapor
vapor venenoso poison vapor
vaporización vaporization
vaporizador de Lundergardh Lundergardh vaporizer
vaporizar vaporize
varilla agitadora stirring rod
variscita variscite
vaso de laboratorio beaker
vasopresina vasopressin
vauquelinita vauquelinite
vecinal vicinal
vehículo vehicle
velocidad velocity, rate
velocidad de combustión combustion rate
velocidad de desintegración decay rate
velocidad de electrón electron velocity
velocidad de fisión fission rate
velocidad de formación rate of formation
velocidad de migración migration velocity
velocidad de propagación propagation rate
velocidad de reacción reaction rate, reaction velocity
velocidad de reacción específica specific reaction rate
velocidad de sedimentación sedimentation rate

velocidad mínima de ionización minimum ionizing speed
velocidad molecular molecular velocity
veneno poison
veneno catalítico catalytic poison
veneno de serpiente snake venom
venenoso poisonous
venturi venturi
veratraldehído veratraldehyde
veratrilideno veratrylidene
veratrilo veratryl
veratroilo veratroyl
veratrol veratrole
verbenona verbenone
verde brillante brilliant green
verde de benzaldehído benzaldehyde green
verde de bromocresol bromocresol green
verde de cobalto cobalt green
verde de cobre copper green
verde de cromo chrome green
verde de naftol naphthol green
verde de París Paris green
verde esmeralda emerald green
verde imperial imperial green
verde malaquita malachite green
verde Scheele Scheele green
verde ultramarino ultramarine green
verde Victoria Victoria green
vermiculita vermiculite
vernolepina vernolepin
verxita verxite
vetivona vetivone
vía pathway
vibración vibration
vibración atómica atomic vibration
vibración molecular molecular vibration
vida media mean life
vidrio glass
vidrio cerámico glass-ceramic
vidrio de agua water glass
vidrio de bórax borax glass
vidrio de borosilicato borosilicate glass
vidrio de fosfato phosphate glass
vidrio de plomo lead glass
vidrio de seguridad safety glass
vidrio de uranio uranium glass
vidrio esmerilado ground glass

vidrio líquido liquid glass
vidrio metálico metal glass
vidrio óptico optical glass
vidrio protector protective glass
vidrio rojo red glass
vidrio soluble soluble glass
villiaumita villiaumite
vinagre vinegar
vinilacetileno vinylacetylene
vinilacetonitrilo vinylacetonitrile
vinilación vinylation
vinilamina vinylamine
vinilbenceno vinylbenzene
vinilbutiléter vinyl butyl ether
vinilcarbazol vinylcarbazole
vinilcetona vinyl ketone
vinilciclohexeno vinylcyclohexene
vinileno vinylene
vinilestireno vinylstyrene
viniléter vinyl ether
viniletileno vinylethylene
viniletiléter vinyl ethyl ether
viniletilhexiléter vinylethylhexyl ether
viniletilpiridina vinylethylpyridine
vinilideno vinylidene
vinilimina vinylimine
vinilisobutiléter vinyl isobutyl ether
vinilmetilcetona vinyl methyl ketone
vinilmetiléter vinyl methyl ether
vinilo vinyl
vinílogo vinylog
vinilpiridina vinylpyridine
vinilpirrolidona vinylpyrrolidone
viniltolueno vinyltoluene
viniltriclorosilano vinyltrichlorosilane
vino wine
vioformo vioform
violantrona violanthrone
violeta de antraceno anthracene violet
violeta de cobalto cobalt violet
violeta de metilo methyl violet
violeta de París Paris violet
violeta neutra neutral violet
viomicina viomycin
virus virus
viscocidad superficial surface viscocity
viscoplástico viscoplastic
viscosa viscose
viscosidad viscosity

viscosidad intrínseca intrinsic viscosity
viscosimetría viscosimetry
viscosímetro viscometer, viscosimeter
viscoso viscid
visualizador analógico analog display
vitamina vitamin
vitelina vitellin
vítreo vitreous
vitrificación vitrification
vitriolo vitriol
vitriolo azul blue vitriol
vitriolo blanco white vitriol
vitriolo de cinc zinc vitriol
vitriolo verde green vitriol
vivianita vivianite
volátil volatile
volatilidad volatility
volatilización volatilization
volatilizar volatilize
volframio wolfram
voltaíta voltaite
voltzina voltzite
volumen volume
volumen atómico atomic volume
volumen crítico critical volume
volumen de combinación combining volume
volumen de retención retention volume
volumen específico specific volume
volumen incompresible incompressible volume
volumen molal molal volume
volumen molar molar volume, mole volume
volumen molar parcial partial molar volume
volumen molecular molecular volume
volumen molecular-gramo gram-molecular volume
volumen nuclear nuclear volume
volumenómetro volumenometer
volumetría titration
volumétrico volumetric
volúmetro volumeter
vulcanita vulcanite
vulcanización vulcanization
vulcanizar vulcanize
vulpinita vulpinite

W

wagnerita wagnerite
warfarina warfarin
warfarina sódica sodium warfarin
wavellita wavellite
whisky whiskey
whitneyita whitneyite
willemita willemite
witherita witherite
wohlerita wöhlerite
wolframato wolframate
wolframato de plomo lead wolframate

wolframato magnésico magnesium wolframate
wolframato potásico potassium wolframate
wolframato sódico sodium wolframate
wolframio wolfram
wolframita wolframite
wollastonita wollastonite
wulfenita wulfenite
wurtzita wurtzite

X

xantano xanthan
xantato xanthate
xantato de almidón starch xanthate
xantato de amilo amyl xanthate
xantato de celulosa cellulose xanthate
xantato de pentilo pentyl xanthate
xantato de propilo propyl xanthate
xantato potásico potassium xanthate
xantato sódico sodium xanthate
xantenilo xanthenyl
xanteno xanthene
xantenol xanthenol
xantilio xanthylium
xantilo xanthyl
xantina xanthine
xanto xantho
xantocromo xanthochromium
xantofila xanthophyll
xantogenato sódico sodium xanthogenate
xantona xanthone
xantopterina xanthopterin
xantosiderita xanthosiderite
xantosina xanthosine
xantoxileno xanthoxylene
xenato xenate
xenilamina xenylamine

xenilo xenyl
xenobiótico xenobiotic
xenol xenol
xenolita xenolite
xenón xenon
xenotima xenotime
xerogel xerogel
xerografía xerography
xi xi
xilano xylan
xileno xylene
xileno de almizcle musk xylene
xilenodiol xylenediol
xilenol xylenol
xilenosulfonato sódico sodium
    xylenesulfonate
xilideno xylidene
xilidina xylidine
xilileno xylylene
xililo xylyl
xilita xylite
xilitol xylitol
xilocaína xylocaine
xiloilo xyloyl
xilol xylol
xilosa xylose

Y

yeso gypsum
yeso de París plaster of Paris
yodado iodized
yodato iodate
yodato amónico ammonium iodate
yodato bárico barium iodate
yodato cálcico calcium iodate
yodato de cadmio cadmium iodate
yodato de cesio cesium iodate
yodato de cinc zinc iodate
yodato de litio lithium iodate
yodato de plata silver iodate
yodato de plomo lead iodate
yodato de rubidio rubidium iodate
yodato magnésico magnesium iodate
yodato potásico potassium iodate
yodato sódico sodium iodate
yodeosina iodeosin
yodilo iodyl
yodimetría iodimetry
yodipamida iodipamide
yodisán iodisan
yodita iodite
yodo iodine
yodoacetato de etilo ethyl iodoacetate
yodoalcano iodoalkane
yodoanilina iodoaniline
yodobehenato cálcico calcium
    iodobehenate
yodobenceno iodobenzene
yodoetano iodoethane
yodoetileno iodoethylene
yodoformo iodoform
yodóforo iodophor
yodohipurato sódico sodium
    iodohippurate
yodometano iodomethane
yodometría iodometry
yodométrico iodometric
yodonio iodonium
yodopropano iodopropane
yodosilo iodosyl
yodoso iodoso

yodosobenceno iodosobenzene
yodosuccinimida iodosuccinimide
yodoxi iodoxy
yodoxilo iodoxyl
yoduro iodide
yoduro alumínico aluminum iodide
yoduro amónico ammonium iodide
yoduro antimonioso antimonous iodide
yoduro arsenioso arsenous iodide
yoduro áurico auric iodide
yoduro bárico barium iodide
yoduro bárico mercúrico mercuric
    barium iodide
yoduro cálcico calcium iodide
yoduro ceroso cerous iodide
yoduro cobaltoso cobaltous iodide
yoduro crómico chromic iodide
yoduro cúprico cupric iodide
yoduro cuproso cuprous iodide
yoduro de acetilo acetyl iodide
yoduro de alilo allyl iodide
yoduro de antimonio antimony iodide
yoduro de arsénico arsenic iodide
yoduro de azufre sulfur iodide
yoduro de bencilo benzyl iodide
yoduro de benzoilo benzoyl iodide
yoduro de berilio beryllium iodide
yoduro de bismuto bismuth iodide
yoduro de bromo bromine iodide
yoduro de butilo butyl iodide
yoduro de cadmio cadmium iodide
yoduro de cesio cesium iodide
yoduro de cianógeno cyanogen iodide
yoduro de cinc zinc iodide
yoduro de cinc potásico potassium zinc
    iodide
yoduro de cobalto cobalt iodide
yoduro de cobre copper iodide
yoduro de estaño tin iodide
yoduro de estroncio strontium iodide
yoduro de etilo ethyl iodide
yoduro de fosfonio phosphonium iodide
yoduro de galio gallium iodide

528

yoduro de hexadecilo hexadecyl iodide
yoduro de hidrógeno hydrogen iodide
yoduro de hierro iron iodide
yoduro de isopropilo isopropyl iodide
yoduro de litio lithium iodide
yoduro de manganeso manganese iodide
yoduro de metileno methylene iodide
yoduro de metilmagnesio
    methylmagnesium iodide
yoduro de metilo methyl iodide
yoduro de neodimio neodymium iodide
yoduro de níquel nickel iodide
yoduro de nitrógeno nitrogen iodide
yoduro de octilo octyl iodide
yoduro de oro gold iodide
yoduro de paladio palladium iodide
yoduro de pentilo pentyl iodide
yoduro de plata silver iodide
yoduro de platino platinum iodide
yoduro de plomo lead iodide
yoduro de propilo propyl iodide
yoduro de rubidio rubidium iodide
yoduro de silicio silicon iodide
yoduro de talio thallium iodide
yoduro de telurio tellurium iodide
yoduro de timol thymol iodide
yoduro de titanio titanium iodide
yoduro de uranio uranium iodide
yoduro de vinilo vinyl iodide

yoduro de zirconio zirconium iodide
yoduro estánnico stannic iodide
yoduro estannoso stannous iodide
yoduro férrico ferric iodide
yoduro ferroso ferrous iodide
yoduro gálico gallic iodide
yoduro magnésico magnesium iodide
yoduro manganoso manganous iodide
yoduro mercúrico mercuric iodide
yoduro mercúrico bárico barium
    mercuric iodide
yoduro mercúrico de plata silver
    mercury iodide
yoduro mercúrico potásico potassium
    mercuric iodide
yoduro mercurioso mercurous iodide
yoduro niqueloso nickelous iodide
yoduro paladioso palladous iodide
yoduro platinoso platinous iodide
yoduro potásico potassium iodide
yoduro potásico de bismuto bismuth
    potassium iodide
yoduro potásico de cadmio cadmium
    potassium iodide
yoduro potásico de cinc zinc potassium
    iodide
yoduro sódico sodium iodide
yoduro talioso thallous iodide
yohimbina yohimbine

Z

zafiro sapphire
zahína sorghum
zaratita zaratite
zeaxantina zeaxanthin
zeína zein
zeolita zeolite
zeolita en jaula cage zeolite
zimasa zymase
zimohexasa zymohexase
zimólisis zymolysis
zimoquímica zymochemistry
zimosa zymose
zimosis zymosis
zimurgia zymurgy
zinc zinc
zingibereno zingiberene
zinnwaldita zinnwaldite
zippeíta zippeite
ziram ziram
zircón zircon
zirconato zirconate

zirconato bárico barium zirconate
zirconato cálcico calcium zirconate
zirconato de estroncio strontium zirconate
zirconato de litio lithium zirconate
zirconato de tetrabutilo tetrabutyl zirconate
zirconato de tetraisopropilo tetraisopropyl zirconate
zirconato magnésico magnesium zirconate
zirconia zirconia
zircónico zirconic
zirconilo zirconyl
zirconio zirconium
zirlita zirlite
zoaleno zoalene
zoisita zoisite
zooquímica zoochemistry
zwitterion zwitterion